Optimising bio-based fertilisation for a more sustainable agriculture

It is widely recognised that agriculture is a significant contributor to global warming and climate change. Agriculture needs to reduce its environmental impact and adapt to current climate change whilst still feeding a growing population, i.e. become more 'climate-smart'. Burleigh Dodds Science Publishing is playing its part in achieving this by bringing together key research on making the production of the world's most important crops and livestock products more sustainable.

Based on extensive research, our publications specifically target the challenge of climate-smart agriculture. In this way we are using 'smart publishing' to help achieve climate-smart agriculture.

Burleigh Dodds Science Publishing is an independent and innovative publisher delivering high quality customer-focused agricultural science content in both print and online formats for the academic and research communities. Our aim is to build a foundation of knowledge on which researchers can build to meet the challenge of climate-smart agriculture.

For more information about Burleigh Dodds Science Publishing simply call us on +44 (0) 1223 839365, email info@bdspublishing.com or alternatively please visit our website at www.bdspublishing.com.

Related titles:

Achieiving sustainable crop nutrition
Print (ISBN 978-1-78676-312-9); Online (ISBN 978-1-78676-314-3, 978-1-78676-315-0)

Biostimulants for sustainable crop production
Print (ISBN 978-1-78676-336-5); Online (ISBN 978-1-78676-338-9, 978-1-78676-339-6)

Improving nitrogen use efficiency in crop production
Print (ISBN 978-1-80146-470-3); Online (ISBN 978-1-80146-471-0, 978-1-80146-472-7)

Improving soil health
Print (ISBN 978-1-78676-670-0); Online (ISBN 978-1-78676-672-4, 978-1-78676-673-1)

Understanding and utilising soil microbiomes for a more sustainable agriculture
Print (ISBN 978-1-80146-474-1); Online (ISBN 978-1-80146-475-8, 978-1-80146-476-5)

Chapters are available individually from our online bookshop:
https://www.bdspublishing.com/webshop/

climate•SMART•publishing
IN AGRICULTURAL SCIENCE

BURLEIGH DODDS SERIES IN AGRICULTURAL SCIENCE
NUMBER 180

Optimising bio-based fertilisation for a more sustainable agriculture

Edited by Professor Ruben Sakrabani, Cranfield University, UK

burleigh dodds
SCIENCE PUBLISHING

Published by Burleigh Dodds Science Publishing Limited
82 High Street, Sawston, Cambridge CB22 3HJ, UK
www.bdspublishing.com

Burleigh Dodds Science Publishing, 1518 Walnut Street, Suite 900, Philadelphia, PA 19102-3406, USA

First published 2026 by Burleigh Dodds Science Publishing Limited

Library of Congress Control Number: 2026942312

British Library Cataloguing in Publication Data
A catalogue record for this book is available from the British Library

ISBN 978-1-83545-186-1 (print)
ISBN 978-1-83545-187-8 (ePub)
ISBN 978-1-83545-188-5 (PDF)
ISSN 2059-6936 (print)
ISSN 2059-6944 (online)

DOI: 10.19103/AS.2025.0166

Typeset by Deanta Global Publishing Services, Dublin, Ireland

Contents

Part 2 Using biostimulants and promoting beneficial microorganisms in soil

Series list

Title	Series number
Achieving sustainable cultivation of maize - Vol 1 From improved varieties to local applications Edited by: Dr Dave Watson, CGIAR Maize Research Program Manager, CIMMYT, Mexico	001
Achieving sustainable cultivation of maize - Vol 2 Cultivation techniques, pest and disease control Edited by: Dr Dave Watson, CGIAR Maize Research Program Manager, CIMMYT, Mexico	002
Achieving sustainable cultivation of rice - Vol 1 Breeding for higher yield and quality Edited by: Prof. Takuji Sasaki, Tokyo University of Agriculture, Japan	003
Achieving sustainable cultivation of rice - Vol 2 Cultivation, pest and disease management Edited by: Prof. Takuji Sasaki, Tokyo University of Agriculture, Japan	004
Achieving sustainable cultivation of wheat - Vol 1 Breeding, quality traits, pests and diseases Edited by: Prof. Peter Langridge, University of Adelaide, Australia	005
Achieving sustainable cultivation of wheat - Vol 2 Cultivation techniques Edited by: Prof. Peter Langridge, University of Adelaide, Australia	006
Achieving sustainable cultivation of tomatoes Edited by: Dr Autar Mattoo, USDA-ARS, USA and Prof. Avtar Handa, Purdue University, USA	007
Achieving sustainable production of milk - Vol 1 Milk composition, genetics and breeding Edited by: Dr Nico van Belzen, International Dairy Federation (IDF), Belgium	008
Achieving sustainable production of milk - Vol 2 Safety, quality and sustainability Edited by: Dr Nico van Belzen, International Dairy Federation (IDF), Belgium	009
Achieving sustainable production of milk - Vol 3 Dairy herd management and welfare Edited by: Emeritus Prof. John Webster, University of Bristol, UK	010
Ensuring safety and quality in the production of beef - Vol 1 Safety Edited by: Prof. Gary Acuff, Texas A&M University, USA and Prof. James Dickson, Iowa State University, USA	011
Ensuring safety and quality in the production of beef - Vol 2 Quality Edited by: Emeritus Prof. Michael Dikeman, Kansas State University, USA	012
Achieving sustainable production of poultry meat - Vol 1 Safety, quality and sustainability Edited by: Prof. Steven C. Ricke, University of Arkansas, USA	013
Achieving sustainable production of poultry meat - Vol 2 Breeding and nutrition Edited by: Prof. Todd Applegate, University of Georgia, USA	014

Achieving sustainable production of poultry meat - Vol 3 015
Health and welfare
Edited by: Prof. Todd Applegate, University of Georgia, USA

Achieving sustainable production of eggs - Vol 1 016
Safety and quality
Edited by: Prof. Julie Roberts, University of New England, Australia

Achieving sustainable production of eggs - Vol 2 017
Animal welfare and sustainability
Edited by: Prof. Julie Roberts, University of New England, Australia

Achieving sustainable cultivation of apples 018
Edited by: Dr Kate Evans, Washington State University, USA

Integrated disease management of wheat and barley 019
Edited by: Prof. Richard Oliver, Curtin University, Australia

Achieving sustainable cultivation of cassava - Vol 1 020
Cultivation techniques
Edited by: Dr Clair Hershey, formerly International Center for Tropical Agriculture (CIAT), Colombia

Achieving sustainable cultivation of cassava - Vol 2 021
Genetics, breeding, pests and diseases
Edited by: Dr Clair Hershey, formerly International Center for Tropical Agriculture (CIAT), Colombia

Achieving sustainable production of sheep 022
Edited by: Prof. Johan Greyling, University of the Free State, South Africa

Achieving sustainable production of pig meat - Vol 1 023
Safety, quality and sustainability
Edited by: Prof. Alan Mathew, Purdue University, USA

Achieving sustainable production of pig meat - Vol 2 024
Animal breeding and nutrition
Edited by: Prof. Julian Wiseman, University of Nottingham, UK

Achieving sustainable production of pig meat - Vol 3 025
Animal health and welfare
Edited by: Prof. Julian Wiseman, University of Nottingham, UK

Achieving sustainable cultivation of potatoes - Vol 1 026
Breeding improved varieties
Edited by: Prof. Gefu Wang-Pruski, Dalhousie University, Canada

Achieving sustainable cultivation of oil palm - Vol 1 027
Introduction, breeding and cultivation techniques
Edited by: Prof. Alain Rival, Center for International Cooperation in Agricultural Research for Development (CIRAD), France

Achieving sustainable cultivation of oil palm - Vol 2 028
Diseases, pests, quality and sustainability
Edited by: Prof. Alain Rival, Center for International Cooperation in Agricultural Research for Development (CIRAD), France

Achieving sustainable cultivation of soybeans - Vol 1 029
Breeding and cultivation techniques
Edited by: Prof. Henry T. Nguyen, University of Missouri, USA

Achieving sustainable cultivation of soybeans - Vol 2 030
Diseases, pests, food and other uses
Edited by: Prof. Henry T. Nguyen, University of Missouri, USA

Acknowledgements

We wish to acknowledge the following for their help in reviewing particular chapters:

- Chapter 1: Professor Andrew Whitmore, Rothamsted Research, UK and Professor Muhammad Riaz, Government College University Faisalabad (GCUF), Pakistan
- Chapter 2: Professor Urzula Norton, University of Wyoming, USA
- Chapter 3: Jean Bonhotal, Director Waste Management Institute - Cornell University, USA
- Chapter 4: Dr Paulo Pagliary, University of Minnesota, USA
- Chapter 5: Dr Mehdi Sharifi, Agriculture and Agri-Food Canada, Canada and Dr Massimo Monteleone, University of Foggia, Italy
- Chapter 6: Dr Sudharsan Varma, University of Arkansas, USA
- Chapter 7: Professor Shabtai Bittmann and Ms Margaret McComb, Agriculture and Agri-Food Canada, Canada and Professor Meredith Niles, Brown University, USA
- Chapter 8: Dr Helen Amorim, University of Arkansas, USA
- Chapter 9: Dr Francisco Moreno-Valencia, Institute of Sciences - Meritorious Autonomous University of Puebla (ICUAP-BUAP), Mexico
- Chapter 10: Emeritus Professor Louise Nelson, University of British Colombia, Canada and Dr Raffaela Balestrini, National Research Council of Italy (CNR), Italy
- Chapter 11: Dr Astrid Oberson, Institute of Agricultural Sciences - ETH Zurich, Switzerland
- Chapter 12: Dr Radha Prasanna, Indian Council of Agricultural Research (ICAR), India
- Chapter 15: Professor Gabriel Acien, University of Almeria, Spain

Introduction

While mineral fertilisers have played a central role in supporting global food production, their widespread use has also contributed to environmental challenges, including significant greenhouse gas (GHG) emissions from their production and application, degradation of soil health and environmentally-damaging nutrient losses. The transition toward more sustainable and resilient agricultural systems has intensified the search for alternatives to conventional fertilisation practices. In this context, bio-based fertilisation (encompassing organic amendments and other biologically derived inputs) offers a promising pathway toward more circular, efficient and ecologically-sound nutrient management.

This volume brings together recent advances in the development, optimisation and application of bio-based fertilisation strategies. It provides a comprehensive overview of approaches aimed at improving nutrient use efficiency, enhancing soil functionality and strengthening plant-soil-microbe interactions. The chapters are organised into two thematic parts: advances in soil amendments, followed by the use of biostimulants and beneficial microorganisms.

Part 1 Advances in soil amendments

Retaining crop residues as soil cover after harvesting is seen as an important technique in improving soil health. Chapter 1 examines the optimisation of crop residues as soil amendments, highlighting their role in maintaining soil organic matter, improving soil structure, and supporting nutrient cycling. Crop residues (including straws, tops, stalks, leaves, and shoots) represent a critical component of sustainable agriculture systems. Their effective management lies at the heart of practices such as reduced tillage, increased crop diversity, and the maintenance of living roots throughout the year. However, as the chapter explains, residue management is shaped by, and needs to take account of, a range of agronomic, economic, logistical, environmental and edaphoclimatic factors.

Cover crops have become a well-accepted management practice for improving soil health. However, many farmers and farm advisors are unfamiliar with the best strategies for using cover crops in a way that accelerates soil health improvements. Chapter 2 describes how selection of cover crop species and mixes, timing of cover crop seeding and termination, management of residue and tillage, and cover crop grazing can impact overall soil health improvement. Relevant data from the Sustainable Agriculture Research and Education (SARE) and Conservation Technology Information Center (CTIC) National Cover

Crop Survey in the United States is included to illustrate farmer experiences with cover crops. New technology used with cover crop management such as virtual fencing for cover crop grazing are described. The chapter concludes by assessing different levels of cover crop adoption, each with progressively more potential for soil health improvement.

Chapter 3 explores the use of vermicompost and biochar, focusing on their complementary effects on soil fertility, microbial activity and carbon stabilisation. As the chapter points out, vermicompost, generated through the joint and synergistic action of earthworms and microorganisms, is widely recognised as a nutrient-rich, biologically-active, humus-like organic material with the potential to enhance soil health and crop yield by supplying essential nutrients, fostering beneficial microbial communities and improving plant resilience against pests and diseases. The chapter also explores the benefits in using biochar, a carbon-rich material produced through the pyrolysis of organic matter (OM). These include the capacity to enhance soil organic carbon (improving soil health and promoting soil carbon sequestration), improve fertiliser quality, promote plant growth and reduce GHG emissions by helping to promote a more circular bioeconomy. Biochar addition also accelerates composting by increasing microbial diversity and shortening the composting cycle, thus yielding an enhanced final compost product. The chapter therefore also introduces emerging research on biochar–vermicompost combinations as next-generation soil amendments for improving soil quality and nutrient availability.

Building on themes in other chapters in Part 1, Chapter 4 discusses the potential of organo-mineral fertilisers (OMF), which integrate the agronomic benefits of mineral nutrients with the soil-enhancing properties of organic matter. OMF involve co-formulating at least one organic and one mineral material into a novel fertiliser product. The chapter reviews the range of organic and inorganic materials used in the production of OMF (including vermicompost, biochar and anaerobic digestate), treatment of organic feedstock to ensure safety and optimise efficacy, as well as product formulation issues. It also summarises research assessing the effects of OMF on crop growth and soil health as well as their wider environmental impact.

The process of anaerobic digestion is a sustainable means of converting manure and other organic biomass in biogas plants into biogas. Digestate is a nutrient-rich by-product of anaerobic digestion. Chapter 5 examines the value of anaerobic digestate as a soil amendment, emphasising its role in nutrient recycling and circular bioeconomy systems. As the chapter explains, digestate contributes to soil structure improvement and can act as a prebiotic, supporting beneficial plant–microbe interactions. The chapter also considers the relative benefits of using digestate in whole, liquid, and solid forms. It discusses methods to optimise the nutrient use efficiency of digestate, reduce nutrient pollution to the environment, optimise recovery of nutrients from digestate, addresses ecotoxicological hazards associated with digestate use and highlights key areas for future research.

Poultry litter is a nutrient-rich by-product of poultry production, widely used to enhance soil fertility and reduce dependence on synthetic fertilisers. Its high content of nitrogen, phosphorus, potassium and organic matter can support soil structure, nutrient cycling and crop productivity. Chapter 6 evaluates the optimisation of poultry litter as a soil amendment. It addresses risks such as pathogens and antibiotic resistance, and treatment methods to control microbial risks (such as composting, pelletisation and heat drying as well as use of chemical amendments). The chapter also discusses key issues relating to soil application of poultry litter, such as the risk of pathogen persistence in soil and consequent contamination of fresh produce as well as managing impact on soil microbiota and nutrient quality.

Slurry is a significant by-product of livestock agriculture. It can be a valuable source of nutrients if managed efficiently. Ensuring that slurry is used as an effective resource is also essential to avoiding pollution of land, water and air and to creating a more sustainable circular bioeconomy. However, despite the existence of many decision support tools and other advice, many livestock farmers still do not engage with the good practice guidance available or manage slurries and manures effectively. Chapter 7 explores research this problem and ways it can be addressed. It describes the SLURRY-MAX Project in the UK which investigated farmer use of slurry management decision support tools, farmers' social practices around slurry, The chapter explores solutions for enhancing the attractiveness, usability and take-up of decision support tools to optimise slurry management for both the benefit of farmers and the wider environment.

As global food demand increases, sustainable alternatives to synthetic fertilisers are becoming increasingly important. Chapter 8 explores the optimisation of insect-derived frass as a soil amendment. Frass is a by-product of insect farming, composed of excreta, exuviae, and uneaten feed, and is gaining attention as a nutrient-rich organic input. This chapter highlights its agronomic potential, including its nutrient composition, availability, and effects on soil health and plant productivity.

Part 2 Using biostimulants and promoting beneficial microorganisms in soil

Chapter 9 examines plant growth-promoting bacteria (PGPB) as inoculants, focusing on their ability to enhance nutrient uptake, stimulate plant growth, and improve resilience to environmental stress. Despite their potential, the effectiveness of PGPB can be inconsistent under field conditions. The chapter explores the key factors influencing their success, including environmental conditions and root colonisation efficiency. It critically synthesises existing literature, identifies key knowledge gaps, technical bottlenecks, and highlights emerging solutions. Particular emphasis is placed on strain selection, advanced formulation technologies, and appropriate integration into sustainable farming systems, all of which are essential to achieve field-ready PGPB-based biostimulants.

Chapter 10 investigates strategies to understand and promote native populations of plant growth-promoting rhizobacteria (PGPR) in soils. While much research has focused on inoculants, the majority of PGPR in agroecosystems are indigenous. Harnessing and enhancing these indigenous populations is central to sustainable agriculture yet remains less developed than strain-based bioproduct approaches. This chapter synthesises current knowledge on the ecology, assembly and functioning of indigenous PGPR and evaluates management strategies to promote in-situ persistence. The chapter examines drivers of PGPR community structure, including soil properties, plant genotype, root exudation, land use, and agricultural inputs, and ways to characterise them using both culture-dependent and culture-independent methods. Agronomic practices that promote healthy microbiomes are also assessed. The chapter reviews key challenges such as context dependence, inconsistent field responses, ecological risks, and key knowledge gaps linking PGPR community shifts to optimising soil health and productive, resilient cropping systems.

Chapter 11 evaluates phosphate-solubilising microorganisms (PSM) and their role in improving crop nutrient use efficiency. Phosphorus (P) is essential for plant growth, yet its availability is often limited due to soil interactions. This necessitates new strategies to utilise existing soil P reserves. PSM can play a crucial role by mobilising soil-bound P so it is available to plants, supporting both soil microbial communities and plant P demand. Testing PSM inoculants involves *in vitro* screening, controlled environment trials, and field tests to ensure P solubilisation enhances crop productivity in practice. However, as the chapter points out, translating positive effects from controlled environments to field conditions remains challenging due to factors like soil colonization, pH and nutrient availability. Optimising the role of PSM in agriculture will involve tailored farming practices, using microbial consortia and/or applying artificial microbial community selection methods. The chapter also identifies future research needs to optimise PSM effectiveness.

Due to their morphological and physiological adaptations, cyanobacteria are an important component of soils and abundant in terms of biomass and species richness. They are, for example, a major component of the formation of biological soil crust (BSC) communities in drylands, where they play a fundamental role in biochemical cycling. They enhance the productivity and fertility of terrestrial ecosystems through carbon and nitrogen fixation, oxygen and biomass production. Chapter 12 explores the diversity of arable soil cyanobacteria and functional characteristics in improving soil structure and stability, carbon sequestration and organic matter production, nitrogen fixation, phosphate solubilisation, micronutrient mobilisation and translocation in plants. The chapter highlights their potential role as biofertilisers in sustainable agronomy, restoring degraded soil, improving soil fertility and health, as well as their benefits for agricultural production and broader environmental protection.

Amino acids used in biostimulants are derived through chemical synthesis or hydrolysis of plant and animal proteins and may be applied as free amino acids or as hydrolysed protein mixtures. Chapter 13 examines amino acid-based biostimulants, focusing on their physiological effects on plants. These products enhance enzymatic and metabolic processes, improving productivity, crop quality and stress tolerance. The chapter reviews the various mechanisms of action of amino acids. It goes on to summarise research on the impact of applying amino acid-based biostimulants on growth, yield, nutrient content (including flavonoid and phenolic content) and quality of different crops.

As Chapter 14 points out, sustaining global food security requires enhancing soil health by increasing levels of soil organic matter (SOM). Judicious use of mineral fertilisers can increase microbial diversity and biomass, which sustains or enhances SOM and soil health. However, synthetic nitrogen (N) applied above the economic optimum rate often increases soil acidity, reduces bacterial diversity and function, which reduces SOM through increased nitrification and mineralization of native SOM. Depending on inorganic fertiliser rate and percolation potential (rainfall regime), increasing soil salinity also impacts microbial functions, further degrading soil health. As the chapter shows, long-term integrated management practices that can both build and maintain SOM include residue management, diverse crop rotations, organic amendments and supplemental N to enhance residue carbon decomposition and conversion to SOM. The resultant increase in SOM enhances native nutrient supply though increased mineralisable organic nutrients and exchangeable nutrients, reducing dependency on nutrient inputs.

The final chapter in the book, Chapter 15, provides an integrated framework for assessing algal biostimulants. It starts with the characterisation of algal biostimulants, their mechanisms of action and composition. Using a series of case studies on the use of both microalgal and macroalgal in various crops, the chapter then explore ways of optimising the use of these products in the field to optimise nutrient use efficiency, and product quality and yields. Examples include use of macroalgal biostimulants to improve lettuce and barley yields as well as optimise olive cultivation, together with integration of microalgal biostimulants with nitrogen fertilisation in durum wheat cultivation.

Together, the contributions in this volume highlight the diversity, potential and complexity of bio-based fertilisation strategies. They emphasise the importance of integrating organic resource management with advances in microbial inoculants and biostimulant technologies to improve nutrient efficiency while maintaining soil health and ecosystem services. At the same time, they acknowledge ongoing challenges, including incomplete understanding of mechanisms of action, context dependency and variability in field performance, and the need for further research to translate scientific advances into scalable, farm-level practices.

Part 1

Advances in soil amendments

Chapter 1

The use of crop residues as soil amendments in farming

Marcelo Valadares Galdos, University of Leeds, UK; and Macdara O'Neill, Agreena, Denmark

1 Introduction

Crop residues, defined here as straws, tops, stalks, leaves and shoots of crops left after the main food or feed component is harvested, account for more than half of the total phytomass harvested annually. It has been estimated that the annual amount of residues produced by the main agricultural crops exceeds 5 billion tons of dry matter (Shinde et al., 2022), and there has been much discussion on alternative uses for what has been called a 'second harvest'. Almost three-quarters of the crop residues produced annually come from cereals, followed by sugar crops, legumes and oil crops (Singh and Rengel, 2007).

Crop residue management is at the heart of sustainable agriculture systems, alongside reducing soil disruption by tillage, increasing crop diversity and maintaining live roots year-round. The effect of protecting soils from wind

http://dx.doi.org/10.19103/AS.2025.0166.01

and water erosion losses by leaving crop residues on the soil surface is widely recognized. The positive effects of mulching on soil moisture retention have been highlighted as an increasingly important component of climate-smart agriculture, particularly considering the increased incidence and duration of severe drought events in several parts of the world. Leaving residues on the field has nutrient cycling benefits, as plant material can be a significant source of nutrients for subsequent crops. Crop residues also represent a relevant source of organic material for maintaining or increasing soil carbon content. The quality, composition and management of crop residues directly and indirectly affect soil chemical, physical and biological functions, which in turn have an impact on a wide range of ecosystem services (Fig. 1). The main agronomic and environmental benefits of crop residues as soil amendments will be presented in this chapter, as well as potential tradeoffs related to other uses. The chapter

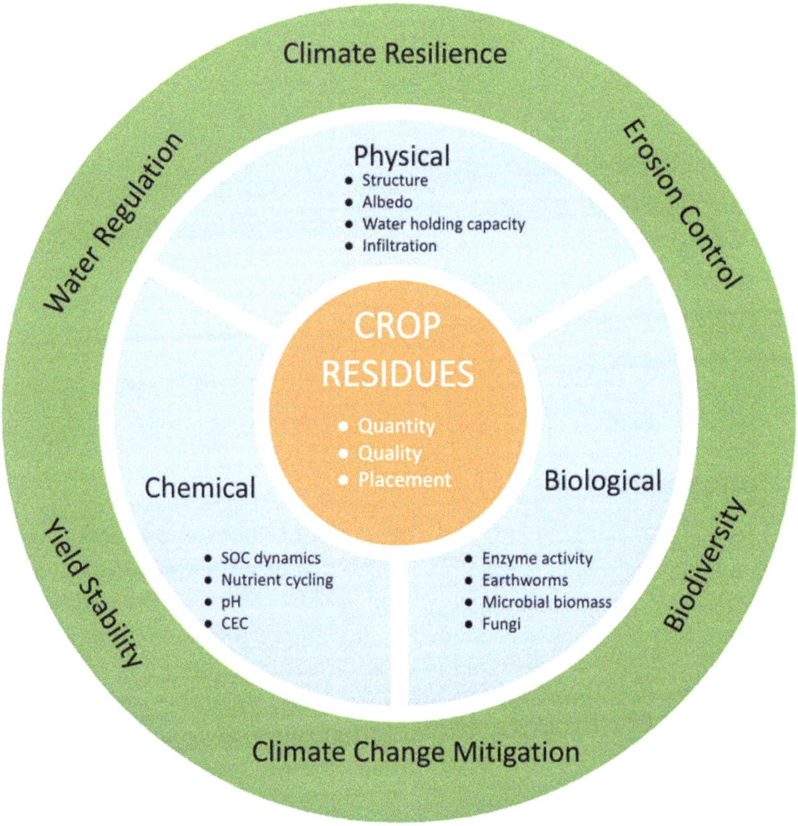

Figure 1 A conceptual framework of how crop residues and their management (inner circle) influence soil physical, chemical and biological processes (middle ring), which in turn support wider ecosystem services and agricultural sustainability (outer ring).

also includes a case study with process-based modelling estimates for soil carbon sequestration rates when farmers change from removing to maintaining crop residues, with data from hundreds of fields in six European countries.

2 Crop residue management

The post-harvest management of crop residues is highly influenced by agronomic, cultural, economic, logistical, environmental and edaphoclimatic factors. Smerald et al. (2023) calculated that approximately 33% of crop residue is used for livestock purposes, 44% left on the field, 6% is burnt on the field and 16% for other purposes including bioenergy, construction materials and mushroom composting. The diversity of uses for post-harvest plant material is reflected in the terminology used, ranging from more negative terms such as waste, litter and trash, to co-products and by-products in the other extreme. Depending on the context, crop residues are considered a nuisance to be rid of, or a valuable resource for soil conservation, nutrient cycling and other uses. Despite the multiple benefits of maintaining crop residues on fields, farmers are faced with competing demands for this material after harvest. For example, feeding animals accounts for approximately 67% of residue use, and 10–25% of agricultural residues are used for cooking in Bangladesh (Islam et al., 2020). These off-field uses for crop residues have been linked to significant challenges in upscaling Conservation Agriculture, leading to soil degradation and decreasing crop yields (Sarker et al., 2022).

Crop residue burning is still a widespread practice in specific crops and regions, leading to annual emissions of 3.4 MtC as soot into the atmosphere, besides CO_2, CH_4, CO, N_2O, NO_x and SO_2 emissions (Bond et al., 2004). Residues are primarily burned on the field but are also used as cooking fuel in households in several regions of the world. The most common arguments for in-field burning residues are to facilitate seedbed preparation for the subsequent crop and to remove biomass that might harbour pests and diseases. Sugarcane tops and leaves account for approximately one-sixth of crop residues produced annually (Smil, 1999), and pre-harvest burning has been a common practice to facilitate manual harvest and transport operations. Due to environmental and economic reasons including lack of manual labour, increased labour costs, sustainability certification schemes, public pressure driven by health concerns and environmental legislation, there are ongoing burning phase-out programmes in the main sugarcane-growing regions, with the gradual replacement of manual harvest with burning by mechanized harvest without burning (Galdos et al., 2013). Similar burning phase-out programmes have been implemented in rice and other crops globally.

Biomass from crop residues has been considered a significant source of feedstock for bioenergy, including combustion, gasification, anaerobic digestion and lignocellulosic ethanol production. Most assessments for the use of crop residues as an energy source account for different potential removal rates based on technical, agronomic and economic criteria (Scarlat et al. 2010). Assuming a removal of 25% of the total available agricultural residues for bioenergy generation, Hoogwijk et al. (2003) calculate the global potential of crop residues for bioenergy as ranging from 5 EJ y^{-1} to 27 EJ y^{-1}, out of a total of 33–1135 EJ y^{-1} from all biomass sources including crop residues, dedicated energy crops, forest residues, animal manures, and organic wastes. In a study assessing the theoretical, technical, environmental and sustainable potentials of crop residues for bioenergy in 36 European countries, Scarlat et al. (2019) estimate that 149 Mt DM or 2.6 EJ y^{-1} could be used sustainably after considering environmental aspects, harvesting constraints and other competitive uses. Large-scale use of crop residues for bioenergy production has taken place in countries such as the US and Brazil, based on maize and sugarcane, respectively.

Converting biomass from crop residues into biochar through pyrolysis – potentially with bio-oil and gas as co-products – has emerged as a promising strategy for sustainable crop residue management, enabling the safe disposal and valorization of agricultural by-products whilst enhancing soil health and contributing to climate mitigation. Biochar application has been shown to improve soil chemical, physical and biological properties, reduce soil N_2O fluxes, increase crop yields and enhance water-holding capacity (Field et al., 2012). This multifunctionality positions biochar as a key component of circular residue management systems, where carbon-rich waste materials are transformed into stable soil amendments that close nutrient and carbon loops. Recent long-term studies underscore the benefits of sustained, low-rate biochar applications in low-input agroecosystems: Mulatu et al. (2025) demonstrated that continuous use of lignocellulosic and bone-derived biochars, as well as biochar–compost mixtures, over 9 years substantially improved soil aggregation, labile carbon fractions and microbial biomass carbon compared with both mineral-fertilized and untreated soils. Similarly, Gross et al. (2024) reported that biochar contributes to significant and persistent increases in SOC stocks, particularly in loamy soils where physical protection of organic matter enhances carbon stability over decadal timescales. Together, these findings affirm biochar's dual role in improving soil health and fostering long-term carbon sequestration. Nevertheless, several research gaps remain, and comprehensive life-cycle assessments are needed to evaluate its broader environmental footprint, including off-site effects such as nutrient leaching and erosion, as well as the energy efficiency of pyrolysis systems and the economic feasibility of large-scale implementation (Lehmann & Joseph, 2024).

Crop residues are used as feed, both in traditional stubble grazing of harvested grain fields and chopped and fed to animals, sometimes supplemented with nitrogen-rich material. Due to their absorption capacity and local availability, crop residues, in particular cereal straws, have been extensively used as bedding for animals. Straw bedding also facilitates manure management, limiting nutrient leaching and often being returned to fields after composting (Smil, 1999).

Where crop residues are retained on the field, they are either incorporated into the soil or left on the soil surface. Tillage and residue management practices are closely interlinked. Conventional tillage involves primary cultivation with moldboard ploughing and secondary cultivation operations such as harrowing or disking for crop establishment. Conventional tillage buries the crop residue enabling its incorporation into the subsoil for decomposition. Conservation tillage systems including minimum tillage and no-tillage systems involve reduced soil disturbance and retention of residue on the surface aimed at reverting soil degradation processes. Retaining residue on the surface as mulch, which includes leaves, roots and stubble of the previous crop, can improve physical, chemical and biological conditions of agricultural soils (Ngosong et al., 2019).

Earthworms are key drivers of soil structure formation, nutrient cycling and organic matter dynamics, acting as biological indicators of soil health and sustainability. Their activity enhances soil porosity, aggregation and nutrient mineralization through burrowing and casting, with casts typically enriched in organic carbon, nitrogen and phosphorus relative to bulk soil (Van Groenigen et al., 2019). Crop residue management strongly influences earthworm abundance and community structure by determining both food availability and the degree of soil disturbance. Global meta-analyses show that reduced and no-tillage systems increase earthworm abundance and biomass by over 100% compared with conventional ploughing, especially for anecic and epigeic species dependent on surface residues (Briones and Schmidt, 2017; Moos et al., 2017). Long-term field experiments also demonstrate that combining no tillage with diverse crop rotations enhances anecic populations and soil bioturbation (Torppa and Taylor, 2022), highlighting the interactive effects of tillage and residue diversity. Residue retention provides both a food source and a protective mulch that moderates soil temperature and moisture, favouring earthworm survival in dry or cold environments (Manono, 2016). Residue quality further shapes community responses: low carbon:nitrogen (C:N) residues such as legumes stimulate short-term activity, whilst cereal straw supports longer-term populations (Akhila and Entoori, 2022). Removing residues or exporting them from the field reduces earthworm abundance and biomass, particularly among endogeic and epi-anecic groups (Lemtiri et al., 2018). Overall, systems that minimize soil disturbance and maintain diverse, high-quality residue inputs promote earthworm abundance and activity, reinforcing soil fertility and ecosystem services central to sustainable crop production.

3 Nutrient cycling

The amendment of crop residues to farmlands is a recommended practice to maintain soil fertility. Crop residues that are retained on fields (either on the soil surface or incorporated) thus act as an important reserve of nutrients for crop production. Mineralization and immobilization processes conducted by soil microorganisms and fauna are essential for making organically bound nutrients in crop residues bioavailable for plant uptake. Mineralization converts organic compounds to inorganic forms whilst immobilization involves the microbial uptake and assimilation of inorganic compounds or their incorporation into organic matter, thus retaining nutrients in the soil (Chen et al., 2014; Cao et al., 2021). Crop residue decomposition is a function of residue quality, soil, climate and the abundance and diversity of microorganisms. Residue quality is primarily determined by its carbon:nitrogen (C:N) ratio and typical quality indices include total N, lignin, cellulose, hemicellulose and polyphenol compounds. High-quality residues with a low C:N ratio (e.g. legumes, brassicas) are decomposed quicker than lower quality, high C:N ratio residues (Nicolardot et al., 2001). Residues with higher N and/or phosphorus (P) contents tend to enhance decomposition, whilst lignin and polyphenol contents reduce decomposition (Constantinides and Fownes, 1994), where soluble material is rapidly decomposed during the initial stages of decomposition followed by the decomposition of the cellulose and lignin. Consequently, crop residues with low C:N ratios (e.g. legumes < 25:1) tend to increase mineralization, whilst those with high C:N ratios (e.g. cereal straw ~80:1) enhance immobilization (Li et al., 2020; Noren et al., 2022). The initial location of crop residues in soil also affects the decomposition dynamics. Holland and Coleman (1987) showed that surface wheat straw residues immobilized more N than did incorporated residues and that microbial N cycling efficiency was greater where residues were allowed to remain on the surface. Chaves et al. (2021) and references therein also found that crop residues decompose more slowly on the surface compared to when they are incorporated; however, no effect of placement is observed for N-rich residues (Abiven and Recous, 2007).

Once decomposed, crop residues supply the soil with some of the major nutrients and micronutrients, existing in both organic and inorganic forms (Singh and Rengel, 2007). Nitrogen is an essential element for plant growth and enhancing crop yield (Vitousek and Howarth, 1991). Crop residues play an important role in supplying N for plant growth, where the organic N can be converted into ammonium (NH_4+) and nitrate (NO_3^-) which are readily available for plant uptake, through the processes of mineralization and nitrification. Breza and Grandy (2025) indicate from their meta-analysis that crop residues produced higher mineralization and immobilization rates relative to organic amendments and synthetic N fertilizer sources, thus potentially increasing soil

N storage and plant N uptake. Torma et al. (2018) investigated the nutrient content of 17 plant residues in Slovakia in a 9-year period. The highest N contents (0.34–0.63%) were found in aboveground and belowground residues of all legumes, aboveground residues of potatoes and belowground residues of alfalfa, clover, soybean and sugar beet.

Phosphorus is an essential element for plant growth. Approximately 30% of the global cropland area is P deficient, meaning croplands require external P fertilization to maintain or increase yields (MacDonald et al., 2011). Global cereal phosphorus use efficiency was estimated to be 16% from 1961 to 2013 (Dhillon et al., 2017) primarily due to inefficiency in P management, low mobility and high fixation of P in soils (Hinsinger, 2001). Soils require an optimum level of plant-available P to optimize crop yields. Soil P exists in inorganic (Pi; Olsen P) and organic (Po) forms. Organic P accounts for 20–80% of the total P in soils and is less mobile than Pi, and it is Po that is derived from mineralization of crop residue by microorganisms. Mineralization of Po in residues occurs at inputs > 3 mg g^{-1} or immobilized inputs are less than 3 mg g^{-1} (Damon et al., 2014), although reported threshold values range from 2 mg g^{-1} to 3 mg g^{-1}. The low bioavailability of P and reliance on mineral P fertilizer inputs for crop growth and SOM indicate the importance of residue management in cropping systems. Poulton et al. (2013) observed that the P reserves accumulated 70–120 years previously contributed to the doubling of spring barley receiving N fertilizer on a long-term field trial in the UK. Guo et al. (2018) show that 30-years of long-term crop straw incorporation led to elevated Pi in the 0–20 cm layer with a simultaneous 43–72% increase in PUE 43% with mineral fertilization plus 3750 kg ha^{-1} wheat straw treatment.

Potassium (K) is a key element for maintaining crop yields playing an essential role in regulating water transport and plant photosynthesis (Römheld and Kirkby, 2010; Zörb et al., 2014). Retention of residues has been demonstrated to increase crop K uptake, and thus, greater removal of crop residues post harvest will have the opposite effect for subsequent crops (Goulding et al., 2021). Whitbread et al. (2000) reported that 80% of K was contained in the straw and also reported a K balance of +8 kg ha^{-1} when straw was retained and –102 kg ha^{-1} when straw was removed, for a wheat cropping system in Australia. Lupwayi et al. (2006) reported that the amount of K in crop residues (wheat, pea, canola and red clover) ranged from 25 kg ha^{-1} to 121 kg ha^{-1} and that approximately 65–99% of K was released under zero and conventional tillage, respectively. These values are in agreement with Torma et al. (2018), who calculated that 100–120 kg K is available in the soil for the following crop without any additional fertilization. Wang et al. (2024) reported a mean K content of 95 kg ha^{-1} for the residue of four main cereal crops globally (barley, maize, rice-paddy and wheat), whilst also highlighting the importance of replenishing K by returning residue to the soil.

Crop residue return is also key to management ecological stoichiometry, defined as the balance between C and other elements such as N, P in soil and how this balance affects soil properties and ecosystem functioning. In a global synthesis of 682 data pairs, Liu et al. (2023) reported that crop residue inputs increase the soil C, N and P contents by 10.5–12%, 7.63–9.2%, 2.62–5.13%, respectively, and that topsoil C, C:N and C:P ratios tended to saturate after 13–16 years of crop residue return.

The beneficial effects of residue as an organic amendment are determined via improvements to crop yield. Crop residue management directly influences overall soil fertility (Sarkar et al., 2024), and understanding the interactive effects of residue and mineral fertilizer will be of critical importance for the adoption of sustainable agricultural management practices. A meta-analysis conducted across 20 long-term experiments in Europe by Hijbeek et al. (2017) indicates a non-significant (+1.4% ± 1.6) increase in crop yields attributed to organic inputs. However, when further disaggregated, it was shown that spring sown crops can benefit (3.4% ± 2.6) in comparison to winter sown crops. Piccoli et al. (2020) found that 55 years of residue incorporation provided a nutritional effect to maize crops and to wheat and sugar beet crops when combined with poultry manure, thereby reducing the need for fertilization. Similar to Hijbeek et al. (2017), spring sown crops were found to benefit from residue amendments, either via improving soil structure and facilitating root growth or the coordination of crop residue mineralization with spring-summer crop growth. Albano et al. (2023) investigated the impact of different organic amendments and mineral fertilizer rates on a winter wheat rotation. The results indicated that plots receiving a combination of organic amendments and mineral fertilizer required lower nitrogen rates to achieve optimum yields. Dămătîrcă et al. (2023) show that for a maize monocropping system, crop residue incorporation combined with organic fertilization increased C and N to 90 cm depth. By contrast, crop residue removal with mineral fertilization depleted soil organic matter to 90 cm depth. Numerous studies also illustrate the importance of residue retention enhancing the availability of P and K for plant utilization after decomposition (Reddy et al., 2025). For example, Gupta et al. (2024) found that retaining rice and wheat residues was equivalent to 30–60 kg P_2O_5 ha^{-1} in terms of grain yield.

Overall, crop residues constitute a significant source of P and K for subsequent crops, where their removal reduces soil P and K nutrient availability and negatively impacts soil nutrient dynamics. Although mineral fertilizers remain necessary to meet crop nitrogen (N) requirements, recommended N fertilizer rates can be adjusted to account for the N contributed by crop residues. One consideration is related to the legacy effect exerted by crop residue that has been shown to impact the dry matter and N grain yield of a crop >5 years after initial sowing (Pullens et al., 2021). Understanding the synergies between residue and fertilizer supports the implementation of more sustainable nutrient management practices.

4 Soil organic matter

Soil organic matter (SOM) contains the largest terrestrial pool of organic carbon (OC) holding approximately 2344 Gt of OC globally, with 54% or 1500 Gt of OC stored in the first 1 m (Jobbágy and Jackson, 2000), more than the vegetation and atmosphere combined (Lehmann and Kleber, 2015). Soil organic matter is formed through the partial decomposition and transformation of plant inputs by soil organisms. These inputs consist of aboveground and belowground residue plus root exudates which vary in their decomposition rates and are further incorporated and stabilized into long-term OM storage.

Recent work has shown that particulate organic matter (POM) and mineral-associated organic matter (MAOM) are the primary fractions governing soil organic matter formation and carbon storage in agricultural soils (Cotrufo and Lavellee, 2022). These fractions are defined operationally: POM consists of low-density, coarse organic particles that are not water-extractable, and can occur either freely in soil or physically protected within aggregates. In contrast, MAOM comprises higher-density organic matter associated with silt- and clay-sized minerals. Dissolved organic matter (DOM) represents the water-extractable fraction of organic matter. According to the Microbial Efficiency–Matrix Stabilization (MEMS) framework, MAOM forms through sorption of plant-derived DOM alongside microbial by-products and necromass (Cotrufo et al., 2013).

Across a range of laboratory and field studies, residue-derived carbon (C) is consistently partitioned into particulate organic matter (POM) and mineral-associated organic matter (MAOM) in proportions that depend on residue type, placement, and soil conditions. Root residues generally yield greater soil C retention than aboveground residues and tend to contribute disproportionately to POM (Fulton-Smith and Cotrufo, 2019; Gale and Cambardarella, 2000). Incorporation of residues into soil enhances SOC formation relative to surface application (Leichty et al., 2020). Short-term studies show that a substantial fraction of residue C enters MAOM even within weeks to months, although POM often dominates early C accrual, particularly in systems receiving fresh plant materials (Córdova et al., 2018; Chen et al., 2024). Long-term field experiments indicate that MAOM represents the majority of total SOC, where, for example, Just et al. (2023) reported that MAOM carbon represented 76±11% (mean ± SD) of the total OC concentrations across 25 Central European long-term field experiments (23 arable and 2 grassland sites).

Crop residue management strongly influences the balance of POM-C and MAOM-C formation and thereby regulates SOC accumulation across different cropping systems and soil environments. A global meta-analysis of 323 field experiments showed that harvesting crop residue reduced SOC stocks by 11% and that the majority of the SOC stock depletion occurred across the whole soil profile within the first 10 years of residue removal (Alvarez, 2024). Ingwersen et al. (2024) also highlighted that ca. 3–4 tC $ha^{-1}y^{-1}$ of straw is required to maintain the SOC stock for the wheat-maize cropping rotation in Germany. Thus, crop residue is critical for the long-term maintenance of SOM in cropland systems.

5 Soil aggregates

Soil structure is the arrangement of particles in soil and the particles of sand, silt and clay, bound together into aggregates of various sizes by organic and inorganic materials. Aggregate stability is critical soil property playing a key role in regulating hydrology, aeration, biological activity and soil fertility (Dai et al., 2025). Aggregates are typically grouped into different size classes with a hierarchy consisting of macroaggregates and microaggregates (Sarker et al., 2018). Soil organic matter contributes to aggregate formation via numerous processes including the decomposition of residues, animal excreta, root exudates, microbial by-products, microbial necromass and fungal hyphae (de Goede et al., 2025; Liang et al., 2019). Inorganic materials such as clay minerals, iron and aluminium oxides (Fe-Al) and carbonates act as the binding agents for soil aggregate formation (Six et al., 2004). Soil aggregation is thus an important process not only for soil structure but also for SOC storage through the physical protection of organic matter from decomposers (Six and Paustian, 2014).

Aggregate formation and stability are strongly influenced by cropland management practices. Macroaggregate formation is primarily driven by plant-derived carbon (Gunina and Kuzykayov, 2014). Husain and Dijkstra (2023) observed a negative relationship between plant residue amendments with microaggregates and silt and clay fractions but an increase related to the formation of macroaggregates. Similarly, De Gryze et al. (2005) observed that aggregate formation increased linearly with increasing amounts of wheat residue at a rate of 12.0 ± 1.24 g aggregates g^{-1} residue added. Coppens et al. (2006) reported that the initial location of oilseed rape crop residue did not significantly affect the incorporation of residue-C in soil aggregates (only 10–12% of carbon applied), as new C entered the aggregate fractions mainly as soluble C. Sonsri and Wattanabe (2023) demonstrated that high molecular weight components such as polysaccharides and water extractable organic matter (a proxy source of dissolved organic matter) promoted macroaggregate formation and aggregate stability. The formation of macroaggregates facilitates the development of microaggregates and the spatial organization of aggregates determines the pore network, which is critical for the movement of liquid, gaseous and colloidal particles in the soil (Totsche et al., 2010; Totsche et al., 2018). Consequently, crop residues and their management play a central role in soil aggregate formation, with important implications for soil organic carbon (SOC) sequestration and overall soil functioning.

6 Soil moisture

Crop residues also play an important role in modulating soil moisture. Soil moisture is a key indicator of water availability and is a limiting factor for crop yields. Soil moisture deficits can have profound effects on crop productivity such that agricultural practices that promote water conservation are adopted to mitigate against the effects of agricultural drought. Crop residues help regulate water loss, where residue left on the surface to mulch can reduce evaporation from bare soil. The adoption of non-inversion tillage combined with mulching in drier climates is specifically aimed at improving water use efficiency.

Crop residues act as thermal insulators on the soil surface where the mulch tends to lower soil temperatures and increase soil moisture contents, principally due to changes in surface radiative fluxes (Horton et al., 1996). In the US Midwest, Klocke et al. (2009) showed that a soil surface covered with 25–75% with corn stover exhibited slight reductions in evaporation compared with bare soil in the absence of a crop canopy. Evaporation was reduced by 50–65% when the soil was covered with 100% corn stover, compared with evaporation from bare soil with no shading. Flerchinger et al. (2003) examined the effects of residue type and configuration on heat and water transfer in three locations in the US. The authors show a contrast in evaporation from highest to lowest between flat wheat residue and bare soil. In their meta-analysis, Ranaivoson et al. (2017) found that: (i) 8 t ha^{-1} of residues were needed to decrease soil water evaporation by about 30% in Conservation Agriculture systems compared to no-till bare soil and (ii) 2 t ha^{-1} residue was required to achieve the maximum effect on soil water infiltration, water runoff and soil loss control. In a canola cropping system, Ward et al. (2015) observed that soils directly underneath crop or crop residue rows absorbed 82–93% of rainfall. However, in the inter-row spaces, tilled soil absorbed 19–30% of event rainfall, compared with 58–78% for inter-row spaces in the no-tilled soil, highlighting the dual importance of no tillage plus crop residue in aiding water conservation.

7 Soil erosion

Agricultural practices such as ploughing, sowing, harvesting, as well as residue removal due to baling or burning can lead to increased risk of soil erosion. Retaining crop residue on the surface in the form of stubble or mulching can minimize soil erosion via rain or wind. Effective control of soil erosion depends on minimizing the impact of raindrops and the velocity of running water on the soil surface (Andraski et al., 1985; Timm et al., 2002). This involves protecting the soil surface by means of a cover crop or mulch to prevent the rain from striking the soil directly, avoiding practices that either compact the soil (thus reducing infiltration and increasing runoff) or pulverize it excessively, thus making the soil more erodible (Hillel, 2003). Bare soils are susceptible to wind erosion, where continuous soil cover either with crops or residues can reduce the effects significantly.

Tebrugge and During (1999) show that surface cover by crop residues and higher aggregate stability under no-tillage protected soil fertility by avoiding surface sealing and erosion in an 18-year study in Germany. In their review of 66 studies, Klopp and Blanco-Canqui (2022) found that residue removal rates >50% adversely affected soil physical properties (reduced wet aggregate stability, dry aggregate stability and water retention). To nullify these effects, around 4 t ha^{-1} of residue must be retained on the field. Nunes et al. (2020) showed that mulching showed a soil erosion reduction efficiency of 60% with results varying by climatic and physical scenarios. Fan et al. (2023) conducted a global meta-analysis based on 421 runoff and 512 soil loss observations from 90 publications to evaluate the effectiveness of mulching in reducing runoff and soil loss across a variety of variables (i.e. mulch type, mulch coverage and application rates, rainfall intensity, land use, soil texture, slope gradient and slope length). The authors show that mulching significantly reduced runoff and soil loss by 47.4% and 76.2%, respectively. It is also suggested that mulch coverage should be at least 60% to balance economic costs and ecological benefits.

8 Soil greenhouse gas emissions

Crop residue management is considered important for sustaining soil fertility and SOM levels in agricultural soils. Exactly how this residue is managed, comprising plant shoots, roots, plus post-harvest straw and stubble can have large impacts on long-term C and greenhouse gas (GHG) budgets (Buysse et al., 2013).

Although primarily examined for their effect on SOM accumulation, the rate of residue decomposition (i.e. mineralization) and associated CO_2 emissions can be used to evaluate its potential for contributing to soil organic matter (Johnson et al., 2007). Ntonta et al. (2022) examined data assessing cumulative CO_2 emissions of 17 crops from 394 trials globally during 0–30, 0–90 and 0–120 days and to relate the results with residue quality (C, N and lignin concentrations) and selected soil properties (texture, pH and SOC concentration). Legumes exhibited the highest CO_2 emissions per gram of C added (1003 mg CO_2-C g^{-1}C after 120 days) followed by grasses (947), oilseed crops (944) and cereals

(846) for all time periods. Maize residues emitted the least C-CO_2 (86, 275 and 495 mg CO_2-C g^{-1}C) for all durations, followed by cotton and sunflower, each characterized by their relatively higher lignin contents.

Crop residues, the biomass left on the soil after harvest of the main crop parts, can also be a direct and indirect source of N_2O emissions, which results in climate impacts (Galdos et al., 2023). The fallow period is an important period of the cropping cycle contributing to 44% of the annual total N_2O emissions globally (Shang et al., 2024). The residue remaining is typically left distributed on the soil surface to decompose and is eventually incorporated into the soil via ploughing. Mineralization of residue occurs during this period in the absence of crop photosynthesis leading to potential increases in GHG emissions. Garnier et al. (2024) reported the highest N_2O emissions during fallow periods were repeatedly observed after a legume crop (alfalfa, green bean and faba bean) due to asynchrony between the biological N fixation and the release of mineral N during residue decomposition (Basche et al., 2014). However, Maier et al. (2025) reported lower post-harvest N_2O emission in pea and maize compared to rapeseed and winter wheat, attributed to dry and warm (pea) plus cool and wet (maize) environmental conditions suppressing denitrification activity in the soil (Butterbach-Bahl et al., 2013). In their meta-analysis, Chen et al. (2013) also noted that residue effects were negatively associated with C:N ratios, but generally residue amendment could not reduce soil N_2O emissions, even for C:N ratios well above *ca.* 30, the threshold for net N immobilization. Residue effects were also comparable to, if not greater than, those of synthetic N fertilizers. In addition, residue effects on soil N_2O emissions were positively related to the amounts of residue C input as well as residue effects on soil CO_2 respiration. Furthermore, most significant and stimulatory effects occurred at 60–90% soil water-filled pore space and soil pH 7.1–7.8. Stimulatory effects were also present for all soil textures except sand or clay content ≤10%, which is in agreement with Badagliacca et al. (2018) that soil characteristics control the magnitude of N_2O production.

In their review, Abalos et al. (2022) found that, on average, crop residue incorporation increased soil N_2O emissions by 43% compared to residue removal; however, crop residues led to both increases and reductions in N_2O emissions. Residue maturity class can be used to predict the emission magnitude, whereby immature residues strongly stimulated N_2O emissions whereas mature had marginal effects on N_2O emissions. Recently, Olesen et al. (2023) also highlight the challenges with estimating N_2O from crop residue and suggest distinguishing N_2O EFs between mature (straw) and immature (cover crops, grass, legumes and vegetables). Essich et al. (2020) found that the removal of sugar beet residue reduced N_2O emissions by 87–95% in conventionally tilled soils. A meta-analysis conducted by the same authors with 176 datasets of rain-fed arable crops found that residue removal resulted in a

11% reduction in N_2O emissions overall. Other long-term studies examining crop residue incorporation revealed a 12-fold increase in N_2O emissions on average, but also an average 7% increase in SOC concentration (Lehtinen et al., 2014); however, other studies have found no significant effects of crop residue retention or incorporation on N_2O release (Malhi and Lemke, 2007).

Valkama et al. (2024) reviewed whether different OM inputs could mitigate N_2O emissions over 50 field experiments encompassing arable cropping systems on mineral soils in 15 European countries. Inputs of crop residues tended to decrease N_2O emissions per unit area, but not significantly, as also observed in previous studies incorporating cereal residues (Rahman et al., 2024; Sanz-Cobena et al., 2014). Residue quality, i.e. legumes with low C:N ratios by contrast, tends to stimulate N_2O emissions (Basche et al., 2014).

Crop residue can also contribute to methane (CH_4) emissions from agricultural systems. Crop residue burning is largely practised by farmers in China, India and Russia, whereas the practice of stubble burning is largely banned in Europe. In China, about 20–30% of post-harvest crop residues are burned by farmers resulting in emissions of CO_2 and CH_4 and particulates to the atmosphere (Yan et al., 2006; Chen et al., 2019). Rice cultivation can be a significant source of CH_4 systems. Rice straw is considered a low-quality animal feed due to its low protein and high silica content (Mandal et al., 2004). The straw is either burned, incorporated into the soil following cultivation or left on the field surface. Rice straw is the primary source of C for CH_4 production during the early growth period of rice plants (Watanabe et al., 1994), and consequently, residue incorporation leads to increased CH_4 emissions under flooded conditions. Cordeiro et al. (2024) found that soil incorporation of residues increased CH_4 emissions relative to open-burning in eastern India. Moreover, wetter sites had 2.8–4.0 times greater CH_4 emissions, on average, than dry sites when rice residues were returned to the soil. Sander et al. (2014) reported higher CH_4 emissions during the fallow period, in agreement with earlier studies showing lower emissions compared to drained fields. Methane emissions in rice-based systems could outweigh the benefits from carbon sequestration achieved through building SOC with crop residue recycling at least in the short term (Jiang et al., 2019); thus, residue management is of key importance to the net GHG budgets in comparison to other cereal-based systems. The positive effects of crop residue often outweigh the negative impacts related to GHGs, where numerous measures can be adopted to reduce the magnitude of residue-induced emissions (Abalos et al., 2022).

9 Albedo effects

Land cover and land management are important factors governing the energy balance of the surface and could therefore play an important role in mitigating climate change. Albedo, the ratio of reflected to incoming shortwave, is one of the most important biogeophysical drivers of climate at local and regional scales (Myhre and Myhre, 2003). Increased surface albedo can lead to a negative radiative forcing and a subsequent cooling effect on the climate (Song, 1999). The albedo of the surface can be influenced by vegetation type, canopy cover, plant phenology, solar angle and soil texture, which in turn influences surface energy partitioning (Bright et al., 2015). Thus there is increasing interest in understanding the interaction between land management, surface albedo and climate change.

In the context of croplands, albedo can be influenced by crop type (encompassing growing season length) and management practices such as tillage, mulching and cover cropping (Sieber 2022; Davin et al., 2014). Increases in albedo have been found for reduced tillage and no-tillage systems compared to conventional tillage systems due to increased crop residue cover in the latter (Lobell et al., 2006). Lei et al. (2024) observed that the no-tillage 5-year rotations of corn-soybean–winter wheat exhibited a higher albedo compared to conventional tillage rotations. The authors attributed the effect to the retention of residue in the no-tillage treatments, as also measured by Sieber et al. (2022). Sieber et al. (2022) also evaluated the inclusion of albedo in the life cycle assessment of a crop, given its role in cooling the climate. The authors found that the increased albedo countered the GHG impact (GWP_{100}) from production of inputs and field operations by 17–47%. Moreover, all crops displayed a net cooling impact on global mean surface temperature on shorter timescales due to albedo (3–12 years under different crops), but a net warming impact on longer timescales due to GHG emissions (e.g. N_2O). Interestingly, Davin et al. (2014) show that a no-tillage system can lead to reduced evaporative cooling during heatwaves, thereby counteracting an albedo-induced cooling effect associated with surface crop residue retention. Thus, residue management in addition to vegetative cover could play an important role in cropland albedo management.

10 Case study

The AgreenaCarbon project is a Pan-European soil carbon programme assisting farmers transition from conventional to regenerative agriculture practices. Regenerative agriculture describes holistic farming practices that improve soil fertility and enhance the storage of SOC whilst maintaining or increasing yields. The project activities that are monitored on farms include: reduction in tillage intensity, leaving crop residues on the field, planting cover or catch crops during the post-harvest/fallow period and reducing synthetic fertilizer application and/or replacing synthetic with organic amendments.

For this case study we describe the impacts of crop residue management on SOC sequestration rates associated with the project. A subset of 558 fields across 6 countries (Bulgaria, Denmark, Estonia, Spain, Ukraine and the United Kingdom) were examined in 2022 where only residue management practice was improved (i.e. changed from removed to mulched). Crop yields were reported by farmers, and crop residue production was calculated from the reported crop yield using the yield-to-residue equations found in the IPCC 2019 Refinement guidelines. The amount of residue was subsequently used for the 'biomass input' parameter in the Rothamsted Carbon model (RothC, Coleman and Jenkinson, 1996) using the following equation:

$$In_t = \sum_{1}^{c}\left(Intercept_{c,t} + Yield_{c,t} \times DMF_{c,t} \times Slope_{c,t}\right) \times Litter\,C$$

where In_t is the input of carbon from crop residue for a field at time step t, c is crop type, $Yield_{c,t}$ is the grain/seed yield per crop type, $DMF_{c,t}$ is the dry matter fraction for specific crop types, $Slope_{c,t}$ is a parameter for estimating aboveground residue (IPCC, 2019), $Intercept_{c,t}$ is a parameter for estimating aboveground residue (IPCC, 2019) and *LitterC* is the carbon content of plant residue (assumed to be 0.44).

The modelled results indicate a mean SOC sequestration rate (+ SD) of 0.27 ± 0.19 tC $ha^{-1}y^{-1}$ attributed to mulching crop residue, and the values are compared with those reported for straw residue management in the literature (Fig. 2). Smith (2008) reported that management practices such as no-till or full residue retention can result in C sequestration rates in arable soils of 0.3–0.8 tC $ha^{-1}y^{-1}$ for about two decades. Powlson et al. (2011) reported SOC% changes ranging from -3.7% to 37.5% in 23 long-term experiments whilst examining the effects of straw incorporation vs. straw removal. The authors also reported maximum increases of 10.2–22.0% in topsoil OC at 2 UK sites after 7–22 years of straw incorporation. Lugato et al. (2014) reported averaged SOC change rates for a cereal straw incorporation scenario for EU-27 of 0.1 and 0.04 tC $ha^{-1}yr^{-1}$. Minasny et al. (2017) reported that residue incorporation could increase SOC accumulation rates by ~0.35 tC $ha^{-1}yr^{-1}$. Lessman et al. (2022) reported an average SOC stock change of 0.21 tC $ha^{-1}yr^{-1}$. Haas et al. (2022) used the bio-geochemical models CERES-EGC (Gabrielle et al., 2006a, 2006b) and the model framework LandscapeDNDC (Haas et al., 2022) to quantify the long-term residue impact of SOC sequestration and N_2O emissions for the EU-27 croplands. They show that 100% residue return leads for ca. 20 years to C sequestration rates of 0.6% y^{-1} equating to 0.35 and 0.21 tC $ha^{-1}y^{-1}$ (left on the surface or incorporated, respectively). This gives a residue soil C-sequestration potential ranging from 14 tCO_2e ha^{-1} to 30 tCO_2e ha^{-1} (0–30 cm). However, Piccoli et al. (2024) reported an SOC storage of 6 tC ha^{-1} in the 0–60 cm soil profile following 127 t ha^{-1} residue-derived C input, lower than that of Haas et al. (2022). The authors calculated a conversion rate of 4.7% of residue-C to SOC

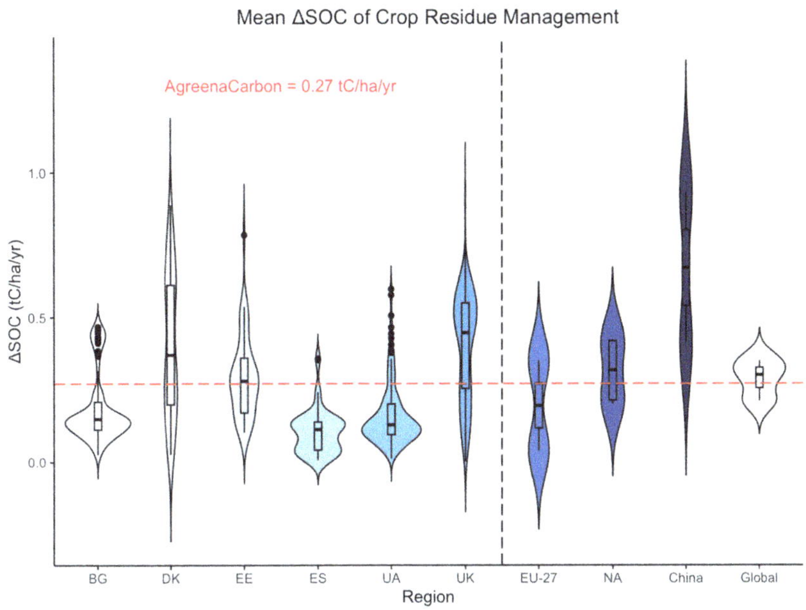

Figure 2 A violin plot of the modelled SOC sequestration rates for fields mulching crop residue in the AgreenaCarbon programme (Bulgaria, BG; Denmark, DK; Estonia, EE; Spain, ES; Ukraine, UA; and United Kingdom, UK) in comparison to estimates derived from the literature (right of the dashed line) for the EU-27 (Haas et al., 2022; Lugato et al., 2014), North America and Canada (Smith et al., 2012), China (Rui and Zhang, 2010; Wang et al., 2020) and Global (Smith, 2008; Minasny et al., 2017; Lessman et al., 2022).

for this long-term (55 years) residue incorporation experiment. Schjønning et al. (2018) showed a 13% increase in SOC content due to residue incorporation at three sites in Denmark in the 0–30 cm soil layer. By contrast, the negative effects of crop residue removal on soil carbon may outweigh the effects of residue retention. Smith et al. (2012) estimated SOC changes from crop residue removal in Canada and Midwest US of 0.423, 0.417, 0.201 and 0.218 tC $ha^{-1}y^{-1}$ for the CENTURY, DAYCENT, DNDC and CAMPBELL models, respectively. Liska et al. (2014) predicted that the removal of 6 t $ha^{-1}y^{-1}$ corn residue for biofuel could decrease regional net SOC by an average of 0.47–0.66 tC $ha^{-1}y^{-1}$ over 5 to 10 years.

The differences in SOC are attributed to numerous factors including climate zone, management practices, annual crop yields, soil type and soil texture. Residue retention, especially surface applied, will increase SOC levels and contribute to improved soil fertility and climate resilience. Haas et al. (2022) suggest appropriate fertilizer application rates can assist with achieving GHG mitigation with crop residue management attributed to the N_2O or CH_4 emissions with the increased availability of N and C for microorganisms

(Paustian et al., 2016). The net cooling effect associated with crop albedo and residue management (surface cover vs. bare soil) may also be explored in order to understand the full extent of residues on the GHG balance of arable cropping systems (Desmet et al., 2025).

11 Conclusion and future trends

This chapter discusses the multifaceted role of crop residues in agriculture, emphasizing their significance for sustainable farming practices. It outlines how residues represent a substantial portion of annual phytomass harvested globally and can be perceived either as a nuisance or as a valuable resource. Crop residue management influences plant nutrient availability, soil structure, hydrology, organic matter dynamics, greenhouse gas emissions and surface albedo. The general consensus in the literature is that leaving residues on fields enhances soil health by reducing erosion, retaining moisture, improving nutrient cycling and increasing soil carbon content. However, competing demands for crop residues including other uses such as cooking, building materials, bioenergy, animal feed and bedding pose a challenge for Conservation Agriculture and require site-specific assessments of the optimal amount to be sustainably removed from fields. A case study illustrates the potential of crop residue mulching to sequester soil carbon, comparing modelled rates with literature values, and highlights the complexities involved in managing residues for optimal environmental and agronomic outcomes.

According to Panagos et al. (2025), ca. 60% of European soils are described as unhealthy, implying that the soils are degraded in terms of their physical, chemical and biological condition. The ability of soils to carry out multi-functional processes related to food production, water storage, carbon sequestration and support biological activity diminishes with increasing degradation (Schulte et al., 2014). Crop residue management can play a central role in alleviating issues related to soil degradation such as reducing soil erosion, SOC losses and GHG losses associated with inefficient nutrient utilization (Panagos et al., 2025). Moreover, crop residues are involved in three measures (organic fertilization via residue incorporation, permanent soil cover and conservation tillage) that when implemented contribute to agricultural soil functioning (Strauss et al., 2023). With the introduction of the EU soil health monitoring law, it's expected that sustainable crop residue management will play a significant role in the future protection of soil fertility and health.

In an effort to optimize the climate mitigation and nutrient cycling components of residue management, there is emerging research on the application of biopreparations to accelerate the decomposition of straw on the field (Rowinska et al. 2024). These microbial and enzymatic biopreparations can be useful when there are large quantities of residues and when climate conditions are not favourable for crop residue decomposition (Rusakova, 2016). However, the widespread use of biopreparations for crop residue decomposition faces standardization and regulatory challenges that are common to biological products in general.

Crop residue decomposition is influenced by a wide range of site-specific parameters including precipitation, air and soil temperature, soil moisture, soil texture, C:N ratio of residues, lignin content, etc. Process-based models, simulating key processes in the crop/soil/atmosphere interface and integrating climate, soil and management factors, can be useful in assessing sustainable uses of crop residues. This approach has been applied to identify the optimal amount of residue to be sustainably removed from subfield (Muth Jr et al. 2012) to continental scales (Monforti et al. 2015). Scenario analysis and modelling approaches are particularly useful to account for the impact of climate change and alternative management practices on biogeochemical processes related to crop residue decomposition, nutrient cycling, carbon dynamics, plant growth and crop yields. Leveraging remote sensing, geospatial analysis and machine learning to optimize crop residue utilization in a spatially explicit manner, whilst considering various factors such as soil, climate, harvesting equipment, logistics, agronomy and policy (Haase et al. 2016, Yang et al. 2025), represents a growing research area enabled by increased data availability and analytic capacity.

12 Where to look for further information

The following articles provide a good overview of the subject:

- Smil, V. (1999), 'Crop residues: agriculture's largest harvest'. *BioScience*, 49, 299–308.
- Turmel, M.-S., et al. (2015), 'Crop residue management and soil health: a systems analysis'. *Agricultural Systems*, 134, 6–16.

Key research in this area can be found at the following organizations:

- Rothamsted Research (www.rothamsted.ac.uk).
- Kellogg Biological Station Long-term Agroecosystem Research Site (https://www.canr.msu.edu/ltar/).
- Joint Research Centre – European Commission (https://commission.europa.eu/about/departments-and-executive-agencies/joint-research-centre_en).
- French Agricultural Research Centre for International Development (CIRAD) (https://www.cirad.fr/en).
- Bangladesh Agricultural Research Institute (BARI) (https://bari.gov.bd).

Major international research projects:

- Global Dataset of crop residues from cereals: A global dataset for the production and usage of cereal residues in the period 1997–2021 | Scientific Data.
- AgriLoop is an EU-funded project developing processes to create innovative high-value products from agricultural waste through sustainable supply chains. https://www.agriloop-project.eu/.

13 References

Abalos, D., et al. (2022), 'Predicting field N_2O emissions from crop residues based on their biochemical composition: a meta-analytical approach'. *Science of the Total Environment*, 812, 152532. https://doi.org/10.1016/j.scitotenv.2021.152532.

Abiven, S. and Recous, S. (2007), 'Mineralisation of crop residues on the soil surface or incorporated in the soil under controlled conditions'. *Biol Fertil Soils*, 43, 849–852. https://doi.org/10.1007/s00374-007-0165-2.

Akhila, A. and Entoori, K. (2022), 'Role of earthworms in soil fertility and its impact on agriculture: a review'. *International journal of fauna and biological studies*, 9(3), 55–63. https://doi.org/10.22271/23940522.2022.v9.i3a.907.

Albano, X., et al. (2023), 'Effect of different organic amendments on actual and achievable yields in a cereal-based cropping system'. *Journal of Soil Science and Plant Nutrition*, 23(2), 2122–2137.

Alvarez, R. (2024), 'A quantitative review of the effects of residue removing on soil organic carbon in croplands'. *Soil and Tillage Research*, 240, 106098. https://doi.org/10.1016/j.still.2024.106098.

Andraski, B. J., et al. (1985), 'Effects of tillage and rainfall simulation date on water and soil losses'. *Soil Science Society of America Journal*, 49(6), 1512–1517. https://doi.org/10.2136/sssaj1985.03615995004900060036x.

Badagliacca, G., et al. (2018), 'Long-term no-tillage application increases soil organic carbon, nitrous oxide emissions and faba bean (Vicia faba L.) yields under rain-fed Mediterranean conditions'. *Science of the Total Environment*, 639, 350–359. https://doi.org/10.1016/j.scitotenv.2018.05.157.

Basche, A. D., et al. (2014), 'Do cover crops increase or decrease nitrous oxide emissions? A meta-analysis'. *Journal of Soil and Water Conservation*, 69, 471–482. https://doi.org/10.2489/jswc.69.6.471.

Bond, T. C., et al. (2004), 'A technology based global inventory of black and organic carbon emissions from combustion'. *Journal of Geophysical Research*, 109, 1–43. https://doi.org/10.1029/2003JD003697.

Breza, L. C. and Grandy, A. S. (2025), 'Organic amendments tighten nitrogen cycling in agricultural soils: a meta-analysis on gross nitrogen flux'. *Frontiers in Agronomy*, 7, 1472749. https://doi.org/10.3389/fagro.2025.1472749.

Bright, R. M., et al. (2015), 'Radiative forcing bias of simulated surface albedo modifications linked to forest cover changes at northern latitudes'. *Biogeosciences*, 12(7), 2195–2205. https://doi.org/10.5194/bg-12-2195-2015.

Briones, M. J. I. and Schmidt, O. (2017), 'Conventional tillage decreases the abundance and biomass of earthworms and alters their community structure in a global meta-analysis'. *Global change biology*, 23(10), 4396–4419. https://doi.org/10.1111/gcb.13744.

Butterbach-Bahl, K., et al. (2013), 'Nitrous oxide emissions from soils: how well do we understand the processes and their controls?'. *Philosophical Transactions of the Royal Society B: Biological Sciences*, 368(1621), 20130122. https://doi.org/10.1098/rstb.2013.0122.

Buysse, P., et al. (2013), 'Fifty years of contrasted residue management of an agricultural crop: impacts on the soil carbon budget and on soil heterotrophic respiration'. *Agriculture, Ecosystems and Environment,* 167, 52–59. https://doi.org/10.1016/j.agee.2013.01.006.

Cao, Y., et al. (2021), 'Organic-C quality as a key driver of microbial nitrogen immobilization in soil: a meta-analysis'. *Geoderma*, 383, 114784. https://doi.org/10.1016/j.geoderma.2020.114784.

Constantinides, M. and Fownes, J.H. (1994) Nitrogen mineralization from leaves and litter of tropical plants: relationship to nitrogen, lignin and soluble polyphenol concentrations. *Soil Biology and Biochemistry*, 26, 49–55. https://doi.org/10.1016/0038-0717(94)90194-5.

Córdova, S., et al. (2018), 'Plant litter quality affects the accumulation rate, composition and stability of mineral-associated soil organic matter.' *Soil Biology and Biochemistry* 125 (2018): 115–124. https://doi.org/10.1016/j.soilbio.2018.07.010.

Chaves, B., et al. (2021), 'The combination of residue quality, residue placement and soil mineral N content drives C and N dynamics by modifying N availability to microbial decomposers'. *Soil Biology and Biochemistry*, 163, 108434. https://doi.org/10.1016/j.soilbio.2021.108434.

Chen, B., et al. (2014), 'Soil nitrogen dynamics and crop residues: a review'. *Agronomy for sustainable development*, 34(2), 429–442. https://doi.org/10.1007/s13593-014-0207-8.

Chen, H., et al. (2013), 'Soil nitrous oxide emissions following crop residue addition: a meta-analysis'. *Global Change Biology,* 19(10), 2956–2964. https://doi.org/10.1111/gcb.12274.

Chen, J., et al. (2019), 'To burn or retain crop residues on croplands? An integrated analysis of crop residue management in China'. *Science of the Total Environment*, 662, 141–150. https://doi.org/10.1016/j.scitotenv.2019.01.150.

Chen, S., et al. (2024), 'Changes in aggregate-associated carbon pools and chemical composition of topsoil organic matter following crop residue amendment in forms of straw, manure and biochar in a paddy soil'. *Geoderma*, 448, 116967. https://doi.org/10.1016/j.geoderma.2024.116967.

Coleman, K. and Jenkinson, D.S. (1996) RothC-26.3 – a model for the turnover of carbon in soil. In: Powlson, D.S., Smith, P. and Smith, J.U. (eds) Evaluation of Soil Organic Matter Models. NATO ASI Series I, Vol. 38. Springer, Berlin, pp. 237–246.

Coppens, F., et al. (2006), 'Impact of crop residue location on carbon and nitrogen distribution in soil and in water-stable aggregates'. *European Journal of Soil Science*, 57(4), 570–582. https://doi.org/10.1111/j.1365-2389.2006.00825.x.

Cordeiro, E. U., et al. (2024), 'The fate of rice crop residues and context-dependent greenhouse gas emissions: Model-based insights from Eastern India'. *Journal of Cleaner Production*, 435, 140240. https://doi.org/10.1016/j.jclepro.2023.140240.

Cotrufo, M. F., et al. (2013), 'The Microbial Efficiency-Matrix Stabilization (MEMS) framework integrates plant litter decomposition with soil organic matter stabilization: do labile plant inputs form stable soil organic matter?'. *Global Change Biology*, 19(4), 988–995. https://doi.org/10.1111/gcb.12113.

Cotrufo, M. F. and Lavallee, J. M. (2022), 'Soil organic matter formation, persistence and functioning: a synthesis of current understanding to inform its conservation and regeneration'. *Advances in Agronomy*, 172, 1–66. https://doi.org/10.1016/bs.agron.2021.11.002.

Dai, W., et al. (2025), 'Soil aggregate stability and erodibility as influenced by soil amendments and winter cover crop in upland soils'. *Soil Science Society of America Journal*, 89(1), e70022. https://doi.org/10.1002/saj2.70022.

Dămătîrcă, C., et al. (2023), 'Residue incorporation and organic fertilisation improve carbon and nitrogen turnover and stabilisation in maize monocropping'. *Agriculture, Ecosystems and Environment*, 342. https://doi.org/10.1016/j.agee.2022.108255.

Damon, P. M., et al. (2014), 'Crop residue contributions to phosphorus pools in agricultural soils: a review'. *Soil Biology and Biochemistry*, 74, 127–137. https://doi.org/10.1016/j.soilbio.2014.03.003.

Davin, E. L., et al. (2014), 'Preferential cooling of hot extremes from cropland albedo management'. *Proceedings of the National Academy of Sciences*, 111(27), 9757–9761. https://doi.org/10.1073/pnas.1317323111.

de Goede, S. P., et al. (2025), 'Fungal-mediated soil aggregation as a mechanism for carbon stabilization'. The ISME Journal, wraf074. https://doi.org/10.1093/ismejo/wraf074.

De Gryze, S., et al. (2005), 'A quantification of short-term macroaggregate dynamics: influences of wheat residue input and texture'. *Soil Biology and Biochemistry*, 37(1), 55–66. https://doi.org/10.1016/j.soilbio.2004.07.024.

Desmet, F., et al. (2025), 'Accounting for albedo changes in carbon farming schemes'. *Science of The Total Environment*, 1002, 180631. https://doi.org/10.1016/j.scitotenv.2025.180631.

Dhillon, J., et al. (2017), 'World phosphorus use efficiency in cereal crops'. *Agronomy Journal*, 109(4), 1670–1677. https://doi.org/10.2134/agronj2016.08.0483.

Essich, L., et al. (2020), 'Is crop residue removal to reduce N_2O emissions driven by quality or quantity? A field study and meta-analysis'. *Agriculture*, 10(11), 546. https://www.mdpi.com/2077-0472/10/11/546.

Fan, D., et al. (2023), 'The effectiveness of mulching practices on water erosion control: a global meta-analysis'. *Geoderma*, 438, 116643. https://doi.org/10.1016/j.geoderma.2023.116643.

Field, J. L., et al. (2012), 'Distributed biochar and bioenergy coproduction: a regionally specific case study of environmental benefits and economic impacts'. *Global Change Biology Bioenergy* 5, 177–191. https://doi.org/10.1111/gcbb.12032.

Flerchinger, G. N., et al. (2003), 'Effects of crop residue cover and architecture on heat and water transfer at the soil surface'. *Geoderma*, 116(1–2), 217–233. https://doi.org/10.1016/S0016-7061(03)00102-2.

Fulton-Smith, S. and Cotrufo, M. F. (2019), 'Pathways of soil organic matter formation from above and belowground inputs in a Sorghum bicolor bioenergy crop'. *GCB Bioenergy* 11(8), 971–987. https://doi.org/10.1111/gcbb.12598.

Gabrielle, B., et al. (2006a), Simulation of nitrous oxide emissions from wheat-cropped soils using CERES. Nutrient Cycling in Agroecosystems, 74, 133–146. https://doi.org/10.1007/s10705-005-5771-5.

Gabrielle, B., et al. (2006b), Process-based modeling of nitrous oxide emissions from wheat-cropped soils at the sub-regional scale. Global Biogeochemical Cycles, 20, GB4018. https://doi.org/10.1029/2005GB002618.

Galdos, M. V., et al. (2013), 'Trends in global warming and human health impacts related to Brazilian sugarcane ethanol production considering black carbon emissions'. *Applied Energy* 104, 576–582. https://doi.org/10.1016/j.apenergy.2012.11.002.

Galdos, M. V., et al. (2023), 'Multi-experiment assessment of soil nitrous oxide emissions in sugarcane'. *Nutrient Cycling in Agroecosystems* 127, 375–392. https://doi.org/10.1007/s10705-023-10321-w.

Gale, W. J. and Cambardella, C. A. (2000), 'Carbon dynamics of surface residue–and root-derived organic matter under simulated no-till'. *Soil Science Society of America Journal*, 64(1), 190–195. https://doi.org/10.2136/sssaj2000.641190x.

Garnier, J., et al. (2024), 'Six years of nitrous oxide emissions from temperate cropping systems under real-farm rotational management'. *Agricultural and Forest Meteorology*, 354, 110085. https://doi.org/10.1016/j.agrformet.2024.110085.

Goulding, K., et al. (2021), 'Outputs: potassium losses from agricultural systems'. In Murrell, T. S., et al. (Eds.), *Improving potassium recommendations for agricultural crops*. Springer Nature, Cham, Switzerland, pp. 75–98.

Gross, A., et al. (2024), Long-term biochar and soil organic carbon stability – evidence from field experiments in Germany. Science of the Total Environment, 954, 176377. https://doi.org/10.1016/j.scitotenv.2024.176377.

Gunina, A. and Kuzyakov, Y. (2014), 'Pathways of litter C by formation of aggregates and SOM density fractions: Implications from 13C natural abundance'. *Soil Biology and Biochemistry*, 71, 95–104. https://doi.org/10.1016/j.soilbio.2014.01.011.

Guo, Z., et al. (2018), 'Long-term straw incorporation benefits the elevation of soil phosphorus availability and use efficiency in the agroecosystem'. *Spanish Journal of Agricultural Research*, 16(3), e1101-e1101. https://doi.org/10.5424/sjar/2018163-12857.

Gupta, R. K., et al. (2024), 'Interactive effects of long-term management of crop residue and phosphorus fertilization on wheat productivity and soil health in the rice–wheat'. *Scientific Reports*, 14(1), 1399. https://doi.org/10.1038/s41598-024-51399-8.

Haas, E., et al. (2022), 'Long term impact of residue management on soil organic carbon stocks and nitrous oxide emissions from European croplands'. *Science of the Total Environment*, 836, 154932. https://doi.org/10.1016/j.scitotenv.2022.154932.

Haase, M., et al. (2016), 'GIS-based assessment of sustainable crop residue potentials in European regions'. *Biomass and Bioenergy* 86, 156–171. https://doi.org/10.1016/j.biombioe.2016.01.020.

Hijbeek, R. V., et al. (2017), 'Do organic inputs matter–a meta-analysis of additional yield effects for arable crops in Europe'. *Plant and Soil*, 411(1), 293–303. https://doi.org/10.1007/s11104-016-3031-x.

Hillel, D. (2003), *Introduction to environmental soil physics*. Academic Press, San Diego, CA.

Hinsinger, P. (2001), 'Bioavailability of soil inorganic P in the rhizosphere as affected by root-induced chemical changes: a review'. *Plant and Soil*, 237(2), 173–195. https://doi.org/10.1023/A:1013351617532.

Hoogwijk, M., et al. (2003), 'Exploration of the ranges of the global potential of biomass for energy'. *Biomass and Bioenergy*, 25(2), 119–133. https://doi.org/10.1016/S0961-9534(02)00191-5.

Holland, E. A. and Coleman, D. C. (1987), 'Litter placement effects on microbial and organic matter dynamics in an agroecosystem'. *Ecology*, 68(2), 425–433. https://doi.org/10.2307/1939274.

Horton, R., et al. (1996), 'Crop residue effects on surface radiation and energy balance'. *Theoretical and Applied Climatology*, 54, 27–37. https://doi.org/10.1007/BF00863556.

Husain, H. and Dijkstra, F. A. (2023), 'The influence of plant residues on soil aggregation and carbon content: a meta-analysis'. *Journal of Plant Nutrition and Soil Science*, 186(2), 177–187. https://doi.org/10.1002/jpln.202200297.

Ingwersen, J., et al. (2024), 'Harvest residues: a relevant term in the carbon balance of croplands?'. *Agricultural and Forest Meteorology*, 349, 109935. https://doi.org/10.1016/j.agrformet.2024.109935.

IPCC (2019), 2019 Refinement to the 2006 IPCC Guidelines for National Greenhouse Gas Inventories. Calvo Buendia, E., Tanabe, K., Kranjc, A., Baasansuren, J., Fukuda, M., Ngarize, S., Osako, A., Pallmans, Y., Plevin, S. and Wiedmann, S. (eds). Intergovernmental Panel on Climate Change, Geneva, Switzerland.

Islam, M., et al. (2020), 'Utilization of crop residues in rural household of Bangladesh'. *Progressive Agriculture*, 31(3): 164–177. https://doi.org/10.3329/pa.v31i3.52119.

Jiang, Y., et al. (2019), 'Acclimation of methane emissions from rice paddy fields to straw addition'. *Science Advances*, 5(1), DOI: 10.1126/sciadv.aau903.

Jobbágy, E. G. and Jackson, R. B. (2000), 'The vertical distribution of soil organic carbon and its relation to climate and vegetation'. *Ecological Applications*, 10(2), 423–436. https://doi.org/10.1890/1051-0761(2000)010[0423:TVDOSO]2.0.CO;2.

Johnson, J.M.F., et al. (2007) Chemical composition of crop biomass impacts its decomposition. Soil Science Society of America Journal, 71(1), 155–162. https://doi.org/10.2136/sssaj2005.0213.

Just, C., et al. (2023), 'Soil organic carbon sequestration in agricultural long-term field experiments as derived from particulate and mineral-associated organic matter'. *Geoderma*, 434, 116472. https://doi.org/10.1016/j.geoderma.2023.116472.

Klocke, N. L., et al. (2009), 'Soil water evaporation and crop residues'. *Transactions of the ASABE*, 52(1), 103–110. doi: 10.13031/2013.25951.

Klopp, H. W. and Blanco-Canqui, H. (2022), 'Implications of crop residue removal on soil physical properties: a review'. *Soil Science Society of America Journal*, 86(4), 979–1001. https://doi.org/10.1002/saj2.20418.

Lehmann, J. and Joseph, S. (2024), 'Biochar for environmental management: an introduction'. In Lehmann, J. and Joseph, S. (Eds.), *Biochar for environmental management - Science, technology and implementation*. Routledge, New York.

Lehmann, J. and Kleber, M. (2015), 'The contentious nature of soil organic matter'. *Nature*, 528(7580), 60–68. https://doi.org/10.1038/nature16069.

Lehtinen, T., et al. (2014), 'Effect of crop residue incorporation on soil organic carbon and greenhouse gas emissions in European agricultural soils'. *Soil Use and Management*, 30(4), 524–538. https://doi.org/10.1111/sum.12151.

Lei, C., et al. (2024), 'Albedo of crops as a nature-based climate solution to global warming'. *Environmental Research Letters*, 19(8), 084032. https://iopscience.iop.org/article/10.1088/1748-9326/ad5fa2/meta.

Leichty, S., et al. (2020), 'Less efficient residue-derived soil organic carbon formation under no-till irrigated corn'. *Soil Science Society of America Journal*, 84(6), 1928–1942. https://doi.org/10.1002/saj2.20136.

Lemtiri, A., et al. (2018), 'Short-term effects of tillage practices and crop residue exportation on soil organic matter and earthworm communities in silt loam arable soil'. In Muñoz, M. Á. and Zornoza, R. (Eds.), *Soil management and climate change: Effects on organic carbon, nitrogen dynamics, and greenhouse gas emissions* (pp. 53–71). Academic Press, New York. https://doi.org/10.1016/B978-0-12-812128-3.00005-7.

Lessman, M., et al. (2022), 'Global variation in soil carbon sequestration potential through improved cropland management'. *Global Change Biology*, 28(3), 1162–1177. https://doi.org/10.1111/gcb.15954.

Li, F., et al. (2020), 'Carbon and nitrogen mineralization differ between incorporated shoots and roots of legume versus non-legume based cover crops'. *Plant and Soil*, 446, 243–257. https://doi.org/10.1007/s11104-019-04358-6.

Liang, C., et al (2019), 'Quantitative assessment of microbial necromass contribution to soil organic matter'. *Global Change Biology*, 25(11), 3578–3590. https://doi.org/10.1111/gcb.14781.

Liska, A. J., et al. (2014), Biofuels from crop residue can reduce soil carbon and increase CO_2 emissions. Nature Climate Change, 4, 398–401. https://doi.org/10.1038/nclimate2187.

Liu, J., et al. (2023), 'Crop residue return sustains global soil ecological stoichiometry balance'. *Global Change Biology*, 29, 2203–2226. https://doi.org/10.1111/gcb.16584.

Lobell, D.B., et al. (2006) Biogeophysical impacts of cropland management changes on climate. Geophysical Research Letters, 33, L06708. https://doi.org/10.1029/2005GL025492.

Lugato, E., et al. (2014), 'Potential carbon sequestration of European arable soils estimated by modelling a comprehensive set of management practices'. *Global Change Biology*, 20(11), 3557–3567. https://doi.org/10.1111/gcb.12551.

Lupwayi, N. Z., et al. (2006), 'Potassium release during decomposition of crop residues under conventional and zero tillage'. *Canadian Journal of Soil Science*, 86(3), 473–481. https://doi.org/10.4141/S05-049.

MacDonald, G. K., et al. (2011), 'Agronomic phosphorus imbalances across the world's croplands'. *Proceedings of the National Academy of Sciences*, 108(7), 3086–3091. https://doi.org/10.1073/pnas.1010808108.

Maier, R., et al. (2025), 'Large nitrous oxide emissions from arable soils after crop harvests prior to sowing'. *Nutrient Cycling in Agroecosystems*, 130, 162–175. https://doi.org/10.1007/s10705-024-10395-0.

Malhi, S. S. and Lemke, R. (2007), 'Tillage, crop residue and N fertilizer effects on crop yield, nutrient uptake, soil quality and nitrous oxide gas emissions in a second 4-yr rotation cycle'. *Soil and Tillage Research*, 96(1–2), 269–283. https://doi.org/10.1016/j.still.2007.06.011.

Mandal, K. G., et al. (2004), 'Rice residue-management options and effects on soil properties and crop productivity'. *Journal of Food Agriculture and Environment*, 2, 224–231.

Manono, B. (2016), 'Agro-ecological role of earthworms (Oligochaetes) in sustainable agriculture and nutrient use efficiency: a review'. *Journal of Agriculture and Ecology Research International*, 8(1), 1–18.

Minasny, B., et al. (2017), 'Soil carbon 4 per mille'. *Geoderma*, 292, 59–86. https://doi.org/10.1016/j.geoderma.2017.01.002.

Monforti, F., et al. (2015), 'Optimal energy use of agricultural crop residues preserving soil organic carbon stocks in Europe'. *Renewable and Sustainable Energy Reviews,* 44, 519–529. https://doi.org/10.1016/j.rser.2014.12.033.

Moos, J. H., et al. (2017), 'Reduced tillage enhances earthworm abundance and biomass in organic farming: a meta-analysis'. *Applied Agricultural and Forestry Research,* 67, 123–128.

Mulatu, G., et al. (2025), Long-term biochar-based amendments enhance soil aggregation and litter decomposition in low-input cropping systems. European Journal of Soil Science, e70171. https://doi.org/10.1111/ejss.70171.

Muth Jr, D. J., et al. (2012), 'Modeling sustainable agricultural residue removal at the subfield scale'. *Agronomy Journal*, 104, 970–981. https://doi.org/10.2134/agronj2012.0024.

Myhre, G. and Myhre, A. (2003), 'Uncertainties in radiative forcing due to surface albedo changes caused by land-use changes', *Journal of Climate*, 16, 1511–1524. https://doi.org/10.1175/1520-0442(2003)016<1511:UIRFDT>2.0.CO;2.

Ngosong, C., et al. (2019). Mulching: a sustainable option to improve soil health. Soil fertility management for sustainable development. https://doi.org/10.1007/978-981-13-5904-0_11.

Nicolardot, B., et al. (2001), 'Simulation of C and N mineralisation during crop residue decomposition: a simple dynamic model based on the C: N ratio of the residues'. *Plant and Soil*, 228(1), 83–103. https://doi.org/10.1023/A:1004813801728.

Norén, I. S., et al. (2022), 'Reference values for arable crop residues: organic matter and C: N ratio'. Wageningen Research, Report WPR-OT 928. https://doi.org/10.18174/566378.

Ntonta, S., et al. (2022), 'Crop residues differ in their decomposition dynamics: Review of available data from world literature'. *Geoderma*, 419, 115855. https://doi.org/10.1016/j.geoderma.2022.115855.

Nunes, M. R., et al. (2020), 'Tillage intensity effects on soil structure indicators–A US meta-analysis'. *Sustainability*, 12(5), 2071. https://www.mdpi.com/2071-1050/12/5/2071.

Olesen, J. E., et al. (2023), 'Challenges of accounting nitrous oxide emissions from agricultural crop residues'. *Global Change Biology*, 29(24), 6846–6855. https://doi.org/10.1111/gcb.16962.

Panagos, P., et al. (2025), 'A soil monitoring law for Europe'. *Global Challenges*, 9, 2400336. https://doi.org/10.1002/gch2.202400336.

Paustian, K., et al. (2016), 'Climate-smart soils'. *Nature*, 532(7597), 49–57. https://doi.org/10.1038/nature17174.

Piccoli, I., et al. (2020), 'Crop yield after 5 decades of contrasting residue management'. *Nutrient Cycling in Agroecosystems*, 117(2), 231–241. https://doi.org/10.1007/s10705-020-10067-9.

Piccoli, I., et al. (2024), 'Low efficacy of different crop residue management on C and N stocks after five decades'. *Soil and Tillage Research*, 244, 106224. https://doi.org/10.1016/j.still.2024.106224.

Poulton, P. R., et al. (2013), 'Plant-available soil phosphorus. Part I: the response of winter wheat and spring barley to Olsen P on a silty clay loam'. *Soil Use and Management*, 29(1), 4–11. https://doi.org/10.1111/j.1475-2743.2012.00450.x.

Powlson, D. S., et al. (2011), 'Implications for soil properties of removing cereal straw: results from long-term studies'. *Agronomy Journal*, 103, 279–287. https://doi.org/10.2134/agronj2010.0146s.

Pullens, J. W. M., et al. (2021), 'Legacy effects of soil fertility management on cereal dry matter and nitrogen grain yield of organic arable cropping systems'. *European Journal of Agronomy*, 122, 126169. https://doi.org/10.1016/j.eja.2020.126169.

Rahman, M. S., et al. (2024), 'Crop residues integration with nitrogen rates reduces yield-scaled nitrous oxide emissions and improves maize yield and soil quality'. *Journal of Integrative Environmental Sciences*, 21, 2310856. https://doi.org/10.1080/1943815X.2024.2310856.

Ranaivoson, L., et al. (2017), 'Agro-ecological functions of crop residues under conservation agriculture: a review'. *Agronomy for Sustainable Development*, 37, 1–17. https://doi.org/10.1007/s13593-017-0432-z.

Reddy, M. S. L., et al. (2025), 'Organic and inorganic nutrient management in combination with rice residue retention influence the productivity and soil health of zero tillage wheat (Triticum Aestivum L.)'. *Journal of Soil Science and Plant Nutrition*, 1–17. https://doi.org/10.1007/s42729-025-02555-0.

Römheld, V. and Kirkby, E. A. (2010), 'Research on potassium in agriculture: needs and prospects'. *Plant and Soil*, 335, 155–180. https://doi.org/10.1007/s11104-010-0520-1.

Rowinska, P., et al. (2024), 'Biopreparations for the decomposition of crop residues'. *Microbial Biotechnology*, 17, e14534. https://doi.org/10.1111/1751-7915.14534.

Rui, W. and Zhang, W. (2010), 'Effect size and duration of recommended management practices on carbon sequestration in paddy field in Yangtze Delta Plain of China: a meta-analysis'. *Agriculture, ecosystems and environment*, 135(3), 199–205. https://doi.org/10.1016/j.agee.2009.09.010.

Rusakova, I. (2016), 'Study effect of microbial inoculants on decomposition of barley straw. Scientific papers series management', *Economic Engineering in Agriculture and Rural Development*, 16, 467–470.

Sander, B. O., et al. (2014), 'Methane and nitrous oxide emissions from flooded rice fields as affected by water and straw management between rice crops'. *Geoderma*, 235, 355–362. https://doi.org/10.1016/j.geoderma.2014.07.020.

Sanz-Cobena, A., et al. (2014), 'Do cover crops enhance N_2O, CO_2 or CH_4 emissions from soil in Mediterranean arable systems?' *Science of the Total Environment*, 466, 164–174. https://doi.org/10.1016/j.scitotenv.2013.07.023.

Sarkar, S., et al. (2024), 'Retention vs incorporation of cereal residues on soil health: a comprehensive review'. *Communications in Soil Science and Plant Analysis*, 55(12), 1883–1902.

Sarker, J. R., et al. (2018), 'Carbon and nutrient mineralisation dynamics in aggregate-size classes from different tillage systems after input of canola and wheat residues'. *Soil Biology and Biochemistry*, 116, 22–38. https://doi.org/10.1016/j.soilbio.2017.09.030.

Sarker, M. R., et al. (2022), 'Conservation tillage and residue management improve soil health and crop productivity-Evidence from a rice-maize cropping system in Bangladesh'. *Frontiers in Environmental Science*, 10, 969819. https://doi.org/10.3389/fenvs.2022.969819.

Scarlat, N., et al. (2010), 'Assessment of the availability of agricultural crop residues in the European Union: potential and limitations for bioenergy use'. *Waste Management* 30, 1889–1897. https://doi.org/10.1016/j.wasman.2010.04.016.

Scarlat, N., et al. (2019), 'Integrated and spatially explicit assessment of sustainable crop residues potential in Europe'. *Biomass and Bioenergy*, 122, 257–269. https://doi.org/10.1016/j.biombioe.2019.01.021.

Schjønning, P., et al. (2018), 'The role of soil organic matter for maintaining crop yields: evidence for a renewed conceptual basis'. *Advances in Agronomy*, 150, 35–79. https://doi.org/10.1016/bs.agron.2018.03.001.

Schulte, R. P. O., et al. (2014), Functional land management: a framework for managing soil-based ecosystem services for the sustainable intensification of agriculture. Environmental Science & Policy, 38, 45–58. https://doi.org/10.1016/j.envsci.2013.10.002.

Shang, Z., et al. (2024), 'Global cropland nitrous oxide emissions in fallow period are comparable to growing-season emissions'. *Global Change Biology*, 30(2), e17165. https://doi.org/10.1111/gcb.17165.

Shinde, R., et al. (2022), 'Management of crop residues with special reference to the on-farm utilization methods: a review'. *Industrial Crops and Products*, 181, 114772. https://doi.org/10.1016/j.indcrop.2022.114772.

Sieber, P., et al. (2022), 'Albedo impacts of current agricultural land use: Crop-specific albedo from MODIS data and inclusion in LCA of crop production'. *Science of the Total Environment*, 835, 155455. https://doi.org/10.1016/j.scitotenv.2022.155455.

Singh, B. and Rengel, Z. (2007), 'The role of crop residues in improving soil fertility'. In: Marschner, P. and Rengel, Z. (Eds.), *Nutrient cycling in terrestrial ecosystems*. Soil Biology, vol 10. Springer, Berlin, Heidelberg. https://doi.org/10.1007/978-3-540-68027-7_7.

Six, J. and Paustian, K. (2014), 'Aggregate-associated soil organic matter as an ecosystem property and a measurement tool'. *Soil Biology and Biochemistry*, 68, A4–A9. https://doi.org/10.1016/j.soilbio.2013.06.014.

Six, J., et al. (2004), 'A history of research on the link between (micro) aggregates, soil biota and soil organic matter dynamics'. *Soil and Tillage Research*, 79(1), 7–31. https://doi.org/10.1016/j.still.2004.03.008.

Smerald, A., et al. (2023), 'A global dataset for the production and usage of cereal residues in the period 1997–2021'. *Scientific Data*, 10(1), 685. https://doi.org/10.1038/s41597-023-02587-0.

Smil, V. (1999), 'Crop residues: agriculture's largest harvest'. *BioScience*, 49, 299–308. https://doi.org/10.2307/1313613.

Smith, P. (2008), 'Land use change and soil organic carbon dynamics'. *Nutrient Cycling in Agroecosystems*, 81, 169–178. https://doi.org/10.1007/s10705-007-9138-y.

Smith, W. N., et al. (2012), 'Crop residue removal effects on soil carbon: measured and inter-model comparisons'. *Agriculture, Ecosystems and Environment*, 161, 27–38. https://doi.org/10.1016/j.agee.2012.07.024.

Song, J. (1999), 'Phenological influences on the albedo of prairie grassland' and crop fields. *International Journal of Biometeorology*, 42, 153–157. https://doi.org/10.1007 /s004840050099.

Sonsri, K. and Watanabe, A. (2023), 'Insights into the formation and stability of soil aggregates in relation to the structural properties of dissolved organic matter from various organic amendments'. *Soil and Tillage Research*, 232, 105774. https://doi.org/10.1016/j.still.2023.105774.

Strauss, V., et al. (2023), 'Sustainable soil management measures: a synthesis of stakeholder recommendations'. *Agronomy for Sustainable Development*, 43(1), 17. https://doi.org/10.1007/s13593-022-00864-7.

Tebrügge, F. and Düring, R. A. (1999),' Reducing tillage intensity–a review of results from a long-term study in Germany'. *Soil and Tillage Research*, 53(1), 15–28. https://doi.org/10.1016/S0167-1987(99)00073-2.

Timm, L. C., et al. (2002), 'Water balance of a sugarcane crop: quantitative and qualitative aspects of its measurement'. *Revista Brasileira de Engenharia Agrícola e Ambiental*, 6, 57–62. https://doi.org/10.1590/S1415-43662002000100011.

Torma, S., et al. (2018), 'Residual plant nutrients in crop residues–an important resource'. *Acta Agriculturae Scandinavica, Section B–Soil and Plant Science*, 68(4), 358–366. https://doi.org/10.1080/09064710.2017.1406134.

Torppa, K. A. and Taylor, A. R. (2022), 'Alternative combinations of tillage practices and crop rotations can foster earthworm density and bioturbation'. *Applied Soil Ecology*, 175, 104460. https://doi.org/10.1016/j.apsoil.2022.104460.

Totsche, K. U., et al. (2010), 'Biogeochemical interfaces in soil: the interdisciplinary challenge for soil science'. *Journal of Plant Nutrition and Soil Science*, 173(1), 88–99. https://doi.org/10.1002/jpln.200900105.

Totsche, K. U., et al. (2018), 'Microaggregates in soils'. *Journal of Plant Nutrition and Soil Science*, 181(1), 104–136. https://doi.org/10.1002/jpln.201600451.

Valkama, E., et al. (2024), 'Effectiveness of soil management strategies for mitigation of N_2O emissions in European arable land: a meta-analysis'. *European Journal of Soil Science*, 75(3), e13488. https://doi.org/10.1111/ejss.13488.

Van Groenigen, J. W., et al. (2019), 'How fertile are earthworm casts? A meta-analysis'. *Geoderma*, 338, 525–535. https://doi.org/10.1016/j.geoderma.2018.11.001.

Vitousek, P. M. and Howarth, R. W. (1991), 'Nitrogen limitation on land and in the sea: how can it occur?'. *Biogeochemistry*, 13, 87–115. https://doi.org/10.1007/BF00002772.

Wang, X., et al. (2020), 'Effects of residue returning on soil organic carbon storage and sequestration rate in China's croplands: a meta-analysis'. *Agronomy*, 10(5), 691. https://www.mdpi.com/2073-4395/10/5/691.

Wang, C., et al. (2024), 'Soil potassium depletion in global cereal croplands and its implications'. *Science of The Total Environment*, 907, 167875. https://doi.org/10.1016/j.scitotenv.2023.167875.

Ward, P. R., et al. (2015), 'Impact of crop residue retention and tillage on water infiltration into a water-repellent soil'. *Biologia*, 70(11), 1480–1484. https://doi.org/10.1515/biolog-2015-0170.

Watanabe, A., et al. (1994), 'Effect of rice straw application on CH_4 emission from paddy fields'. *Soil Science and Plant Nutrition*, 40(3), 497–504. http://dx.doi.org/10.1080/00380768.1994.10413327.

Whitbread, A. M., et al. (2000), 'Managing legume leys, residues and fertilisers to enhance the sustainability of wheat cropping systems in Australia: 2. Soil physical fertility and carbon'. *Soil and Tillage Research*, 54(1–2), 77–89. https://doi.org/10.1016/S0167-1987(99)00113-0.

Yan, X., et al. (2006), 'Bottom-up estimate of biomass burning in mainland China'. *Atmospheric Environment*, 40(27), 5262–5273. https://doi.org/10.1016/j.atmosenv.2006.04.040.

Yang, L., et al. (2025), 'Applications of remote sensing for crop residue cover mapping', *Smart Agricultural Technology*, 11, 100880. https://doi.org/10.1016/j.atech.2025.100880.

Zörb, C., et al. (2014), 'Potassium in agriculture–status and perspectives'. *Journal of Plant Physiology*, 171(9), 656–669. https://doi.org/10.1016/j.jplph.2013.08.008.

Chapter 2

Optimizing cover crops for soil health

Robert L. Myers, University of Missouri, USA

1 Introduction

Cover crops are commonly defined as conservation plants used to cover the soil, particularly at times when no cash crop is being grown. Cover crops provide soil protection, seedbed improvement and soil enhancement between periods of normal crop production. They are known by many names that reflect their diverse functions and management contexts, including green manures, living mulches, under-sown or companion crops, smother crops, rest crops and catch crops. Regardless of terminology, their shared role is to occupy ecological niches left vacant by cash crops, thereby restoring biological activity, protecting soil resources and enhancing system function and resilience (Clark, 2007; Myers, 2018a).

A growing body of research now documents the multifunctional benefits of cover crops across a wide range of agroecosystems. These benefits include improved nutrient cycling and retention, enhanced soil biological activity and biodiversity, as well as suppression of weeds, pests and diseases. Cover crops can also buffer cropping systems against climate variability by reducing erosion

http://dx.doi.org/10.19103/AS.2025.0166.02
Published by Burleigh Dodds Science Publishing Limited, 2026.

during extreme rainfall events while increasing water infiltration and retention to improve soil resilience during drought (Snapp et al., 2023).

Cover crops can enhance macronutrient cycling within agricultural systems in various ways. These include biological nitrogen fixation, nutrient scavenging, improvement of soil health processes that regulate nutrient transformations and nutrient supply to subsequent crops, as well as reduction of nutrient losses from agroecosystems, including reduced leaching and runoff as well as mitigation of nitrous oxide (N_2O) emissions (Peng, 2026).

Given the importance of soil biology to overall soil health, cover crops can positively influence soil microbial communities, soil fauna and overall biological functioning (Wander, 2026). Root exudates from cover crops, residue inputs and rhizosphere processes help to shape microbial diversity and activity. Increased biological activity contributes to improved soil structure, enhanced nutrient mineralization, disease suppression and greater resilience to environmental stress. More broadly, cover crops can enhance both above- and below-ground biodiversity in agricultural landscapes, including arthropods which play pivotal roles in food webs, nutrient cycling, pest suppression and pollination (Prischman-Voldeth, 2026). Complementing this, cover crops can also help to suppress pests. Cover cropping can be used as a biodiversity-driven pest management strategy that supports natural enemies, disrupts pest life cycles and, in some cases, exerts bio-fumigation effects (Irvin, 2026). Cover crops can also serve as foundational components of integrated weed management (IWM) systems, particularly in the face of increasing herbicide-resistant weed populations, by imposing ecological stressors that reduce weed fitness and population growth (Wallace, 2026).

Given the potential benefits of cover crops, the Conservation Technology Information Center (CTIC) has, since 2012, conducted a National Farmer Cover Crop Survey with support from the USDA Sustainable Agriculture Research and Education (SARE) program and the American Seed Trade Association (ASTA). In several of these surveys, farmers were asked to identify their top reasons for using cover crops. Soil health has consistently been the number one reason (CTIC, 2023).

While it has been clear for several years that farmers are motivated to try or keep using cover crops because of soil health benefits, there have also been many questions raised by farmers and farm advisers regarding the best ways to use cover crops in support of soil improvements. This chapter addresses several cover crop selection and management strategies.

As one might expect, getting a good solid stand of cover crops and growing them for a longer period are core aspects of using cover crops for soil health purposes. However, there are other factors that have an impact as well, including the species of cover crops being used, whether mixes of cover crop species are part of the strategy and how cover crop residue is managed. Use of cover crops for grazing is also discussed, as an increasing number of farmers are adopting that strategy, both for economic and environmental reasons.

2 Enhancing the duration of cover crop growth

A common trend among farmers gaining experience with cover crops is to grow them for a longer period to maximize the benefits from the cover crop (Belfry and Van Eerd, 2016; Noland et al., 2018). For winter annual cover crops, this means both earlier fall planting and later spring termination. For summer cover crops, this means not only early establishment with fast-growing mixes but also including winter annuals that can extend the period of living roots until the approximate time of cash crop planting the following year.

2.1 Earlier seeding

With winter cover crops, there are a number of techniques used to get earlier seeding, including methods shown in Fig. 1. Most common is using aerial seeding or a high clearance broadcast spreader, such as a commercial spray rig modified for seeding, to get the cover crop seeded prior to harvest of the cash crop (Licht, 2019). This overseeding practice is most often done with corn (*Zea mays*), because the corn ear is harvested well-above any newly emerging cover crops and because there is typically a month-long period when overseeding can be done with corn as the leaves dry down and light penetration to the soil surface improves.

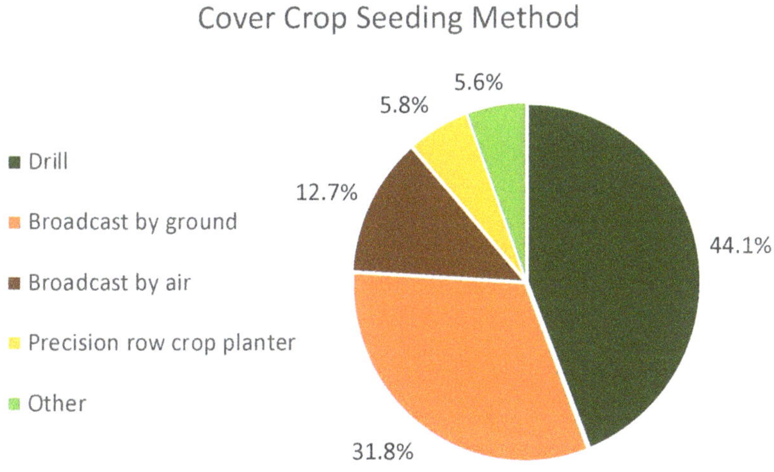

Figure 1 Cover crop seeding method used by farmer respondents in the 2022–3 SARE/CTIC National Cover Crop Survey. Other methods were specified individually by some respondents.

Published by Burleigh Dodds Science Publishing Limited, 2026.

Aerial seeding can be done by helicopter or drone but is most often done by airplane. The prop wash from helicopters can make it more difficult to evenly spread cover crop seed, although experienced pilots can make it work. The use of drones for cover crop seeding until recently was primarily experimental, but commercial use of drones for cover crop seeding has started to take off in the last few years. Drone seeding has the advantage of being able to be done in a more precise spread pattern than by airplane (White, 2024) and reduces risk for a pilot navigating around utility wires and other obstacles. Drone seeding works best for cover crops with a lower seeding rate, like brassicas and legumes, but is being done with higher volume seeds like winter cereals. Any of the aerial seeding methods mentioned here have the advantage of being done when the ground is too wet for equipment to be driven on the ground.

Use of high clearance equipment has gradually taken hold in some areas. Some ag retailers and some larger grain farmers have invested in kits that allow them to adapt high clearance spray rigs for cover crop seeding, adding a seed box and air seeder to equipment that may have boom arms covering 90 feet or more. Such modified spreaders can seed up to 500 acres a day if fast-loading seed tenders are available and efficient distribution patterns. The high clearance seeders have the advantage over aerial seeding of being able to be used on a windy day and having little risk of getting cover crop seed in unintended areas. Wet soil conditions, however, can delay the use of a high clearance seeder at a time when it would otherwise be advantageous to get seed broadcast onto the moist soil.

While interseeding cover crops into standing corn has worked fairly well (Fig. 2), it does face challenges (Sedghi et al., 2023). First and foremost is that any type of broadcast seeding requires timely rainfall or at a minimum, already moist soil conditions, which may be inconsistent with running equipment through a field (impossible if the soil is excessively wet). Cover crop seed broadcast onto a dry surface with no rain may sit there for weeks, gradually being eaten by insects, birds and small mammals such as field mice or voles.

Soybeans (*Glycine max*) can be more challenging to interseed cover crops into during late narrow window of about a week that can work, when leaves are turning yellow but before they have fallen (Klopp et al., 2024). If the leaves are still green when interseeding, too much shade may prevent the cover crop seedlings from growing. If seeding is too late, after soybean leaves have fallen and covered the soil surface, the cover crop seed may just sit on top of soybean leaves and not sprout due to lack of seed-to-soil contact.

Another potential challenge with late summer interseeding (as shown in Fig. 2) is the impact of residual herbicides (Cornelius and Bradley, 2017). Brassica cover crops are particularly sensitive to many broadleaf residual herbicides, which is unfortunate because brassicas establish well when broadcast-seeded

Figure 2 Hagie Cover Crop Interseeder being used to seed cover crops into corn. Photo credit: Rob Myers.

and benefit from earlier seeding. Winter cereal grasses, such as cereal rye and triticale, are less likely to be affected by residual herbicides and legumes in general are intermediate in sensitivity.

By careful planning with residual herbicide selections, or using organic approaches that avoid herbicides, some farmers have pushed the boundaries with cover crop interseeding to interplant the cover crops in early summer or even in some cases, at the same time as the cash crop planting. Use of high clearance drills is one method of early seeding that has allowed farmers to drill in cover crops into corn that is knee high. Figure 3 shows about 14% of surveyed farmers tried early interseeding, compared to 33% who did seeding later before harvest. Success with such early seeding has depended on timely summer rainfalls and careful selection of the cover crops being used; some are more tolerant of growing in the shade of a tall cash crop like corn while others will lack persistence in such conditions.

To help avoid the excessive shading experienced by cover crops interseeded into standing corn or other cash crops, a few farmers have experimented with wide rows of corn; most often 60-inch (152 cm) row spacing compared to the more typical 30-inch (76 cm) row spacing on corn. These unusually wide row spacings allow adequate sunlight to get down to the shorter cover crops and keep them growing. In some cases, the farmers use livestock to graze both the corn and cover crops together in late summer, in other cases the corn is harvested by running the corn head above the cover crop canopy, then after corn harvest the cover crops may be grazed at that point.

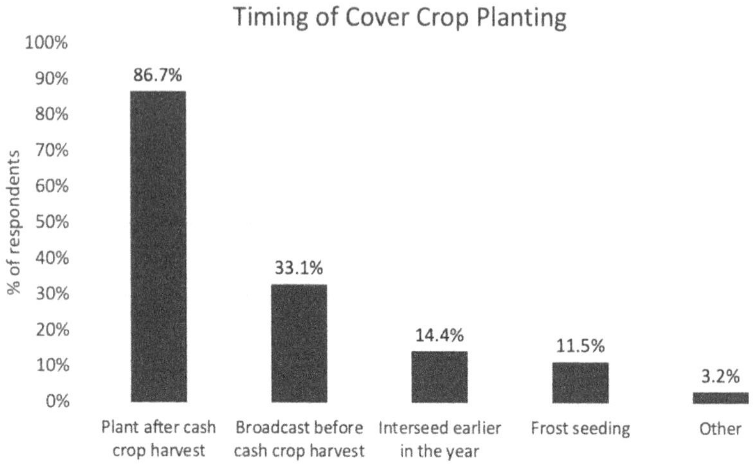

Figure 3 Timing of cover crop planting reported by farmer respondents in the 2022–3 SARE/CTIC National Cover Crop Survey.

Two other ways to facilitate earlier seeding of cover crops are to use shorter season cash crop varieties or plant cash crop species that allow earlier harvest. For example, some farmers choose earlier maturity varieties of corn, soybeans or other summer crops without significant yield reduction, allowing them to have an extra one to three weeks to get the cover crops growing. However, it pays to experiment with multiple early season varieties and carefully evaluate variety test reports to find some that will work without significant yield loss, if possible.

Even earlier cover crop seeding can be allowed by using a completely different cash crop species, such as adding a small grain to a corn and soybean rotation. For many US Midwest farmers, the most common option is to add winter wheat (*Triticum aestivum*) back to the corn and soybean rotation, allowing them to get cover crops started in mid-summer instead of mid-to-late fall. The challenge with this approach is making the winter wheat profitable. Often the best way to do that is to plant a summer cover crop mix that can be grazed, preferably more than once in the fall. Careful planning of a summer cover crop cocktail with some fast-growing, high-quality grazing species can allow for late summer grazing. If cool-season species are mixed in, it may be possible to get a late fall grazing and even another grazing pass in spring.

2.2 Later termination

While earlier establishment of cover crops can aid over-wintering success of fall-planted species and maximum growth of summer species, another way to boost cover crop benefits is with later termination. This approach mainly applies to over-wintering cover crops like winter cereals, certain legumes and canola or rapeseed.

Timing of cover crop termination is a frequent point of discussion among farmers and farm advisers and has even been influenced by policy considerations, such as earlier rules on crop insurance (Fig. 4). Current crop insurance guidelines are considerably more flexible on termination timing but do vary by region (Connor et al., 2022). Most novice cover crop users are encouraged to terminate any over-wintering cover crops early to simplify the spring planting of their cash crops.

Planting a spring crop into cover crops that have been terminated when they are only a few inches tall is considerably easier than planting into tall, thick cover crop residue, especially for a farmer not used to setting up a planter for heavy cover crop residue conditions.

Many farmers are now more experienced with planting into heavy amounts of cover crop biomass, which has made late termination of cover crops much more viable (Silva, 2024). This late termination can benefit soil health by providing not only more biomass to decompose but also prolonging the period of living roots from the cover crops. The key is for a farmer to figure how to successfully establish their cash crop in a taller cover crop, which may include making changes to their planter set up.

Some farmers will switch. closing wheels on their cash crop planter, from one with a round rubber wheel to a metal spiked wheel (Quinn, 2024). Others will remove no-till coulters and lift up trash cleaning devices on their planter that might otherwise be used in high residue situations (Balkcom et al., 2024). Planting into a cover crop that is still living is referred to as 'planting green' and has become increasingly more popular with experienced cover crop

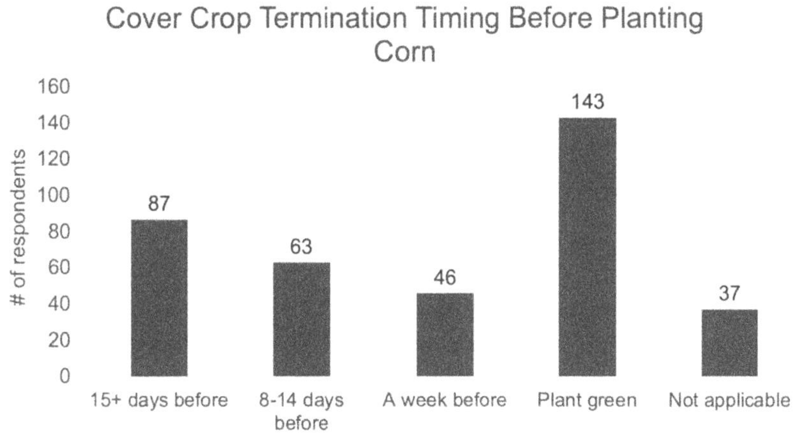

Figure 4 Termination timing used by farmer respondents in the 2022–3 SARE/CTIC National Cover Crop Survey.

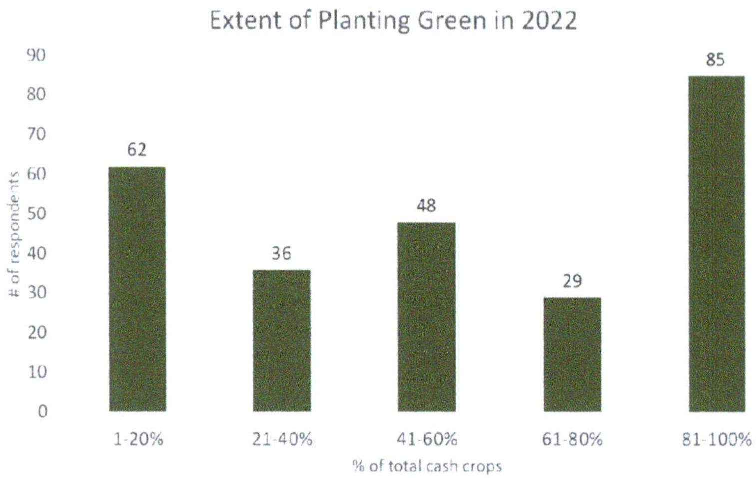

Figure 5 Adoption of planting green technique by farmer respondents in the 2022–3 SARE/CTIC National Cover Crop Survey.

Figure 6 Trial of soybeans being planted green into cereal rye cover crop. Photo credit: Rob Myers.

farmers (see Figs. 5 and 6). Planting green and ecosystem services in row crop production systems (Blanco and Jhala, 2024). The greater amount of cover crop biomass and longer cover crop growth period can help considerably with weed control (example in Fig. 6), can help soil moisture be moderated in a wet spring, and speed soil health improvements.

3 Getting a good stand of cover crops

One of the more popular topics of cover crop presentations, publications and videos is discussing all they ways that cover crops can be seeded and how to be improve the odds of getting a good solid stand of cover crops. Planting date, as described previously, is a big factor in establishment success, but so is the planting method. It's well known that broadcast seeding of cover crops, while popular as a speedy and economical method, is not as likely to provide consistent success as direct seeding with a grain drill or other planter. Some cover crop species, such as annual ryegrass, are relatively easy to broadcast establish, while large-seeded cover crops like peas almost have to be direct seeded.

Many large grain farmers who have experience with cover crops will use more than one strategy on their farm, maybe drilling cover crop mixes while broadcast seeding something like cereal rye that can survive a late start. Labor is often a constraint to being able to run a grain drill or planter while dealing with fall harvest, but either hiring a part-time employee or service company can allow direct seeding to get accomplished.

Other aspects of getting a good stand of cover crops include purchasing good-quality seed with known germination percentage and seeding at an appropriate rate. Those who are grazing cover crops or using them for weed suppression may seed at a high rate per acre, while other farmers may use half as much that are just wanting some living roots to help with soil biology and reducing erosion.

4 Choice of cover crop species, varieties and mixtures

Equally important to cover crop planting dates and termination timing is the choice of appropriate cover crop species, varieties and mixes for the season, soil type and cropping system. There can be big differences in growth and biomass accumulation among cover crop species. However, there is still much we need to learn about how individual cover crop species affect soil health and even the potential for different varieties to influence selected soil health benefits. It has been observed that some cover crop varieties will produce more biomass, have different maturity dates or potentially impact soil nitrogen. Some legume varieties have shown better ability to fix nitrogen than others, and it's reasonable to assume that nitrogen sequestration may differ somewhat among selected varieties and species. Since biomass is the most studied aspect of cover crop species and variety comparisons, it's worth looking at how performance can vary among the various types of cover crops. Figure 7 shows the relative popularity of cover crop species among surveyed farmers.

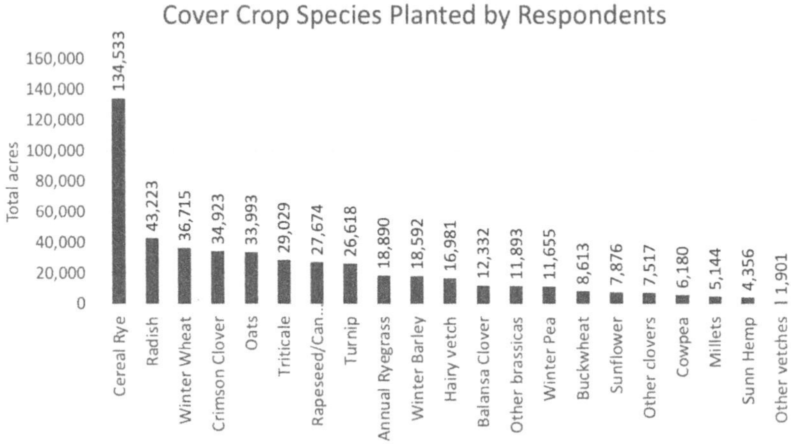

Figure 7 Cover crop species used by respondents in the 2022–3 SARE/CTIC National Cover Crop Survey.

4.1 Cool-season cover crops

For farmers who want a cover crop that will winter kill but still produce good fall growth, oats and radishes have been the two most popular choices. When planted on the same day in early or mid-fall, oats will normally produce more above-ground biomass than over-wintering grass cover crops like cereal rye, winter wheat or annual ryegrass. This rapid fall growth makes oats popular for fall grazers of cover crops and they are also an effective winter-kill option. Radishes and other Brassicas can also put on an impressive amount of fall growth if planted in early fall, but their fall growth falls off sharply if planted in the latter part of the fall. For farmers that graze cover crops, turnips have been a popular fall choice, worth mentioning in the soil health context because of the benefits of grazing for soil health.

Moving to cover crops that can over-winter, the choices are more varied and depend on the growing region, since winter conditions can really affect success of species like winter barley, canola/rapeseed and many legumes (particularly the less hardy clovers).

For maximum biomass, many farmers turn first to the hardier winter cereals, namely cereal rye, triticale and winter wheat. None will produce an extensive amount of fall biomass, but when allowed to grow long enough in the spring, can produce multiple tons of spring biomass per acre. By comparison, the over-wintering legumes or Brassicas produce less biomass than the winter cereals, but can play a role as complementary components of cover crop mixes. While tricky to balance the seeding rates just right, a well-designed mix of one or more winter cereals with one or more legumes and possibly brassicas can

Published by Burleigh Dodds Science Publishing Limited, 2026.

Figure 8 Cover crop mix of cereal rye, hairy vetch and crimson clover. Photo credit: Rob Myers.

produce not only a large amount of spring biomass but really boost soil health because of the extra biodiversity (see example mix in Fig. 8).

4.2 Warm-season cover crops

Likewise, summer cover crop selection is a bit of a choice between a species that puts on prolific growth, like sorghum-sudan or a forage sorghum, versus a diverse mix of warm-season cover crops. Total biomass may not necessarily be higher with a mix, but the mix will be adaptable to a wider range of growing conditions than a single species and will generally benefit soil health more because of the diversity of plant species grown. Figure 7 illustrates how about 45% of farmers surveyed have increased their use of cover crop mixes compared to single species.

Among the summer grass cover crops, various options from the sorghum family predominate, ranging from sorghum-sudan, sudangrass and forage sorghums. These sorghum options grow fast, can get 7 to 10 feet tall depending on variety and moisture, and accumulate a large amount of biomass, even in dry summers. Some of the millet species, particularly pearl millet and foxtail millet, also do very well in dry conditions, but don't generate quite as much biomass as the sorghum options (Myers, 2018b). The millet options are a very diverse group of plants from various regions of the world, representing several different genera. 'Millets' is actually a vague term not unlike saying cereal grain to describe crops as different as wheat and oats. The term millet applies to a variety of annual grass crops with small, rounded seed. While pearl millet and foxtail millet do well in dry conditions, there are other millets that tolerate wetter conditions, particularly Japanese millet and browntop millet. A summer cover crop mix that includes a variety of millet options and provides some versatility in case the summer turns wetter or drier than expected. Also, the millets grow well with various sorghum options and are mixed with summer legumes and other broadleaves.

Among the warm-season legume options, sunn hemp is the cover crop species that puts on the most summer growth, particularly from a height standpoint (Fig. 9). Sunn hemp, more of a tropical crop, can get 3 m tall when planted in June, providing effective nitrogen fixation as well as being a good grazing option. Cowpeas also fit in well with other summer cover crops and, depending on variety, can produce a good amount of growth. The most popular option for cowpeas as a cover crop is iron clay cowpeas, which are a vining type that can climb on other crops and produce good nitrogen fixation. Cowpeas are members of the *Vigna* genus of plants, which includes mung beans and adzuki beans, sometimes seen as additions to complex summer cover crop cocktail mixes.

Besides grasses and legumes, diverse summer cover crop mixes often include one or more non-leguminous broadleaf options. Sunflowers and buckwheat are the two most common options in this category. Sunflowers can put on an impressive amount of biomass growth when planted in mid-summer and also are known for deep rooting. Buckwheat, while a shallower-rooting plant, forms a fibrous root system that seems to aid soil tilth and is a fast grower as well, though likely to only get about half the height of buckwheat. Other broadleaf warm-season plants are sometimes sold for cover crop mixes,

Figure 9 Sunn hemp cover crop on the left and a mix of cowpeas with pearl millet on the right. Photo credit: Rob Myers.

Published by Burleigh Dodds Science Publishing Limited, 2026.

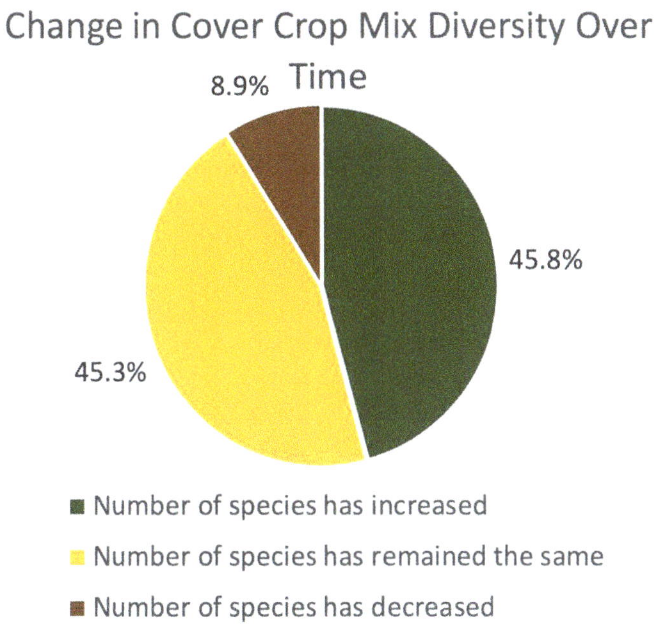

Figure 10 Pattern of use with cover crop mixes reported by farmer respondents in the 2022–3 SARE/CTIC National Cover Crop Survey.

including sesame and okra. As shown in Fig. 10, just under half of farmers surveyed increased their use of cover crop mixes over time.

5 Cover crop root systems

Evaluation of cover crop management and comparison of species and varieties has often centered on above-grown biomass because it's easy to observe, but root growth of the cover crop is likely just as important and maybe even more important. Unfortunately, evaluating root growth of cover crops is tough to do in a systematic way because of the time involved in extracting delicate roots from the soil, washing off the soil and measuring root biomass and other root characteristics. Even the process of root decomposition is not well studied, although it is currently believed that decomposing roots contribute more to soil organic matter than above-ground biomass, a fair portion of which oxidizes and goes off into the atmosphere as microbes decompose leaves and stems.

The limited research done on cover crop roots seems to indicate that the winter annual grasses produce the most overall root growth among the annual cover crops (perennial grasses can produce even more root growth but are seldom used as cover crops in annual cropping, being limited to use in perennial systems like orchards and vineyards).

Table 1 Carbon to nitrogen ratios in cropping systems

Material	C:N ratio
Rye straw	82:1
Wheat straw	80:1
Oat straw	70:1
Corn stover	57:1
Rye cover crop (anthesis)	37:1
Rye cover crop (vegetative)	26:1
Ideal microbial diet	24:1
Legume hay	17:1
Hairy vetch cover crop	11:1
Soil microbes (average)	8:1

Adapted from NRCS, 2011.

6 Carbon-to-nitrogen (C:N) ratio of cover crops

When considering the role of above-ground and below-ground biomass from cover crops for soil organic matter and health, it is worth considering the relative carbon-to-nitrogen (C:N) ratio of various cover crop species and even cover crop growth stages.

For soil microbes to break down plant residues (leaves, stems, roots, etc.), they need to have a carbon-to-nitrogen (C:N) ratio of 24:1. If the plant material has a higher C:N ratio, then it takes longer for the microbes to break down the plant residue, and N must be accessed from the soil fraction, reducing the available soil nitrogen for cash crops. While vegetative cereal rye plants are close to that ideal C:N ratio and decompose readily, mature rye plants have a much higher C:N ratio that can be 82:1 as shown in Table 1. This is helpful for situations where a long-persisting rye mulch is desired in a field, such as for weed control and helping reduce soil moisture evaporation but means that relatively little of that above-ground biomass is being incorporated into the soil fraction. Adding a legume and/or brassica to a mix with a winter cereal like rye can contribute to more rapid breakdown of a large amount of spring cover crop biomass, allowing more of it to be incorporated into the soil organic matter in a way that can contribute to soil health more quickly.

7 Additional factors affecting persistence of cover crop residue

The most important factor for cover crop residue persistence is the C:N ratio of the plant biomass. However, contact of the residue with the soil is also important. In years past, when cover crops were often managed as green manure crops with a primary focus on soil fertility, the cover crops were tilled into the soil before planting the cash crop, most often with a moldboard plow as the first tillage step before additional seedbed preparation such as disking and harrowing. Traditionally, farmers were most comfortable with a tilled seedbed for planting their cash crops. While the tillage of the residue can lead to faster incorporation into the soil organic matter fraction, the tillage has to be done in a way that works for the amount of residue present. For example, disking a thick layer of cover crop residue can lead to more problems than it solves when it comes to spring cash crop planting. Thus, experienced cover crops today are much more likely to no-till plant into cover crop residue or in some regions will use a strip-till system. If a farmer is no-tilling and allowing the cover crop to get sizable, some of these farmers like to roll the cover crop to get a thicker residue mat for weed control and also to facilitate soil contact with the residue that breaks more of it down in the following weeks, as opposed to leaving the residue standing more upright (note that most cash crop planters at least partially knock down taller cover crop plants).

Another option, rarely used in row cropping but occasionally used in speciality crop operations, is to mow the cover crop. The mowing chops up the cover crop residue and leaves it lying close to the soil surface in a way that can speed up residue breakdown compared to leaving the cover crop standing. A flail mower in particular can provide an even layer of mowed residue. If the mowing is intended to terminate the cover crop, as opposed to simply chopping up an already dead cover crop (such as from spraying), it is important that the cover crop already be in a reproductive (flower or seeding) growth stage. Most cover crops, if mowed while still vegetative (pre-flowering), will simply regrow, much like a mowed lawn. For cover crops that are already flowering and still alive, it's best to mow them as close to the ground as possible if the goal is to terminate the cover crop while mowing.

8 Combining cover crops with no-till

Many of the soil health benefits of cover crops are significantly enhanced when combining cover crops with minimal soil disturbance, particularly in no-till cropping systems (Mbuthiaa, 2015). Tillage has a negative impact on mycorrhizal soil fungi and over time, can diminish soil aggregate structure. Tillage can also quickly diminish soil macropores that cover crops would otherwise help foster; greater numbers of macropores in the soil are important for improved rainfall infiltration and contribute to root growth. Another downside of tillage for soil health is that tillage can cause increased oxidation of soil organic matter (Wright, 1999).

9 The role of grazing cover crops

Some recent research has shown that grazing cover crops is a way to more rapidly improve soil health, at least in some environments (Simon, 2024). From an agroecological perspective, this makes sense, because our soil microbial communities evolved with grazing animals, such as buffalo on the US prairie. Soil microbes seem to respond favorably to manure, urine and saliva from the grazing animals.

Benefits from grazing cover crops go beyond just the presence of the grazing animal in the field, although that is a very important benefit. The grazing is perhaps the most profitable way to utilize cover crops (Franzluebbers and Stuedemann, 2014) especially in their first year of use (Myers et al., 2019). Since grazing makes cover crops profitable, it also makes it much more justifiable from an economic sense to lengthen the time the cover crops are present in the field (Golden, 2016), because then the farmer is getting more forage as well as extra soil health benefits from the longer period of cover crop growth. It also becomes more viable to use a higher seeding rate of cover crops and to make use of cover crop mixes, since a moderately higher seed cost can pay off in better grazing returns.

There is still a significant amount to learn about how cover crop grazing impacts soil health. Does it benefit some soil microbial groups more than others, such as bacteria more than fungi? Is it important to graze a field two to three times per year or does the biggest benefit happen just with a single grazing window each year? Is there an advantage to having more than one species of grazing animal, such as rotating or combining grazing sheep or goats with cattle? How should the grazing be managed to avoid soil compaction issues?

Cover crop grazing has been steadily increasing in many regions, especially where farmers have both livestock and row crops. However, many farmers still limit their use of grazing animals to perennial pastures because they have no fencing around their crop fields. Virtual fencing is a new technology that offers great potential to expand the use of cover crop grazing (Dozler and Myers, 2026).

Virtual fencing has been used commercially for the last few years in New Zealand, Australia and parts of Europe. The technology has recently come to the US (sales really got underway from a few companies in 2025 and are expected to expand significantly in the next few years for both pasture and cover crop grazing). The companies currently offering livestock virtual fencing technology all use the same basic approach, which is to include the following components on a collar worn on each adult grazing animal (young animals are not normally collared, partly due to rapid growth in neck size).

- GPS device to locate the animal's position relative to the prescribed boundary;

- Communication device that transmits the animal's location either via a cell phone signal or to an on-farm tower that then further relays information between animal collars and the herd manager;
- Sound generator to emit an audible tone or series of sounds that alerts the grazing animal it is approaching the boundary of the prescribed grazing area (paddock);
- Electric pulse device that can provide a momentary shock to the grazing animal if it goes past the warning zone (auditory cues) and is at the prescribed boundary;
- Long-lasting battery to power the electronics listed above; and
- Small solar panels to recharge the battery.

Typically, all the electronics are in one device that hangs from the collar, but in the case of the current Halter company product, their small solar panels are on top of the collar, and the collar also can vibrate, providing another cue to the animal as to where they should be grazing (Fig. 11).

There are several benefits to virtual fencing, with the largest being the labor savings that comes from being able to eliminate regular movement of physical electric fencing in order to set up a management grazing system

Figure 11 Beef cattle wearing the Halter virtual fencing product. Photo credit: Kaitlyn Dozler.

(Meehan, 2024). Since there is essentially no extra labor to design frequent grazing moves, it becomes possible to more intensely and efficiently make use of cover crop forage (or pasture).

Another benefit of virtual fencing is being able to monitor the location and movement of individual animals (Melo et al., 2024). If any animals get outside of the boundary, perhaps chased by loose dogs or coyotes, the herd manager is notified. Typically, these animals return to the herd without manager intervention since they do not receive an electric shock when rejoining the herd, but the manager has the option of redrawing the paddock boundary to re-enclose any animals outside the previous boundary. If an animal is not moving for an extended period, potentially from illness, the herd manager can be notified and will check on the animal.

Given that the boundaries of the prescribed grazing areas are drawn on a phone, tablet or computer, it is easy to design more irregular boundaries that follow the shape of streams or other sensitive areas, compared to trying to string electric fencing, more typically in straighter lines. It also makes it easier to design access to in-field watering systems.

There are also some limitations with virtual fencing. Most notably, it is still recommended by companies selling these systems that the herd be constrained by a physical perimeter fence, especially near busy roadways. Thus, if cattle were in a 40-acre crop field for cover crop grazing with virtual fencing collars, it would likely involve having an electric fence or other physical fence around the perimeter of the 40 acres. However, all movement of the cattle for management-intensive grazing, such as grazing one-acre paddocks with each daily move, could be managed through the virtual fencing system (some physical movement of water tanks might be needed).

Also, in areas without good cell reception, the cattle are still restrained by the GPS signal, but the herd manager needs some method of wireless connection to the collars both to locate the animals and update boundary information recognized by the collars. Since the normal cell signal might not be feasible, that is where an on-farm tower (typical cost $3000–5000) needs to be used, and then the tower uses Starlink or some other method to connect to the internet. Finally, there is the question of economics; more data is still needed on how economical these virtual fencing collars will be for grazing in various scenarios. Clearly, someone using the collars year-round would be able to make them pay more easily than someone just grazing cover crops for a few weeks.

Fortunately, some of the companies allow farmers using their collars to just pay a monthly service fee on the collars (covering the cost of network support) based on the months of actual use.

10 Conclusion

There are several ways that cover crops can be managed and used to more effectively build soil health as identified. To summarize this information, it's perhaps helpful to lay out a progression of how a given farmer might expand or intensify their cover crop use over time in way that accelerates soil health improvements. Not every farmer will follow the steps outlined below, but this outline depicts what some of the more advanced cover croppers have done with their row crop farming systems. Each cover crop 'level' below depicts a progressively greater level of potential soil health benefit.

Level 1: Cover crops adopted for the first time on a field or farm, starting with a fall-planted single species cover crop that winter kills, such as oats or radishes.

Level 2: Farmer progressed to use of a cover crop that over-winters, most often a winter cereal such as winter wheat or cereal rye. Cover crop is usually still a single species planted relatively late in the fall (often broadcast) and terminated early in the spring while still only a few inches tall.

Level 3: To get more reliable cover crop establishment, the farmer either switches from late fall broadcasting of the cover crop to either drilling it in right after grain harvest or broadcasting the seed into the standing cash crop with an airplane, drone or high clearance seeder.

Level 4: To maximize cover crop benefits, the farmer starts terminating the cover crop later in the spring (usually trying this first with soybeans, maybe a year or two later with corn). This later termination may be in conjunction with moving back the planting date of the cash crop or simply coincide with when the cash crop would typically be planted.

Level 5: As the farmer sees the benefits of getting more growth out of the cover crop for a longer period, they start experimenting with ways to get the cover crops going even earlier in the year, most often by using an earlier maturity cash crop variety that allows earlier harvest followed by cover crop drilling. Alternatively, they may add an earlier maturity cash crop species such as winter wheat or another small grain to the crop rotation in some fields, allowing them to get several additional weeks of cover crop growth following harvest of the small grain. For those using a small grain with early harvest, they are normally switching to use of a summer cover crop cocktail with exclusively or primarily warm-season cover crops that grow fast in the warmer days of summer (a few cool-season species may be mixed in, but with the recognition they won't grow much until the cooler days of fall). By the time a farmer gets to level 5, they are often cutting back on fertilizer use and/or herbicide and pesticide use.

Level 6: For individuals using cover crops with their own livestock, they may start experimenting with cover crop grazing on some crop fields. Farmers may start cover crop grazing because they are driven primarily by the idea of further boosting their soil health. They may even be farmers who haven't had any livestock on their farm for years but are motivated to re-integrate livestock because they've become convinced of the benefits of soil health grazing by hearing about other farmers' experiences.

Level 7: Relatively few farmers are at this more experimental phase, but there are some farmers who go beyond all the steps to do very unconventional steps like 60-inch (152 cm) corn rows with diverse cover crops planted at the same time as corn, then grazing the whole field, either with the corn grazed too or right after the corn is combined. These are also farms where a diversity of livestock is being grazed combined with more diversified crop rotations. These farmers are typically trying to maximize biological diversity with improved soil health as a strong priority and then figure out how to capitalize on improved farm resiliency from an economic standpoint. Economic benefits may extend to doing more direct marketing of their farm products to gain value from their regenerative management methods, rather than selling into conventional commodity markets.

Overall, it's clear that many farmers do go through some sort of progression of growing cover crops for longer in fall and/or spring, using more diverse mixtures of cover crops over time, and focusing increasingly on how to maximize soil health. These farmers are seeing benefits in the resiliency of their fields to challenging weather situations, and many have reported benefiting financially. The Soil Health Institute has done an extensive number of economic evaluations of soil health practices using the case study method of actual farmer experience (SHI, 2021, 2023). While they continue to add to the types of cropping systems and regions being studied, their initial set of 100 focused on corn and soybean farmers using soil health practices (particularly no-till and cover crops). They documented that 85% of the farmers had an increase in net income for corn from the use of soil health practices. The average net benefit for corn from using soil health practices was $51.60 per acre. Similarly, 88% of the farmers reported an increase in net income for soybeans, and over all the farms, the average net economic benefit for soybeans from using soil health practices (including cover crops) was $44.89 per acre. In sum, cover crops are an effective tool for helping improve soil health and can contribute to economic profitability.

11 Where to look for further information

The regional cover crop councils in the US have all developed websites with a variety of cover crop publications and decision tools. They can be found online by searching for 'cover crop council' with the relevant regional name, which is Midwest, Northeast, Southern or Western. The USDA Sustainable Agriculture Research and Education (SARE) program has extensive information on cover

Published by Burleigh Dodds Science Publishing Limited, 2026.

crops and soil health, including bulletins, books, fact sheets, videos, research reports, a cover crop image library with free cover crop photos and illustrations, and other information. See: https://www.sare.org/resources/cover-crops/.

Additional cover crop information is available through the Center for Regenerative Agriculture at University of Missouri, see: https://cra.missouri.edu/topics/cover-crops/.

The Soil Health Institute has a variety of information related to soil health, including soil health measurement, soil health economics and other topics. See: www.soilhealthinstitute.org.

12 References

Balkcom, K., et al. (2024). *How do I get good seed/soil contact in cover crop residue?* Southern Cover Crop Council Website. https://southerncovercrops.org/cover-crop-resource-guide/row-crops/mountains-ridge-valley-piedmont/planting-cash-crops-in-cover-crop-residue/how-do-i-get-good-seed-soil-contact-in-cover-crop-residue/.

Belfry, K.D. and Van Eerd, L.L. (2016). Establishment and impact of cover crops intersown into corn. *Crop Science*, 56: 1245–1256. https://doi.org/10.2135/cropsci2015.06.0351.

Blanco, H. and Jhala, A.J. (2024). Planting green and ecosystem services in row crop production systems. *CABI Reviews*, 19: 1. https://doi.org/10.1079/cabireviews.2024.0029.

Clark, A. (2007). *Managing cover crops profitably Third Edition*. USDA-SARE (Sustainable Agriculture Research and Education). University of Maryland.

Connor, L., et al. (2022). Crop insurance participation and cover crop use: Evidence from Indiana county-level data. *Applied Economic Perspectives and Policy*, 44(4): 2181–2208. https://doi.org/10.1002/aepp.13206.

Cornelius, C.D. and Bradley, K.W. (2017). Carryover of common corn and soybean herbicides to various cover crop species. *Weed Technology*, 31: 21–31.

CTIC. (2023). *Report of the 2022–23 cover crop survey*. Joint publication of the Conservation Technology Information Center and the North Central Region Sustainable Agriculture Research and Education Program.

Dozler, K. and Myers, R. (2026). *Virtual fence 101: An introduction to virtual fencing technology*. Technical publication. Center for Regenerative Agriculture, University of Missouri.

Franzluebbers, A. and Stuedemann, J.A. (2014). Crop and cattle production responses to tillage and cover crop management in an integrated crop–livestock system in the southeastern USA. *European Journal of Agronomy*, 57: 62–70.

Golden, L., et al. (2016). *Cover crops for grazing use in Idaho* (University of Idaho Extension Bulletin No. 901). University of Idaho Extension.

Irvin, N.A., et al. (2026). Cover crops for arthropod pest management. In S. W. Duiker and A. Clark (Eds.), *Cover crops for sustainable soil management*. Burleigh Dodds Science Publishing.

Klopp, H., et al. (2024). *Inter-seeding cover crops into soybean*. South Dakota State University Extension.

Licht, M. (2019). *A look at cover crop seeding methods*. Iowa State University Extension.

Mbuthiaa, L.W., et al. (2015). Long term tillage, cover crop, and fertilization effects on microbial community structure, activity: Implications for soil quality. *Soil Biology and Biochemistry*, 89: 24–34.

Meehan, M., et al. (2024). Effectiveness of virtual fence in North Dakota grazing systems. AS2241-6. 2024 *North Dakota Livestock Research Report*.

Melo, J., et al. (2024). There will always be collars in my future: exploring US ranchers' and company representatives' perspectives of virtual fencing for livestock. *Smart Agricultural Technology*. https://doi.org/10.1016/j.atech.2024.100632.

Myers, R.L. (2018a). Use of cover crops to promote soil health. In D. Reicosky (Ed.), *Managing soil health for sustainable agriculture Volume 2: Monitoring and management*. Burleigh Dodds Science Publishing.

Myers, R.L. (2018b). *Growing millets for grain, forage or cover crop use* (Extension Publication No. G4164). University of Missouri Extension. https://extension.missouri.edu/publications/g4164

Myers, R.L., et al. (2019). *Cover crop economics*. USDA-SARE Technical Bulletin.

Noland, R.L., et al. (2018). Establishment and function of cover crops interseeded into corn. *Crop Science*, 58: 863–873.

NRCS. (2011). *Carbon to nitrogen ratios in cropping systems*. Prepared by the USDA NRCS East National Technology Support Center, Greensboro, NC, in cooperation with North Dakota NRCS.

Peng, Y., et al. (2026). The role of cover crops in optimizing nutrient cycles. In S. W. Duiker and A. Clark (Eds.), *Cover crops for sustainable soil management*. Burleigh Dodds Science Publishing.

Quinn, D. (2024). Trials evaluate after-market closing wheels for corn. *Indiana Prairie Farmer*. https://www.farmprogress.com/corn/trials-evaluate-after-market-closing-wheels-for-corn.

Sedghi, N., et al. (2023). Aerial interseeding and planting green to enhance nitrogen capture and cover crop biomass carbon. *Journal of Soil and Water Conservation*, 78(3): 282–298.

Silva, T.S., et al. (2024). Cereal rye cover crop termination at or before soybean planting has minimal effect on soybean yield across the Midwestern US. *Field Crops Research*, 312: 109393.

Simon, L.M., et al. (2024). Cover crop grazing effects on soil properties in no-tillage dryland cropping systems in the Central Great Plains. *Agriculture, Ecosystems & Environment*, 374: 109140. https://doi.org/10.1016/j.agee.2024.109140.

Snapp, S.S., et al. (2023). Cover crops for soil health. In W. R. Horwath (Ed.), *Improving soil health*. Burleigh Dodds Science Publishing.

Soil Health Institute. (2021). *Economics of soil health systems on 100 farms*. https://soilhealthinstitute.org/our-work/.

Soil Health Institute. (2023). *Nationwide study on 30 U.S. farms shows positive economic impact of soil health management systems*. https://soilhealthinstitute.org/news-events/nationwide-study-on-30-u-s-farms-shows-positive-economic-impact-of-soil-health-management-systems/.

Wallace, J.M. and Wellman, L. (2026). Revisiting the contribution of cover crops to weed management in Conservation Agriculture. In S. W. Duiker and A. Clark (Eds.), *Cover crops for sustainable soil management*. Burleigh Dodds Science Publishing.

Published by Burleigh Dodds Science Publishing Limited, 2026.

Wander, M. (2026). The role of cover crops in improving soil biological health. In S. W. Duiker and A. Clark (Eds.), *Cover crops for sustainable soil management*. Burleigh Dodds Science Publishing.

White, J. (2024). *Drone cover crop seeding success*. Brownfield Ag Network. https://www.brownfieldagnews.com/news/drone-cover-crop-seeding-success/.

Wright, S.F., et al. (1999). Changes in aggregate stability and concentration of glomalin during tillage management transition. *Soil Science Society of America Journal*, 63: 1825–1829.

Chapter 3

Combined use of vermicompost and biochar to enhance soil health and crop yield in sustainable agriculture

Parveen Fatemah Rupani, Cranfield University, UK; Inés María Alonso-Crespo, María Gómez-Brandón and Jorge Domínguez, University of Vigo, Spain; and Ruben Sakrabani, Cranfield University, UK

1 Introduction

Soil plays a vital role on Earth as a living ecosystem that sustains plants, animals, and humans (Lehmann et al., 2020). Nonetheless, human-induced pressures on soil health have threatened and compromised long-term food security (FAO, 2023). Limited resources have also resulted in the mismanagement and improper disposal of organic waste in the environment, accelerating soil degradation due to the accumulation of waste in landfills (Zhang et al., 2022b). Maintaining nutrient-rich topsoil has become essential to counteract the numerous challenges derived from rapid population growth, economic development, urbanisation, and increased food demand (Amundson et al., 2015; Zeiss et al., 2022). The use of organic fertilisers with a carbon-to-nitrogen (C to N) ratio of 42:1 is a promising solution that would help to address this crisis by immobilising nutrients such as nitrogen, reducing leaching losses and, above all, preserving soil health (Cameron et al., 2013; Chivenge

http://dx.doi.org/10.19103/AS.2025.0166.03

et al., 2011; Gentile et al., 2009). The role of earthworms in organic solid waste management has been well established since first highlighted by Darwin (1892). Vermicomposting technology has been expanded over the years aiming to produce a valuable bio-product called vermicompost (Domínguez, 2023; Ismail, 2004). Vermicompost, generated through the joint and synergistic action of earthworms and microorganisms, is widely recognised as a nutrient-rich, biologically active humus-like organic material with the potential to enhance soil health and crop yield by supplying essential nutrients, fostering beneficial microbial communities, and improving plant resilience against pests and diseases (Gómez-Brandón et al., 2020). The transformation of organic waste into vermicompost takes place under aerobic conditions as a result of the feeding, casting (deposition of faeces), and tunnelling activities of earthworms (Sanchez-Hernandez and Domínguez, 2017).

Biochar is a carbon-rich material produced through the pyrolysis of organic matter (OM). This material was first used as a soil amendment in the late twentieth century, when researchers began to investigate the properties of *Terra Preta*, a black soil with a high charcoal content and recognised its ability to enhance soil fertility and carbon sequestration. The term 'biochar' became widely recognised by the early 2000s, driven by the work of pioneers such as Johannes Lehmann, who emphasised the role of biochar in improving soil health, reducing greenhouse gas (GHG) emissions and addressing environmental challenges. Nowadays, biochar is considered a potentially valuable material for sustainable soil management and climate change mitigation (Li et al., 2020; Mosa et al., 2023; Siedt et al., 2021; Vijay et al., 2021). Indeed, the capacity of biochar to enhance soil organic carbon, improve fertiliser quality and promote plant growth has been demonstrated (Agegnehu et al., 2017). Based on these benefits, biochar addition also accelerates composting by increasing microbial diversity and shortening the composting cycle, thus yielding an enhanced final compost product (Chen et al., 2017).

Taking this a step further, researchers have begun exploring biochar–vermicompost combinations as a next-generation soil amendment for enhancing soil quality and for improving nutrient availability and ecological balance while aligning with circular economy principles and advancing net-zero waste processes (Antileo-Mellado et al., 2024; EL-Mogy et al., 2024; Li et al., 2024). In this context, the present chapter considers the synergistic effects of biochar and vermicompost in mitigating the adverse impacts of organic waste on soil health and in promoting soil restoration. This chapter also examines how earthworms respond to the presence of biochar and whether biochar affects their growth and activity. These critical questions address the potential challenges that earthworms may face in contact with biochar and the eventual benefits of incorporating biochar in the final vermicompost product (Fig. 1).

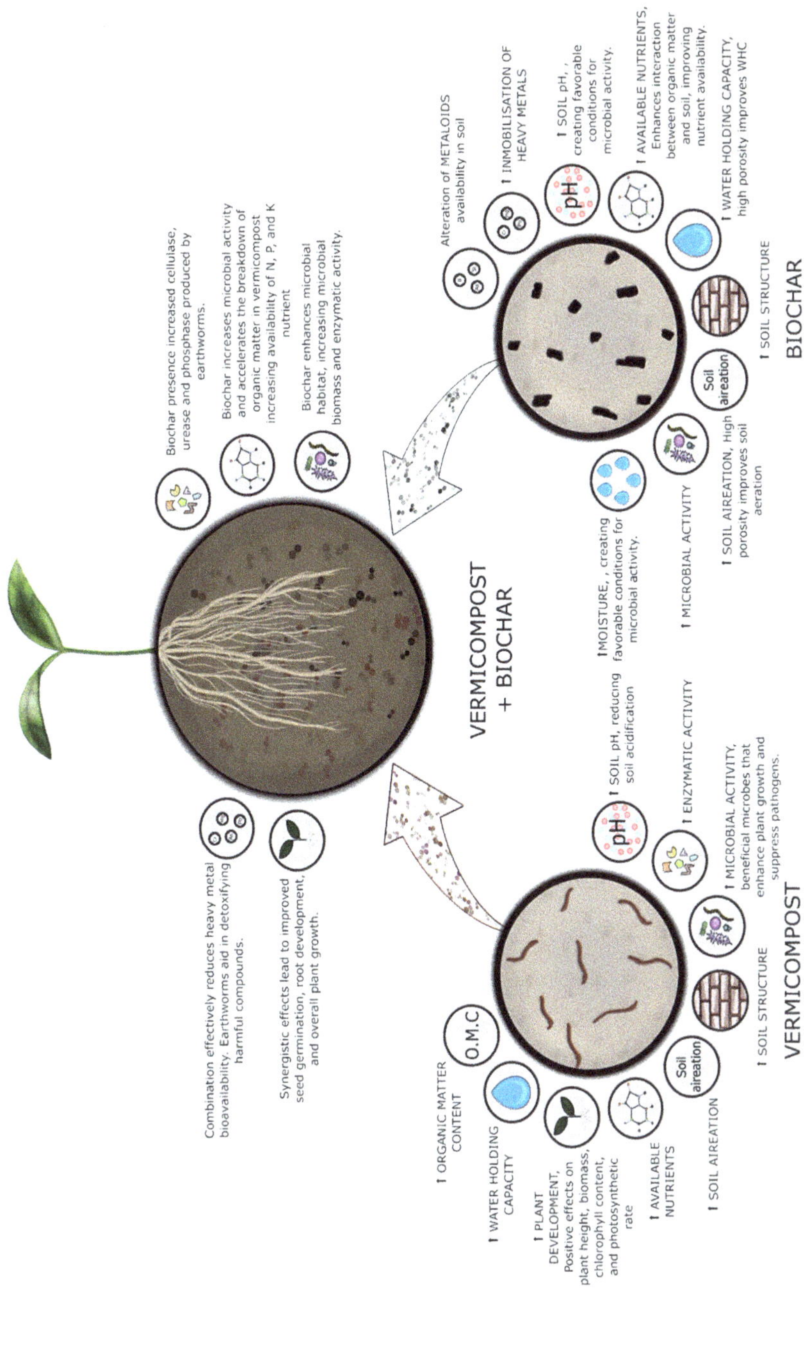

Figure 1 The figure illustrates the biochar–vermicompost interaction.

2 The role of biochar in composting

Biochar has been shown to improve compost quality and mitigate GHG emissions (Agyarko-Mintah et al., 2017). It has also been found that biochar produced at 550°C reduces CO_2 emissions by 50% relative to that generated at other pyrolysis temperatures (Lévesque et al., 2018). There are three plausible reasons for this: (1) Biochar acquires a more stable carbon structure at 550°C, which allows it to retain more carbon in a form resistant to microbial degradation, thus increasing carbon storage in the soil and reducing CO_2 release into the atmosphere. (2) The higher surface area and porosity of biochar produced at this temperature enhance the capacity of the material to absorb CO_2 and other gases. These properties also promote soil aeration and water retention, further decreasing GHG emissions from the soil. (3) Finally, pyrolysis at 550°C strikes a balance between maximising biochar yield and minimising production of GHGs (e.g. methane and CO_2), thus optimising environmental benefits.

Biochar interacts with microbes and, subsequently, with nutrient cycling in several ways. On the one hand, biochar slows down nitrification by adsorbing ammonium (NH_4^+) on its large, porous surface. This reduces the availability of NH_4^+ to nitrifying bacteria, which convert the cation to nitrate (NO_3^-), thus helping to reduce N_2O emissions. On the other hand, biochar increases soil pH and moisture, indirectly affecting microbial activity and also helping to slow down the nitrification process (Yang et al., 2017). Biochar also further accelerates carbon turnover through different mechanisms. First, it induces a 'priming effect' by providing a habitat for soil microorganisms that decompose carbon sources, thereby enhancing the interactions between soil microbes and organic matter (Rasul et al., 2022) and favouring decomposition and carbon cycling. Second, the highly porous surface of biochar interacts with soil organic matter (SOM), facilitating the adsorption and desorption of carbon materials (Kalu et al., 2024). This process makes carbon more bioavailable to microbes, which then transform the carbon into simpler compounds like CO_2 or more stable microbial-derived compounds like amino sugars and microbial necromass. Third, biochar improves soil aeration and moisture regulation, thus creating favourable conditions for aerobic microbial activity and carbon cycling (Gougoulias et al., 2014; Razzaghi et al., 2020). The high porosity of the biochar can also improve the physical properties of soil, including the water-holding capacity, due to the dissociation of H^+ ions and leaching of basic cations such as calcium (Ca), magnesium (Mg), potassium (K), and sodium (Na). Table 1 summarises the effect of H^+ ions on soil properties. The application of biochar also significantly increases soil pH, creating alkaline conditions that help to immobilise soil metals, further contributing to soil health and sustainability (Neina, 2019; Qiu et al., 2020).

Table 1 Effects of H^+ ions dissociation on soil properties

Process	Effect on soil chemistry	Implication for soil function
Dissociation of H^+ ions from biochar	Increases soil acidity (lowers pH)	Promotes cation exchange and mobilisation of nutrients
Displacement of basic cations (Ca^{2+}, Mg^{2+}, K^+, Na^+)	Enhances the leaching of these cations from exchange sites	Alters nutrient availability and buffering capacity
Increased cation exchange capacity (CEC)	Facilitates nutrient retention and exchange	Improves fertiliser efficiency and microbial access to ions
Mobilisation of Ca^{2+} and Mg^{2+}	May improve soil structure and flocculation	Enhances aeration and water infiltration
Leaching of K^+ and Na^+	Can reduce salinity and ionic imbalance in some soils	Supports root development and microbial activity

Source: Gougoulias et al., 2014; Razzaghi et al., 2020.

Another important factor to consider when using biochar as a soil amendment is its effect on seed germination, which varies depending on the feedstock and the application rate. While some studies have reported that biochar inhibits seed germination (Gasco et al., 2016), others have shown that applying biochar at rates of 10–50 t/ha improved seed germination and plant growth (Solaiman et al., 2012).

For instance, biochar can act as a biostimulant for growth of lentils but can also be phytotoxic to lettuce (Gasco et al., 2016), and it may also have no effect on plant growth, as reported by Alburquerque et al. (2014), after the application of more than 100 t/ha of biochar. Similarly, Bargmann et al. (2013) reported that biochar generally had no effect on the germination of spring barley (*Hordeum vulgare*). This lack of effect, however, does not preclude potential risks. Biochar can exhibit phytotoxicity, often attributed to the presence of phenolic compounds. Moreover, the phytotoxic effects on plants depend on the biochar and seeds considered. The variable effects of biochar on plant growth and germination highlight the complex interactions with different plants and biochar types, which extend to its influence on soil properties and metal dynamics. Biochar amendments greatly alter metal availability in soils (Antonangelo et al., 2021; Bilias et al., 2021; Xu et al., 2011). For instance, Wang et al. (2020) observed that arsenic concentration in rice straw biochar (RSBC)-amended soil increased to 29.96 mg/kg, while cadmium (Cd) levels decreased to 0.25 mg/kg in both RSBC and sewage sludge biochar (SSBC)-amended soils relative to the control treatment. These authors also reported that plant biomass increased by 37.91% in SSBC-amended soil and that RSBC boosted plant growth. Moreover, RSBC reduced arsenic concentration in plants and increased phosphorus (P) uptake, while SSBC reduced Cd accumulation in plants by 30.63% (Wang et al., 2020).

3 Earthworms and their role in vermicomposting

Over the last decades, vermicomposting has become a well-established bioconversion process that ensures the circular management of organic waste in a sustainable, eco-friendly, and low-impact manner (Domínguez, 2023). This bio-oxidative process is driven by the combined action of detritivorous earthworms, microorganisms, and other fauna within the decomposer community, thus accelerating the breakdown and stabilisation of organic matter and modifying its physical, chemical, and biological properties (Domínguez, 2004). In particular, the epigeic earthworm species *Eisenia fetida* (Savigny, 1826) and *Eisenia andrei* (Bouché, 1972) and the tropical species *Eudrilus eugeniae* (Kinberg, 1867) have been extensively used in vermicomposting systems due to their short life cycles, high reproductive rates, and high rate of consumption, digestion, and assimilation of organic matter. Earthworm feeding and digestion processes fragment and aerate the input material and alter microbial populations and communities, thereby affecting organic matter quality and nutrient availability, enhancing microbial activity, and accelerating the decomposition process (Bhat et al., 2018; Domínguez, 2004; Gómez-Brandón and Domínguez, 2014). The resulting vermicompost is a nutrient-rich, well-decomposed, and biologically active product (Aira and Domínguez, 2016), with beneficial effects on a wide variety of agronomic and horticultural crops (Blouin et al., 2019; Lazcano and Domínguez, 2011).

The chemical and (micro)biological properties of the final vermicompost and its potential value as an organic amendment are inextricably linked to the nature of the initial organic matter, as well as to the effects of earthworms on the decaying of these substrates and associated microbial communities through gut- and cast-associated processes, GAPs and CAPs, respectively (Gómez-Brandón et al., 2021; González et al., 2024). Indeed, the type and quality of the organic feedstock largely determine the transformations that occur during processing and, together with earthworm-driven mechanisms, shape the composition, dynamics, and performance of the system. These factors also influence the duration of the ageing period required before a stable end product is obtained. To this end, García-Sánchez et al. (2017) monitored the vermicomposting process of different feedstocks including horse manure, apple pomace, grape pomace, and digestate over time, that is after 60, 120, 180, and 240 days of processing with *E. andrei*. All materials exhibited neutral to alkaline pH throughout the process. Electrical conductivity (EC) remained largely unchanged (between 0.1 and 1.6 mS cm^{-1}), between days 60 and 240. Stabilisation of the materials was achieved after 240 days, as reflected by a decrease or stabilisation of the organic carbon content. Aligned with this, microbial activity also declined after 240 days, corresponding with a lower earthworm activity, which peaked at 120 days. Moreover, phytotoxicity tests indicated that apple pomace and digestate had reached a stable, nutrient-rich, and safe state, while horse manure and grape pomace did not reach the same level of maturation. Plaza et al. (2008) analysed the products obtained from a prolonged process of about eight months of vermicomposting with *E. andrei* using cattle manure and a mixture of olive pomace and cattle manure. The authors reported that the pH remained alkaline in cattle manure treatment and increased to 6.6 to 8.6 in the mixture. While the authors reported total organic carbon (TOC) was decreased in both organic sources, the mixture one kept more carbon. The C/N ratio decreased significantly in both cases, reflecting stabilisation and humification of the materials. Total nitrogen increased in the case of the mixture, while in the cattle manure, it decreased. Fernández-Gómez et al. (2010) conducted a pilot-scale experiment using a metal vermireactor with an initial layer containing sheep manure as habitat for the earthworms (*E. fetida*). The reactor was fed with tomato-fruit waste paste for 5 months (+2 months of maturation), and the chemical and enzymatic parameters were monitored over the course of the process. The results showed that the earthworm biomass increased after 90 days; followed by a decline due to an increase in pH level, EC, and ammonium concentration in the mixture. The temporal patterns of β-glucosidase, protease, and urease activities appeared to be related to earthworm growth and OM stabilisation. Taken together, the authors suggested that the continuous-feeding vermicomposting process could be a suitable management system for processing vegetable wastes. Hřebečková et al. (2019) compared vermicompost from household biowaste, malthouse sludge + agricultural waste, and grape marc. The process lasted more than 12 months, during which fresh layers of organic sources were added every 2 weeks. The pH stabilised in neutral ranges for all materials, and the EC decreased toward the end of the process. The authors reported that microbial biomass and activity were concentrated in the upper, fresher layers, where earthworm density was also the highest. Likewise, the maximum enzymatic activity occurred in these upper layers, whereas deeper, older layers showed reduced activity. To further investigate these spatial dynamics, Gómez-Brandón et al. (2021) evaluated the changes in vermicompost derived from grape marc using *E. andrei* for 42 weeks (294 days), sampling every 3 weeks in a continuously fed system, where fresh material was added weekly in layers. The main objective of this study was to compare the evolution of the initial feedstock with and without earthworms; only results from the treatments with earthworms are considered here. They reported that pH was weak-alkaline across the layers, while C/N ratio increased or remained stable, likely due to the high lignocellulosic content of the residues. By the end of the experiment, macro- and micronutrients content increased, while microbial biomass and activity decreased with depth, with fungi predominating in the upper, fresher layers (Table 2).

Table 2 Summary of initial and final values of physicochemical and biological parameters

Organic source	Maximum time of vermicomposting	pH initial	pH final	Electrical conductivity ($mS\ cm^{-1}$) initial	Electrical conductivity ($mS\ cm^{-1}$) final	Organic carbon/OM (TC/TOC/DOC) initial	Organic carbon/OM (TC/TOC/DOC) final	Nitrogen (TN, $N\text{-}NO_3^-$, NH_4^+-$mgkg^{-1}$-TKN) initial	*Nitrogen (TN (%), $N\text{-}NO_3^-$, NH_4^+-$mgkg^{-1}$-TKN) final*	C/N initial	C/N final	Microbial activity initial	Microbial activity final	Study
Cattle manure (CM)	*8 months (≈244 days)*	8.6	8.5	760	143	TOC 212	TOC 88	TN (gkg^{-1}) 14.0	TN (gkg^{-1}) 11.0	15	8	NA	Humification: TEC, FAC, HAC, HR, HI increased; FTIR spectra: decrease in aliphatic and N-containing groups, enrichment of aromatic structures; E4/E6 ratio decreased approaching soil HA; fluorescence: decrease in RFI, increase in EEWP	Plaza et al., 2008
Cattle manure + olive pomace (CM + OP)	*8 months (≈244 days)*	6.6	8.6	321	249	TOC 483	TOC 292	TN (gkg^{-1}) 12.8	TN (gkg^{-1}) 20.0	38	15	NA	Humification: TEC, FAC, HAC, HR, HI increased; FTIR and fluorescence similar trends as above, indicating high degree of humification and maturity	Plaza et al., 2008
Sheep manure + tomato-fruit wastes	5 months + 60 days maturation (210 days)	8.3	10.4	168	4.4	TOC (gkg^{-1}) 139	TOC (gkg^{-1}) 148	*TKN (gkg^{-1}) 12.1 NH_4^+ 348*	*TKN (gkg^{-1}) 12.9 NH_4^+ 2561*	11.4	11.4	Dehydrogenase: peak at day 30, sharp increase 90–120; β-glucosidase: peak day 90; Protease: increase 0–60, decline day 90, rise 90–120; Urease: stable 0–60, increase 60–90	Dehydrogenase: stable after day 120; β-glucosidase decline with earthworm mortality; Protease: decline post day 120; Urease: decreased after maturation; bacterial/fungal DGGE: high diversity, community shifts	Fernández-Gómez et al., 2010

Horse manure (HM)	240 days	7.8	7.1	1.4	1.1	TOC 40 %	DOC (gkg^{-1}) 3.3	TN 2.7 %	*N-NO_3^- 735NH_4^+ 215*	NA	NA	Dehydrogenase activity peak at 120 d; β-glucosidase decline at 120 d; Protease: increase 0–60 d, decline 90–120 d; Phosphatase: increase 240 d; o-diphenol oxidase peak 120 d	Dehydrogenase decline at 180–240 d; Protease increase at 240 d; β-glucosidase low; phosphatase stable; o-diphenol oxidase decline at 240 d	García-Sánchez et al., 2017
Apple pomace (AP)	240 days	7.9	8	1.3	0.4	TOC 40 %	DOC (gkg^{-1}) 2.6	TN 2.5 %	N-NO_3^- 56.2NH_4^+ 163.7	NA	NA	Dehydrogenase peak 120 d; β-glucosidase stable; Protease 60–120 d increase; phosphatase high; o-diphenol oxidase peak 120 d	Dehydrogenase decline at 180–240 d; Protease decline at 240 d; β-glucosidase low; phosphatase remained stable; o-diphenol oxidase decline at 240 d	García-Sánchez et al., 2017
Grape pomace (GP)	241 days	7.7	9.2	1.2	1	TOC 42 %	DOC (gkg^{-1}) 3.6	TN 2.1 %	N-NO_3^- 36.6NH_4^+ 162.3	NA	NA	Dehydrogenase peak 120 d; β-glucosidase 60–120 d increase; Protease 120 d peak; phosphatase stable; o-diphenol oxidase low	Dehydrogenase decline at 180–240 d; β-glucosidase decline at 180–240 d; protease decline at 240 d; phosphatase stable; o-diphenol oxidase increase at 240 d	García-Sánchez et al., 2017
Household biowaste	>12 months (365 days)	5.4	7.45	NA	0.23	NA	NA	NA	NA	26.7	NA	NA	Enzymatic activity varies: higher β-D-glucosidase, acid phosphatase, chitinase, cellobiohydrolase, aminopeptidases in intermediate layers; higher earthworm biomass in top layers	Hřebečková et al., 2019

(*Continued*)

Table 2 *Continued*

Organic source	Maximum time of vermicomposting	pH initial	pH final	Electrical conductivity (mS cm^{-1}) initial	Electrical conductivity (mS cm^{-1}) final	Organic carbon/OM (TC/TOC/DOC) initial	Organic carbon/OM (TC/TOC/DOC) final	Nitrogen (TN, N-NO_3^-, NH_4^+-mgkg^{-1}-TKN) initial	*Nitrogen (TN (%), N-NO_3^-, NH_4^+ -mgkg^{-1}-TKN) final*	C/N initial	C/N final	Microbial activity initial	Microbial activity final	Study
Malthouse sludge + agricultural waste (20:80 vol)	>12 months (365 days)	8.3	8.28	0.20	0.14	NA	NA	NA	NA	6	NA	NA	Enzymatic activity: moderate levels of β-D-glucosidase, lipase, and aminopeptidase; earthworms only present in top layers	Hřebečková et al., 2019
Grape marc	>12 months (365 days)	7.6	8.68	0.41	0.14	NA	NA	NA	NA	21.4	NA	NA	Earthworm biomass and enzymatic activity (β-D-glucosidase, acid phosphatase, chitinase, and lipase) were higher in top layers	Hřebečková et al., 2019
Digestate (DG): 50% slurry, 40% corn silage, 10% hay)	242 days	8.5	8.1	3.5	1.6	TOC 36 %	DOC (gkg^{-1}) 2.5	TN 2.2 %	N-NO_3^- 1170.5 NH_4^+ 159.6	NA	NA	Dehydrogenase peak at 120 d; β-glucosidase, protease, and o-diphenol oxidase were low; while phosphatase remained stable	Dehydrogenase decline at 180–240 d; decline in β-glucosidase and protease at 240 d; while, phosphatase remain stable; and o-diphenol increased	García-Sánchez et al., 2020
Grape marc var. Albariño	*42 weeks (≈294 days)*	8.23	8.5	0.37	0.35	TC (g) 112	TC (g) 154	TN (g) 4.93	TN (g) 5.50	23.3	28.5	Upper layers: high fungal activity, lower bacterial activity; high β-D-glucosidase in intermediate layers; high lipase; moderate acid phosphatase; moderate cellobiohydrolase and aminopeptidases.	Lower layers: reduction in fungal activity; β-D-glucosidase and other enzymes; bacteria remain present but overall activity was lower; stabilisation of organic matter.	Gómez-Brandón et al., 2021

Grape marc var. Mencía	*42 weeks (≈294 days)*	7.93	7.55	0.27	0.22	TC (g) 159	TC (g) 233	TN (g) 6.34	TN (g) 8.93	25.2	26.4	Upper layers: high earthworm biomass, increased fungal activity; high β-D-glucosidase in intermediate layers; moderate lipase and acid phosphatase; cellobiohydrolase and aminopeptidases present.	Lower layers: decrease in enzymatic activity; lower fungal activity; bacteria predominate; stabilisation of organic matter.	Gómez-Brandón et al., 2021
Common conclusions	Similar evaluation periods.	Most substrates had a slightly alkaline initial pH, except for household waste. During the vermicomposting process, the pH tends to stabilise or increase slightly in some cases (AP, sheep manure + tomato waste) or decrease moderately (HM, Mencía grape marc). However, all pH values remain in an alkaline to slightly alkaline range.		The EC varies depending on the original organic source. It generally decreases at the end of the study, indicating salt leaching and stabilisation of the vermicompost.		The organic carbon content is high at the start of the process. During vermicomposting, it decreases due to the mineralisation and degradation of organic matter, with a tendency towards stabilisation.		The nitrogen content varies depending on the organic source. During vermicomposting, nitrogen is preserved or increases slightly in some types of waste (e.g. grape marc, CM + OP), while in others it may decrease due to leaching or volatilisation.		The initial values are high, indicating a high proportion of available carbon. The C/N ratio decreases in most cases, indicating the maturity and stabilisation of the organic matter.		Dehydrogenase peaks at 120 days, then declines. ß-glucosidase, protease, phosphatase, and other enzymes show intermediate increases followed by a decline towards the end of the process. Spatial distribution may vary with the upper or fresher layers showing greater earthworm and enzyme activity.		

Physicochemical and biological parameters: pH, electrical conductivity, organic carbon, nitrogen forms, C/N ratio, and microbial/enzymatic activity reported in different vermicomposting studies.
The table presents mean values either before the start of the experiments or from the first available measurement, and the corresponding values at the end of the experimental period. When data are not available, cells are indicated as NA (Not Available).
Abbreviations: TOC: Total Organic Carbon; DOC: Dissolved Organic Carbon; TC: Total Carbon; OM: Organic Matter; TN: Total Nitrogen; TKN: Total Kjeldahl Nitrogen; N-NO_3^-: Nitrate Nitrogen; NH^{4+}: Ammonium; EC: Electrical Conductivity; C/N: Carbon-to-Nitrogen Ratio; TEC: Total Extractable Carbon; FAC: Fulvic Acid Carbon; HAC: Humic Acid Carbon; HR: Humification Ratio; HI: Humification Index; FTIR: Fourier Transform Infrared Spectroscopy; E4/E6: Absorbance ratio at 465 nm and 665 nm (humification indicator).

Nonetheless, apart from aforementioned properties, there is increasing evidence indicating that the potential usefulness of vermicompost lies in its biological components (Gopalakrishnan et al., 2011; Rehman et al., 2023; Vambe et al., 2023). These authors reported that vermicompost and its liquid derivatives (known as vermicompost teas) harbour a broad spectrum of bacterial genera, including *Bacillus*, *Enterobacter*, *Pseudomonas,* and *Streptomyces* spp., which are capable of inhibiting the growth of plant pathogenic fungi such as members of the genera *Phytophthora* and *Pythium* species. The presence of beneficial plant-growth promoting microorganisms (PGPMs) makes vermicompost of great agronomic value fuelling its use as a natural biofertiliser (Ma et al., 2022; Rehman et al., 2023). Recent studies have also shown that vermicompost derived from grape marc and leguminous shrubs have higher functional diversity and more genes putatively involved in metabolic processes that are potentially beneficial for plant growth and development than their respective parent materials (Domínguez et al., 2019; Gómez-Brandón et al., 2020). A similar trend has been observed regarding the microbial richness of bacterial communities, which was enhanced during the vermicomposting of grape marc (Domínguez et al., 2019; Kolbe et al., 2019). Earthworms also facilitate the replacement of these microbial communities in the vermicompost by introducing the microbiome from their digestive system into their casts, thus enhancing fungal populations, reducing bacterivorous nematodes and increasing mineralisation rates, CO_2 flux, and labile C and N pools in earthworm density-dependent processes (Aira et al., 2008). Earthworm activity can stimulate initial microbial growth and enzyme activity, with the response varying depending on the slurry rate and the age of the layer, indicating a relationship between microbial biomass, enzyme activity, and substrate availability (Aira et al., 2007).

During the early stages of vermicomposting, the levels of N_2O can increase as an intermediate product of fluctuating oxygen levels. Thus, N_2O emissions have been found to be higher in the vermicompost than in the control feedstock without earthworms (Lv et al., 2020). This study reported that both nitrification and denitrification contributed to N_2O production in the early stages, but nitrification became the dominant pathway as the process continued. Earthworms stimulated nitrifying bacteria, leading to higher *amoA* gene copies and increased N_2O emissions via nitrification. Additionally, denitrification occurred primarily in the earthworm gut, where conditions inhibited the *nosZ* gene, further contributing to N_2O production. Overall, nitrification was responsible for 55.8 to 88.7% of N_2O emissions, driven by the presence of earthworms. Altogether, it points to vermicomposting as a valuable and environmentally friendly approach for sustainable agriculture and environmental conservation. The most significant advantage of the process lies in its dual role in organic waste management and the generation of a product that contributes to soil fertility and health (Ambasta and Kumari, 2013; Vyas et al., 2022).

4 Effects of vermicompost on soil

In intensive cropping systems, the excessive use of nitrogenous fertilisers has been shown to accelerate soil acidification (Zhang et al., 2022a). Owing to its generally alkaline properties, vermicompost can potentially mitigate soil acidification and represents an environmentally sustainable alternative to mineral fertilisers. When added to acidic soils, vermicompost can act as a buffer, increasing the soil pH to more neutral values (Toor et al., 2024; Zhang et al., 2020). Adjusting the soil pH improves the solubility of some nutrients such as P, thus enhancing P availability to plants (Hinsinger, 2001; Lauber et al., 2009), as found after the addition of vermicompost (Zhang et al., 2020). Beneficial soil microorganisms, such as nitrogen-fixing bacteria and organic matter decomposers, also tend to thrive in slightly alkaline or almost neutral soils (Lauber et al., 2009). Soil pH also influences the dispersion of clays and aggregation of particles. As the pH increases, the negative charge on clay particles also increases, causing the particles to repel each other, which benefits the structural stability of soils with high clay dispersion. This process promotes the formation of large aggregates, especially in soils with high carbonate content (Bronick and Lal, 2005). Moreover, adding vermicompost, which is rich in OM, can favour soil microbial activity by enhancing microbial respiration (Bronick and Lal, 2005; García-Sánchez et al., 2017).

The changes that take place in soil following the application of vermicompost can enhance plant performance. For instance, Xu and Mou (2016) reported that variations in soil parameters, such as increased pH, EC and nutrient content and improved soil structure and water holding capacity, were accompanied by an increase in several plant traits, including plant height, number and area of leaves, chlorophyll content and photosynthetic rate, after the addition of vermicompost. Similarly, Arancon et al. (2006) observed that vermicompost has a positive effect on plant growth and strawberry yield. This was accompanied by a significant increase in soil microbial biomass, enzyme activity and available nutrients, as well as an improvement in soil structure, water-holding capacity and a reduction in soil bulk density. Based on a meta-analysis of 68 articles, Blouin et al. (2019) estimated that vermicompost increased plant yields by 26%, total biomass by 13%, shoot biomass by 78%, and root biomass by 57%. The maximum yields were obtained when vermicompost represented between 30% and 50% of the soil volume and when cattle manure was used as feedstock. Although the meta-analysis had some limitations – such as potential over-representation of studies with large effect sizes (e.g. substantial increases in biomass or yield) and variability in experimental setups – the authors concluded that vermicompost has significantly positive effects on plant growth. Furthermore, vermicompost application has also been reported to be an effective strategy for suppressing plant diseases (Vambe et al., 2023; Yatoo et al., 2021). Overall, these different findings indicate that vermicompost is a valuable amendment for improving soil and plant health and, in turn, increasing the abundance and diversity of microbial communities.

Despite the aforementioned advantages of vermicomposting and the potential benefits of vermicompost, there are some challenges that affect its efficiency. The slow breakdown of lignocellulosic materials (e.g. straw), which are complex and resistant to decomposition by earthworms and the associated microorganisms, and the presence of heavy metals can have harmful effects on earthworms. These negative aspects of vermicomposting could be addressed by combining vermicompost with other types of amendments (Tang et al., 2017; Zhang and Sun, 2014). To this date, the inclusion of biochar in vermicomposting has been reported to be a promising way of overcoming these limitations (EL-Mogy et al., 2024).

5 Combining biochar and vermicompost

The capacity of biochar to promote the degradation of pollutants and reduce GHG emissions has been demonstrated and attributed to properties such as its carbonaceous structure, high cation exchange capacity, and porous surface area, which help to enhance microbial activity and water retention (Li et al., 2024; Paul et al., 2020). These characteristics also favour soil aggregation and increase enzymatic activity, leading to faster decomposition and improved vermicompost quality (Hijikata et al., 2015).

Qiang et al. (2018) showed that adding biochar, particularly at a rate of 6%, during the vermicomposting process increased the activity of cellulase, urease, and phosphatase enzymes excreted by *E. fetida*. Roy et al. (2022) reported that the availability of essential nutrients such as N, P, and K increased by approximately 1.5–2.0 times in exfoliated biochar-based vermicompost beds. This improvement was attributed to the increased surface area, which provided an environment more conducive to microbial activity and earthworm colonisation. As a result, microbial biomass carbon and microbial respiration levels increased significantly, along with a 6 to 12-fold acceleration in dehydrogenase and fluorescein diacetate (FDA) hydrolysing enzyme activities in biochar-based vermibeds.

Earthworms are known to play a vital role in eliminating toxic compounds and heavy metals from their environment (Chen and Hu, 2019; Sinha et al., 2010). For instance, Roy et al. (2022) showed that *E. fetida* played an essential role in reducing arsenic (As^{3+}) levels in biochar vermibeds. The bioavailable arsenic fractions were reduced by 25–40%, while the recalcitrant arsenic fractions increased by 8–15%. This demonstrates the effectiveness of *E. fetida* in transforming arsenic into less bioavailable forms. The performance of *E. fetida* was particularly notable in the exfoliated biochar combined with cow dung (12.5% exfoliated biochar to 87.5% cow dung feedstock), making this combination highly effective for arsenic removal and nutrient recovery. Furthermore, some studies have demonstrated that biochar produced from rice husks significantly reduces the bioavailability and mobility of cadmium in contaminated soils (Cui et al., 2016; Oranratmanee and Tungittiplakorn, 2021). The porous structure and high cation exchange capacity of biochar enhance the adsorption and stabilisation of cadmium ions, thus preventing their uptake by plants. To enhance the bioremediation strategy, activation of biochar with extracellular enzymes is emerging as a potential alternative to chemical or physical activation, which is often expensive and generates chemical waste. In this process, microorganisms act as a source of enzymes that adhere to the biochar surface. Recent studies have highlighted the role of soil macrofauna, particularly earthworms, in enhancing enzyme adsorption onto biochar (Sanchez-Hernandez, 2018; Sanchez-Hernandez et al., 2019). Hence, the inclusion of biochar during the vermicomposting process, with the aim of removing pollutants, has also been investigated. Bošković et al. (2021) evaluated the addition of biochar during the vermicomposting process using *E. andrei* and studied the influence of the degradation rate of fungicides. These authors reported that mixing biochar with vermicompost changed the fate and bioaccumulation of the fungicides.

The addition of biochar to the vermicomposting process has been shown to improve humification processes, resulting in high-quality vermicompost with enriched fulvic acid content, which is beneficial for plant growth (Antileo-Mellado et al., 2024; Hou et al., 2023; Zhang et al., 2023). Moreover, the combination of vermicompost and biochar enhances CO_2 production and microbial respiration, leading to improved seed germination and plant growth (Dawar et al., 2022; EL-Mogy et al., 2024). In this regard, Zhang et al. (2023) reported that the combined application of vermicompost and biochar resulted in a higher yield of continuously cropped pepper grown in karst areas than yields obtained with each amendment alone. This indicates that the organic amendments had a synergistic effect on crop productivity that may be related to an improved nutrient status and microbial environment (Rivelli and Libutti, 2022; Wang et al., 2021).

6 Toxicity of biochar to earthworms

Despite the emerging benefits of biochar, the potentially adverse effects on earthworms have raised concerns (Odinga et al., 2021; Sanchez-Hernandez et al., 2019; Wu et al., 2023). Biochar contains ash and combustion by-products that can have harmful effects on earthworm growth, causing oxidative stress and inhibiting the vermicomposting process (Elliston and Oliver, 2020; Zhang et al., 2019). Studies have shown that biochar can contain toxic compounds, such as polycyclic aromatic hydrocarbons (PAHs), which may inhibit earthworm growth and cause oxidative stress (Pokharel et al., 2020; Zhang et al., 2017). The toxicity of biochar varies depending on feedstock type, pyrolysis temperature, and posterior treatment. For instance, cow dung biochar was found to be more toxic than corncob or sewage sludge biochar (Intani et al., 2018). However, thermal treatment of biochar can reduce its vermitoxicity by lowering the PAH content, as demonstrated by Buss et al. (2015).

Careful consideration must therefore be given to the strategy of including biochar in the vermicomposting process, to minimise the potential drawbacks. Recent studies have aimed to explore the vermitoxicity of biochar produced from various types of feedstock – animal (cow dung), plant (corncob), and sewage sludge, at different pyrolysis temperatures (350°C, 550°C, and 750°C) (Shi et al., 2021). The toxicity was evaluated using a combination of acute toxicity tests, neutral red retention time (NRRT) assay, and oxidative stress response assays. The test results revealed that cow dung biochar was more toxic to earthworms than biochar derived from corncob and sewage sludge. The toxicity level was measured by determining the LC50 and determining weight loss in the earthworms. Cow dung biochar caused significant disruption to lysosomal membrane stability, indicating cellular stress. Additionally, oxidative stress was indicated by disturbed antioxidant enzyme activity and elevated malondialdehyde levels, highlighting the vermitoxicity of cow dung biochar. Interestingly, the study demonstrated that thermal treatment can mitigate the toxic effects of biochar. By subjecting biochar to higher pyrolysis temperatures, the LC50 increased and earthworm weight loss decreased. This suggests that PAHs, which are by-products of biochar production, may be key contributors to the toxicity. The use of the NRRT assay provided further insights, offering a dose-response relationship that could be instrumental in assessing biochar toxicity. The findings highlight the importance of feedstock selection and thermal treatment in biochar production to minimise the harmful effects on earthworms, which are vital for soil health. These results emphasise the need for further research to understand the comprehensive ecological impacts of biochar and to develop guidelines for its safe application in agricultural practices (Shi et al., 2021).

Bošković et al. (2021) examined fungicide bioaccumulation in earthworms (*E. andrei*) in biochar-amended soils and found that bioaccumulation decreased over time but varied with soil properties. Equilibrium was reached faster (3–7 days) in more absorbent soils, while in less absorbent sandy soils it was not reached until around 21 days, indicating that biochar delays equilibrium in some types of soil. Wang et al. (2020) used high-throughput sequencing to examine the effects of biochar amendments on the gut bacteria of the earthworms (*E. fetida*). The analysis revealed high concentrations of metals in earthworm tissues, with earthworm gut bacterial composition dominated by Firmicutes (39.36%), Actinobacteria (23.69%), and Proteobacteria (13.02%); and different from that of the surrounding soil. Additionally, rice straw biochar (RSBC) had a more pronounced effect on the earthworm gut microbiome than sewage sludge biochar (SSBC). It is necessary to clarify the above process as *E. fetida* or *E. andrei* are not soil earthworms. However, to perform the toxicity test, the earthworms were bred separately in the lab with different substrates such as manure, garden waste, and peat, as in the experiment conducted by Bošković (2021). Before testing, adult earthworms with developed clitellum (~200 mg weight) were acclimatised in experimental soils moistened to 50% water-holding capacity with added manure (7 mg $g_{soil\text{-}dw}^{-1}$) for three days (Wang et al., 2020).

By contrast, some studies have reported no significant effect on earthworms when added to the biochar-amended soil. For example, Whalen and Benslim (2021) stated that wood-based biochar, applied as a soil amendment alone or in combination with dairy cattle slurry, had no effect on earthworm populations in agricultural fields on a dairy farm in southern Quebec (Canada). The study authors concluded that the changes in earthworm populations can be attributed to environmental stress, such as flooding and other unfavourable weather conditions. Similarly, Huang et al. (2020) reported no significant negative effects on earthworm survival rates in heavy metal-contaminated soil + biochar. However, these authors observed that the initial reduction in earthworm growth was consistent with that found in a previous study by Zhang et al., (2019), in which a decrease in earthworm biomass during the first 28 days of the vermicomposting process was reported. Interestingly, other studies showed that the presence of *E. fetida* led to an increase in antioxidant enzyme activity in the initial 28 days and subsequently declined (Liu et al., 2014; Panzarino et al., 2016). These findings are consistent with those of Sanchez-Hernandez et al., (2019), who reported that the enzyme activities of earthworms (*Lumbricus terrestris*) decreased significantly in the soil including 1% biochar for 30 days.

Several studies have explored methods of mitigating biochar toxicity, including water or solvent washing and thermal treatment. Kołtowski and Oleszczuk (2015) reported a significant reduction in the PAH content of biochar when subjected to drying at temperatures of 100°C to 300°C. Intani et al. (2018) also demonstrated that heat-treated biochar reduced phytotoxicity in ryegrass and improved plant biomass. Such findings highlight the importance of treating biochar to reduce its toxicity before application to agricultural soils.

7 Conclusion and future trends in research

Earthworms serve as effective bioindicators of soil pollution and play a significant role in soil remediation. The distinct differences between soil microbial communities and those in the earthworm gut highlight the gut's unique habitat and its influence on soil microbial dynamics. Vermicomposting remains a promising bioremediation strategy, and its integration with biochar offers a synergistic approach to improving soil quality and mitigating environmental impacts.

Remediation of metal-contaminated soils remains complex due to their heterogeneous physicochemical properties and the persistent risk of metal re-mobilisation. In this context, combining biochar with vermicompost presents a sustainable solution, enhancing contaminant stabilisation, boosting microbial diversity, and improving soil structure and nutrient retention. Their combined use not only improves soil health but also enhances climate resilience by increasing carbon sequestration and buffering crops against drought stress.

Ultimately, bioremediation should aim for holistic soil restoration, addressing biological, chemical, and physical dimensions of soil health. Vermicompost, enriched with beneficial microbes, plays a pivotal role in this transformation. Future research should focus on optimising biochar and vermicompost formulations to minimise toxicity and deepen our understanding of their combined effects on soil ecosystems. Future studies should also explore shifts in microbial functional groups and their roles in nutrient cycling, pathogen suppression, and long-term soil resilience. Advancing this knowledge will be key to developing resilient, regenerative agricultural systems.

8 Acknowledgements

This study was partly co-funded by the Spanish Ministerio de Ciencia e Innovación (CPP2023-010797), the Xunta de Galicia (ED431C 2022/07) and the European Union's HORIZON-MISS-2023-SOIL-01 research and innovation programme (LivingSoiLL_10157502).

9 References

Agegnehu, G., et al. 2017. The role of biochar and biochar-compost in improving soil quality and crop performance: a review. *Applied Soil Ecology*, 119, 156–170.

Agyarko-Mintah, E., et al. 2017. Biochar lowers ammonia emission and improves nitrogen retention in poultry litter composting. *Waste Management*, 61, 129–137.

Aira, M., et al. 2007. Earthworms strongly modify microbial biomass and activity triggering enzymatic activities during vermicomposting independently of the application rates of pig slurry. *Science of the Total Environment*, 385(1–3), 252–261.

Aira, M., et al. 2008. Detritivorous earthworms directly modify the structure, thus altering the functioning of a microdecomposer food web. *Soil Biology and Biochemistry*, 40(10), 2511–2516.

Aira, M., et al. 2016. Characterization of the bacterial communities of casts from Eisenia andrei fed with different substrates. *Applied Soil Ecology*, *98*, pp.103-111.

Alburquerque, J.A., et al. 2014. Effects of biochars produced from different feedstocks on soil properties and sunflower growth. *Journal of Plant Nutrition and Soil Science*, 177(1), 16–25.

Ambasta, S.K. and Kumari, S. 2013. Earthworm: a natural ecofriendly organism for recycling of organic residues and improvement of soil health by Vermicompost. *International Journal of Applied Sciences and Biotechnology*, 1(4), 171–175.

Amundson, R., et al. 2015. Soil and human security in the 21st century. *Science*, 348(6235), 1261071.

Antileo-Mellado, S., et al. 2024. Effect of biochar on vermicompost production: chemical, biochemical, and biological properties. *Agronomy*, 14(3), 615.

Antonangelo, J.A., et al. 2021. The roles of co-composted biochar (COMBI) in improving soil quality, crop productivity, and toxic metal amelioration. *Journal of Environmental Management*, 277, 111443.

Arancon, N., et al. 2006. Influences of vermicomposts on field strawberries: part 2. Effects on soil microbiological and chemical properties. *Bioresource Technology*, 97(6), 831–840.

Bargmann, I., et al. 2013. Hydrochar and biochar effects on germination of spring barley. *Journal of Agronomy and Crop Science*, 199(5), 360–373.

Bhat, S.A., et al. 2018. Bioremediation and detoxification of industrial wastes by earthworms: vermicompost as powerful crop nutrient in sustainable agriculture. *Bioresource Technology*, 252, 172–179.

Bilias, F., et al. 2021. Towards a soil remediation strategy using biochar: effects on soil chemical properties and bioavailability of potentially toxic elements. *Toxics*, 9(8), 184.

Blouin, M., et al. 2019. Vermicompost significantly affects plant growth: a meta-analysis. *Agronomy for Sustainable Development*, 39, 1–15.

Bošković, N., et al. 2021. Conazole fungicides epoxiconazole and tebuconazole in biochar amended soils: degradation and bioaccumulation in earthworms. *Chemosphere*, 274, 129700.

Bronick, C.J. and Lal, R. 2005. Soil structure and management: a review. *Geoderma*, 124(1–2), 3–22.

Buss, W., et al. 2015. Inherent organic compounds in biochar–their content, composition and potential toxic effects. *Journal of Environmental Management*, 156, 150–157.

Cameron, K.C., et al. 2013. Nitrogen losses from the soil/plant system: a review. *Annals of Applied Biology*, 162(2), 145–173.

Chen, Y., et al. 2017. Influence of biochar on heavy metals and microbial community during composting of river sediment with agricultural wastes. *Bioresource Technology*, 243, 347–355.

Chen, Z. and Hu, S. 2019. Heavy metals distribution and their bioavailability in earthworm assistant sludge treatment wetland. *Journal of Hazardous Materials*, 366, 615–623.

Chivenge, P., et al. 2011. Comparison of organic versus mineral resource effects on short-term aggregate carbon and nitrogen dynamics in a sandy soil versus a fine textured soil. *Agriculture, Ecosystems & Environment*, 140(3–4), 361–371.

Cui, L., et al. 2016. Continuous immobilization of cadmium and lead in biochar amended contaminated paddy soil: a five-year field experiment. *Ecological Engineering*, 93, 1–8.

Darwin, C., 1892. The formation of vegetable mould, through the action of worms, with observations on their habits. J. Murray.

Dawar, K., et al. 2022. Maize productivity and soil nutrients variations by the application of vermicompost and biochar. *PLoS One*, 17(5), e0267483.

Domínguez, J. 2004. State-of-the-art and new perspectives on vermicomposting research. in: *Earthworm ecology*, CRC Press, pp. 401–424.

Domínguez, J. 2023. State-of-the-art and new perspectives on vermicomposting research: 18 years of Progress. in: *Vermicomposting for sustainable food systems in Africa*, Springer, pp. 27–44.

Domínguez, J., et al. 2019. Changes in the composition and function of bacterial communities during vermicomposting may explain beneficial properties of vermicompost. *Scientific Reports*, 9(1), 9657.

EL-Mogy, M.M., et al. 2024. Integration of biochar with vermicompost and compost improves agro-physiological properties and nutritional quality of greenhouse sweet pepper. *Agronomy*, 14(11), 2603.

Elliston, T. and Oliver, I.W. 2020. Ecotoxicological assessments of biochar additions to soil employing earthworm species Eisenia fetida and Lumbricus terrestris. *Environmental Science and Pollution Research*, 27(27), 33410–33418.

FAO. 2023. *The state of food security and nutrition in the world 2023*. FAO.

Fernández-Gómez, M.J., et al. 2010. Continuous-feeding vermicomposting as a recycling management method to revalue tomato-fruit wastes from greenhouse crops. *Waste Management*, 30(12), 2461–2468.

García-Sánchez, M., et al. 2017. Stabilization of different starting materials through vermicomposting in a continuous-feeding system: changes in chemical and biological parameters. *Waste Management*, 62, 33–42.

Gasco, G., et al. 2016. Relation between biochar properties and effects on seed germination and plant development. *Biological Agriculture & Horticulture*, 32(4), 237–247.

Gentile, R., et al. 2009. Managing N availability and losses by combining fertilizer-N with different quality residues in Kenya. *Agriculture, Ecosystems & Environment*, 131(3–4), 308–314.

Gómez-Brandón, M. and Domínguez, J. 2014. Recycling of solid organic wastes through vermicomposting: microbial community changes throughout the process and use of vermicompost as a soil amendment. *Critical Reviews in Environmental Science and Technology*, 44(12), 1289–1312.

Gómez-Brandón, M., et al. 2020. Vermicomposts are biologically different: microbial and functional diversity of green vermicomposts. *Earthworm Assisted Remediation of Effluents and Wastes*, 125–140.

Gómez-Brandón, M., et al. 2021. Changes in the nutrient dynamics and microbiological properties of grape marc in a continuous-feeding vermicomposting system. *Waste Management*, 135, 1–10.

González, A.B.M., et al. 2024. Earthworms and microplastics: transport from sewage sludge to soil, antibiotic-resistant genes, and soil remediation. in: *Earthworm technology in organic waste management*, Elsevier, pp. 207–229.

Gopalakrishnan, S., et al. 2011. Evaluation of actinomycete isolates obtained from herbal vermicompost for the biological control of Fusarium wilt of chickpea. *Crop Protection*, 30(8), 1070–1078.
Gougoulias, C., et al. 2014. The role of soil microbes in the global carbon cycle: tracking the below-ground microbial processing of plant-derived carbon for manipulating carbon dynamics in agricultural systems. *Journal of the Science of Food and Agriculture*, 94(12), 2362–2371.
Hijikata, N., et al. 2015. Suitability of biochar as a matrix for improving the performance of composting toilets. *Waste Management & Research*, 33(4), 313–321.
Hinsinger, P. 2001. Bioavailability of soil inorganic P in the rhizosphere as affected by root-induced chemical changes: a review. *Plant and Soil*, 237(2), 173–195.
Hou, S., et al. 2023. Role of vermicompost and biochar in soil quality improvement by promoting Bupleurum falcatum L. nutrient absorption. *Soil Use and Management*, 39(4), 1600–1617.
Hřebečková, T., et al. 2019. Changes of enzymatic activity during a large-scale vermicomposting process with continuous feeding. *Journal of Cleaner Production*, 239, 118127.
Huang, C., et al. 2020. Role of biochar and Eisenia fetida on metal bioavailability and biochar effects on earthworm fitness. *Environmental Pollution*, 263, 114586.
Intani, K., et al. 2018. Phytotoxicity of corncob biochar before and after heat treatment and washing. *Sustainability*, 11(1), 30.
Ismail, Z.I. 2004. New machine set to boost palm kernel cake industry. *Business Times*.
Kalu, S., et al. 2024. Biochar reduced the mineralization of native and added soil organic carbon: evidence of negative priming and enhanced microbial carbon use efficiency. *Biochar*, 6(1), 7.
Kolbe, A.R., et al. 2019. Bacterial succession and functional diversity during vermicomposting of the white grape marc Vitis vinifera v. Albariño. *Scientific Reports*, 9(1), 7472.
Kołtowski, M. and Oleszczuk, P. 2015. Toxicity of biochars after polycyclic aromatic hydrocarbons removal by thermal treatment. *Ecological Engineering*, 75, 79–85.
Lauber, C.L., et al. 2009. Pyrosequencing-based assessment of soil pH as a predictor of soil bacterial community structure at the continental scale. *Applied and Environmental Microbiology*, 75(15), 5111–5120.
Lazcano, C. and Domínguez, J. 2011. The use of vermicompost in sustainable agriculture: impact on plant growth and soil fertility. *Soil Nutrients*, 10(1–23), 187.
Lehmann, J., et al. 2020. The concept and future prospects of soil health. *Nature Reviews Earth & Environment*, 1(10), 544–553.
Lévesque, V., et al. 2018. Mitigation of CO_2, CH_4 and N_2O from a fertigated horticultural growing medium amended with biochars and a compost. *Applied Soil Ecology*, 126, 129–139.
Li, J., et al. 2024. Optimizing biochar addition strategies in combined processes: Comprehensive assessment of earthworm growth, lignocellulose degradation and vermicompost quality. *Bioresource Technology*, 406, 131031.
Li, Y., et al. 2020. Soil, biochar, and nitrogen loss to runoff from loess soil amended with biochar under simulated rainfall. *Journal of Hydrology*, 591, 125318.
Liu, T., et al. 2014. The cytotoxic and genotoxic effects of metalaxy-M on earthworms (Eisenia fetida). *Environmental Toxicology and Chemistry*, 33(10), 2344–2350.

Lv, B., et al. 2020. Elucidating the role of earthworms in N_2O emission and production pathway during vermicomposting of sewage sludge and rice straw. *Journal of Hazardous Materials*, 400, 123215.
Ma, H., et al. 2022. Vermicompost improves physicochemical properties of growing medium and promotes plant growth: a meta-analysis. *Journal of Soil Science and Plant Nutrition*, 22(3), 3745–3755.
Mosa, A., et al. 2023. Biochar as a soil amendment for restraining greenhouse gases emission and improving soil carbon sink: Current situation and ways forward. *Sustainability*, 15(2), 1206.
Neina, D. 2019. The role of soil pH in plant nutrition and soil remediation. *Applied and Environmental Soil Science*, 2019(1), 5794869.
Odinga, E.S., et al. 2021. Occurrence, formation and environmental fate of polycyclic aromatic hydrocarbons in biochars. *Fundamental Research*, 1(3), 296–305.
Oranratmanee, M. and Tungittiplakorn, W. 2021. Application of biochar for cadmium stabilization in contaminated paddy soil. *E3S Web of Conferences. EDP Sciences*. pp. 02021.
Panzarino, O., et al. 2016. Rank-based biomarker index to assess cadmium ecotoxicity on the earthworm Eisenia andrei. *Chemosphere*, 145, 480–486.
Paul, S., et al. 2020. Biogenic stabilization and heavy metal immobilization during vermicomposting of vegetable waste with biochar amendment. *Journal of Hazardous Materials*, 390, 121366.
Plaza, C., et al. 2008. Organic matter humification by vermicomposting of cattle manure alone and mixed with two-phase olive pomace. *Bioresource Technology*, 99(11), 5085–5089.
Pokharel, P., et al. 2020. Biochar increases soil microbial biomass with changes in extra- and intracellular enzyme activities: a global meta-analysis. *Biochar*, 2, 65–79.
Qiang, G.X., et al. 2018. Bamboo biochar amendment improves the growth and reproduction of Eisenia fetida and the quality of green waste vermicompost. https://doi.org/10.1016/j.ecoenv.2018.03.023.
Qiu, Z., et al. 2020. Remediation of cadmium-contaminated soil with biochar simultaneously improves biochar's recalcitrance. *Environmental Pollution*, 256, 113436.
Rasul, M., et al. 2022. Biochar-induced priming effects in soil via modifying the status of soil organic matter and microflora: a review. *Science of the Total Environment*, 805, 150304.
Razzaghi, F., et al. 2020. Does biochar improve soil water retention? A systematic review and meta-analysis. *Geoderma*, 361, 114055.
Rehman, S.u., et al. 2023. Vermibiochar: a novel approach for reducing the environmental impact of heavy metals contamination in agricultural land. *Sustainability*, 15(12), 9380.
Rivelli, A.R. and Libutti, A. 2022. Effect of biochar and inorganic or organic fertilizer co-application on soil properties, plant growth and nutrient content in Swiss chard. *Agronomy*, 12(9), 2089.
Roy, S., et al. 2022. Assessing the arsenic-saturated biochar recycling potential of vermitechnology: insights on nutrient recovery, metal benignity, and microbial activity. *Chemosphere*, 286, 131660.
Sanchez-Hernandez, J.C. 2018. Biochar activation with exoenzymes induced by earthworms: a novel functional strategy for soil quality promotion. *Journal of hazardous materials*, 350, 136–143.

Sanchez-Hernandez, J.C. and Domínguez, J. 2017. Vermicompost derived from spent coffee grounds: assessing the potential for enzymatic bioremediation. in: *Handbook of coffee processing by-products*, Elsevier, pp. 369–398.

Sanchez-Hernandez, J.C., et al. 2019. Biochar and earthworms working in tandem: Research opportunities for soil bioremediation. *Science of the Total Environment*, 688, 574–583.

Shi, Z., et al. 2021. Effects of biochar and thermally treated biochar on Eisenia fetida survival, growth, lysosomal membrane stability and oxidative stress. *Science of The Total Environment*, 770, 144778.

Siedt, M., et al. 2021. Comparing straw, compost, and biochar regarding their suitability as agricultural soil amendments to affect soil structure, nutrient leaching, microbial communities, and the fate of pesticides. *Science of the Total Environment*, 751, 141607.

Sinha, R.K., et al. 2010. Earthworms–the environmental engineers: review of vermiculture technologies for environmental management and resource development. *International Journal of Global Environmental Issues*, 10(3–4), 265–292.

Solaiman, Z.M., et al. 2012. Biochars influence seed germination and early growth of seedlings. *Plant and Soil*, 353, 273–287.

Tang, R., et al. 2017. Metabolic responses of Eisenia fetida to individual Pb and Cd contamination in two types of soils. *Scientific Reports*, 7(1), 13110.

Toor, M.D., et al. 2024. Vermicompost rate effects on soil fertility and morpho-physio-biochemical traits of lettuce. *Horticulturae*, 10(4), 418.

Vambe, M., et al. 2023. Potential role of vermicompost and its extracts in alleviating climatic impacts on crop production. *Journal of Agriculture and Food Research*, 12, 100585.

Vijay, V., et al. 2021. Review of large-scale biochar field-trials for soil amendment and the observed influences on crop yield variations. *Frontiers in Energy Research*, 9, 710766.

Vyas, P., et al. 2022. Vermicomposting with microbial amendment: implications for bioremediation of industrial and agricultural waste. *BioTechnologia*, 103(2), 203.

Wang, F., et al. 2021. Biochar and vermicompost improve the soil properties and the yield and quality of cucumber (Cucumis sativus L.) grown in plastic shed soil continuously cropped for different years. *Agriculture, Ecosystems & Environment*, 315, 107425.

Wang, H.-T., et al. 2020. The effect of biochar on soil-plant-earthworm-bacteria system in metal (loid) contaminated soil. *Environmental Pollution*, 263, 114610.

Whalen, J.K. and Benslim, H. 2021. Earthworm populations are stable in temperate agricultural soils receiving wood-based biochar. *Pedosphere*, 31(3), 398–404.

Wu, X., et al. 2023. Earthworms enhance the bioremediation of tris (2-butoxyethyl) phosphate-contaminated soil by releasing degrading microbes. *Journal of Hazardous Materials*, 452, 131303.

Xu, C. and Mou, B. 2016. Vermicompost affects soil properties and spinach growth, physiology, and nutritional value. *HortScience*, 51(7), 847–855.

Xu, R.-K., et al. 2011. Adsorption of methyl violet from aqueous solutions by the biochars derived from crop residues. *Bioresource Technology*, 102(22), 10293–10298.

Yang, D., et al. 2017. Potential benefits of biochar in agricultural soils: a review. *Pedosphere*, 27(4), 645–661.

Yatoo, A.M., et al. 2021. Sustainable management of diseases and pests in crops by vermicompost and vermicompost tea. A review. *Agronomy for Sustainable Development*, 41(1), 7.
Zeiss, R., et al. 2022. Challenges of and opportunities for protecting European soil biodiversity. *Conservation Biology*, 36(5), e13930.
Zhang, F., et al. 2020. Influences of a vermicompost application on the phosphorus transformation and microbial activity in a paddy soil. *Soil & Water Research*, 15(4), 19–9210.
Zhang, L. and Sun, X. 2014. Effects of rhamnolipid and initial compost particle size on the two-stage composting of green waste. *Bioresource Technology*, 163, 112–122.
Zhang, M., et al. 2017. Effects of straw and biochar amendments on aggregate stability, soil organic carbon, and enzyme activities in the Loess Plateau, China. *Environmental Science and Pollution Research*, 24, 10108–10120.
Zhang, M., et al. 2023. Effects of biochar and vermicompost on growth and economic benefits of continuous cropping pepper at karst yellow soil region in Southwest China. *Frontiers in Plant Science*, 14, 1238663.
Zhang, Q. et al. 2019. Effects of biochar on the earthworm (Eisenia foetida) in soil contaminated with and/or without pesticide mesotrione. *Science of The Total Environment*, 671, 52–58.
Zhang, Y. et al. 2022a. Soil Acidification caused by excessive application of nitrogen fertilizer aggravates soil-borne diseases: Evidence from literature review and field trials. *Agriculture, Ecosystems & Environment*, 340, 108176.
Zhang, Z. et al. 2022b. Environmental impacts of hazardous waste, and management strategies to reconcile circular economy and eco-sustainability. *Science of The Total Environment*, 807, 150856.

Chapter 4

The potential for utilising organo-mineral fertilisers in agriculture

Otobong Mfon Isong, Cranfield University, UK; Kiran Pavuluri, International Fertilizer Development Centre, USA; and Parveen Fatemeh Rupani and Ruben Sakrabani, Cranfield University, UK

1 Introduction

An increasing global population reaching up to 10 billion by 2050 (United Nations, 2019) will result in ever-greater demand for food. Nearly 50% of food supply currently relies on the use of mineral fertilisers to meet crop nutrient demand (Rickson et al., 2015). Mineral fertilisers have been used to increase crop yield and improve soil fertility since the mid-nineteenth century. Their popularity in agriculture is attributed to a range of benefits (Haifa Group, 2022). First, they are a reliable and predictable source of the nutrients required by plants. Their nutrient composition is usually tailored in specific doses for different plants and soil conditions. Second, mineral fertilisers are generally fast-acting, providing immediate nutrition for plants. Additionally, they are available in various physical states, can be transported over long distances, are easy to use, and historically were relatively affordable.

However, long-term and excessive dependence on the application of mineral fertilisers has had negative effects on soil and the wider environment.

http://dx.doi.org/10.19103/AS.2025.0166.04

These include negative effects on soil microbial communities, decreased soil quality and pollution through nutrient losses via run-off, leaching, volatilisation or denitrification (Gomiero et al., 2011; Kour et al., 2020; Li et al., 2022; Ren et al., 2020). Given the complex nature of nutrient uptake from the time of application of mineral fertilisers to root absorption through plant growth stages, nutrient losses to the environment are inevitable (Tejada et al., 2005). Leaching of nutrients, e.g. leads to a myriad of environmental problems, including eutrophication of water bodies and a reduction in marine life activity. This also results in significant economic losses to the farmer, as essential nutrients are lost rather than being used by the plants. Studies show that over 50% of the nitrogen (N) found in waterbodies is linked to the application of mineral fertilisers in agriculture (Hansen et al., 2000; Meissner et al., 1998; Owens et al., 2000; Sogbedji et al., 2000).

Mineral fertilisers are also characterised by a high carbon footprint derived from their production and application, accounting for up to 2% of global greenhouse gas (GHG) emissions and 11% of agricultural emissions contributing to climate change (Menegat et al., 2022). The IPCC (2014, 2022) identifies the use of N fertiliser as a major source of anthropogenic non-carbon dioxide (CO_2) greenhouse gas emissions. Nitrous oxide (N_2O), methane (CH_4), and CO_2 are the most common greenhouse gas emissions, although ammonia (NH_3) emissions are also significant. According to 2022 IPCC estimates, agriculture is responsible for more than 80% of anthropogenic N_2O emissions and 70% of anthropogenic NH_3 emissions, primarily caused by the application of livestock manure and inorganic fertiliser.

Governments are now increasingly legislating to mitigate these climate change impacts. Policy drivers such as the European Union's Carbon Border Adjustment Mechanism seek to impose additional taxes on high-energy commodities such as fertilisers (Bux et al., 2024). Growing geopolitical instability caused by events such as the 2020 COVID-19 pandemic, the Russian invasion of Ukraine in 2022, and the 2026 conflict between Iran, Israel, and the United States have disrupted the flow of key commodities such as fertilisers and dramatically increased their price (Naja et al., 2026).

There is now a renewed focus on the use of organic or bio-based fertilisers to offset the challenges associated with over-reliance on mineral fertilisers. Some of the advantages of organic fertilisers include potentially improved crop quality, increased soil carbon storage, and improved soil physical, biological, and chemical health (Cui et al., 2018; Du et al., 2022; Li et al., 2020; Liu et al., 2019; Zhou et al., 2015). Nevertheless, organic fertilisers often have less predictable, relatively low, and variable concentrations of key nutrients such as nitrogen (N), phosphorus (P), and potassium (K) compared to tailored formulations in mineral fertilisers. This variability can have negative effects on crop yields (compared to mineral fertilisers) and may require the use of substantial amounts to achieve the required levels of nutrition. This then adds to the cost and may then lead to problems such as leaching into neighbouring watercourses.

This is where there is an opportunity for organo-mineral fertilisers (OMFs) as an alternative nutrient source (Uddin et al., 2025). The aim of this chapter is to provide an overview of the role of OMFs as an alternative source of bio-based fertiliser in agriculture, highlighting both their advantages and disadvantages.

2 Organo-mineral fertilisers

The European Union (2019) defines organo-mineral fertilisers (OMFs) as liquid or solid compounds obtained by co-formulating at least one organic and one inorganic material with a declarable amount of primary nutrients, although no more than 16% of N by mass. This blend of organic and inorganic fertilisers is designed to combine the advantages of each component while minimising the effects of sole dependence on one or more fertiliser types.

OMFs are designed to provide plants with essential nutrients for growth, while also improving soil health and fertility. The organic component of an OMF is typically derived from a natural source, such as plant residue, animal manure, compost, digestate or other organic materials. Potential additional sources include recycled industrial waste products that would otherwise have been disposed of (Deeks et al., 2013; de Moraes et al., 2020; Mumbach et al., 2020). OMFs produced from urban industrial waste are also proving to be an efficient alternative to mineral fertilisation (Magela et al., 2019a). In this respect, OMF fits within the broader circular economy concept to enable products previously considered waste to become valuable inputs for multiple production processes.

Organic substances used in OMF contain various beneficial elements, including N, P, K, and micronutrients, which are essential for plant growth (Oliveira et al., 2017; Moreira et al., 2021). The organic matter in these fertilisers may also help to improve soil structure, moisture retention, and nutrient-holding capacity. The mineral component of an OMF consists of inorganic nutrients, usually in the form of mineral salts. These minerals are added to the organic matter to supplement the nutrient content and ensure a balanced and readily available nutrient supply for plants. Common mineral nutrients found in these fertilisers include N, P, and K, as well as secondary nutrients such as calcium (Ca), magnesium (Mg), sulphur (S), and micronutrients. One key benefit of OMF is that, as well as topping up nutrient content, the mineral content of OMF can potentially provide highly concentrated fast-release nutrients for immediate plant uptake, while the organic component releases additional nutrients more slowly for longer-term plant needs.

Moreira et al. (2021) have highlighted that the quality of the compounds used in OMF formulation influences nutrient availability and mineralisation. Additionally, product modifications are essential for variables such as different soil types and climatic conditions. Each product should ideally be specific to particular crop needs, soil, and climatic conditions (Mazeika et al., 2021). Proper soil testing and nutrient management practices are important to ensure appropriate formulation, ensure application, and to prevent nutrient imbalances. However, the pioneering work of Burak and Sakrabani (2023), Pawlett et al. (2015), Deeks et al (2013), and Antille et al (2017) have shown that, if appropriately designed and used, the application of OMF in field-scale trials does not compromise crop yield compared to mineral fertilisers.

2.1 Advantages of OMFs

OMFs have several important benefits:

1. Nutrient supply: OMFs provide essential nutrients, including N, P, K, and micronutrients, to plants. These nutrients are crucial for plant growth, development, and productivity. By ensuring an adequate nutrient supply of these nutrients over time, OMFs can help optimise crop yields and quality (Bastos et al., 2021; Nascimento et al., 2021).
2. Slow-release properties: As noted, OMFs often have slow-release characteristics due to the organic component. This slow-release feature ensures a more gradual and sustained nutrient release, reducing the risk of nutrient leaching, runoff, and gaseous emissions. It helps plants receive a continuous supply of nutrients over an extended period, leading to better nutrient use efficiency and minimising the potential for nutrient imbalances or wastage (Semida et al., 2014). This promotes nutrient use efficiency of the inorganic component, as the organic matter helps retain the mineral element in the soil until it is needed by the crop. The rate of nutrient release is governed by a range of variables, including the balance of organic and inorganic material in the final product, physical structure (e.g. pore size), and other variables (Wen et al., 2017; Kim et al., 2018; Liu et al., 2019; An et al., 2020; Lustosa Filho et al., 2020)
3. Soil fertility improvement: The organic component of OMFs helps improve soil fertility. Organic matter enhances soil structure, promotes water retention, and increases the soil's ability to hold and release nutrients in a way that optimises crop nutrient uptake (Oliveira et al., 2017, Moreira et al., 2021; Kumar et al., 2022). It also supports beneficial microbial activity, which further improves soil health and nutrient cycling. Over time, the use of OMFs can lead to healthier, more productive soils.
4. Enhanced plant resilience: The organic matter in OMFs helps improve soil moisture retention and nutrient availability. This, in turn, enhances plant resilience to environmental stresses, such as drought, heat, and

disease. Healthy soils with improved fertility provide a favourable environment for plant growth, enabling them to better withstand adverse conditions (Semida et al., 2014; Li et al., 2022).

5. Environmental sustainability: OMFs contribute to sustainable agricultural practices. They reduce the dependence on mineral fertilisers, which (as has been discussed) can have negative environmental impacts. By incorporating organic materials, OMFs promote recycling and waste management, utilising agricultural by-products or organic waste. They also help preserve soil health, reduce erosion, and support biodiversity (Jha et al., 2023; Szulc et al., 2021).
6. Reduced chemical inputs: By using OMFs, farmers can reduce their reliance on mineral fertilisers. This can have several advantages, including reduced environmental pollution and decreased health risks associated with the handling and application of mineral fertilisers. OMFs offer a potentially more sustainable and environmentally friendly alternative (Mazeika et al., 2021).

2.2 Disadvantages of OMFs

While OMFs offer several benefits, there are some limitations associated with their use:

1. Cost: OMFs can be more expensive compared to mineral fertilisers. The production and processing of organic materials, as well as the additional steps required to blend them with mineral nutrients, can increase the cost of these fertilisers.
2. Availability: Availability of OMFs may be limited in certain regions or markets. The production and distribution infrastructure for these fertilisers is not as developed or widespread as that of conventional fertilisers. Farmers may face challenges in accessing a consistent supply of OMFs, especially in remote areas.
3. Nutrient content variability of organic fraction: While the overall nutrient content of OMFs can be fixed, the amount of mineral fertiliser that needs to be added can vary depending on the organic materials used, the manufacturing process, and blending ratios (Bagreev et al., 2001; Deeks et al., 2013).
4. Nutrient specificity: OMFs may not provide all the required nutrients in the desired ratios for specific crops or soil conditions. Although they can supply essential macronutrients, such as N, P, and K, they may not always contain secondary or micronutrients in sufficient quantities. Supplemental fertilisation or soil amendments may still be needed to address any nutrient deficiencies or imbalances.

5. Decomposition rate: The organic materials in OMFs undergo decomposition over time, releasing nutrients gradually. However, the decomposition rate can be influenced by environmental factors, such as temperature, moisture, and microbial activity. In some cases, the release of nutrients from organic materials may not align with the crop's nutrient demands, leading to suboptimal nutrient availability during critical growth stages.
6. The moisture content of feedstocks used in formulating OMF is high (in some cases up to 85% in biosolids), which needs to be reduced before it can be used in the final product. This is both costly and energy-intensive. Some attempts have been made to reduce the moisture content using waste heat, which is more environmentally friendly.
7. Application logistics: The application of OMFs may require specialised equipment or techniques, depending on their physical form and consistency. For example, some organic materials may need to be properly incorporated into the soil to maximise nutrient release and minimise nutrient loss. Adequate training and equipment may be necessary for farmers to apply these fertilisers effectively.

While there has been no comprehensive study which has addressed the costs of producing and using OMFs compared to mineral fertilisers, a review by Rodrigues et al. (2021) using sewage sludge as a case study highlights its potential cost-effectiveness if properly managed. The cost of production of the organic fraction is about eight times lower than the costs associated with mineral fertilisers at US$28.92/mt and US$121.33/mt. This suggests that the production cost would be significantly lower after enrichment with nutrients from inorganic sources. In Brazil, the cost of producing OMFs from waste materials (US$136–144/mt) was found to be significantly lower than the costs associated with landfilling (US$70.00/mt) (Borges et al., 2015). Further studies are required to quantify the costs and savings of using OMFs in agriculture compared to mineral fertilisation. It is important for farmers to consider these factors when deciding whether to use OMFs or not. As noted, tailoring fertiliser choices to specific crop requirements, soil, and climatic conditions is essential for achieving optimal nutrient management and sustainable agricultural practices.

3 Organic and inorganic materials used in the production of organo-mineral fertilisers

Several materials have been studied for use as organic and inorganic components of OMFs, with varying effects on soil and plants. Some of these are discussed in the following sections.

3.1 Livestock manure

In Europe, an estimated 148 million pigs, 88 million cattle, and 1428 million chickens generate over 1000 million mt of manure annually (Bernal et al., 2015). This output, which is a rich source of organic and inorganic forms of N and P, has significant potential as an organic fertiliser. However, the composition of manure and its nutrient content are highly variable and determined by factors such as animal type and age, feeding practice, volume, and concentration of livestock production (Buckwell and Nadeu, 2016; Kominko et al., 2019). Manure is a significant source of GHG and NH_3 emissions in agriculture. Inappropriate storage, handling, and application of manure are linked to soil and air pollution, eutrophication, and groundwater pollution (Hou et al., 2018). If poorly handled, manure can also be a significant source of pathogens and other contaminants. Well-managed incorporation into OMF provides the potential to offset some of these problems.

3.2 Poultry litter

Poultry litter is obtained as a by-product from growing chickens in a confined space. It is composed of a variety of flooring materials, including wood shavings, grain husks, and cobs, which become mixed with feed and bird faecal matter, as well as skin and feathers shed by birds (Angelo et al., 1997; Chen and Jiang, 2014; Wilkinson et al., 2011). Poultry litter has long been used as an organic fertiliser to supply macro- and micronutrients to plants (Nascimento et al., 2021). Due to its abundance and relative simplicity in handling. as well as affordability, poultry litter is the most popular organic material used in OMF manufacture (Rogeri et al. 2016).

However, its composition is variable and dependent upon multiple factors, including bird species, the number of birds per square metre, their diet, and housing conditions (Ritz and Merka, 2009). Direct application of this material to the soil without further treatment can be a risk to animals, plants, and humans, as it can harbour pathogens and other contaminants. Furthermore, its high moisture content (40–70%) and low nutrient content (5–15 kilograms [kg] of N per metric ton [mt]) compared to mineral fertilisers (30–200 kg/mt) reduce its efficacy, requiring large amounts to be transported and then spread onto fields. This increases both costs and the potential for gaseous emissions and leaching into water courses.

These disadvantages can effectively be offset by using poultry litter in the production of OMF, where the mineral component can improve overall nutrient content. The resultant fertiliser has been found to promote crop yields and protein content of potato, rapeseed, and wheat straw. In addition, an increase in soil-available phosphorus and potassium has been observed (Mazeika et al., 2016; Mazeika et al., 2021; Nascimento et al., 2021).

3.3 Vermicompost

The process of vermicomposting is a biochemical, eco-friendly method of producing high-value agricultural manure using earthworms as shredders of organic matter, converting it into a stable product. Different substrates, such as municipal and household wastes, can be utilised to produce vermicompost (Kumar et al., 2022). Organic matter undergoes a series of biological, chemical, and physical changes because of the digestive enzymes and microorganisms present in the tract of earthworms (Ali and Nazir, 2021; Singh et al., 2020). The final product is a nutrient-rich humic substance capable of supporting both plant growth and soil microbial activity (Olle, 2019). The latter is an essential component of soil health, which can further promote plant growth and health (Atteya et al., 2021). However, like most organic manures, vermicompost is low in the essential macronutrients required by plants compared to inorganic fertiliser (Sannigrahi, 2016). Vermicompost can be enriched with specific nutrients using mineral additions (Molina et al., 2013; Ndegwa and Thompson, 2001).

3.4 Biochar

Biochar is a carbon-rich material produced through the pyrolysis of organic feedstock in an anaerobic environment. Although it has a long history of use, its availability has increased more recently as a co-product of bioenergy production. Primary feedstocks include wood products (Domingues et al., 2017), agricultural crop residues (Tripathi et al., 2016), poultry litter, swine manure, or municipal wastes such as sewage sludge (Liu et al., 2018). Feedstocks with a higher lignin content (e.g. pine or eucalyptus) are characterised by a lower nutrient content (N, P, K, and Ca), while feedstock with a higher cellulose content results in biochar with a significantly higher concentration of volatile compounds and nutrients (Tripathi et al., 2016). Enaime et al. (2020) noted that pyrolysis conducted at temperatures above 500°C results in biochar that is also able to absorb organic pollutants due to its large surface area, high volume of micropores, and high hydrophobicity. Below 500°C, the resulting biochar is more efficient for rendering inorganic pollutants immobile.

Depending on the type and quality of feedstock, as well as processing temperatures and oxygen concentration, biochar can achieve a high nutrient content. A high concentration of macro- and micropores enables biochar to absorb nutrients and then release them gradually (Soinne et al., 2014; Gao et al., 2016; Kizito et al., 2019; Wali et al. 2020), as well as, more broadly, contribute to improved soil aeration and water retention (Downie et al., 2009; Troeh and Thompson, 2005). Biochar can also contribute to improved carbon sequestration and recycling within a circular economy (Lehmann et al., 2003;

Van Zwieten et al., 2014; Puga et al., 2020; Zhang et al., 2021; Ndoung et al., 2021).

3.5 Sewage sludge

Sewage sludge is obtained from the treatment of wastewater and sewage. The process of treatment aims at separating the solid component from the liquid component, with the latter treated and, once made safe, returned to reservoirs (de Moraes et al., 2020). The resultant concentrated sludge, which is rich in organic matter and nutrients (de Moraes et al., 2020), is treated (e.g. via anaerobic digestion) to make it safe for use in agriculture and referred to as a biosolid. Biosolids typically consist of around 40% organic matter, 4% N, and 2% P, with macro- and micronutrients present in smaller quantities, while liming increases the concentration of Ca and Mg (Bueno et al., 2011). Sewage sludge in countries such as the UK is predominantly treated using anaerobic digestion, resulting in a nearly pathogen-free digestate which is subject to further post-treatment to obtain a biosolid.

3.6 Anaerobic digestate

The process of anaerobic digestion is a sustainable means of converting manure and other organic biomass in biogas plants into biogas and digestate (Tambone et al., 2010, Möller and Müller, 2012). Anaerobic digestion works through four stages (hydrolysis, acidogenesis, acetogenesis and methanogenesis), each carried out by specialised groups of microorganisms responsible for breaking down biomass and producing specific compounds as end-products (Harirchi et al., 2022, Fagbohungbe et al., 2019). Digestate is a nutrient-rich by-product of anaerobic digestion. While a portion of the organic compounds from the initial feedstock is converted to biogas during the process, the mineral fraction (containing nitrogen, phosphorus, and potassium) remains largely intact as digestate (Seleiman et al., 2017).

Whole, unprocessed digestate can be applied directly to land. However, it is commonly separated into liquid and solid fractions to make it easier to transport and apply, using various mechanical techniques, sometimes with added coagulants and flocculants to improve efficiency. The liquid fraction of digestate is characterised by high N content and low dry matter and can be applied to nearby farmland or further upgraded via ammonia stripping and scrubbing or membrane filtration. The solid fraction is stable and rich in carbon and phosphorus, can be composted, dried and pelletised, or subjected to pyrolysis, gasification, or hydrothermal carbonisation, making it particularly suitable as a feedstock for OMF (European Biogas Association, 2024; Möller and Müller, 2012).

3.7 Sugarcane vinasse

Several co-products from sugarcane cultivation are used in the manufacture of alcohol. These include solid by-products, including filter cake and bagasse, as well as liquid effluents, such as molasses and vinasse. Vinasse is seen as a significant source of distillery waste and a significant pollutant. Its high N and K content make it an essential ingredient for OMF, though variable components of organic and mineral components require careful handling to avoid potential nutrient imbalances in soil and potential leaching (Da Silva et al., 2007; Bastos et al., 2021).

3.8 Inorganic components of OMFs

Mineral compounds are used in the production of OMFs to enrich the fraction derived from organic wastes. They help to increase the primary nutrient content of the product to obtain the correct ratio necessary for plant growth and productivity. They also serve as blending and disinfecting agents and increase the bioavailability of nutrients. Inorganic components of OMFs include the following:

- Urea, bentonite clay, sepiolite, and phosphate rock (Chew et al., 2020; Shi et al., 2020).
- Phosphate fertiliser (bone meal and triple superphosphate). This has a high P concentration, high carbon retention, and enhanced metal stabilisation in soils (Zhao et al., 2016).
- Phosphoric acid and magnesium oxide (Carneiro et al., 2021).
- Potassium sulphate, potassium nitrate, ammonium nitrate, ammonium sulphate, and diammonium phosphate (Kominko et al., 2019).
- Magnesium chloride ($MgCl_2$): This is characterised by a high P adsorption capacity and contributes to plant growth and accumulation of P, e.g. in maize (Nardis et al., 2021).

The choice of mineral addition is dependent upon the desired result for a crop or soil and the potential interactions that may occur between soil, plant, and climatic conditions.

4 Production of organo-mineral fertilisers: treatment of organic feedstocks and product formulation

Various techniques, such as anaerobic digestion and composting, can be used in the production of OMFs. These are determined by the type of organic and inorganic feedstock and the different chemical interactions that occur to obtain a balanced product capable of delivering multiple agronomic benefits (Bouhia et al., 2022). The options for treating organic amendments can be broadly classified into two categories: thermal and microbiological.

4.1 Thermal treatment

Thermal treatment is the most popular method of treating the waste materials used in OMF (Szulc et al., 2021). Advantages of thermal treatment include decontamination, moisture reduction, and volume reduction. Thermal processing can potentially be done through incineration, gasification, or pyrolysis. While these occur at various temperatures, they all aim at reducing the moisture content of the organic material and then producing either dried manure or concentrated ash, which can then be used in OMF formulations. However, incineration requires very high amounts of energy and produces harmful emissions (including dioxins and furans), while gasification produces very little solid material. This leaves pyrolysis, a thermochemical process that decomposes organic materials in anaerobic conditions (without combustion taking place). Factors that affect the quality of OMFs produced through pyrolysis include temperature (Biederman and Harpole, 2013; Enaime et al., 2020), type of feedstock used (Ndoung et al., 2021), and process time (Karim et al., 2017; Mosa et al., 2018). As noted earlier, pyrolysis conducted at temperatures above 500°C results in biochar-based OMFs, which are additionally useful in absorbing organic pollutants because of high surface area, volume of micropores, and high hydrophobicity. Below 500°C, the resulting biochar is more efficient for rendering inorganic pollutants immobile.

4.2 Microbiological treatments: vermicomposting, composting and anaerobic digestion

The microbiological treatment of organic materials involves the transformation of waste materials using living microorganisms. This may occur either with (aerobic) or without (anaerobic) oxygen. If properly managed, the resulting product is safe, stable, rich in organic compounds and humus, and does not emit an unpleasant odour (Awasthi et al., 2020; Ayilara et al., 2020; Lohri et al., 2017; Sayara et al., 2020). The quality of the product, however, depends on multiple factors, including the quality of the feedstock used in composting, pH, moisture content, and aeration. Moisture and pH are crucial, as they affect the activity of microorganisms and the conditions of the composting process. A moisture content below 30% limits bacterial activity and above 65% results in potentially uncontrolled anaerobic conditions and emission of odours. The recommended optimal humidity is 50–60% (Jain et al., 2019; Petric et al., 2012; Yeh et al., 2020).

As noted earlier, vermicomposting is a biochemical, eco-friendly method of producing high-value agricultural manure that uses earthworms as shredders of organic matter, converting it into a stable product. California Red earthworms (*Eisenia fetida*) are usually used for vermicomposting due to characteristics such as a high reproductive capacity and ability to adapt to varied environmental conditions, such as pH, temperature, and moisture. They have been used, e.g. to produce vermicompost from dairy cattle manure, poultry droppings, food industry waste, and sugarcane vinasse (Das et al., 2021; Guimarães et al., 2017). Choice of suitable earthworms is highly dependent on the temperature of the region, the biomass to be used, and their life span. Organic matter undergoes a series of biological, chemical, and physical changes because of the digestive enzymes and microorganisms present in the gut of earthworms (Ali and Nazir, 2021; Singh et al., 2020). The product is a nutrient-rich humic substance (Olle, 2019; Atteya et al., 2021). However, like most organic manures, vermicompost is low in some essential macronutrients (Sannigrahi, 2016), requiring mineral additions to create a balanced fertiliser (Molina et al., 2013; Ndegwa and Thompson, 2001).

Composting is the process of decomposing organic matter through the action of microorganisms in an aerobic environment at temperatures up to 70°C (Das et al., 2021; Guimarães et al., 2017). This allows for sanitisation, degradation, and concentration into a nutrient-dense product (Guimarães et al. (2017). While relatively simple and cheap, some nutrients are lost and the nutrient profile of the resulting product is variable and may not match crop nutrition requirements (Das et al., 2021).

Composting and vermicomposting are often used together in producing OMFs. For example,Guimarães et al. (2017) have identified their potential use in producing a P-rich fertiliser that reduces dependence on acidified phosphate fertilisers. In this case, the manure was pre-composted to eliminate potential fermentation and enhance colonisation by earthworms. However, moisture and temperature must be maintained at specific levels to prevent either dehydration or oversaturation. The survival and activity of earthworms are also dependent on the proper combination of organic wastes and mineral additions that do not limit their activity by altering the pH or carbon-to-nitrogen (C/N) ratios or become toxic in higher concentrations (Das et al., 2021). The timing of additions also affects these variables, final nutrient content, and, thus, the quality of the final OMF product (Molina et al., 2013).

As noted, the process of anaerobic digestion is a sustainable means of converting manure and other organic biomass in biogas plants into biogas and digestate. Digestate is available in solid and liquid forms, both of which have beneficial applications as fertiliser for crops or amendments for soil (Tambone et al., 2010; Möller and Müller, 2012). As noted earlier, anaerobic digestion works through four stages:

- Hydrolysis: Hydrolytic bacteria use enzymes to break down complex biological matter such as proteins, polysaccharides, and fats into simpler blocks, including fatty acids, amino acids, and sugars (Mani et al., 2016). The end-products of hydrolysis are soluble in the liquid state of the digestate and serve as substrates for the next phase.

- Acidogenesis: The end-products of hydrolysis are then converted into short-chain organic acids (Strazzera et al., 2018).
- Acetogenesis: These acids are further broken down by acetogenic bacteria into hydrogen, acetic acid, and carbon dioxide (Chen et al., 2008).
- Methanogenesis: Finally, methanogenic microorganisms generate methane and carbon dioxide, known as biogas, leaving anaerobic digestate as residue (Uddin and Wright, 2022; Demirel and Scherer, 2008).

Each is carried out by specialised groups of microorganisms responsible for breaking down biomass and producing specific compounds as end-products (Harirchi et al., 2022; Fagbohungbe et al., 2019).

4.3 OMF product formulation

The importance of further processing to produce OMF can be seen in the case of digestate. Direct application of raw digestate is possible but poses risks: residual methane, ammonia, hydrogen sulphide, and volatile fatty acids (VFAs) can cause phytotoxicity and odorous gas emissions that limit viability without prior treatment (Chojnacka and Moustakas, 2024). Producing OMFs that meet regulatory standards, therefore, typically requires supplementary processing, including stabilisation, conditioning, or thermal treatment. Chojnacka and Moustakas (2024) suggest that digestate may be conditioned with sulphuric and phosphoric acid, neutralised, and corrected in composition using ash from digestate thermal processing (e.g. fly ash containing oxides of Ca, K, Mg, Zn, and Mn), with additional formulants. The porous physical structure and chemical composition of digestate-derived pellets enable sustained nutrient mobilisation over time, supporting both short- and long-term plant nutrition (Valentinuzzi et al., 2020). The resulting fertilisers compare well in performance with mineral fertilisers (Koszel and Lorencowicz, 2015).

Key general requirements for OMF product formulation include (Klyosov and Orekhovskaya, 2021; Bouhia et al., 2022):

- Simplicity, cost-effectiveness, and a low energy requirement;
- Versatility: suitability for a range of feedstocks, e.g. with differing moisture levels; and
- Product quality: the ability to create a consistent product with a sufficient shelf life.

Granulation is particularly suitable due to the relatively mild conditions associated with the process, such as temperature and pressure. These are sufficient to eliminate pathogens but retain nutrients in a stable form that has

Figure 1 OMF in the form of granules (left) (Deeks et al., 2013) or pellets (right) (Burak and Sakrabani, 2023).

a long shelf life and is easy to apply using existing equipment (Fig. 1). As a relatively simple process, it also reduces capital requirements and operational costs (Sharara et al., 2018; Mazeika et al., 2021).

5 Assessing the effects of organo-mineral fertilisers on crops and soil

The effectiveness of the OMF application to the soil can be determined by the resultant effects on crop yield and growth. Plant growth is heavily influenced by soil conditions, such as pH, cation exchange capacity (CEC), organic matter content, and nutrient availability (Hadroug et al., 2019; Zhu et al., 2014). However, the interaction between OMF and soil health is complex. For example, adjusting soil pH can potentially increase the bioavailability of soil nutrients and, thus, OMF efficacy.

Bouhia et al. (2022) provide a synthesis of the impact of OMF applications on crop yields. Tables 1 and 2 summarise a range of studies of OMF by feedstock and mineral addition and reported effects on crop yields and overall soil health. These tables are not an exhaustive list but provide an overview of the mixed effects of OMF, depending on the context. Figure 2 suggests proposed mechanisms on how OMF may influence the physical, chemical, and biological properties of soil, again depending on the local context.

6 Assessing the wider environmental impacts of organo-mineral fertilisers

OMFs have been designed to both maintain crop yields and reduce environmental impacts associated with the use of inorganic fertilisers (Vakal et al., 2021). Being slow-release fertilisers, given their organic component, they can reduce the risk of nutrient losses. Vitale et al. (2017) observed lower

Table 1 Effects of organo-mineral fertilisers by feedstock material

Organic feedstock category	Feedstock material	Mineral addition	Effects	References
Crop residue				
	Maize stalk	Triple superphosphate, diatomite, and urea	• Higher amounts of N, P, Ca, and silicon and higher CEC in the biochar fertiliser • Increased maize growth and yield • Decrease in cadmium toxicity	Chen et al. (2021)
	Wastes of peanut shell and sugarcane bagasse	$MgCl_2$ and $CaCl_2$	• Increase in P adsorption • Increase in plant nutrients and reduced metal(loid)s	Fang et al. (2020)
	Dry wheat straw	Urea, bentonite clay, rock phosphate, Fe_2O_3, and $FeSO_4.7H_2O$	• Increase in soil Eh (potential difference) • Increase in the Eh between the rhizosphere and the root membrane • Increase plant growth and microbial population	Chew et al. (2020)
	Corn straw	$MgCl_2$ solution	• Slow release of nitrate and ammonium • Improved soil moisture retention • Improved maize growth and N use efficiency	Khajavi-Shojaei et al. (2020)
	Cotton straw	$Mg_3(PO_4)_2$ and bentonite coating with sodium alginate (NaAlg), starch, cellulose, chitosan, and maltodextrin	• Slower release of nutrients • Superior degradability	An et al. (2020)
	Maize residue	Dorowa phosphate rock (DPR)	• Increase in P, N, and C content	Tumbure et al. (2020)
	Cotton straw	K_3PO_4 and bentonite	• Better P and K slow release	An et al. (2020)

(Continued)

Table 1 *Continued*

Organic feedstock category	Feedstock material	Mineral addition	Effects	References
	Palm leaf waste	Solution containing 200 mg of P per litre	• Immobilisation of heavy metals • Increase in soil P • Improved maize growth	Ahmad et al. (2018)
	Urban green waste	Urea, clay minerals of bentonite, and sepiolite	• Slower release of N and increase in maize growth • Reduced N leaching	Shi et al. (2020)
Animal Products				
	Poultry litter (PLB) and coffee husk (CHB)	Phosphoric acid, magnesium oxide, triple superphosphate (TSP)	• Slower P release • Increased soil P • Increased plant growth	Carneiro et al. (2021)
	Pelletised poultry manure	5% TSP	• Reduced iron (Fe), zinc (Zn), copper (Cu), and manganese (Mn) in soils • Increased N, P, and K uptake • Lower Fe, Zn, Cu, and Mn concentrations in plants • Increased plant growth • Reduced Ca and Mg concentrations in leaf	Gunes et al. (2015)
	Poultry litter	TSP, phosphoric acid (H_3PO_4), and magnesium oxide (MgO)	• Promoted the accumulation of both labile and moderately labile P • Higher biomass yield of Marandu grass	Lustosa Filho et al. (2020)
	Poultry litter, pig manure, and sewage sludge	Magnesium chloride ($MgCl_2$)	• Promoted P adsorption from aqueous solution increased plant yields	Nardis et al. (2021)

Table 2 Effects of organo-mineral fertilisers on plants and soil

Feedstock	Crop	Soil type	Soil properties	Crop properties	References
Maize straw biochar	Tomato (*Solanum lycopersicum*)	Sandy loam		Superior control of *M. incognita*; significant increase in plant growth	Abdelnabby et al. (2018)
Palm leaf waste biochar	Maize (*Zea mays*)		Increase soil available P; decreased soil labile heavy metal (heavy metals in exchangeable and reducible fractions were transformed to more stable fraction)	Enhanced plant growth parameters (shoot and root lengths and dry matter) and uptake of P	Ahmad et al. (2018)
Poultry litter (PLB) and coffee husk (CHB)	Maize (*Zea mays*), grass beans (*Phaseolus vulgaris*)	Oxisols	Increase soil P content	Increase in crop yields; better plant P uptake	Carneiro et al. (2021)
Pelletised poultry manure	Lettuce			Increased N, P, and K uptake; lower Fe, Zn, Cu, Ca, Mg, and Mn concentrations in leaves; increased plant growth	Gunes et al. (2015)
Poultry litter	Maize (*Zea mays*)	Oxisols	Decreased water-soluble P, thereby causing a slow release of P; improved soil pH and Mg	Dry biomass yields equivalent to TSP treatments	Lustosa Filho et al. (2020)
Poultry litter, pig manure, and sewage sludge	Maize (*Zea mays*)	Oxisols		Promoted plant growth and accumulation of P in maize; increased P and Mg uptake	Nardis et al. (2021)

(*Continued*)

Table 2 *Continued*

Feedstock	Crop	Soil type	Soil properties	Crop properties	References
Urban Green waste	Maize (*Zea mays*)		Improved carbon retention	Improved maize growth and increased N in the maize plant; increase in N use efficiency	Shi et al. (2020)
	Wheat	Alfisol with inceptisol intergradations	Reduced N, P, and K loss; higher N/P ratio (lower eutrophication risk)	Improved N and K content; improved protein content; improved number of spikes per square; increased grain gross protein	Tejada et al. (2005)
Sewage sludge and filter cake	Soybean (*Glycine max* L. Merrill)	Oxisol	Urea, triple superphosphate, potassium chloride; increase in amino acids	Increase in root growth	Silva et al. (2019)

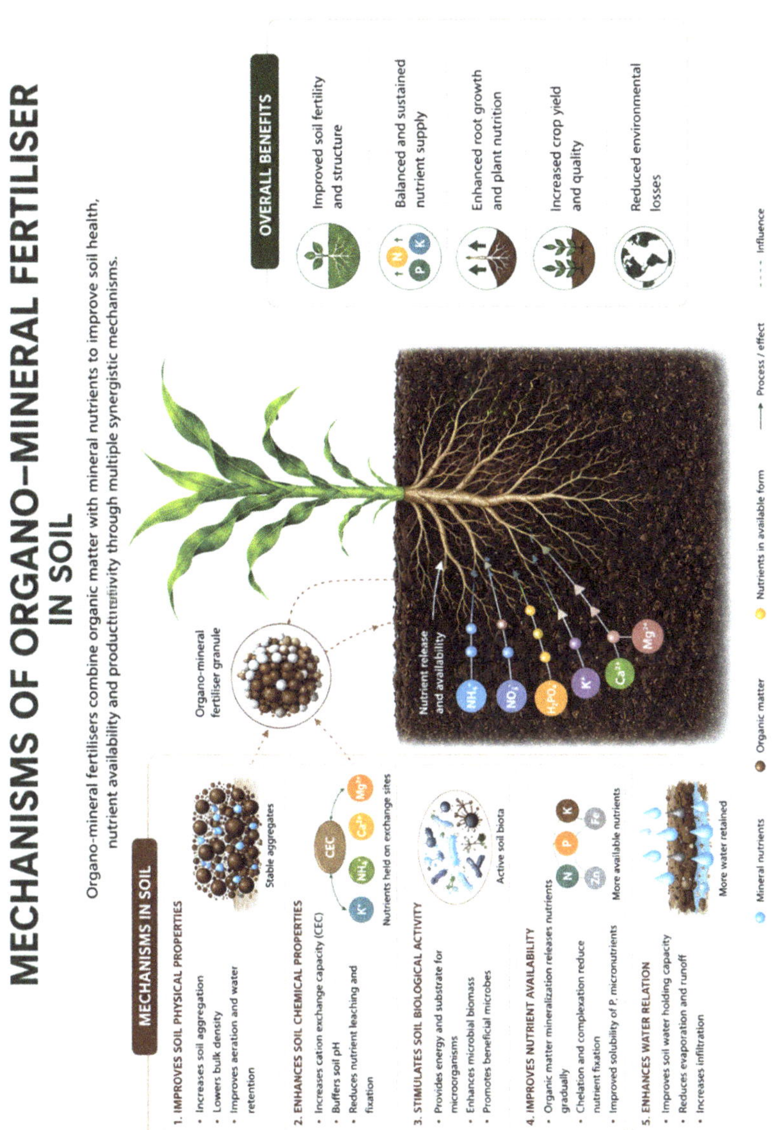

Figure 2 Mechanisms of OMF interactions in soil and impact on plants.

N emissions after the use of OMF because of slow mineralisation of the organic component. This slow release leads to better nutrient efficiency in plants and reduced losses of N to the environment. Based on field trials using biochar-enriched OMF, Chojnacka (2025) reported 18% lower GHG emissions compared to the use of inorganic fertilisers. These results were corroborated by Kulikova et al. (2020) and Zhang et al. (2018), who found that OMF application resulted in a 79.66% reduction in global warming potential compared to mineral fertilisers. Carbon emissions were 10% lower for OMF compared to mineral fertilisers. The risk of aquatic eutrophication was reduced by 60.68% in a study by Zhang et al. (2018). However, more research is needed on potential risks of groundwater contamination from the use of OMF.

It is widely acknowledged that improving soil carbon sequestration (e.g. by increasing soil organic matter (SOM) content) is key to combating climate change and achieving net zero targets by 2050 (Paul et al., 2023). Increased levels of SOM are also associated with improved soil physical, chemical, and biological health. While OMF look promising, as they add stable organic matter to soil that is then decomposed by microbial organisms over time, long-term contributions to carbon sequestration will require a combination of management techniques (Tang et al., 2016). These techniques include the adoption of minimum or no tillage systems, use of cover crops and more diverse rotations, agroforestry systems, and other measures. Adoption of OMF and other measures like these to promote soil carbon sequestration may need incentives such as carbon markets where farmers can apply for tradable carbon credits based on adopting sequestration measures (Brown and Leonard, 2004). However, there is still much to learn about the carbon credit potential of OMF, including how to quantify their impact, potential leakage effects, and how lasting changes may be (Paul et al., 2023).

One way of assessing the overall environmental impact of OMF is via life cycle assessment (LCA) which provides a framework for quantifying potential environmental impacts of products throughout their life cycle (Walling et al., 2020). Litskas et al. (2023) has used LCA to review the environmental impact of producing various types of fertiliser, including an OMF based on pig, cattle, and poultry manure. The LCA was based on the ability of a fertiliser to supply 1 kg of N to a given crop. The LCA considered categories related to acidification, atmospheric emissions, ecotoxicity potential, eutrophication, human health, and resource use.

When compared with mineral and organic fertilisers, the OMF product was observed to have the highest environmental impact across all categories, including carbon emissions, an indicator of its global warming potential. This result may be attributed to the low N content of the OMF product. These findings can be criticised as not taking sufficient account of emissions from mineral fertilisers and not accounting for wider benefits such as carbon storage

(Ioannidou et al., 2022). Further research should also include other life cycle stages, such as transportation, field application, and soil processes, to obtain a complete picture of how the use of OMF impacts the environment. Egas et al. (2023) has also identified problems such as inconsistencies in choice of functional unit, feedstock, and technology used in manufacture. There is a need for further research and greater standardisation in assessing the environmental impact of OMF.

7 Conclusion

The use of OMF in agriculture shows promise, with documented benefits for plants and soils. Knowledge gaps still exist around developing standardised doses for various plant species and the resultant effect on plant growth and soil health. In addition, given the multiple sources of organic feedstock and inorganic minerals, it is difficult to perform direct comparisons of the effects of OMF on agricultural systems. To fully validate OMF as a viable option in agriculture, a more holistic approach considering feedstock quality, energy, and transportation needs to be considered in addition to impact on soil health, crop yield, and quality. Most importantly, there seems to be no consensus on the environmental impact of their use, as life cycle assessments differ in their boundaries and functional units. These will present areas for future research on OMF.

8 References and further reading

Abdelnabby, H., et al. (2018). Furfural–biochar-based formulations show synergistic and potentiating effects against *Meloidogyne incognita* in tomato. *Journal of Pest Science*, 91, 203–218.

Ahmad, M., et al. (2018). Phosphorus-loaded biochar changes soil heavy metals availability and uptake potential of maize (*Zea mays* L.) plants. *Chemosphere*, 194, 327–339.

Ali, R. & Nazir, R. (2021). The science of vermicomposting for sustainable development. In Bhat, R. A., et al. (eds.), *Agricultural waste* (pp. 133–150). Apple Academic Press.

Alp, Y. & Sensoy, S. (2023). The effects of different fertiliser applications on some morphological traits in fresh bean. *Yuzuncu Yıl University Journal of Agricultural Sciences*, 33(1), 100–110. https://doi.org/10.29133/yyutbd.1209177.

Ameloot, N., et al. (2013). Short-term CO_2 and N_2O emissions and microbial properties of biochar amended sandy loam soils. *Soil Biology and Biochemistry*, 57, 401–410.

An, X., et al. (2020). Copyrolysis of biomass, bentonite, and nutrients as a new strategy for the synthesis of improved biochar-based slow-release fertilisers. *ACS Sustainable Chemistry & Engineering*, 8(8), 3181–3190.

Angelo, J. C., et al. (1997). Poultry litter material: quality, quantity, and effect on the performance of broiler chickens. *Revista Brasileira de Zootecnia*, 26, 121–30.

Antille, D. L., et al. (2014). Effects of biosolids-derived organomineral fertilisers, urea, and biosolids granules on crop and soil established with ryegrass (*Lolium perenne* L.). *Communications in Soil Science and Plant Analysis*, 45(12), 1605–1621.

Antille, D. L., et al. (2017). Field-scale evaluation of biosolids-derived organomineral fertilisers applied to winter wheat in England. *Agronomy Journal*, 109(2), 654–674.

Atteya, A. K., et al. (2021). Response of *Moringa oleifera* seeds and fixed oil production to vermicompost and NPK fertilisers under calcareous soil conditions. *Plants*, 10(10), 1998.

Awasthi, S. K., et al. (2020). Changes in global trends in food waste composting: research challenges and opportunities. *Bioresource Technology*, 299, 122555.

Ayilara, M. S., et al. (2020). Waste management through composting: challenges and potentials. *Sustainability*, 12(11), 4456.

Bagreev, A., et al. (2001). Pore structure and surface chemistry of adsorbents obtained by pyrolysis of sewage sludge-derived fertiliser. *Carbon*, 39(13), 1971–1979. https://doi.org/10.1016/S0008-6223(01)00026-4.

Bastos, A. V. S., et al. (2021). Immediate and residual effects of mineral and organomineral nitrogen sources associated with concentrated vinasse on maize. *Journal of Soil Science and Plant Nutrition*, 21(2), 1382–1396.

Bebber, D. P. & Richards, V. R. (2022). A meta-analysis of the effect of organic and mineral fertilisers on soil microbial diversity. *Applied Soil Ecology*, 175, 104450.

Bernal, M. P., et al. (2015). *Evaluation of manure management systems in Europe*. http://hdl.handle.net/2117/88745.

Besen, M. R., et al. (2022). The efficiency of stabilized fertilisers in reducing ammonia loss is influenced by liming method. *Communications in Soil Science and Plant Analysis*, 53(12), 1552–1564.

Biederman, L. A. & Harpole, W. S. (2013). Biochar and its effects on plant productivity and nutrient cycling: a meta-analysis. *GCB Bioenergy*, *5*(2), 202–214.

Borges, N. B., et al. (2015). Characterization of residual sand removed from the grit chambers of a wastewater treatment plant and its use as fine aggregate in the preparation of non-structural concrete. *Water Practice and Technology*, 10(1), 164–171.

Bouhia, Y., et al. (2022). Conversion of waste into organo-mineral fertilisers: current technological trends and prospects. *Reviews in Environmental Science and Bio/Technology*, 21(2), 425–446.

Brown, S. & Leonard, P. (2004). Building carbon credits with biosolids recycling. *BioCycle*, 45(9), 25–29.

Bruun, E. W., et al. (2012). Effects of slow and fast pyrolysis biochar on soil C and N turnover dynamics. *Soil Biology and Biochemistry*, 46, 73–79.

Buckwell, A. & Nadeu, E. (2016). *Nutrient recovery and reuse (NRR) in European agriculture: a review of the issues, opportunities, and actions*. RISE Foundation.

Bueno, J. R. P., et al. (2011). Chemical and microbiological attributes of an oxisol treated with successive applications of sewage sludge. *Revista Brasileira de Ciência do Solo*, 35(4), 1461–1470.

Burak, E. & Sakrabani, R. (2023). Novel carbon capture-based organo-mineral fertilisers show comparable yields and impacts on soil health to mineral fertiliser across two cereal crop field trials in Eastern England, *Field Crops Research*, 302(October), 109043.

Bux, C., et al. (2024). Carbon border adjustment mechanism (CBAM) to tackle carbon leakage in the international fertiliser trade. *Sustainability*, 16, 10661.

Carneiro, J. S. D. S., et al. (2021). Long-term effect of biochar-based fertilisers application in tropical soil: agronomic efficiency and phosphorus availability. *Science of the Total Environment*, 760, 143955.

Chew, J., et al. (2020). Biochar-based fertiliser: supercharging root membrane potential and biomass yield of rice. *Science of the Total Environment*, 713, 136431.

Chen, L., et al. (2018). Formulating and optimizing a novel biochar-based fertiliser for simultaneous slow-release of nitrogen and immobilization of cadmium. *Sustainability*, 10(8), 2740.

Chen, Y., et al. (2008). Inhibition of anaerobic digestion process: a review. *Bioresource Technology*, 99(10), 4044–4064. https://doi.org/10.1016/j.biortech.2007.01.057.

Chen, Z. & Jiang, X. (2014). Microbiological safety of chicken litter or chicken litter-based organic fertilisers: a review. *Agriculture*, 4(1), 1–29. https://doi.org/10.3390/agriculture4010001.

Chen, Z., et al. (2021). A novel maize biochar-based compound fertiliser for immobilizing cadmium and improving soil quality and maize growth. *Environmental Pollution*, 277, 116455.

Chojnacka, K. (2025). Global adoption of organo-mineral fertilisers: environmental and policy insights. *Discover Agriculture*, 3, 184.

Chojnacka, K. & Moustakas, K. (2024). Anaerobic digestate management for carbon neutrality and fertilizer use: a review of current practices and future opportunities. *Biomass and Bioenergy*, 180, 106991. https://doi.org/10.1016/j.biombioe.2023.106991.

Cui, X., et al. (2018). Long-term combined application of manure and chemical fertiliser sustained higher nutrient status and rhizospheric bacterial diversity in reddish paddy soil of Central South China. *Scientific Reports*, 8, 16554. https://doi.org/10.1038/s41598-018-34685-0.

Dad, F. P., et al. (2020). Influence of iron-enriched biochar on Cd sorption, its ionic concentration and redox regulation of radish under cadmium toxicity. *Agriculture*, 11(1), 1.

Das, D., et al. (2021). A valorisation approach in recycling of organic wastes using low-grade rock minerals and microbial culture through vermicomposting. *Environmental Challenges*, 5, 100225.

da Silva, M. A., et al. (2007). Use of stillage and its impact on soil properties and groundwater. *Revista Brasileira de Engenharia Agrícola e Ambiental*, 11, 108–114.

Deeks, L. K., et al. (2013). A new sludge-derived organo-mineral fertiliser gives similar crop yields as conventional fertilisers. *Agronomy for Sustainable Development*, 33, 539–549.

Demirel, B. & Scherer, P. (2008). The roles of acetotrophic and hydrogenotrophic methanogens during anaerobic conversion of biomass to methane: a review. *Reviews in Environmental Science and Biotechnology*, 7, 173–190. https://doi.org/10.1007/s11157-008-9131-1.

de Moraes, E. R., et al. (2020). Yield and biometry of fertilized sugar cane with organomineral fertiliser of sewage sludge and biostimulant. *Bioscience Journal*, 36(5), 1564–1576.

Diacono, M. & Montemurro, F. (2011). Long-term effects of organic amendments on soil fertility. *Sustainable Agriculture*, 2, 761–786.

Dimkpa, C. O., et al. (2020). Development of fertilisers for enhanced nitrogen use efficiency – trends and perspectives. *Science of the Total Environment*, 731, 139113.

Domingues, R. R., et al. (2017). Properties of biochar derived from wood and high-nutrient biomasses with the aim of agronomic and environmental benefits. *PLoS One*, 12(5), e0176884.

Dong, D., et al. (2020). An effective biochar-based slow-release fertiliser for reducing nitrogen loss in paddy fields. *Journal of Soils and Sediments*, 20, 3027–3040.

Downie, A., et al. (2009). Physical properties of biochar. In Lehmann, J. & Joseph, S. (eds.), *Biochar for environmental management: science and technology*. Earthscan.

Du, T., et al. (2022). Positive effects of organic fertilisers and biofertilisers on soil microbial community composition and walnut yield. *Applied Soil Ecology*, 175, 104457. https://doi.org/10.1016/j.apsoil.2022.104457.

Egas, D., et al. (2023). Life cycle assessment of bio-based fertilisers production systems: where are we and where should we be heading? *The International Journal of Life Cycle Assessment*, 1–25.

Enaime, G., et al. (2020). Biochar for wastewater treatment – conversion technologies and applications. *Applied Sciences*, 10(10), 3492. https://doi.org/10.3390/app10103492.

European Biogas Association. (2024). *Exploring digestate's contribution to healthy soils* (Position paper). https://www.europeanbiogas.eu/news/exploring-digestates-contribution-to-healthy-soils/.

European Union. (2019). *Regulation 1009 of the European Parliament and of the Council laying down rules on making available on the market of EU fertilizing products and amending Regulations (EC) No 1069/2009 and (EC) No 1107/2009 and repealing Regulation (EC) No 2003/2003*.

Fagbohungbe, M. O., et al. (2019). The effect of acidogenic and methanogenic conditions on the availability and stability of carbon, nitrogen and phosphorus in a digestate. *Journal of Environmental Chemical Engineering*, 7(4), 103138. https://doi.org/10.1016/j.jece.2019.103138.

Fan, D., et al. (2022). Global evaluation of inhibitor impacts on ammonia and nitrous oxide emissions from agricultural soils: a meta-analysis. *Global Change Biology*, 28(17), 5121–5141.

Fang, L., et al. (2020). Use of Mg/Ca modified biochars to take up phosphorus from acid-extract of incinerated sewage sludge ash (ISSA) for fertiliser application. *Journal of Cleaner Production*, 244, 118853.

Farrar, M. B., et al. (2019). Short-term effects of organo-mineral enriched biochar fertiliser on ginger yield and nutrient cycling. *Journal of Soils and Sediments*, 19, 668–682.

Gao, M., et al. (2021). Effects of long-term biochar and biochar-based fertiliser application on brown earth soil bacterial communities. *Agriculture, Ecosystems & Environment*, 309, 107285.

Gao, S., et al. (2016). Locally produced wood biochar increases nutrient retention and availability in agricultural soils of the San Juan Islands, USA. *Agriculture, Ecosystems & Environment*, 233, 43–54.

Gao, Y. & Cabrera Serrenho, A. (2023). Greenhouse gas emissions from nitrogen fertilisers could be reduced by up to one-fifth of current levels by 2050 with combined interventions. *Nature Food*, 4(2), 170–178.

Głąb, T., et al. (2016). Effect of biochar application on soil hydrological properties and physical quality of sandy soil. *Geoderma*, 281, 11–20.
Gomiero, T., et al. (2011). Environmental impact of different agricultural management practices: conventional vs. organic agriculture. *Critical Reviews in Plant Sciences*, 30, 95–124. http://dx.doi.org/10.1080/07352689.2011.554355.
Grobelak, A., et al. (2019). General considerations on sludge disposal, industrial and municipal sludge. In Prasad, M. N. V., et al. (eds.), *Industrial and municipal sludge* (pp. 135–153). Butterworth-Heinemann.
Guimarães, R. N., et al. (2017). Production of phosphate biofertiliser through composting and vermicomposting process. *Comunicata Scientiae*, 8(3), 432–443.
Gunes, A. Y. D. I. N., et al. (2015). Variations in mineral element concentrations of poultry manure biochar obtained at different pyrolysis temperatures, and their effects on crop growth and mineral nutrition. *Soil Use and Management*, 31(4), 429–437.
Guo, X. X., et al. (2020). The role of biochar in organic waste composting and soil improvement: a review. *Waste Management*, 102, 884–899. https://doi.org/10.1016/j.wasman.2019.12.003.
Hadroug, S., et al. (2019). Pyrolysis process as a sustainable management option of poultry manure: characterization of the derived biochars and assessment of their nutrient release capacities. *Water*, 11(11), 2271.
Haifa Group. (2022). *Chemical fertilisers: advantages and disadvantages*. Haifa Group. https://www.haifa-group.com/haifa-blog/chemical-fertilisers-advantages-and-disadvantages.
Halvorson, A. D., et al. (2014). Enhanced-efficiency nitrogen fertilisers: potential role in nitrous oxide emission mitigation. *Agronomy Journal*, 106(2), 715–722.
Hansen, B., et al. (2000). Nitrogen leaching from conventional versus organic farming systems – a systems modelling approach. *European Journal of Agronomy*, 13(1), 65–82.
Harirchi, S., et al. (2022). Microbiological insights into anaerobic digestion for biogas, hydrogen or volatile fatty acids (VFAs): a review. *Bioengineered*, 13(3), 6521–6557.
Hou, Y., et al. (2018). Stakeholder perceptions of manure treatment technologies in Denmark, Italy, the Netherlands, and Spain. *Journal of Cleaner Production*, 172, 1620–1630.
Ioannidou, S. C., et al. (2022). Linking management practices and soil properties to Ecosystem Services in Mediterranean mixed orchards. *Ecosystem Services*, 53, 101378.
IPCC. (2014). Climate change 2014: mitigation of climate change. Available online: https://www.ipcc.ch/report/ar5/wg3/ (accessed on 23 June 2023).
IPCC. (2022). Climate change 2022: mitigation of climate change. Available online: https://www.ipcc.ch/report/ar6/wg3/ (accessed on 23 June 2023).
Jaafar, N. M., et al. (2014). Microscopy observations of habitable space in biochar for colonization by fungal hyphae from soil. *Journal of Integrative Agriculture*, 13(3), 483–490.
Jain, M. S., et al. (2019). Variation in the key indicators during composting of municipal solid organic wastes. *Sustainable Environment Research*, 29(1), 1–8.
Jha, S., et al. (2023). Enriching vermicompost with rock phosphate applied with inorganic fertilisers improves soil biological quality in calcareous soil. *The Pharma Innovation Journal*, 12(3), 4870–4874.

Joseph, S., et al. (2015). Effects of enriched biochars containing magnetic iron nanoparticles on mycorrhizal colonisation, plant growth, nutrient uptake, and soil quality improvement. *Pedosphere*, 25(5), 749–760.

Jung, S., et al. (2019). Strategic use of biochar for CO_2 capture and sequestration. *Journal of CO2 Utilization*, 32, 128–139.

Kacprzak, M., et al. (2017). Sewage sludge disposal strategies for sustainable development. *Environmental Research*, 156, 39–46.

Kannari, N., et al. (2020). Calcium-enriched biochar: pyrolysis of digested sludge, phosphorus recovery, and fertiliser properties. *Journal of the Japan Institute of Energy*, 99, 236–242.

Karhu, K., et al. (2011). Biochar addition to agricultural soil increased CH_4 uptake and water holding capacity – results from a short-term pilot field study. *Agriculture, Ecosystems & Environment*, 140(1–2), 309–313.

Karim, A. A., et al. (2017). Potassium enriched biochar production by thermal plasma processing of banana peduncle for soil application. *Journal of Analytical and Applied Pyrolysis*, 123, 165–172.

Karim, A. A., et al. (2019). Co-plasma processing of banana peduncle with phosphogypsum waste for production of lesser toxic potassium – sulphur rich biochar. *Journal of Material Cycles and Waste Management*, 21, 107–115.

Kasozi, G. N., et al. (2010). Catechol and humic acid sorption onto a range of laboratory-produced black carbons (biochars). *Environmental Science & Technology*, 44(16), 6189–6195.

Khajavi-Shojaei, S., et al. (2020). Synthesis modified biochar-based slow-release nitrogen fertiliser increases nitrogen use efficiency and corn (*Zea mays* L.) growth. *Biomass Conversion and Biorefinery*, 13, 593–601.

Kim, J. A., et al. (2018). A phosphorus-enriched biochar fertiliser from bio-fermentation waste: a potential alternative source for phosphorus fertilisers. *Journal of Cleaner Production*, 196, 163–171.

Kizito, S., et al. (2019). Role of nutrient-enriched biochar as a soil amendment during maize growth: exploring practical alternatives to recycle agricultural residuals and to reduce chemical fertiliser demand. *Sustainability*, 11(11), 3211.

Klyosov, D. N. & Orekhovskaya, A. A. (2021). On the development of technology for obtaining organomineral fertilisers. In *IOP Conference Series: Earth and Environmental Science* (Vol. 723, No. 3, p. 032024). IOP Publishing.

Kominko, H., et al. (2017). The possibility of organo-mineral fertiliser production from sewage sludge. *Waste and Biomass Valorization*, 8, 1781–1791.

Kominko, H., et al. (2019). Closing the nutrient cycle by using organo-mineral fertilisers based on secondary raw materials. Paper presented at the *7th International Conference on Sustainable Solid Waste Management*, Heraklion, Greece.

Koszel, M. & Lorencowicz, E. (2015). Agricultural use of biogas digestate as a replacement fertilizers. *Agriculture and Agricultural Science Procedia*, 7, 119–124. https://doi.org/10.1016/j.aaspro.2015.12.004.

Kour, D., et al. (2020). Microbial biofertilisers: bioresources and eco-friendly technologies for agricultural and environmental sustainability. *Biocatalysis and Agricultural Biotechnology*, 23, 101487. https://doi.org/10.1016/j.bcab.2019.101487.

Kulikova, M. A., et al. (2020). Environmental assessment of the impact of ammophos-based organic fertilisers on the environment. In *IOP Conference Series: Earth and Environmental Science*, Vol. 579, No. 1, p. 012052. IOP Publishing.

Kumar, R., et al. (2022). Combined application of chemical fertiliser and enriched household vermicompost influences maize yield and soil quality under calcareous soil. *The Pharma Innovation Journal*, 11(8), 2069–2075.

Lehmann, J. & Joseph, S. (2015). Biochar for environmental management: an introduction. In Lehmann, J. & Joseph, S. (eds.), *Biochar for environmental management* (pp. 1–13). Routledge.

Lehmann, J., et al. (2003). Nutrient availability and leaching in an archaeological Anthrosol and a Ferralsol of the Central Amazon basin: fertiliser, manure, and charcoal amendments. *Plant and Soil*, 249, 343–357.

Lehmann, J., et al. (2011). Biochar effects on soil biota – a review. *Soil Biology and Biochemistry*, 43(9), 1812–1836.

Li, L., et al. (2020). Planning green infrastructure to mitigate urban surface water flooding risk – a methodology to identify priority areas applied in the city of Ghent. *Landscape and Urban Planning*, 194, 103703. https://doi.org/10.1016/j.landurbplan.2019.103703.

Li, Y., et al. (2022). Organomineral fertiliser application enhances *Perilla frutescens* nutritional quality and rhizosphere microbial community stability in karst mountain soils. *Frontiers in Microbiology*, 13, 1058067. https://doi.org/10.3389/fmicb.2022.1058067.

Liao, J., et al. (2020). Effects of biochar-based controlled release nitrogen fertiliser on nitrogen-use efficiency of oilseed rape (*Brassica napus* L.). *Scientific Reports*, 10(1), 11063.

Litskas, V. D. (2023). Environmental impact assessment for animal waste, organic and synthetic fertilisers. *Nitrogen*, 4(1), 16–25.

Liu, X., et al. (2019). A biochar-based route for environmentally friendly controlled release of nitrogen: urea-loaded biochar and bentonite composite. *Scientific Reports*, 9(1), 9548.

Liu, Z., et al. (2018). Characteristics and applications of biochars derived from wastewater solids. *Renewable and Sustainable Energy Reviews*, 90, 650–664. https://doi.org/10.1016/j.rser.2018.02.040.

Lohri, C. R., et al. (2017). Treatment technologies for urban solid biowaste to create value products: a review with focus on low-and middle-income settings. *Reviews in Environmental Science and Bio/Technology*, 16, 81–130.

Lustosa Filho, J. F., et al. (2020). Aging of biochar-based fertilisers in soil: effects on phosphorus pools and availability to *Urochloa brizantha* grass. *Science of The Total Environment*, 709, 136028. https://doi.org/10.1016/j.scitotenv.2019.136028.

Magela, M. L. M., et al. (2019a). Application of organomineral fertilisers sourced from filter cake and sewage sludge can affect nutrients and heavy metals in soil during early development of maize. *Australian Journal of Crop Science*, 13(6), 863–873.

Magela, M. L. M., et al. (2019b). Efficacy of organomineral fertilisers derived from biosolid or filter cake on early maize development. *Australian Journal of Crop Science*, 13(5), 662–670.

Mani, S., et al. (2016). Process simulation and modeling: anaerobic digestion of complex organic matter. *Biomass and Bioenergy*, 93, 158–167. https://doi.org/10.1016/j.biombioe.2016.07.018.

Mazeika, R., et al. (2016). Engineered pelletized organo-mineral fertilisers (OMF) from poultry manure, diammonium phosphate and potassium chloride. *ACS Sustainable Chemistry & Engineering*, 4(4), 2279–2285.

Mazeika, R., et al. (2021). Nutrient dynamics and plant response in soil to organic chicken manure-based fertilisers. *Waste and Biomass Valorization*, 12, 371–382.

Meissner, R., et al. (1998). Lysimeter studies in East Germany concerning the influence of set aside of intensively farmed land on the seepage water quality. *Agriculture, Ecosystems & Environment*, 67(2–3), 161–173.

Menegat, S., et al. (2022). Greenhouse gas emissions from global production and use of nitrogen synthetic fertilisers in agriculture. *Scientific Reports*, 12, 14490.

Molina, M. J., et al. (2013). Stabilisation of sewage sludge and vinasse bio-wastes by vermicomposting with rabbit manure using *Eisenia fetida*. *Bioresource Technology*, 137, 88–97.

Möller, K. & Müller, T. (2012). Effects of anaerobic digestion on digestate nutrient availability and crop growth: a review. *Engineering in Life Sciences*, 12(3), 242–257. https://doi.org/10.1002/elsc.201100085.

Moreira, J. G., et al. (2021). Chemical changes in soil with use of pelletized organomineral fertiliser made from biosolids and sugarcane filter cake. *Australian Journal of Crop Science*, 15(1), 67–72.

Mosa, A., et al. (2018). Functionalized biochar derived from heavy metal rich feedstock: phosphate recovery and reusing the exhausted biochar as an enriched soil amendment. *Chemosphere*, 198, 351–363.

Mumbach, G. L., et al. (2020). Agronomic efficiency of organomineral fertiliser in sequential grain crops in southern Brazil. *Agronomy Journal*, 112(4), 3037–3049.

Naja, F., et al. (2026). Food security amid the US Iran war: a food system analysis and a framework for coordinated multilevel action, *Global Food Security*, 49, 100919.

Nardis, B. O., et al. (2021). Phosphorus recovery using magnesium-enriched biochar and its potential use as fertiliser. *Archives of Agronomy and Soil Science*, 67(8), 1017–1033.

Nascimento, A. L., et al. (2020). Chemical attributes of sewage sludges: relationships to sources and treatments, and implications for sludge usage in agriculture. *Journal of Cleaner Production*, 258, 120746. https://doi.org/10.1016/j.jclepro.2020.120746.

Nascimento, C. O., et al. (2021). Phosphorus diffusion and agronomic efficiency of chicken litter organomineral fertilisers improved with binder materials. *Waste and Biomass Valorization*, 12, 3765–3772.

Ndegwa, P. M. & Thompson, S. A. (2001). Integrating composting and vermicomposting in the treatment and bioconversion of biosolids. *Bioresource Technology*, 76(2), 107–112.

Ndoung, O. C. N., et al. (2021). A scoping review on biochar-based fertilisers: enrichment techniques and agro-environmental application. *Heliyon*, 7(12), e08473.

Nguyen, T. T. N., et al. (2018). The effects of short term, long term, and reapplication of biochar on soil bacteria. *Science of the Total Environment*, 636, 142–151.

Nielsen, S., et al. (2014). Comparative analysis of the microbial communities in agricultural soil amended with enhanced biochars or traditional fertilisers. *Agriculture, Ecosystems & Environment*, 191, 73–82.

Oh, T. K., et al. (2014). Utilization of biochar impregnated with anaerobically digested slurry as slow-release fertiliser. *Journal of Plant Nutrition and Soil Science*, 177(1), 97–103.

Oka, Y., et al. (2007). Enhanced nematicidal activity of organic and inorganic ammonia-releasing amendments by *Azadirachta indica* extracts. *Journal of Nematology*, 39(1), 9.

Oliveira, D. P., et al. (2017). Organic matter sources in the composition of pelletized organomineral fertilisers used in sorghum crops. *African Journal of Agricultural Research*, 12(32), 2574–2581.

Olle, M. (2019). Vermicompost, its importance and benefit in agriculture. *Journal of Agricultural Science*, 2, 93–98. https://doi.org/10.15159/jas.19.19.

Osman, A., et al. (2014). Efficacy of three nitrogenous fertilisers urea, ammonium nitrate, ammonium sulphate with or without chitin on suppression of root-knot nematode, *Meloidogyne incognita* population infesting sunflower plant: a field study. *Middle East Journal of Agricultural Research*, 3(4), 900–906.

Owens, L. B., et al. (2000). Lysimeter study of nitrate leaching from a corn-soybean rotation. *Soil Science Society of America Journal*, 29(2), 467–474.

Paul, C., et al. (2023). Carbon farming: are soil carbon certificates a suitable tool for climate change mitigation? *Journal of Environmental Management*, 330, 117142.

Pawlett, M., et al. (2015). Nutrient potential of biosolids and urea derived organo-mineral fertilisers in a field scale experiment using ryegrass (*Lolium perenne* L.). *Field Crops Research*, 175, 56–63.

Petric, I., et al. (2012). Evolution of process parameters and determination of kinetics for co-composting of organic fraction of municipal solid waste with poultry manure. *Bioresource Technology*, 117, 107–116.

Puga, A. P., et al. (2020). Biochar-based nitrogen fertilisers: greenhouse gas emissions, use efficiency, and maize yield in tropical soils. *Science of the Total Environment*, 704, 135375.

Quilliam, R. S., et al. (2013). Life in the 'charosphere'–does biochar in agricultural soil provide a significant habitat for microorganisms? *Soil Biology and Biochemistry*, 65, 287–293.

Ren, H., et al. (2020). Responses of soil enzyme activities and plant growth in a eucalyptus seedling plantation amended with bacterial fertilisers. *Archives of Microbiology*, 202(6), 1381–1396. https://doi.org/10.1007/s00203-020-01849-4.

Rickson R.J., et al. (2015). Input constraints to food production: the impact of soil degradation. *Food Security*, 7(2), 351–364.

Ritz, C. W. & Merka, W. C. (2009). Maximizing poultry manure use through nutrient management planning. https://esploro.libs.uga.edu/esploro/outputs/report/Maximizing-poultry-manure-use-through-nutrient/9949316025802959#file-0.

Rodrigues, M. M., et al. (2021). Sewage sludge as organic matrix in the manufacture of organomineral fertilisers: physical forms, environmental risks, and nutrients recycling. *Journal of Cleaner Production*, 313, 127774.

Rogeri, D. A., et al. (2016). Composition of poultry litter in Southern Brazil. *Revista Brasileira de Ciência do Solo*, 40. https://doi.org/10.1590/18069657rbcs20150421.

Sannigrahi, A. K. (2016). Major constraints in popularising vermicompost technology in eastern India. *Modern Environment Science and Engineering*, 2(2), 123–133.

Sayara, T., et al. (2020). Recycling of organic wastes through composting: process performance and compost application in agriculture. *Agronomy*, 10(11), 1838.

Schmidt, H. P., et al. (2017). Biochar-based fertilization with liquid nutrient enrichment: 21 field trials covering 13 crop species in Nepal. *Land Degradation & Development*, 28(8), 2324–2342.

Seleiman, M. F., et al. (2017). Chemical composition and in vitro digestibility of whole-crop maize fertilized with synthetic fertilizer or digestate and harvested at two maturity stages in Boreal growing conditions. *Agricultural and Food Science*, 26(1), 47.

Semida, W. M., et al. (2014). A novel organo-mineral fertiliser can alleviate negative effects of salinity stress for eggplant production on reclaimed saline calcareous soil. *Acta Horticulturae*, 1034, 493–499. https://doi.org/10.17660/ActaHortic.2014.1034.61.

Sharara, M. A., et al. (2018). Techno-economic optimization of community-based manure processing. *Agricultural Systems*, 161, 117–123.

Shi, W., et al. (2020). Biochar bound urea boosts plant growth and reduces nitrogen leaching. *Science of the Total Environment*, 701, 134424.

Silva, H., et al. (2019). Effects of organomineral fertilizers formulated with biosolids and filter cake on common bean yield crop (*Phaseolus vulgaris* L.). *Australian Journal of Crop Science*. 1566–1571.

Singh, A., et al. (2020). Earthworms and vermicompost: an eco-friendly approach for repaying nature's debt. *Environmental Geochemistry and Health*, 42, 1617–1642.

Sogbedji, J. M., et al. (2000). Nitrate leaching and nitrogen budget as affected by maize nitrogen rate and soil type. *Soil Science Society of America Journal*, 29(6), 1813–1820.

Soinne, H., et al. (2014). Effect of biochar on phosphorus sorption and clay soil aggregate stability. *Geoderma*, 219, 162–167.

Strazzera, G., et al. (2018). Volatile fatty acids production from food wastes for biorefinery platforms: a review. *Journal of Environmental Management*, 226, 278–288. https://doi.org/10.1016/j.jenvman.2018.08.039.

Szulc, W., et al. (2021). Possibilities of using organic waste after biological and physical processing – an overview. *Processes*, 9(9), 1501.

Tambone, F., et al. (2010). Assessing amendment and fertilizing properties of digestates from anaerobic digestion through a comparative study with digested sludge and compost. *Chemosphere*, 81(5), 577–583.

Tang, K., et al. (2016). Carbon farming economics: what have we learned? *Journal of Environmental Management*, 172, 49–57.

Tejada, M., et al. (2005). Effects of application of two organomineral fertilisers on nutrient leaching losses and wheat crop. *Agronomy Journal*, 97(3), 960–967.

Tripathi, M., et al. (2016). Effect of process parameters on production of biochar from biomass waste through pyrolysis: a review. *Renewable and Sustainable Energy Reviews*, 55, 467–481.

Troeh, F. R. & Thompson, L. M. (2005). *Soils and soil fertility* (pp. 489). Blackwell Publishing.

Tumbure, A., et al. (2020). Co-pyrolysis of maize stover and igneous phosphate rock to produce potential biochar-based phosphate fertiliser with improved carbon retention and liming value. *ACS Sustainable Chemistry & Engineering*, 8(10), 4178–4184.

Uddin, M. M. & Wright, M. M. (2022). Anaerobic digestion fundamentals, challenges, and technological advances. *Physical Sciences Reviews*, 8(9), 2819–2837. https://doi.org/10.1515/psr-2021-0068.

Uddin, M. K., et al. (2025). Organo-mineral fertilizers to sustain soil health and crop yield and reduce environmental impact: a comprehensive review. *European Journal of Agronomy*, 162, 127433.

United Nations. (2019). Department of economic and social affairs, population division world population prospects 2019: highlights (ST/ESA/SER.A/423).

Utomo, W. H., et al. (2011). Effect of biochar on the release and loss of nitrogen from urea fertilization. *Journal of Agriculture and Food Technology*, 1(7), 127–132. www.textroad.com.

Vakal, S., et al. (2021). Minimization of soil pollution as a result of the use of encapsulated mineral fertilisers. *Journal of Ecological Engineering*, 22(1), 221–230.

Valentinuzzi, F., et al. (2020). The fertilising potential of manure-based biogas fermentation residues: pelleted vs. liquid digestate. *Heliyon*, 6(2), e03325. https://doi.org/10.1016/j.heliyon.2020.e03325.

Vangeli, S., et al. (2022). Effects of fertiliser type on nitrous oxide emission and ammonia volatilization in wheat and maize crops. *Soil Use and Management*, 38(4), 1519–1531.

Van Zwieten, L., et al. (2014). An incubation study investigating the mechanisms that impact N_2O flux from soil following biochar application. *Agriculture, Ecosystems & Environment*, 191, 53–62.

Vitale, L., et al. (2017). Fertiliser type influences tomato yield and soil N_2O emissions. *Plant, Soil and Environment*, 63(3), 105–110.

Wali, F., et al. (2020). Formulation of biochar-based phosphorus fertiliser and its impact on both soil properties and chickpea growth performance. *Sustainability*, 12(22), 9528.

Walling, E. & Vaneeckhaute, C. (2020). Greenhouse gas emissions from inorganic and organic fertiliser production and use: a review of emission factors and their variability. *Journal of Environmental Management*, 276, 111211.

Wen, P., et al. (2017). Microwave-assisted synthesis of a novel biochar-based slow-release nitrogen fertiliser with enhanced water-retention capacity. *ACS Sustainable Chemistry & Engineering*, 5(8), 7374–7382.

Wilkinson, K., et al. (2011). Effect of heating and aging of poultry litter on the persistence of enteric bacteria. *Poultry Science*, 90(1), 10–18. https://doi.org/10.3382/ps.2010-01023.

Yang, W., et al. (2021). Influence of biochar and biochar-based fertiliser on yield, quality of tea and microbial community in an acid tea orchard soil. *Applied Soil Ecology*, 166, 104005.

Yeh, C. K., et al. (2020). Optimizing food waste composting parameters and evaluating heat generation. *Applied Sciences*, 10(7), 2284.

Zhang, B., et al. (2010). Does microbial habitat or community structure drive the functional stability of microbes to stresses following re-vegetation of a severely degraded soil? *Soil Biology and Biochemistry*, 42(5), 850–859.

Zhang, D., et al. (2016). Biochar helps enhance maize productivity and reduce greenhouse gas emissions under balanced fertilization in a rainfed low fertility inceptisol. *Chemosphere*, 142, 106–113. https://doi.org/10.1016/j.chemosphere.2015.04.088.

Zhang, P., et al. (2020). Beneficial effects of biochar-based organic fertiliser on nitrogen assimilation, antioxidant capacities, and photosynthesis of sugar beet (*Beta vulgaris* L.) under saline-alkaline stress. *Agronomy*, 10(10), 1562.

Zhang, S., et al. (2021). Contrasting short-term responses of soil heterotrophic and autotrophic respiration to biochar-based and chemical fertilisers in a subtropical Moso bamboo plantation. *Applied Soil Ecology*, 157, 103758.

Zhang, W., et al. (2018). Dynamics of soil fertility and maize growth with lower environment impacts depending on a combination of organic and mineral fertiliser. *Journal of Soil Science and Plant Nutrition*, 18(2), 556–575.

Zhang, X., et al. (2015). Managing nitrogen for sustainable development. *Nature*, 528(7580), 51–59.

Zhang, Y., et al. (2014). Community structure and elevational diversity patterns of soil Acidobacteria. *Journal of Environmental Sciences*, 26(8), 1717–1724.

Zhao, L., et al. (2016). Copyrolysis of biomass with phosphate fertilisers to improve biochar carbon retention, slow nutrient release, and stabilize heavy metals in soil. *ACS Sustainable Chemistry & Engineering*, 4(3), 1630–1636.

Zhou, C., et al. (2014). Characteristics and the recovery potential of plastic wastes obtained from landfill mining. *Journal of Cleaner Production*, 80, 80–86. https://doi.org/10.1016/j.jclepro.2014.05.083.

Zhou, J., et al. (2015). Influence of 34-years of fertilization on bacterial communities in an intensively cultivated black soil in northeast China. *Soil Biology and Biochemistry*, 90, 42–51. https://doi.org/10.1016/j.soilbio.2015.07.005.

Zhou, Z., et al. (2019). Soil microbial community structure shifts induced by biochar and biochar-based fertiliser amendment to karst calcareous soil. *Soil Science Society of America Journal*, 83(2), 398–408.

Zhu, L., et al. (2014). Effects of different tillage and straw return on soil organic carbon in a rice-wheat rotation system. *PLoS One*, 9(2), e88900.

Chapter 5

Optimizing use of anaerobic digestate as a soil amendment in agriculture

Christina van Midden, Agri-Food and Biosciences Institute, UK

1 Introduction

Anaerobic digestate is a nutrient-dense slurry generated from biogas production. It is a mixture of water, partially digested organic matter, inorganic compounds, microbial biomass and necromass, proteins, lipids, carbohydrates, and minerals. A key characteristic of digestate is its low dry matter content, typically ranging between 3–10%. Anaerobic digestate is one of two products produced by anaerobic digestion, the other being biogas, which is used for heat and electricity. Digestate and biogas are generated under controlled microbial decomposition of biodegradable materials, such as organic waste products from households, food and drink processing, animal manures, crop residues, sewage sludge and biosolids, as well as purpose-grown energy crops such as silage maize and forage grass. As not all the organic material is decomposed during anaerobic digestion, digestate has been proposed as a source of organic matter to improve or maintain agricultural soil productivity.

http://dx.doi.org/10.19103/AS.2025.0166.05

Digestate contains a range of bioavailable macro- to micro-nutrients, making it a source of nutrients for plant production (Möller and Müller, 2012). Additionally, digestate contains bioactive substances such as phytohormones, amino acids, vitamins, monosaccharides and humic substances which promote beneficial microbial-plant interaction and enhance both plant growth and resilience against biotic and abiotic stresses (Möller and Müller, 2012, Sani et al., 2024). As a source of organic matter, digestate can positively impact soil structure, nutrient availability, and microbial activity by increasing soil organic matter (Ragályi et al., 2025), which in turn benefits plant production.

Due to its ability to transform organic wastes into valuable products, anaerobic digestion technology can offer a significant role in advancing circular bioeconomy. Anaerobic digestion enables the recovery and reuse of carbon as energy and the recovery and reuse of nutrients as a fertilizer (Figure 1). This aligns with circular economy principles by closing resource loops and minimizing waste. Furthermore, the energy generated by biogas contributes to energy security by reducing reliance on fossil fuel imports. A component of the circular economy acknowledges the importance of environmental sustainability. Biogas plants contribute toward mitigating climate change, as they reduce greenhouse gas emissions from organic waste disposal by capturing and reusing the gases (Fagerström et al, 2018). In addition, this technology can reduce soil and groundwater pollution by reducing the amount of organic material going to landfill (Obaideen et al., 2022) and reducing the volume of animal slurries. The latter contributes to alleviating the issue of excess animal manure application in livestock-intensive areas and its subsequent environmental pollution (Nag et al., 2019, Massini et al., 2025). Finally, anaerobic digestion technology can

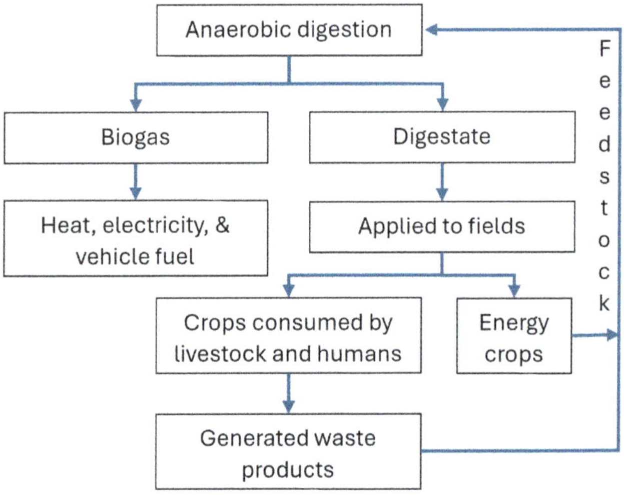

Figure 1 Nutrient cycle in food production system incorporating anaerobic digestion.

contribute to economic sustainability by creating new jobs and diversifying farm incomes.

This chapter reviews current research knowledge by first outlining the process of producing digestate and how this impacts the characteristics of it. Secondly, it will compare the effectiveness of digestate as a replacement fertilizer for synthetic and mineral fertilizers. As there are concerns of nutrient pollution regarding digestate use, the third section reviews research into methods to reduce nutrient loss and optimize the nutrient use efficiency of digestate, while the fourth provides an overview of nutrient recovery technologies. The fifth section discusses the applicability of using digestate to build soil organic matter. The sixth section addresses concerns of contaminants in the digestate and how this could impact food security and environmental health. The penultimate section discusses future research areas that are needed to ensure the sustainable use of digestate.

2 The process of producing quality anaerobic digestate

The fundamental process of anaerobic digestion to produce biogas involves creating an oxygen-free environment to induce methane and carbon dioxide production by anaerobic microbes during the decomposition of organic matter (Al Seadi et al., 2008). The biogas is then either upgraded to biomethane that can be used in gas grids or converted to electricity. There are four successive phases to methane production (Figure 2). During the hydrolysis phase, microorganisms break down complex compounds in the feedstock into simple compounds, the speed of which is determined by the complexity of the compound. Lignin, a complex structure, is highly resistant to decomposition, whereas the simpler structured hemicellulose rapidly decomposes (Molinuevo-Salces et al., 2013). In the acidogenesis stage, the products of hydrolysis are absorbed into microbial cells and broken down to form acidic products. Crucially, the microbes carrying out the transformation use up the remaining oxygen. During the acetogenesis phase, the products from the previous phase are transformed into the volatile fatty acids needed by methanogenic microbes to produce methane, which is carried out in phase four (methanogenesis).

The conditions to achieve anaerobic digestion are flexible, so long as the unit is sealed. The temperature of the digestate can be mesophilic (35–40°C) or thermophilic (50–60°C). The latter has a higher potential to increase methane production, whereas mesophilic digestion is energetically cheaper because the liquid does not require as much heating (Gebreeyessus and Jenicek, 2016). Feedstock can be fed into the digester either continually or in batches, which allows for the manipulation of moisture levels (Al Seadi et al., 2008). Batch feeding reduces the amount of liquid in the digestate, which

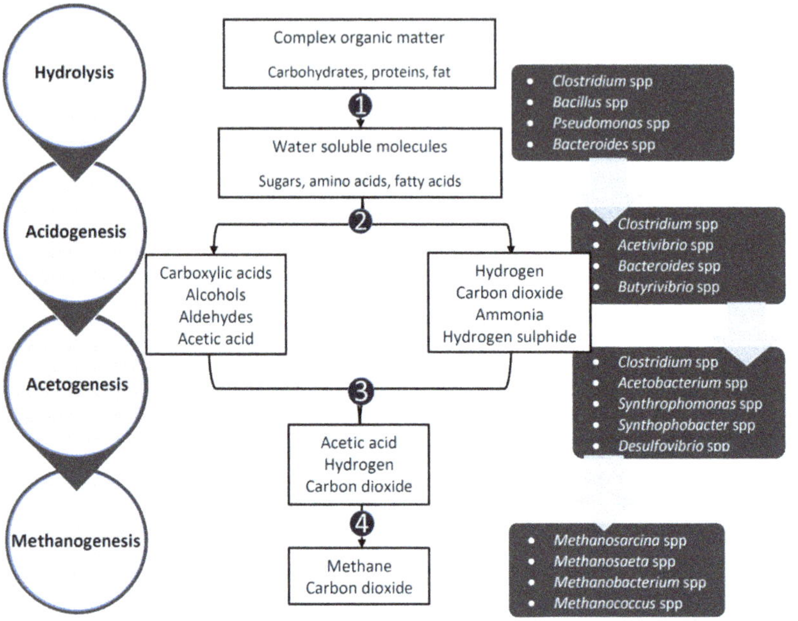

Figure 2 Flow diagram of the four stages of biogas production through anaerobic digestion and the microbial groups involved.

reduces dewatering costs, but results in fluctuating methane production. In contrast, continuous feeding minimizes the repeated heating costs associated with feedstock addition and ensures more stable methane production, though it increases dewatering costs. The process of biogas production itself can be made more efficient if there are connected digester tanks tailored with optimal conditions for the microorganisms carrying out the separate stages of methane production (De Gioannis et al., 2017).

Once the biogas has been extracted, the digestate is ready for post-processing. Anaerobic digestate can be applied directly to fields in the form that it is in (whole digestate), so long as its feedstock was of a sanitary origin, such as agricultural wastes (Al Seadi and Lukehurst, 2012). This system works well for on-farm digesters, where the farmer can add their own animal and/or crop wastes, or purpose-grown energy crops such as maize or ryegrass, into the tank and spread the digestate onto their fields. For off-farm digesters, particularly those that handle food or sewage waste, the resulting digestate will likely need to meet quality assurance schemes set by governmental legislation, such as the EU's ECN–QAS or the UK's BSI PAS110 schemes. Satisfying the conditions of these standards usually involves post-processing of the digestate to remove, for example, biological contaminants such as pathogens by pasteurization or reduce chemical and physical contaminants such as heavy metals and plastics, to acceptable levels prior to land application (see Section 6 for further detail on contaminants).

After removing contaminants to meet the standards required by quality assurance schemes, the digestate is still a liquid-rich substance that is bulky to store and expensive to transport. Digestate can be stored in tanks or lagoons but requires a cover to prevent greenhouse gas emissions (Al-Seadi et al 2008). In temperate countries, provinces or states, there can be restrictions that ban or limit the spreading of organic manures such as digestate to land during winter months (e.g. DAERA, 2019, MDA, 2024, OMAFRA, 2024). This requires storage facilities to have the capacity to store digestate during this closed spreading period. To reduce transport costs and storage pressure, whole digestate is separated into solid (also known as fiber) and liquid fractions using screw presses, decantation centrifuges and belt filters (Nowak and Czekała, 2024). The solid fraction can be stored in an open heap, whereas the liquid needs to be stored in lagoons, preferably with a gas-tight seal to reduce greenhouse gas emissions and ammonia volatilization during storage (Liebetrau et al., 2013) and allow for the capture of methane produced during storage (Al-Seadi et al., 2008). The solid fraction can be transported in trailers, whereas the liquid and whole digestate require specialized tankers.

The separation of digestate into the two fractions has consequences for the nutrient and organic matter distribution. Between 70–80% of the nitrogen and potassium remains in the liquid fraction, while 55–65% of phosphorus and the dry organic matter remain in the solid fraction (Fuchs and Drosg, 2013). This is due to the chemical nature of the nutrients, as phosphorus often precipitates as struvite minerals during the digester process and settles to become part of the solid fraction (Marti et al., 2008), while nitrogen and potassium remain water soluble (Aguirre-Villegas et al., 2019). However, this separation is dependent on the efficiency of the separation technology (Figure 3). Low efficient separation results in a greater proportion of organic matter, organic nitrogen and P, S, Ca and Mg moving into the liquid fraction, whereas a high-efficiency separator allows the majority of this organic matter and these nutrients to move into the solid fraction (Guilayn et al., 2019). Based on separation efficiencies, Tambone et al. (2017) suggested that the solid fraction be used as NP organic fertilizer due to having a similar N:P ratio to poultry manure and sewage sludge, while the liquid fraction could be used as a substitute for mineral nitrogen fertilizer.

3 Using digestate as a fertilizer

The process of anaerobic digestion removes carbon from the organic feedstock but not the nutrients. This produces a digestate containing the range of macro to micronutrients that were originally present in the feedstock, as such the nutrient characterization of digestate is dependent on

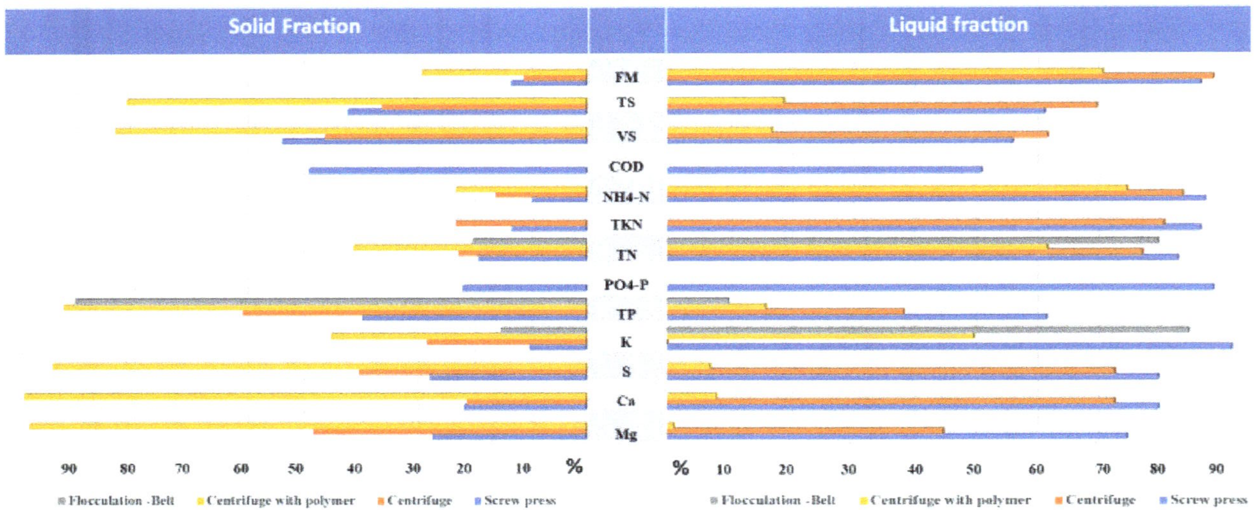

Figure 3 The distribution of the main digestate properties after applying different solid-liquid separation techniques on a weight basis (Barampouti et al., 2020). Fm = fresh matter, TS = total solids, VS = volatile solids, COD = chemical oxygen demand, NH4-N = ammonium nitrogen, TKN = total Kjeldahl nitrogen, TN = total nitrogen, PO4-P = phosphate, TP = total phosphorus, K = potassium, S = sulfur, Ca = calcium, Mg = magnesium.

its feedstock (Table 1). During digestion, the organic forms of nutrients are broken down and transformed into bioavailable forms. For example, between 15–82% of the organic nitrogen bound within feedstock is mineralized into ammonium (Bareha et al., 2018). An exception to this is phosphorus, where the transformed bioavailable P is absorbed to organic matter particles or ions (namely magnesium, iron, aluminum, or calcium) in the digestate. This reduces the concentration of plant-available P to total P in the digestate compared to ingestate (Ran et al., 2023, Tuszynska et al., 2021). This effect is, however, highly dependent on the feedstock, particularly its content of phosphorus-binding compounds. Despite this, digestate application to soil increases concentrations of soil available phosphorus as well as nitrogen and potassium (Slepetiene et al., 2023, Rolka et al., 2024), providing a source of nutrients to support plant growth. Furthermore, digestates contain a mixture of bioactive substances such as phytohormones, humic acids, vitamins and monosaccharides, which can improve plant growth (Möller and Müller, 2012, Scaglia et al., 2017).

The use of liquid and whole digestate as either a partial or whole replacement for synthetic nitrogen fertilizers has been shown to produce equivalent yields for cereals (Buligon et al., 2023, Doyeni et al., 2021, Du et al., 2019, Grillo et al., 2021, Pastorelli et al., 2021, Ran et al., 2022, Šimon et al., 2015), fodder or energy crops such as grasses and silage maize (Brychkova et al., 2024, Luo et al., 2022, Montemurro et al., 2015, Riva et al., 2016, Samoraj et al., 2022, Tilvikienė et al., 2018, Tsachidou et al., 2019, Vaneeckhaute et al., 2013, Walsh et al., 2018) and vegetable crops (Alburquerque et al., 2012a, Barzee et al., 2019, Elbashier et al., 2018, Stoknes et al., 2016, Weimers et al., 2022, Yagüe and Lobo, 2024). The reason why digestate produces these equivalent yields is that between 60–90%

Table 1 Characteristics of digestate based on feedstock. Please note that it is more common for digestors to mix feedstock types (co-digest) to improve performance of digestors by enhancing the breakdown of the organic matter and biomethane yields (Al Seadi and Lukehurst, 2012)

Characteristic	Energy crops (a)	Animal manure (b)	Food waste (c)
DM (%)	4.3 - 8.2	2.2–7.8	1.9–7.8
OM (%dmb)	68.5–86.9	70–75	61.8–80.8
TC (% dmb)	36.2–41.1	35–43	36.2–39.1
TN (% dmb)	0.54–6.8	5.4–11.51	14.1–16.9
N-NH4 (% TN)	49.4–60.2	55–76	21.3–89
TP (% dmb)	0.9–3.5	1.7–1.8	1.5
TK (% dmb)	0.2–6.2	3.6–6.4	4.1

DM = dry matter, OM = organic matter, TC = total carbon, TN = total nitrogen, NH4-N = ammonium nitrogen, TP = total phosphorus, TK= total potassium, dmb = dry matter basis
a) Häfner et al., 2022, Pokoj et al., 2015, Robles-Aguilar et al., 2019, Tambone et al., 2017, Voća et al., 2005.
b) Häfner et al., 2022, Mortola et al., 2019, Voća et al., 2005, Zhang et al., 2020.
c) Häfner et al., 2022, Nicholson et al., 2017, Tampio et al., 2016, Zhang et al., 2012.

of the total nitrogen in liquid and whole digestate is in the form of bioavailable ammonium (Czekała, 2022). As the rate of nitrogen fertilization is based on total nitrogen content, the high ammonium content in digestate means that most of the nitrogen applied in digestate is immediately available for plant uptake. In order to use liquid digestate as a phosphorus replacement fertilizer, it requires biological treatment to partially remove nitrogen to prevent excess nitrogen loading. This treated digestate produces comparable crop yields to mineral P fertilizer (Vanden Nest et al., 2015).

The concentration of phosphorus in the solid fraction enables it to successfully substitute mineral P fertilizer (Vanden Nest et al., 2015). When solid digestate was used as a synthetic nitrogen substitute, it decreased yields of field-grown vegetables (Horta and Carneiro, 2021, Liedl et al., 2006). This is due to the low bioavailability of nitrogen in the solid digestate, as between 54–90% of the total nitrogen is organic nitrogen (Czekała, 2022), which requires mineralization by microorganisms to become bioavailable. As such, nitrogen supply to crops was limited, particularly in the early growth stages. However, over a 3-year period Jurgutis et al. (2021b) measured slightly higher yields in grasslands when applying solid digestate to meet the same nitrogen content as liquid digestate. Cavalli et al. (2016) grew maize followed by ryegrass without applying fertilizers between the two crops and measured an increase in ryegrass yield in the solid digestate amended plots compared to liquid digestate or synthetic nitrogen. This is due to the legacy effect from the residual nitrogen in the soil that accumulated due to low plant uptake of the solid digestate supplied nitrogen within the first year of application. This results in more nitrogen being available for the following crop.

Solid digestate can also be used in growing media in horticulture, as a source of nutrients and physical substrate (Asp and Bergstrand, 2024, Crippa et al., 2013, Dubský et al., 2019). The concentration of ammonium nitrogen determines the amount of digestate that can substitute peat (Dubský et al., 2019, Caspersen et al., 2023) or be used in combination with other substrates (Nesse et al., 2018) to produce equivalent yields to plants grown only on peat.

4 Improving the fertilizer use efficiency of digestate

Nutrient use efficiency of a fertilizer denotes the plant capacity to effectively use fertilizer supplied nutrients for growth. The availability of nutrients to the plant is influenced by the amount that is either immobilized in the soil or lost. Immobilization is the process whereby nutrients are absorbed either by soil biota uptake, chemical fixation or physical occlusion and become unavailable to plants in the immediate term (Figure 4). Nutrients can be lost from the soil via gaseous, surface run-off, or leaching routes (Figure 4). These losses can have environmental consequences including the eutrophication of water bodies,

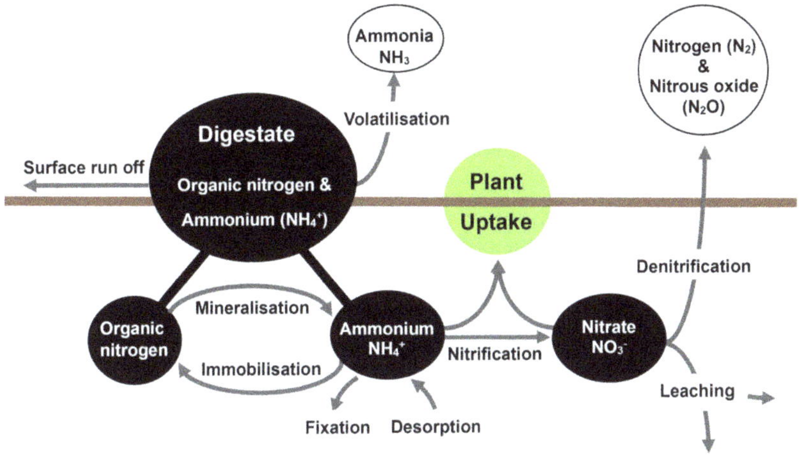

Figure 4 Dynamics of nitrogen supplied by digestate in the soil and loss pathways.

contributing to climate change, and the acidification of terrestrial and aquatic ecosystems via acid rain (Nkoa, 2013). However, these consequences result from the inappropriate use of any fertilizer, but the following methods can be taken to mitigate these problems when applying digestates and improve the overall nutrient use efficiency of using digestate.

The key factor when applying digestate, as with any form of fertilizer, is timing. Schwager et al. (2016) and Nicholson et al. (2017) measured a reduction in nitrogen losses when they applied in spring versus autumn application. Over-winter crop growth is limited and nitrogen demand is very low, therefore the nitrogen is vulnerable to leaching losses. As such, it is recommended to only apply liquid or whole digestate to an actively growing crop. However, the solid fraction can be used as a basal fertilizer before crop sowing, as the nutrients contained are mainly in organic forms that take time to mineralize into the bioavailable forms that are vulnerable to being leached. Further improvements for nitrogen use efficiencies of fertilizers in general can be attained by applying multiple smaller doses to match the different nitrogen demanding stages of the crop as it grows (Hu et al., 2021). While digestate can be applied in splits (Brychkova et al., 2024, Doyeni et al., 2021), the type of crop needs to be considered, as Wilson et al. (2021) recorded mechanical damage when applying digestate to maize at later crop growing stages.

A second key recommendation is to apply digestate at a nutrient rate that matches what the crop requires, to avoid surplus nutrients building up in the soils and being vulnerable to loss. This can be improved by variable rate application, which is a precision agriculture technique that applies fertilizer at different rates across a field, rather than a uniform rate, to account for heterogeneous nutrient concentrations across a field (Piccoli et al., 2023). However, applying digestate to meet crop nitrogen requirements can result in greater than plant-needed amounts of phosphorus or potassium being added to the soil. Therefore, careful monitoring of soil is required to ensure concentrations do not exceed crop optimal conditions. This may mean that digestate application needs to be mixed with mineral fertilizers to ensure only the correct amount of nutrients is added for what the crop requires.

When calculating the rate of nutrient to apply, a soil nutrient analysis is important to understand the nutrient balance in the field to determine how much fertilizer is needed to satisfy crop requirement. Further knowledge of soil properties is key to understanding how much nutrients are needed. For example, the Agriculture and Horticulture Development Board's (AHDB) nutrient management guide RB209 (2023) recommends less nitrogen for crops grown in sandy textured soils compared to clay or silt textured soils, as crop uptake efficiency of nitrogen is higher in sandy soils. The nutrient content of digestate is variable, due to the different organic materials that can be used to make it (Table 1; Risberg et al., 2017). Therefore, a nutrient analysis of the digestate that includes the concentration of bioavailable forms such as ammonium (Sharifi et al., 2019), prior to use is important to ensure the correct amount is applied to meet plant requirements. These measures of soil and digestate properties ensure that nutrients are applied at the required rate.

Surface run-off moves nutrients from the field into nearby watercourses. To avoid run-off, a guide by WRAP (2012) states that soil conditions, topography and upcoming weather need to be considered. Applying digestate to frozen, waterlogged, or snow-covered soil prevents digestate from infiltrating the soil, resulting in either immediate run-off or run-off during thaw. Conversely, applying to fields containing drains where the soil is deeply cracked causes digestate to preferentially flow into the drains and subsequently into waterbodies, rather than infiltrate the soil. Run-off is generated while applying during rainy weather or when heavy rain is forecast within the next few days following application. Lastly, digestate should not be applied to steep slopes as gravity will cause the liquid material to move down the slope, leaving the upper slope under-fertilized, with most of the digestate collecting either at the slope bottom or leaving the field.

While the previous methods improve the overall nutrient use efficiency of digestate, there are some nitrogen-specific issues that come with digestate application. The plant-available ammoniacal nitrogen in the digestate constitutes around 60–80% of the total nitrogen (Makádi, 2012). This nitrogen is easily volatilized, losing between 35–60% of the total applied nitrogen when surface applied (Tiwary et al., 2015), causing a deterioration in air quality and subsequent environmental and human health concerns (Nkoa, 2013). Most of the volatilized nitrogen occurs within 6 hours of application (Nicholson et al., 2017) and rapid incorporation following application significantly reduces ammonia loss.

As not all agricultural systems support tillage, low emission slurry spreading techniques can be used to reduce ammonia volatilization. Precision application, or low emission slurry spreading equipment (LESSE), minimizes the contact that digestate has with air compared to the traditional splash plate method. There are several LESSE systems: trailing hose, trailing shoe, and injection. Trailing hose, also known as dribble bar, is a boom set with hoses that distributes digestate close to the ground in narrow bands. Trailing shoe is similar, but the key difference is that the digestate is deposited via a metal 'shoe' at the end of the hoses, which are designed to apply bands of digestate on the soil surface below the crop canopy. Injectors place the digestate beneath the soil surface by cutting slits into the soil and depositing the digestate directly into the slits. These injectors can be either open or close slotted. The use of LESSE reduces ammonia losses between 40–77% (Nicholson et al., 2018, Riva et al., 2016). The ammonia reduction efficiency of this equipment is based on reducing contact between air and the digestate, as such injection reduces emissions more than trailing shoe (Nicholson et al., 2018). However, the choice of equipment is affected by soil conditions such as stone and clay content (Akoumianaki, 2022). While the key aim of LESSE is to reduce atmospheric nitrogen losses, it has also demonstrated a reduction in phosphorus losses from run-off (McConnell et al., 2013). This is attributed to the reduction in slurry-rainfall contact area, reducing the likelihood of phosphorus dissolution and solubilizing into the water run-off.

Another method to reduce ammonia volatilization is to acidify the digestate, since pH is a key driver of ammonia emissions. Digestate is inherently high in pH, usually between pH 7 and 8, whereas acidification usually targets a reduction to pH 5.5 (Fangueiro et al., 2015). Acidification can be achieved in one of two ways: addition of acid or plasma treatment. Strong acids, mainly sulfuric, are added directly to the digestate, and have demonstrated reductions of nitrogen loss via ammonia emissions between 35–95% (Eihe et al., 2019, Sánchez-Rodríguez et al., 2018, Wagner et al., 2021). Plasma treatment is a technology developed by N2 Applied, which uses air and electricity to generate an air plasma where oxygen and nitrogen combine to form a reactive nitrogen gas. This gas is subsequently absorbed in the slurry as nitrate and nitrite, which lowers the pH of the digestate, while also doubling its readily available N content compared to non-acidified (N2 Applied, 2023). Plasma-treated digestate reduced ammonia emissions from field application by between 27–57% (Pedersen et al., 2024, Rollett et al., 2025).

Once in the soil, ammonium is transformed by nitrifying bacteria into nitrite and then nitrate (Figure 4). Nitrate is used as an energy source by denitrifying bacteria and transformed into N_2 and N_2O, the latter of which is a potent greenhouse gas with a global warming potential 273 times greater than that of CO_2 (IPCC, 2021). Nitrate itself is highly mobile in the soil and vulnerable to being leached. To prevent these gaseous and leaching losses, chemicals or physical materials can be added to the digestate. Nitrification inhibitors are chemical compounds that suppress the activity of nitrifying bacteria, thereby slowing the conversion of ammonium into nitrate. Ammonium is less mobile in the soil compared to nitrate, as it has a greater affinity for sorption to soil constituents, which reduces its vulnerability to being leached. The addition of nitrification inhibitors to digestate has been shown to reduce N_2O emissions (Kesenheimer et al., 2021, Sánchez-Rodríguez et al., 2018, ten Huf and Olfs, 2020, Severin et al., 2016) and nitrate leaching (Giacometti et al., 2020).

Physical materials that can be added to digestate to reduce nitrogen losses include biochar, glycerol and zeolites. Biochar is a carbon-rich material made from organic biomass waste by pyrolysis under oxygen-free or limited conditions. Biochar has a highly porous structure and an array of functional groups with a strong affinity for inorganic ions, which causes it to absorb and fix nutrients (Ding et al., 2016). Applying digestate with biochar in mixtures between 1:1 and 120:1 units of digestate per unit of biochar has been shown to reduce N_2O emissions (Dicke et al., 2015, Martin et al., 2015) and nitrate leaching (Bian et al., 2023, Plaimart et al., 2021). However, the effectiveness of biochar as a nutrient absorbent and its subsequent impact on nitrogen loss pathways is influenced by feedstock and pyrolysis temperature (Gai et al., 2014). Additionally, information on guidance for application methods and rates is currently unknown and requires comprehensive field research. These questions include the rate of biochar to apply, whether biochar should be mixed into the digestate before application, or applied separately to the soil before digestate application. In the latter case, the frequency of application is unknown for the biochar to effectively reduce leaching and N_2O emissions from digestate application.

Glycerol is a viscous liquid by-product of the biodiesel industry. When added to soil, it increases soil microbial biomass (De et al., 2022). As microbes require nitrogen to build cellular compounds, they immobilize nitrogen supplied by the digestate in their biomass. Adding glycerol with digestate has the potential to reduce nitrate losses within the first month following application (van Midden et al., 2025). It is worth mentioning that mixing carbon-rich materials with digestate to reduce nitrogen losses could also reduce other nutrient losses, but this has not yet been explored. Lastly, zeolites, an aluminosilicate material, also has the potential to reduce nitrogen losses due to their highly porous and high adsorption properties (Cataldo et al., 2021, Jakkula and Wani, 2018), however, research is required to test their effectiveness with digestate. The research area of adding these materials into digestate to reduce nitrogen losses is in the early stages. Calculations are still required to work out the optimal amounts of material to add with digestate to be of benefit, considering crop demand, soil type and climate.

Methods mentioned previously in this chapter have focused on reducing N_2O emissions by slowing the rate of ammonium nitrogen from the digestate that is available for nitrifying bacteria to convert into nitrate. This mechanism allows plants to uptake more of the nitrate over their growing season, as they cannot utilize all of it at the time of application. Therefore, there is less nitrate available for denitrifying bacteria to convert into N_2O. Jonassen et al. (2022a) propose an alternative mechanism to reduce N_2O emissions by bioengineering the soil microbiome. By enriching the microbial community with non-denitrifying N_2O-respiring bacteria (NNRB) containing the *nosZ* gene, which encodes for the enzyme nitrous oxide reductase, N_2O emissions from soil could be reduced by up to 60% (Hiis et al., 2024). Through a series of laboratory experiments, Jonassen et al. (2022a), Jonassen et al. (2022b), Wang et al. (2023a) and Hiis et al. (2024) have demonstrated proofs of concept to show that digestate is both a suitable substrate to cultivate these microorganisms and a vector to inoculate soil. However, further work is needed to select microorganisms that are both strong sinks for N_2O and can survive and compete in soil in the long term to ensure continued suppression of N_2O emissions.

The high ammoniacal nitrogen content of the digestate can be phytotoxic to plants by impairing germination or damaging plant tissue when the digestate comes into contact with the plant (Lencioni et al., 2016). This is detrimental to the successful establishment and growth of emerging seedlings. When applying digestate at field scale, phytotoxicity can be avoided by applying digestate several days prior to sowing or planting out to avoid digestate coming into contact with germinating seeds or young plants (Alburquerque et al., 2012b), or by band spreading between crop rows, however no study has been found to confirm this. When used as a substrate in growing media, the amount of solid digestate used in the mix is limited by its ammonium content (Caspersen et al., 2023). Composting is a method that can reduce the acute phytotoxic issue and produce a product that slowly releases nitrogen and other nutrients over time. Composting is the degradation of organic matter by microorganisms under aerobic conditions, although to make a more nutrient-rich product, earthworms (Hanc and Vasak, 2014, Krishnasamy et al., 2014) or insects (Cui et al., 2023) can be used. In order to be composted effectively, additional materials such as wood chips, corn stalks, biochar or oyster shells as bulking agents are required (Li et al., 2020, Lu et al., 2020, Weldon et al., 2023, Zeng et al., 2016). Usually, only the solid fraction is composted, but the liquid fraction can be used to water compost piles (Bustamante et al., 2012, Vu et al., 2015a). Compost is more biologically stable than the solid digestate feedstock (Tambone et al., 2015), reducing the phytotoxicity of digestate (Gurmessa et al., 2021, Inbar et al., 1985) and creating a material that is more suitable to use as growing media (Bustamante et al., 2012).

5 Nutrient recovery from digestate

There are circumstances where digestate supply exceeds the amount that can be safely applied to land, regardless of techniques to reduce nutrient pollution pathways. In EU countries, the Nitrate Directive (91/676/EEC) sets a maximum application limit of 250 kg N per ha. This limit is a particular challenge in regions where livestock farms are heavily concentrated, or where there are landless farms. Landless farms are industrial animal units where animal feed is not grown on land surrounding where the animals are housed. If these systems have an integrated anaerobic digester, some or all of the digestate produced would need to be exported. Nutrient recovery technologies can condense nutrients to make materials that are less bulky than digestate and consequently cheaper to store and transport, allowing it to be exported to greater distances. Additionally, these recovered nutrients can be processed into products that are compatible with fertilizer spreaders. This section will provide a brief overview of the main nutrient recovery options and their effectiveness. However, it will not go into detail on how each product affects soil properties and crop performance, as they are no longer classed as digestate and therefore beyond the scope of this chapter.

There are multiple technologies to recover nutrients that have been extensively reviewed (Arellano-Yasaca and Chu, 2025, Barampouti et al., 2020, Shi et al., 2018, Sobhi et al., 2022, Vaneeckhaute et al., 2016, Zielińska and Bułkowska, 2024). Nutrient recovery from liquid digestate can be achieved by vacuum evaporation and membrane filtration to create a liquid concentrate. Ammonia stripping is a process that recovers captured volatilized ammonia by acidic substances such as sulfuric or nitric acid. Ion exchange and struvite precipitation use materials such as biochar, zeolite, resins and clays (for ion exchange), or magnesium and calcium (for struvite), to adsorb and remove nutrients from the digestate. The solid fraction can be converted into pellets to make a P-rich product (Valentinuzzi et al., 2020) or combined with nitrogen to make a NP-rich pellet (Burak and Sakrabani, 2023). Chapter 4 of this book gives further information on organo-mineral fertilizers. Alternatively, the solid fraction can be pyrolyzed (decomposition of organic matter at high temperatures under oxygen-free or limited conditions) to make biochar or incinerated and phosphorus recovered from the ashes. Chapter 3 of this book has more detail on biochar as a soil amendment.

However, many of these techniques are not yet cost-effective, due in part to operating costs, the infrastructure required and implementing with current systems. Additionally, the methods themselves require optimization and scaling up to industry-wide application. For example, membrane technology is vulnerable to fouling and the regeneration frequency of ion-exchange materials is uncertain (Arellano-Yasaca and Chu, 2025, Sobhi et al., 2022). Processes of evaporation, ammonia stripping, and struvite precipitation require reagents to alter the chemical properties of the digestate, such as pH, to ensure the greatest recovery of nutrients occurs, adding to the overall cost of the process (Arellano-Yasaca and Chu, 2025). Ammonia stripping and struvite precipitation are relatively widely established, but work is still required to optimize the end-use product. The ammonia sulfate from stripping is very acidic and requires neutralizing to avoid damaging the crop (Vaneeckhaute et al., 2013), while precipitated struvite is powdery and should be processed into prills or pellets for improved handling (Latifian et al., 2012). Despite these limitations, digestate has great potential as a source of recovered nutrients for agricultural use, which can reduce reliance on inorganic fertilizers. However, there remain challenges that have to be overcome for full-scale implementation of the multiple nutrient recovery technologies described.

6 Effect of anaerobic digestate on soil organic matter

Soil organic matter (SOM) is the sum of all living or once-living materials within the soil, of which approximately 58% is organic carbon (SOC). Soil microorganisms comprise between 1–5% of total soil organic carbon (Jenkinson and Ladd, 1981), but are key to biochemical functioning. SOM has a vital role in regulating a range of soil functions such as nutrient availability, water infiltration and storage, and carbon sequestration. Inputs of organic materials are important to maintain SOM and sustain agricultural productivity, as concentrations in agricultural soils can be low due to factors such as the removal of plant biomass from fields. The effect of an organic material on SOM is influenced by both the quantity and the quality of the organic carbon contained within it (Bhogal et al., 2018).

Digestate contains partially decomposed organic matter, containing a higher concentration of more chemically recalcitrant aliphatic molecules compared to its undigested feedstock (Tambone et al., 2013). Due to this, the authors suggested that digestate could be a better material to sequester carbon in soil compared to undigested material. Furthermore, the remaining proportion of labile carbon in the digestate can be rapidly assimilated by soil microorganisms and converted into necromass, a more stable form of carbon, as the nutrients supplied in the digestate stimulate microbial growth (Reuland et al., 2022). However, the total amount of organic matter supplied by the digestate is influenced by the feedstock and can be between 38–77% of the dry matter (Barampouti et al., 2020). Furthermore, the fraction of digestate impacts the amount of organic matter that it will add to the soil. The liquid fraction typically has a dry matter of less than 4%, the unseparated digestate between 3–10%, and the solid fraction greater than 20%.

Repeated applications of whole or the liquid fraction of digestate between 2 and 6 years have no significant impact on either SOC (Barłóg et al., 2020,

Odlare et al., 2008, Pastorelli et al., 2021, Šimon et al., 2015, Mayerová et al., 2023, Giannini et al., 2023, Jurgutis et al., 2021a) or SOM (Bhogal et al., 2018). A 13-year experiment showed a SOC increase of 11% under digestate fertilization compared to without (Levin et al., 2021). A 10-year study showed that SOC concentrations increased between 0.06–0.2% in 68 farms in the Czech Republic that regularly applied digestate, but that the ability of the soil to sequester the carbon provided by digestate was influenced by soil type and texture (Voltr et al., 2021). In contrast, two separate 10-year field studies measured no effect of annual digestate applications on SOM (Rittl et al., 2023) or SOC (Mora-Salguero et al., 2025). This highlights that the contributions to SOM from whole or liquid digestate are inconsistent within a decade of repeated applications, but increases can be expected over longer timescales. Voltr et al. (2021) concluded that digestate can help build SOM when it is co-applied with additional organic matter, for example, animal manure. This is supported by Villarino et al. (2025) who assessed the impact of applying solid and liquid digestate on 14 fields over 5–12 years and measured an increase in SOC stocks.

Regarding the solid fraction, very few multi-year field experiments are available. The repeated application of solid digestate resulted in a significant increase in SOC (Badagliacca et al., 2020), and SOM (Ivezić et al., 2021) or an increasing (though non-significant) trend in SOC (Jurgutis et al., 2021a) over 2–3 years. However, it is worth noting that changes in SOC concentrations can take between 6 and 10 years to be detected under various rates of carbon inputs, land uses, and soil types (Smith, 2004). Changes in soil microbial carbon can be used as an early indicator for longer-term changes. A review of digestate impact on soil microorganisms found that the solid fraction consistently had a positive influence on microbial biomass within a year of application (van Midden et al., 2023). This shows the potential of the solid fraction as an amendment for increasing soil organic matter within soil.

Soil amendments are required to meet particular criteria set by national or federation governing bodies. A characterization of solid digestates from multiple feedstocks showed that the solid fraction does not meet the standards for a soil amendment, due to high moisture and low organic matter contents (Teglia et al., 2011). Furthermore, applying the solid fraction to a carbon rate may oversupply nutrients as the solid digestate still contains a considerable amount of bioavailable nutrients. Therefore, treatments that result in a moisture content reduction and convert bioavailable nutrients to organic forms, such as air-drying or composting, can be done to improve the soil amending properties of the solid digestate. However, more and longer-term field studies covering a range of climate conditions, soil types, farming systems, digestate feedstocks, application methods and rates, and post-processing technologies are needed to gain a more accurate understanding of how solid digestate application affects SOM dynamics in agricultural soils.

7 Contaminants in anaerobic digestate

While digestate has good fertilizer properties, there are some potential hazards that need to be considered in addition to issues of nutrient pollution mentioned previously. The application of digestate, as with any organic material (e.g. animal manure, compost, biosolids), to soil can result in soil contamination by physical, chemical, and biological means. Physical contamination in digestates made from mixed organic wastes or garden wastes includes materials such as glass, ceramics, metals, stones, wood, paper and plastics (Logan and Visvanathan, 2019, O'Connor et al., 2022), which can accumulate in soils. Plastics can harm life in the soil through the release of toxic substances. Due to their size, microplastics (5 mm - 1 μm) and nanoplastics (< 1 μm) can be directly taken up by soil organisms and plants. The negative effects of plastics on soil and plant life include stunted growth, reduced enzymatic activity and altered microbial community structures, which consequently impact nutrient and carbon cycles (Azeem et al., 2021, Gorde et al., 2024). The presence of impurities in digestate can lead to negative public perception and a reluctance by farmers to utilize it.

Chemical contaminants include heavy metals, salts and organic pollutants. Studies using digestates sourced from sewage, industrial and urban wastes have shown that the resultant digestate can contain high levels of copper, cadmium, nickel, lead and zinc above those set by the relevant governing bodies as acceptable for land application (Bonetta et al., 2014, Coelho et al., 2018, Govasmark et al., 2011), while digestates from manures contain high contents of copper and zinc (Alburquerque et al., 2012b). While these micronutrients are beneficial to plant growth, when in excess they can reduce crop yield by interfering with the metabolic activities of plants and reducing soil microbial functioning (Angon et al., 2024, Singh and Kalamdhad, 2011).

Digestates, particularly those from food wastes, contain high sodium contents (Ries et al., 2023). While sodium is a plant-required nutrient, excessive sodium concentrations damages plants (Kronzucker et al., 2013). Furthermore, soil sodicity, as a result of sodium accumulating in soils, causes soil structural deterioration, reduces water infiltration, and negatively impacts soil biology and subsequently soil biogeochemical cycles (Mohanavelu et al., 2021). The long term application of digestates containing high concentrations of sodium is of particular risk to soils with a naturally high level of sodium (Pawlett and Tibbett, 2015).

Organic pollutants in digestates can include antibiotics, hormones, pesticides, pharmaceuticals and personal care products, phenols, and persistent organic compounds such as PCBs, PAHs, PFAS, phthalates, and dioxin-like compounds (van Midden et al., 2023, O'Connor et al., 2022). These organic pollutants can have an ecotoxicological effect on life in the soil. A few of the effects are changes in the structure and functioning of soil microbial communities, plant growth inhibition, and bioaccumulation in the food web (Cycoń et al., 2019, Gworek et al., 2021, Stefan et al., 2024, Steiner et al., 2024).

Biological contamination of anaerobic digestate is the presence of pathogens, weed seeds, and mycotoxins (O'Connor et al., 2022, Nkoa, 2013, Guan et al., 2024). These pose a risk to both animal and human health and are of particular concern for biogas facilities that supply digestate to multiple farms. Pathogen content can be influenced by digestate processing, with thermophilic digestion reducing pathogen load more so than mesophilic (Jiang et al., 2020). Additional pasteurization of the digestate further sanitizes the digestate (Nag et al., 2019).

The contaminants from digestates can have detrimental effects on soil, plant, animal and human health. However, information regarding the risk of these contaminants in digestate poses to the environment is scarce. Much of the research studying the impacts of these contaminants is from lab-based incubation experiments, which often explore the impacts of these contaminants applied at a higher rate than would be found in the digestate. However, they do provide an indication of what could occur due to the buildup of these contaminants in the soil. This is valuable, as information regarding the cumulative effect of repeated digestate applications on the concentration of these contaminants in the soil and their subsequent impact is even more scarce.

A process to improve the safety of digestate is to remove or reduce the concentration of the contaminants, of which there are a wide range of techniques being studied to do so. These include using biochar (Beljin et al., 2024, Liu et al., 2023) to absorb and remove heavy metals and organic pollutants, advanced oxidation processes to break down organic pollutants (Moradi et al., 2023), microbial cultures to biodegrade compounds (Mateescu et al., 2024), or converting digestate into hydrochar (Wang et al., 2025) or biochar (Wang et al., 2023b) to sanitize the digestate, immobilize heavy metals and reduce organic pollutant contents. However, many of these technologies require further research to optimize the process, scale up from laboratory research to industry level, and would need to be integrated into current systems. Another way to minimize risk is to use digestate that meets the requirements of a quality assurance scheme, such as the EU's ECN–QAS or national certification systems (Al Seadi and Lukehurst, 2012). These schemes and certification systems set a standard for which digestates must pass in order to be considered safe to apply based on current research knowledge.

8 Future trends in research

Considerable research has been conducted to reduce the negative environmental consequences that can occur from nutrient pollution when using digestate, in particular nitrogen. Yet most studies focus on reducing nitrogen loss from one pathway, such as ammonia volatilization, but there are concerns of pollution swapping. For example, injecting digestate directly into the soil is more effective at reducing ammonia volatilization compared to surface band application (Nicholson et al., 2018), but it can result in greater leaching losses (Cameira et al., 2019). Therefore, further research is needed to understand the impacts of ammonia reduction techniques on the non-target nitrogen loss routes. It may be that ammonia reduction techniques need to be used in combination with methods that reduce leaching or N_2O emissions, such as nitrification inhibitors, biochar, or glycerol, but this requires more research to determine. Finally, the impact of these methods on non-target nutrients also needs evaluation.

The risks of non-nutrient pollutants in digestate, as described in the previous section, are not well studied in terms of field experiments. There remains a lack of long-term studies to understand the effect of repeated digestate applications on the accumulation of pollutants that could result in food safety and environmental quality concerns. While this is a risk of applying any material of organic waste origin, the information from this research work is key to developing and updating guidelines and quality assurance schemes to ensure the sustainable and safe use of digestate in agriculture.

Long-term experiments are also necessary to understand the impact of digestate on soil organic matter. A few such studies were found and indicated that the solid fraction has the potential to be used as a soil conditioner. However, more research is required to understand the effect of solid digestate on soil organic matter, and with studies that run longer than 3 years. Furthermore, an understanding of how repeated digestate impacts soil biology, from micro- to macro-organisms, is missing, as highlighted in reviews by Karimi et al. (2022) and van Midden et al. (2023). Biological diversity is of vital importance to soil nutrient and carbon cycles, and therefore the sustainable functioning and productivity of agricultural soils. Long-term studies are a much-needed resource to answer these questions. Karimi et al. (2022) also suggested developing farm networks to identify the impact of digestate over a variety of soil types and farming systems to enable robust conclusions to be drawn.

Regarding the interaction of digestate in soil, the influence of soil properties needs to be accounted for. There is a lack of multi-field studies available to look at the impact of soil properties on the dynamics of digestate fertilization on crops or the accumulation of nutrients, pollutants, and carbon in the soil. A few pot-scale experiments have compared different soil types and measured differences in the impact of digestate application on soil chemical and biological properties (Abubaker et al., 2013, Makádi et al., 2016, Panuccio et al., 2021). Additional studies comparing the effects of different soils would provide valuable knowledge toward identifying the amount of digestate that needs to be applied to ensure crop productivity.

Climate is also a consideration when it comes to understanding the dynamics of digestate when it is applied to soil. Most of the studies found and reviewed in this chapter were based on research conducted in the Global North. The use of small-scale digesters is being promoted in the developing world as a way to deal with organic wastes, with the benefit of generating fuel for cooking (Kinyua et al., 2016, Vu et al., 2015b), therefore reducing pressures on resources such as firewood. Yet there is a lack of research on the fertilizer replacement value of the digestate and how best to utilize digestate in those regions, particularly in tropical and subtropical climates. Research in this area is needed to inform best practice that is relevant to local conditions.

9 Conclusion

Anaerobic digestate is a nutrient dense slurry that is produced during biogas generation. Digestate is commonly separated into liquid and solid fractions, with the solid fraction containing higher organic matter contents than the liquid. As such, solid digestate has the potential to build soil organic matter and sequester carbon in the soil. Digestate in all its forms can be used as a fertilizer due to its high content of plant-available macro- to micro- nutrients. The liquid fraction and unseparated whole digestates can be used to partially or wholly substitute inorganic fertilizers without reducing crop yields. The solid fraction can be used to build soil fertility or be used in a substrate mixture for horticultural use.

The nutrient concentration of digestate is variable, due to the various organic materials that can be used as feedstock. Therefore, a nutrient characterization of the digestate prior to use as a fertilizer is recommended. The nutrient use efficiency from digestate can be optimized by following recommended best practices: using the right equipment (e.g. low emission slurry spreading equipment), at the right time and conditions (e.g. when the crop is actively growing while avoiding saturated or frozen soils), at the right rate of nutrient that the crop needs (include results from a soil nutrient analysis to incorporate field nutrient supply and knowledge of soil type), and in the right place (e.g. avoid steep slopes, place digestate close to plant roots). Further methods to improve the nutrient use efficiency of digestate are to modify it. Acidification reduces ammonia volatilization losses when the digestate is on the soil surface, while materials such as biochar and glycerol can reduce leaching and gaseous losses once the digestate is in the soil.

Due to the nature of the feedstock, digestate can contain materials that are toxic to soil, plant, animal and human health. Setting and adhering to quality assurance schemes can reduce these risks. While digestate is a relatively new organic material compared to animal manures or compost to be used in agriculture, it demonstrates great potential to be used as fertilizer. Yet there are questions related to its long-term use that still need to be addressed, in particular the effect of repeated applications on soil organic carbon and the accumulation of hazardous compounds in the soils. As such, two pathways can be suggested for the use of digestates. Digestates derived from feedstocks considered safe can be considered appropriate for application on agricultural land. In contrast, digestates derived from feedstocks with potential contaminants should not be applied to agricultural land until sufficient research and evidence demonstrate their safety.

Further research is also required to determine the effectiveness of material addition to digestate to reduce nitrogen losses, particularly in combination with methods to reduce ammonia volatilization. Yet the work already done, as identified in this chapter, does provide information on how to manage digestate responsibly. By following the recommendations output of these research studies, it will improve the sustainability of utilizing it as a source of nutrients and organic matter in agriculture.

10 Where to look for further information

10.1 Further reading

A comprehensive introduction to biogas and digestate found in the *Biogas Handbook* by Teodorita Al Seadi, Dominik Rutz, Heinz Prassl, Michael Köttner, Tobias Finsterwalder, Silke Volk and Rainer Janssen is available at: www.lemvigbiogas.com/BiogasHandbook.pdf.

A white paper published by the European Biogas Association (EBA) on 'exploring the potential of digestate in fostering healthy soils and advancing sustainable agricultural practices across Europe' can be downloaded here: www.europeanbiogas.eu/exploring-digestates-contribution-to-healthy-soils/.

Reports from field trials to provide a good practice guide for digestate application commissioned by the Waste & Resources Action Programme (WRAP) can be found here: www.wrap.ngo/resources/report/digestate-and-compost-agriculture-dc-agri-project-reports.

The *Biogas Handbook* (2013) edited by Arthur Wellinger, Jerry Murphy, and David Baxter contains chapters of interest regarding the 'land application of digestate' and 'biogas digestate quality and utilization.' doi.org/10.1533/9780857097415.

10.2 Key organizations/conferences

WBA (World Biogas Association): brings together professionals from the industry (www.worldbiogasassociation.org). It is also worth checking if there are biogas associations at relevant national (e.g. the Anaerobic Digestion and Bioresources Association, UK) or regional (e.g. the European Biogas Association) level.

IUSS (International Union of Soil Sciences) hosts a range of events that include sustainable soil management in agriculture (https://www.iuss.org/).

IFS (International Fertilizer Society) hosts events for professionals interested in any aspect of the fertilizer market, including organic sources (https://fertiliser-society.org/).

RAMIRAN (Recycling of Agricultural, Municipal and Industrial Residues in Agricultural Network) conference in Europe.

ASA/CSSA/SSSA conference brings together sessions on agricultural research in the US.

11 Acknowledgements/declaration of interests

The author of this chapter declares that no funding was received related to this chapter and has no conflicts to declare.

12 References

Abubaker, J., et al. 2013. Bacterial community structure and microbial activity in different soils amended with biogas residues and cattle slurry. *Applied Soil Ecology,* 72, 171–180. https://doi.org/10.1016/j.apsoil.2013.07.002.

Aguirre-Villegas, H. A., et al. 2019. Anaerobic digestion, solid-liquid separation, and drying of dairy manure: measuring constituents and modeling emission. *Science of the Total Environment,* 696, 134059. https://doi.org/10.1016/j.scitotenv.2019.134059.

Akoumianaki, I. 2022. *Moving to more sustainable methods of slurry application: implications for water quality of waterbodies and water protected areas*. Centre of Expertise for Waters.

Al Seadi, T. & Lukehurst, C. 2012. *Quality management of digestate from biogas plants used as fertiliser*. IEA Bioenergy Task 37 Publication.

Al Seadi, T., et al. 2008. *Biogas handbook*. University of Southern Denmark.

Alburquerque, J. A., et al. 2012a. Agricultural use of digestate for horticultural crop production and improvement of soil properties. *European Journal of Agronomy,* 43, 119–128. https://doi.org/10.1016/j.eja.2012.06.001.

Alburquerque, J. A., et al. 2012b. Assessment of the fertiliser potential of digestates from farm and agroindustrial residues. *Biomass and Bioenergy,* 40, 181–189. https://doi.org/10.1016/j.biombioe.2012.02.018.

Angon, P. B., et al. 2024. Sources, effects and present perspectives of heavy metals contamination: Soil, plants and human food chain. *Heliyon,* 10, 28357. https://doi.org/10.1016/j.heliyon.2024.e28357.

Arellano-Yasaca, D. V. & Chu, C. Y. 2025. Insights into nutrients recovery from food waste liquid Digestate: a critical review and systematic analysis. *Waste Manag,* 200, 114743. https://doi.org/10.1016/j.wasman.2025.114743.

Asp, H. & Bergstrand, K.-J. 2024. Nutrient utilization and growth of tomato crops fertilized with solid anaerobic digestate. *Scientia Horticulturae,* 329, 112986. https://doi.org/10.1016/j.scienta.2024.112986.

Azeem, I., et al. 2021. Uptake and accumulation of nano/microplastics in plants: a critical review. *Nanomaterials (Basel)*, 11, 2935. https://doi.org/10.3390/nano11112935.

Badagliacca, G., et al. 2020. Use of solid anaerobic digestate and no-tillage practice for restoring the fertility status of two Mediterranean orchard soils with contrasting properties. *Agriculture, Ecosystems & Environment*, 300. https://doi.org/10.1016/j.agee.2020.107010.

Barampouti, E. M., et al. 2020. Exploring technological alternatives of nutrient recovery from digestate as a secondary resource. *Renewable and Sustainable Energy Reviews*, 134, 110379. https://doi.org/10.1016/j.rser.2020.110379.

Bareha, Y., et al. 2018. Characterization and prediction of organic nitrogen biodegradability during anaerobic digestion: a bioaccessibility approach. *Bioresource Technology*, 263, 425–436. https://doi.org/10.1016/j.biortech.2018.04.085.

Barłóg, P., et al. 2020. Effect of digestate on soil organic carbon and plant-available nutrient content compared to cattle slurry and mineral fertilization. *Agronomy*, 10, 379. https://doi.org/10.3390/agronomy10030379.

Barzee, T. J., et al. 2019. Digestate biofertilizers support similar or higher tomato yields and quality than mineral fertilizer in a subsurface drip fertigation system. *Frontiers in Sustainable Food Systems*, 3, 56. https://doi.org/10.3389/fsufs.2019.00058.

Beljin, J., et al. 2024. Biochar in the remediation of organic pollutants in water: a review of polycyclic aromatic hydrocarbon and pesticide removal. *Nanomaterials (Basel)*, 15. https://doi.org/10.3390/nano15010026.

Bhogal, A., et al. 2018. Improvements in the quality of agricultural soils following organic material additions depend on both the quantity and quality of the materials applied. *Frontiers in Sustainable Food Systems*, 2, 9. https://doi.org/10.3389/fsufs.2018.00009.

Bian, X., et al. 2023. Biochar-enhanced agricultural application of liquid digestate from food waste anaerobic digestion for celery cultivation. *Science of the Total Environment*, 869, 161562. https://doi.org/10.1016/j.scitotenv.2023.161562.

Bonetta, S., et al. 2014. Agricultural reuse of the digestate from anaerobic co-digestion of organic waste: microbiological contamination, metal hazards and fertilizing performance. *Water, Air, & Soil Pollution*, 225, 2046. https://doi.org/10.1007/s11270-014-2046-2.

Brychkova, G., et al. 2024. Use of anaerobic digestate to substitute inorganic fertilisers for more sustainable nitrogen cycling. *Journal of Cleaner Production*, 446, 141016. https://doi.org/10.1016/j.jclepro.2024.141016.

Buligon, E. L., et al. 2023. Fertilizer performance of a digestate from swine wastewater as synthetic nitrogen substitute in maize cultivation: physiological growth and yield responses. *Agriculture*, 13, 565. https://doi.org/10.3390/agriculture13030565.

Burak, E. & Sakrabani, R. 2023. Novel carbon capture-based organo-mineral fertilisers show comparable yields and impacts on soil health to mineral fertiliser across two cereal crop field trials in Eastern England. *Field Crops Research*, 302. https://doi.org/10.1016/j.fcr.2023.109043.

Bustamante, M. A., et al. 2012. Co-composting of the solid fraction of anaerobic digestates, to obtain added-value materials for use in agriculture. *Biomass and Bioenergy*, 43, 26–35. https://doi.org/10.1016/j.biombioe.2012.04.010.

Cameira, M. D. R., et al. 2019. Band application of acidified slurry as an alternative to slurry injection in Mediterranean winter conditions: Impact on nitrate leaching. *Soil and Tillage Research*, 187, 172–181. https://doi.org/10.1016/j.still.2018.12.001.

Caspersen, S., et al. 2023. Nutrient challenges with solid-phase anaerobic digestate as a peat substitute - Storage decreased ammonium toxicity but increased phosphorus availability. *Waste Management,* 165, 128–139. https://doi.org/10.1016/j.wasman.2023.04.032.

Cataldo, E., et al. 2021. Application of zeolites in agriculture and other potential uses: a review. *Agronomy,* 11, 1547. https://doi.org/10.3390/agronomy11081547.

Cavalli, D., et al. 2016. Nitrogen fertilizer replacement value of undigested liquid cattle manure and digestates. *European Journal of Agronomy,* 73, 34–41. https://doi.org/10.1016/j.eja.2015.10.007.

Coelho, J. J., et al. 2018. Physical-chemical traits, phytotoxicity and pathogen detection in liquid anaerobic digestates. *Waste Management,* 78, 8–15. https://doi.org/10.1016/j.wasman.2018.05.017.

Crippa, L., et al. 2013. Utilization of the solid fraction of digestate from anaerobic digestion as container media substrate. *Acta Horticulturae* 1013, 367–373. https://doi.org/10.17660/ActaHortic.2013.1013.45.

Cui, G., et al. 2023. Feasibility of housefly larvae-mediated vermicomposting for recycling food waste added digestate as additive. *Journal of Environmental Sciences,* 128, 150–160. https://doi.org/10.1016/j.jes.2022.07.016.

Cycoń, M., et al. 2019. Antibiotics in the soil environment-degradation and their impact on microbial activity and diversity. *Frontiers in Microbiology,* 10, 338. https://doi.org/10.3389/fmicb.2019.00338.

Czekała, W. 2022. Digestate as a source of nutrients: nitrogen and its fractions. *Water,* 14, 4067. https://doi.org/10.3390/w14244067.

DAERA, (Department of Agriculture, Environment and Rural Affairs), Northern Ireland. 2019. Nutrient Action Programme. Available at: https://www.daera-ni.gov.uk/articles/nutrients-action-programme#toc-0 [Accessed 23/09/2025].

De, M., et al. 2022. Crude glycerol, a biodiesel byproduct, used as a soil amendment to temporarily immobilise and then release nitrogen. *European Journal of Soil Science,* 73. https://doi.org/10.1111/ejss.13241.

De Gioannis, G., et al. 2017. Energy recovery from one- and two-stage anaerobic digestion of food waste. *Waste Management,* 68, 595–602. https://doi.org/10.1016/j.wasman.2017.06.013.

Dicke, C., et al. 2015. Effects of different biochars and digestate on N2O fluxes under field conditions. *Science of the Total Environment,* 524–525, 310–318. https://doi.org/10.1016/j.scitotenv.2015.04.005.

Ding, Y., et al. 2016. Biochar to improve soil fertility: a review. *Agronomy for Sustainable Development,* 36. https://doi.org/10.1007/s13593-016-0372-z.

Doyeni, M. O., et al. 2021. The effectiveness of digestate use for fertilization in an agricultural cropping system. *Plants (Basel),* 10, 1734. https://doi.org/10.3390/plants10081734.

Du, H., et al. 2019. Effects of digested biogas slurry applicationmixed with irrigation water on nitrate leaching during wheat-maize rotation in the North China Plain. *Agricultural Water Management,* 213, 882–893. https://doi.org/10.1016/j.agwat.2018.12.012.

Dubský, M., et al. 2019. Use of solid phase of digestate for production of growing horticultural substrates. *Horticultural Science,* 46, 34–42. https://doi.org/10.17221/221/2016-hortsci.

Eihe, P., et al. 2019. The effect of acidification of pig slurry digestate applied on winter rapeseed on the ammonia emission reduction.
Elbashier, M. M. A., et al. 2018. Effects of anaerobic digestate on Chinese Melon (Cucumis melo L.) yield components, soil properties, and microbial communities under saline irrigation condition. *Communications in Soil Science and Plant Analysis,* 49, 2446–2455. https://doi.org/10.1080/00103624.2018.1510954.
Fagerström, A., et al., 2018.The role of Anaerobic Digestion and Biogas in the Circular Economy, IEABioenergy.
Fangueiro, D., et al. 2015. Acidification of animal slurry–a review. *Journal of Environmental Management,* 149, 46–56. https://doi.org/10.1016/j.jenvman.2014.10.001.
Fuchs, W. & Drosg, B. 2013. Assessment of the state of the art of technologies for the processing of digestate residue from anaerobic digesters. *Water Science and Technology,* 67, 1984–1993. https://doi.org/10.2166/wst.2013.075.
Gai, X., et al. 2014. Effects of feedstock and pyrolysis temperature on biochar adsorption of ammonium and nitrate. *PLoS One,* 9, e113888. https://doi.org/10.1371/journal.pone.0113888.
Gebreeyessus, G. D. & Jenicek, P. 2016. Thermophilic versus mesophilic anaerobic digestion of sewage sludge: a comparative review. *Bioengineering (Basel),* 3, 15. https://doi.org/10.3390/bioengineering3020015.
Giacometti, C., et al. 2020. A nitrification inhibitor, nitrapyrin, reduces potential nitrate leaching through soil columns treated with animal slurries and anaerobic digestate. *Agronomy,* 10, 865. https://doi.org/10.3390/agronomy10060865.
Giannini, V., et al. 2023. Agronomic management strategies to increase soil organic carbon in the short-term: evidence from on-farm experimentation in the Veneto region. *Plant and Soil,* 491, 561–574. https://doi.org/10.1007/s11104-023-06135-y.
Gorde, N. B., et al. 2024. The impact of microplastics on soil ecosystems: a review. *International Journal of Research in Agronomy,* 7, 544–550. https://doi.org/10.33545/2618060X.2024.v7.i12g.2208.
Govasmark, E., et al. 2011. Chemical and microbiological hazards associated with recycling of anaerobic digested residue intended for agricultural use. *Waste Management,* 31, 2577–83. https://doi.org/10.1016/j.wasman.2011.07.025.
Grillo, F., et al. 2021. Agro-environmental sustainability of anaerobic digestate fractions in intensive cropping systems: insights regarding the nitrogen use efficiency and crop performance. *Agronomy,* 11. https://doi.org/10.3390/agronomy11040745.
Guan, D., et al. 2024. A critical review on sustainable management and resource utilization of digestate. *Process Safety and Environmental Protection,* 183, 339–354. https://doi.org/10.1016/j.psep.2024.01.029.
Guilayn, F., et al. 2019. Digestate mechanical separation: efficiency profiles based on anaerobic digestion feedstock and equipment choice. *Bioresource Technology,* 274, 180–189. https://doi.org/10.1016/j.biortech.2018.11.090.
Gurmessa, B., et al. 2021. Post-digestate composting benefits and the role of enzyme activity to predict trace element immobilization and compost maturity. *Bioresource Technology,* 338, 125550. https://doi.org/10.1016/j.biortech.2021.125550.
Gworek, B., et al. 2021. Pharmaceuticals in the soil and plant environment: a review. *Water, Air, & Soil Pollution,* 232, 145. https://doi.org/10.1007/s11270-020-04954-8.

Häfner, F., et al. 2022. Digestate composition affecting N fertiliser value and C mineralisation. *Waste and Biomass Valorization,* 13, 3445–3462. https://doi.org/10.1007/s12649-022-01723-y.
Hanc, A. & Vasak, F. 2014. Processing separated digestate by vermicomposting technology using earthworms of the genus Eisenia. *International Journal of Environmental Science and Technology,* 12, 1183–1190. https://doi.org/10.1007/s13762-014-0500-8.
Hiis, E. G., et al. 2024. Unlocking bacterial potential to reduce farmland NO emissions. *Nature,* 630, 421–428. https://doi.org/10.1038/s41586-024-07464-3.
Horta, C. & Carneiro, J. P. 2021. Use of digestate as organic amendment and source of nitrogen to vegetable crops. *Applied Sciences,* 12, 248. https://doi.org/10.3390/app12010248.
Hu, C., et al. 2021. A global meta-analysis of split nitrogen application for improved wheat yield and grain protein content. *Soil and Tillage Research,* 213, 105111. https://doi.org/10.1016/j.still.2021.105111.
Inbar, Y., et al. 1985. *The use of composted slurry produced by methanogenic fermentation of cow manure as a growth media.* International Society for Horticultural Science (ISHS), Leuven, Belgium, 75–82.
IPCC, 2021: Climate Change 2021: The Physical Science Basis.Contribution of Working Group I to the Sixth Assessment Report of theIntergovernmental Panel on Climate Change [Masson-Delmotte, V., P. Zhai, A.Pirani, S.L. Connors, C. Péan, S. Berger, N. Caud, Y. Chen, L. Goldfarb, M.I.Gomis, M. Huang, K. Leitzell, E. Lonnoy, J.B.R. Matthews, T.K. Maycock, T.Waterfield, O. Yelekçi, R. Yu, and B. Zhou (eds.)]. Cambridge University Press,Cambridge, United Kingdom and New York, NY, USA, 2391 pp. doi:10.1017/9781009157896.
Ivezić, V., et al. 2021. Potential of industrial by-products as liming materials and digestate as organic fertilizer and their effect on soil properties and yield of Alfalfa (Medicago sativa L.). *Sustainability,* 13. https://doi.org/10.3390/su131911016.
Jakkula, V. J. & Wani, S. P. 2018. Zeolites: potenital soil amendments for improving nutrient and water use efficiency and agriculture productivity. *Scientific Reviews and Chemical Communications,* 8, 119.
Jenkinson, D. S. & Ladd, J. 1981. Microbial biomass in soil: measurement and turnover. *Soil biochemistry,* 5, 415–471.
Jiang, Y., et al. 2020. Inactivation of pathogens in anaerobic digestion systems for converting biowastes to bioenergy: a review. *Renewable and Sustainable Energy Reviews,* 120, 109654. https://doi.org/10.1016/j.rser.2019.109654.
Jonassen, K. R., et al. 2022a. Nitrous oxide respiring bacteria in biogas digestates for reduced agricultural emissions. *ISME Journal,* 16, 580–590. https://doi.org/10.1038/s41396-021-01101-x.
Jonassen, K. R., et al. 2022b. A dual enrichment strategy provides soil- and digestatecompetent nitrous oxide-respiring bacteria for mitigating climate forcing in agriculture. *mBio,* 13, e00788-22.
Jurgutis, L., et al. 2021a. The effect of digestate fertilisation on grass biogas yield and soil properties in field-biomass-biogas-field renewable energy production approach in Lithuania. *Biomass and Bioenergy,* 153, 106211. https://doi.org/10.1016/j.biombioe.2021.106211.

Jurgutis, L., et al. 2021b. Towards a full circular economy in biogas plants: sustainable management of digestate for growing biomass feedstocks and use as biofertilizer. *Energies,* 14, 4272. https://doi.org/10.3390/en14144272.
Karimi, B., et al. 2022. Impact of biogas digestates on soil microbiota in agriculture: a review. *Environmental Chemistry Letters,* 20, 3265–3288. https://doi.org/10.1007/s10311-022-01451-8.
Kesenheimer, K., et al. 2021. Nitrification inhibitors reduce N_2O emissions induced by application of biogas digestate to oilseed rape. *Nutrient Cycling in Agroecosystems,* 120, 99–118. https://doi.org/10.1007/s10705-021-10127-8.
Kinyua, M. N., et al. 2016. Review of small-scale tubular anaerobic digesters treating livestock waste in the developing world. *Renewable and Sustainable Energy Reviews,* 58, 896–910. https://doi.org/10.1016/j.rser.2015.12.324.
Krishnasamy, K., et al. 2014. Evaluation of anaerobic digestate as a substrate for vermicomposting. *International Journal of Environment and Waste Management,* 14, 149. https://doi.org/10.1504/ijewm.2014.064084.
Kronzucker, H. J., et al. 2013. Sodium as nutrient and toxicant. *Plant and Soil,* 369, 1–23. https://doi.org/10.1007/s11104-013-1801-2.
Latifian, M., et al. 2012. Struvite-based fertilizer and its physical and chemical properties. *Environmental Technology,* 33, 2691–2697. https://doi.org/10.1080/09593330.2012.676073.
Lencioni, G., et al. 2016. Environmental application and phytotoxicity of anaerobic digestate from pig farming by in vitro and in vivo trials. *International Journal of Environmental Science and Technology,* 13, 2549–2560. https://doi.org/10.1007/s13762-016-1088-y.
Levin, K. S., et al. 2021. Effects of organic energy crop rotations and fertilisation with the liquid digestate phase on organic carbon in the topsoil. *Agronomy,* 11, 1393. https://doi.org/10.3390/agronomy11071393.
Li, Y., et al. 2020. Factors affecting gaseous emissions, maturity, and energy efficiency in composting of livestock manure digestate. *Science of the Total Environment,* 731, 139157. https://doi.org/10.1016/j.scitotenv.2020.139157.
Liebetrau, J., et al. 2013. Analysis of greenhouse gas emissions from 10 biogas plants within the agricultural sector. *Water Science and Technology,* 67, 1370–1379. https://doi.org/10.2166/wst.2013.005.
Liedl, B. E., et al. 2006. Fertilizer potential of liquid and solid effluent from thermophilic anaerobic digestion of poultry waste. *Water Science and Technology,* 53, 69–79. https://doi.org/10.2166/wst.2006.237.
Liu, Z., et al. 2023. Organic contaminants removal and carbon sequestration using pig manure solid residue-derived biochar: A novel closed-loop strategy for anaerobic liquid digestate. *Chemical Engineering Journal,* 471. https://doi.org/10.1016/j.cej.2023.144601.
Logan, M. & Visvanathan, C. 2019. Management strategies for anaerobic digestate of organic fraction of municipal solid waste: Current status and future prospects. *Waste Management and Research,* 37, 27–39. https://doi.org/10.1177/0734242X18816793.
Lu, M.-Y., et al. 2020. Addition of oyster shell to enhance organic matter degradation and nitrogen conservation during anaerobic digestate composting. *Environmental*

Science and Pollution Research, 27, 33732–33742. https://doi.org/10.1007/s11356-020-09460-2.

Luo, H., et al. 2022. Benefits of biobased fertilizers as substitutes for synthetic nitrogen fertilizers: field assessment combining minirhizotron and UAV-based spectrum sensing technologies. *Frontiers in Environmental Science*, 10, 988932. https://doi.org/10.3389/fenvs.2022.988932.

Makádi, M. 2012. Digestate: a new nutrient source - review. In: Kumar, S. (ed.) *Biogas*. IntechOpen.

Makádi, M., et al. 2016. Impact of digestate application on chemical and microbiological properties of two different textured soils. *Communications in Soil Science and Plant Analysis*, 47, 167–178. https://doi.org/10.1080/00103624.2015.1109652.

Marti, N., et al. 2008. Struvite precipitation assessment in anaerobic digestion processes. *Chemical Engineering Journal*, 141, 67–74. https://doi.org/10.1016/j.cej.2007.10.023.

Martin, S. L., et al. 2015. Biochar-mediated reductions in greenhouse gas emissions from soil amended with anaerobic digestates. *Biomass and Bioenergy*, 79, 39–49. https://doi.org/10.1016/j.biombioe.2015.04.030.

Massini, G., et al. 2025. Anaerobic digestion of cattle manure contaminated with an antibiotic mixture: a nature-based solution for environmental management. *Land*, 14. https://doi.org/10.3390/land14020353.

Mateescu, C., et al. 2024. Effectiveness of biological approaches for removing persistent organic pollutants from wastewater: a mini-review. *Microorganisms*, 12. https://doi.org/10.3390/microorganisms12081632.

Mayerová, M., et al. 2023. Long-term application of biogas digestate improves soil physical properties. *Soil and Tillage Research*, 231, 105715. https://doi.org/10.1016/j.still.2023.105715.

McConnell, D. A., et al. 2013. Phosphorus losses from low-emission slurry spreading techniques. *Journal of Environmental Quality*, 42, 446–54. https://doi.org/10.2134/jeq2012.0024.

MDA (Maryland Department of Agriculture), USA. 2015. Nutrient application requirements. Available at: https://mda.maryland.gov/resource_conservation/Documents/nm_manual/1-D1-1-1D1-6.pdf [Accessed 23/09/2025].

Mohanavelu, A., et al. 2021. Irrigation induced salinity and sodicity hazards on soil and groundwater: an overview of its causes, impacts and mitigation strategies. *Agriculture*, 11, 983. https://doi.org/10.3390/agriculture11100983.

Molinuevo-Salces, B., et al. 2013. Anaerobic co-digestion of livestock and vegetable processing wastes: fibre degradation and digestate stability. *Waste Management*, 33, 1332–1338. https://doi.org/10.1016/j.wasman.2013.02.021.

Möller, K. & Müller, T. 2012. Effects of anaerobic digestion on digestate nutrient availability and crop growth: a review. *Engineering in Life Sciences*, 12, 242–257. https://doi.org/10.1002/elsc.201100085.

Montemurro, F., et al. 2015. Yield and performance and soil properties of organically fertilized fodder crops. *Journal of Plant Nutrition*, 38, 1558–1572. https://doi.org/10.1080/01904167.2015.1016177.

Mora-Salguero, D., et al. 2025. Long-term effect of repeated application of pig slurry digestate on microbial communities in arable soils. *Heliyon*, 11, 41117. https://doi.org/10.1016/j.heliyon.2024.e41117.

Moradi, N., et al. 2023. Removal of contaminants of emerging concern from the supernatant of anaerobically digested sludge by O_3 and O_3/H_2O_2: Ozone requirements, effects of the matrix, and toxicity. *Environ Res,* 235, 116597. https://doi.org/10.1016/j.envres.2023.116597.

Mortola, N., et al. 2019. Potential use of a poultry manure digestate as a biofertiliser: evaluation of soil properties and lactuca sativa growth. *Pedosphere,* 29, 60–69. https://doi.org/10.1016/s1002-0160(18)60057-8.

Nag, et al. 2019. Anaerobic digestion of agricultural manure and biomass - Critical indicators of risk and knowledge gaps. *Science of the Total Environment,* 690, 460–479. https://doi.org/10.1016/j.scitotenv.2019.06.512.

Nesse, A. S., et al. 2018. Peat replacement in horticultural growth media: the adequacy of coir, paper sludge and biogas digestate as growth medium constituents for tomato (Solanum lycopersicum L.) and lettuce (Lactuca sativa L.). *Acta Agriculturae Scandinavica, Section B – Soil & Plant Science,* 69, 287–294. https://doi.org/10.1080/09064710.2018.1556728.

Nicholson, F. A., et al. 2017. Nitrogen losses to the environment following food-based digestate and compost applications to agricultural land. *Environmental Pollution,* 228, 504–516. https://doi.org/10.1016/j.envpol.2017.05.023.

Nicholson, F. A., et al. 2018. Precision application techniques reduce ammonia emissions following food-based digestate applications to grassland. *Nutrient Cycling in Agroecosystems,* 110, 151–159. https://doi.org/10.1007/s10705-017-9884-4.

Nkoa, R. 2013. Agricultural benefits and environmental risks of soil fertilization with anaerobic digestates: a review. *Agronomy for Sustainable Development,* 34, 473–492. https://doi.org/10.1007/s13593-013-0196-z.

Nowak, M. & Czekała, W. 2024. Sustainable use of digestate from biogas plants: separation of raw digestate and liquid fraction processing. *Sustainability,* 16, 5461. https://doi.org/10.3390/su16135461.

O'Connor, J., et al. 2022. Physical, chemical, and microbial contaminants in food waste management for soil application: a review. *Environmental Pollution,* 300, 118860. https://doi.org/10.1016/j.envpol.2022.118860.

Obaideen, K., et al. 2022. Biogas role in achievement of the sustainable development goals: evaluation, challenges, and guidelines. *Journal of the Taiwan Institute of Chemical Engineers,* 131. https://doi.org/10.1016/j.jtice.2022.104207.

Odlare, M., et al. 2008. Changes in soil chemical and microbiological properties during 4 years of application of various organic residues. *Waste Management,* 28, 1246–1253. https://doi.org/10.1016/j.wasman.2007.06.005.

OMAFRA (Ontario Ministry of Agriculture, Food and Rural Affairs), Canada. 2024. Nutrient management. https://www.ontario.ca/page/nutrient-management#section-7 [Accessed 23/09/2025].

Panuccio, M. R., et al. 2021. Digestate application on two different soils: agricultural benefit and risk. *Waste and Biomass Valorization,* 12, 4341–4353. https://doi.org/10.1007/s12649-020-01318-5.

Pastorelli, R., et al. 2021. Recycling biogas digestate from energy crops: effects on soil properties and crop productivity. *Applied Sciences,* 11, 750. https://doi.org/10.3390/app11020750.

Pawlett, M. & Tibbett, M. 2015. Is sodium in anaerobically digestated food waste a potential risk to soils. *Sustainable Environment Research,* 25, 235–239.

Pedersen, J., et al. 2024. Effect of slurry separation and air-plasma treatment on NH3 and VOC emissions from field applied biogas digestate and pig slurry to grassland. *Biosystems Engineering,* 247, 257–266. https://doi.org/10.1016/j.biosystemseng.2024.09.014.

Piccoli, I., et al. 2023. A farm-scale sustainability assessment of the anaerobic digestate application methods. *European Journal of Agronomy,* 146, 126811. https://doi.org/10.1016/j.eja.2023.126811.

Plaimart, J., et al. 2021. Coconut husk biochar amendment enhances nutrient retention by suppressing nitrification in agricultural soil following anaerobic digestate application. *Environmental Pollution,* 268, 115684. https://doi.org/10.1016/j.envpol.2020.115684.

Pokoj, T., et al. 2015. Semi-continuous anaerobic digestion of different silage crops: VFAs formation, methane yield from fiber and non-fiber components and digestate composition. *Bioresource Technology,* 190, 201–210. https://doi.org/10.1016/j.biortech.2015.04.060.

Ragályi, P., et al. 2025. Factors influencing the impact of anaerobic digestates on soil properties. *Soil Systems,* 9. https://doi.org/10.3390/soilsystems9030078.

Ran, X., et al. 2023. Comparison of phosphorus species in livestock manure and digestate by different detection techniques. *Science of The Total Environment,* 874, 162547. https://doi.org/10.1016/j.scitotenv.2023.162547.

Ran, Y., et al. 2022. Yield and quality of rice under the effects of digestate application. *Agriculture,* 12, 514. https://doi.org/10.3390/agriculture12040514.

Reuland, G., et al. 2022. Assessment of the carbon and nitrogen mineralisation of digestates elaborated from distinct feedstock profiles. *Agronomy,* 12. https://doi.org/10.3390/agronomy12020456.

Ries, J., et al. 2023. Potential applications of food-waste-based anaerobic digestate for sustainable crop production practice. *Sustainability,* 15, 8520. https://doi.org/10.3390/su15118520.

Risberg, K., et al. 2017. Comparative characterization of digestate versus pig slurry and cow manure - Chemical composition and effects on soil microbial activity. *Waste Management,* 61, 529–538. https://doi.org/10.1016/j.wasman.2016.12.016.

Rittl, T. F., et al. 2023. Anaerobic digestion of dairy cattle slurry–long-term effects on crop yields and chemical soil characteristics. *Organic Agriculture,* 13, 547–563. https://doi.org/10.1007/s13165-023-00447-0.

Riva, C., et al. 2016. Short-term experiments in using digestate products as substitutes for mineral (N) fertilizer: agronomic performance, odours, and ammonia emission impacts. *Science of the Total Environment,* 547, 206–214. https://doi.org/10.1016/j.scitotenv.2015.12.156.

Robles-Aguilar, A. A., et al. 2019. Maize silage digestate application affecting germination and early growth of maize modulated by soil type. *Agronomy,* 9, 473. https://doi.org/10.3390/agronomy9080473.

Rolka, E., et al. 2024. Digestate from an agricultural biogas plant as a factor shaping soil properties. *Agronomy,* 14, 1528. https://doi.org/10.3390/agronomy14071528.

Rollett, A. J., et al. 2025. Ammonia emissions and nitrogen use efficiency of untreated and plasma treated liquid organic materials. *Soil Use and Management,* 41, 70004. https://doi.org/10.1111/sum.70004.

Samoraj, M., et al. 2022. The challenges and perspectives for anaerobic digestion of animal waste and fertilizer application of the digestate. *Chemosphere,* 295, 133799. https://doi.org/10.1016/j.chemosphere.2022.133799.

Sánchez-Rodríguez, A. R., et al. 2018. Advanced processing of food waste based digestate for mitigating nitrogen losses in a winter wheat crop. *Frontiers in Sustainable Food Systems,* 2, 35. https://doi.org/10.3389/fsufs.2018.00035.

Sani, M. N. H., et al. 2024. Harnessing biostimulants from biogas digestates for high-value resource recovery: a review. *Environmental Chemistry Letters,* 23, 139–164. https://doi.org/10.1007/s10311-024-01801-8.

Scaglia, B., et al. 2017. The anaerobic digestion process capability to produce biostimulant: the case study of the dissolved organic matter (DOM) vs. auxin-like property. *Science of the Total Environment,* 589, 36–45. https://doi.org/10.1016/j.scitotenv.2017.02.223.

Schwager, E. A., et al. 2016. Field nitrogen losses induced by application timing of digestate from dairy manure biogas production. *Journal of Environmental Quality,* 45, 1829–1837. https://doi.org/10.2134/jeq2016.04.0148.

Severin, M., et al. 2016. Greenhouse gas emissions after application of digestate: short-term effects of nitrification inhibitor and application technique effects. *Archives of Agronomy and Soil Science,* 62, 1007–1020. https://doi.org/10.1080/03650340.2015.1110575.

Sharifi, M., et al. 2019. Short-term nitrogen dynamics in a soil amended with anaerobic digestate. *Canadian Journal of Soil Science,* 99, 173–181. https://doi.org/10.1139/cjss-2018-0060.

Shi, L., et al. 2018. Nutrient recovery from digestate of anaerobic digestion of livestock manure: a review. *Current Pollution Reports,* 4, 74–83. https://doi.org/10.1007/s40726-018-0082-z.

Šimon, T., et al. 2015. The effect of digestate, cattle slurry and mineral fertilization on the winter wheat yield and soil quality parameters. *Plant, Soil and Environment,* 61, 522–527. https://doi.org/10.17221/530/2015-pse.

Singh, J. & Kalamdhad, A. S. 2011. Effects of heavy metals on soil, plants, human health and aquatic life. *International Journal of Research in Chemistry and Environment,* 1, 15 - 21.

Slepetiene, A., et al. 2023. Solid and liquid phases of anaerobic digestate for sustainable use of agricultural soil. *Sustainability,* 15, 1345. https://doi.org/10.3390/su15021345.

Smith, P. 2004. How long before a change in soil organic carbon can be detected? *Global Change Biology,* 10, 1878–1883. https://doi.org/10.1111/j.1365–2486.2004.00854.x.

Sobhi, M., et al. 2022. Selecting the optimal nutrients recovery application for a biogas slurry based on its characteristics and the local environmental conditions: a critical review. *Sci Total Environ,* 814, 152700. https://doi.org/10.1016/j.scitotenv.2021.152700.

Stefan, E.-M., et al. 2024. Impact of soil organic pollutants on different types of crops - review. *23rd International Scientific Conference Engineering for Rural Development Proceedings.*

Steiner, M., et al. 2024. Effects of pesticides on soil bacterial, fungal and protist communities, soil functions and grape quality in vineyards. *Ecological Solutions and Evidence,* 5, 12327. https://doi.org/10.1002/2688–8319.12327.

Stoknes, K., et al. 2016. Efficiency of a novel "Food to waste to food" system including anaerobic digestion of food waste and cultivation of vegetables on digestate in a bubble-insulated greenhouse. *Waste Management,* 56, 466–476. https://doi.org/10.1016/j.wasman.2016.06.027.

Tambone, F., et al. 2013. Organic matter characterization during the anaerobic digestion of different biomasses by means of CPMAS 13C NMR spectroscopy. *Biomass and Bioenergy,* 48, 111–120. https://doi.org/10.1016/j.biombioe.2012.11.006.

Tambone, F., et al. 2015. Composting of the solid fraction of digestate derived from pig slurry: biological processes and compost properties. *Waste Management,* 35, 55–61. https://doi.org/10.1016/j.wasman.2014.10.014.

Tambone, F., et al. 2017. Solid and liquid fractionation of digestate: mass balance, chemical characterization, and agronomic and environmental value. *Bioresource Technology,* 243, 1251–1256. https://doi.org/10.1016/j.biortech.2017.07.130.

Tampio, E., et al. 2016. Agronomic characteristics of five different urban waste digestates. *Journal of Environmental Management,* 169, 293–302. https://doi.org/10.1016/j.jenvman.2016.01.001.

Teglia, C., et al. 2011. Characterization of Solid Digestates: Part 2, Assessment of the Quality and Suitability for Composting of Six Digested Products. *Waste and Biomass Valorization,* 2, 113–126. https://doi.org/10.1007/s12649-010-9059-x.

ten Huf, M. & Olfs, H.-W. 2020. Effect of the nitrification inhibitor DMPP on nitrous oxide emissions and the stabilization of ammonium following the injection of dairy slurry and digestate in a soil-column experiment. *Journal of Plant Nutrition and Soil Science,* 183, 129–135. https://doi.org/10.1002/jpln.201900025.

Tilvikienė, V., et al. 2018. Effects of 5 years of digestate application on biomass production and quality of cocksfoot (Dactylis glomerata L.). *Grass and Forage Science,* 73, 206–217. https://doi.org/https://doi.org/10.1111/gfs.12306.

Tiwary, A., et al. 2015. Assessment and mitigation of the environmental burdens to air from land applied food-based digestate. *Environmental Pollution,* 203, 262–270. https://doi.org/10.1016/j.envpol.2015.02.001.

Tsachidou, B., et al. 2019. Biogas residues in substitution for chemical fertilizers: a comparative study on a grassland in the Walloon Region. *Science of the Total Environment,* 666, 212–225. https://doi.org/10.1016/j.scitotenv.2019.02.238.

Tuszynska, A., et al. 2021. Phosphorus concentration and availability in raw organic waste and post fermentation products. *Journal of Environmental Management,* 278, 111468. https://doi.org/10.1016/j.jenvman.2020.111468.

Valentinuzzi, F., et al. 2020. The fertilising potential of manure-based biogas fermentation residues: pelleted vs. liquid digestate. *Heliyon,* 6, e03325. https://doi.org/10.1016/j.heliyon.2020.e03325.

Van Midden, C., et al. 2023. The impact of anaerobic digestate on soil life: a review. *Applied Soil Ecology,* 191, 105066. https://doi.org/10.1016/j.apsoil.2023.105066.

Van Midden, C., et al. 2025. Glycerol immobilises anaerobic digestate supplied nitrogen. *Waste and Biomass Valorization*. https://doi.org/10.1007/s12649-024-02876-8.

Vanden Nest, T., et al. 2015. P availability and P leaching after reducing the mineral P fertilization and the use of digestate products as new organic fertilizers in a 4-year field trial with high P status. *Agriculture, Ecosystems & Environment,* 202, 56–67. https://doi.org/10.1016/j.agee.2014.12.012.

Vaneeckhaute, C., et al. 2016. Nutrient recovery from digestate: systematic technology review and product classification. *Waste and Biomass Valorization*, 8, 21–40. https://doi.org/10.1007/s12649-016-9642-x.

Vaneeckhaute, C., et al. 2013. Closing the nutrient cycle by using bio-digestion waste derivatives as synthetic fertilizer substitutes: a field experiment. *Biomass and Bioenergy*, 55, 175–189. https://doi.org/10.1016/j.biombioe.2013.01.032.

Villarino, S. H., et al. 2025. Adding anaerobic digestate to commercial farm fields increases soil organic carbon. *Journal of Agriculture and Food Research*, 21. https://doi.org/10.1016/j.jafr.2025.101942.

Voća, N., et al. 2005. Digested residue as a fertilizer after the mesophilic process of anaerobic digestion. *Plant, Soil and Environment*, 51, 262–266. https://doi.org/10.17221/3584-pse.

Voltr, V., et al. 2021. The soil organic matter in connection with soil properties and soil inputs. *Agronomy*, 11, 779. https://doi.org/10.3390/agronomy11040779.

Vu, Q. D., et al. 2015a. Greenhouse gas emissions from passive composting of manure and digestate with crop residues and biochar on small-scale livestock farms in Vietnam. *Environmental Technology*, 36, 2924–2935. https://doi.org/10.1080/09593330.2014.960475.

Vu, T. K., et al. 2015b. Life cycle assessment of biogas production in small-scale household digesters in Vietnam. *Asian-Australas Journal of Animal Science*, 28, 716–729. https://doi.org/10.5713/ajas.14.0683.

Wagner, C., et al. 2021. Acidification effects on in situ ammonia emissions and cereal yields depending on slurry type and application method. *Agriculture*, 11, 1053. https://doi.org/10.3390/agriculture11111053.

Walsh, J. J., et al. 2018. Repeated application of anaerobic digestate, undigested cattle slurry and inorganic fertilizer N: impacts on pasture yield and quality. *Grass and Forage Science*, 73, 758–763. https://doi.org/https://doi.org/10.1111/gfs.12354.

Wang, J., et al. 2025. Migration and transformation mechanisms of heavy metals and PAHs during hydrothermal treatment of food waste digestate. *Process Safety and Environmental Protection*, 201. https://doi.org/10.1016/j.psep.2025.107569.

Wang, X., et al. 2023a. Using adaptive and aggressive N_2O-reducing bacteria to augment digestate fertilizer for mitigating N_2O emissions from agricultural soils. *Science of the Total Environment*, 903, 166284. https://doi.org/10.1016/j.scitotenv.2023.166284.

Wang, Y., et al. 2023b. Pyrolysis of food waste digestate residues for biochar: pyrolytic properties, biochar characterization, and heavy metal behaviours. *Fuel*, 353. https://doi.org/10.1016/j.fuel.2023.129185.

Weimers, K., et al. 2022. Liquid anaerobic digestate as sole nutrient source in soilless horticulture-or spiked with mineral nutrients for improved plant growth. *Frontiers in Plant Science*, 13, 770179. https://doi.org/10.3389/fpls.2022.770179.

Weldon, S., et al. 2023. Co-composting of digestate and garden waste with biochar: effect on greenhouse gas production and fertilizer value of the matured compost. *Environmental Technology*, 44, 4261–4271. https://doi.org/10.1080/09593330.2022.2089057.

Wilson, M. L., et al. 2021. Dragging a manure hose over corn at early growth stages does not reduce yield. *Agronomy Journal*, 113, 3910–3921. https://doi.org/10.1002/agj2.20797.

WRAP 2012. Using quality anaerobic digestate to benefit crops. Available at: https://www.triedandtested.org.uk/media/vqxdsty0/using-quality-digestate-to-benefit-crops.pdf [Accessed 14th February 2025].

Yagüe, M. R. & Lobo, M. C. 2024. Valorization of organic residues for lettuce production minimizes nitrogen loss in soil in comparison to mineral fertilization. *Journal of Plant Nutrition,* 47, 1680–1696. https://doi.org/10.1080/01904167.2024.2319161.

Zeng, Y., et al. 2016. Improving composting as a post-treatment of anaerobic digestate. *Bioresource Technology,* 201, 293–303. https://doi.org/10.1016/j.biortech.2015.11.013.

Zhang, T., et al. 2020. Phosphorus recovered from digestate by hydrothermal processes with struvite crystallization and its potential as a fertilizer. *Science of the Total Environment,* 698, 134240. https://doi.org/10.1016/j.scitotenv.2019.134240.

Zhang, Y., et al. 2012. Anaerobic digestion of two biodegradable municipal waste streams. *Journal of Environmental Management,* 104, 166–174. https://doi.org/10.1016/j.jenvman.2012.03.043.

Zielińska, M. & Bułkowska, K. 2024. Sustainable management and advanced nutrient recovery from biogas energy sector effluents. *Energies,* 17. https://doi.org/10.3390/en17153705.

Chapter 6

Optimizing poultry litter for soil improvement

Runan Yan, Chinwe Justice-Alucho and Xiuping Jiang, Clemson University, USA

1 Introduction

Poultry litter (PL), a nutrient-rich byproduct of poultry production, is widely utilized as a bio-based soil amendment to improve soil fertility and reduce dependence on synthetic fertilizers. Its high nitrogen, phosphorus, potassium, and organic matter contents support soil structure, nutrient cycling, and crop productivity. According to the USDA NASS Poultry Production and Value: 2024 Summary, approximately 9.33 billion broilers were produced in the United States in 2024, representing a 1% increase from the 9.24 billion broilers produced in 2023 (USDA NASS, 2025). In addition, the report indicates that the total live weight of broilers produced in 2024 reached 61.1 billion pounds, highlighting the substantial scale of broiler production in the United States. Globally, poultry production is projected to rise by ~2% in 2025, with the United States, EU, Brazil, Turkey, and China as major contributors (USDA, 2025). This rapid growth directly translates into increased litter generation, as waste scales proportionally with bird numbers and live weight. Based on the typical litter generation rate of 2.5 lb per bird, this level of production corresponds to the estimated 14~29 million tons of broiler litter generated annually, highlighting the substantial volume of poultry waste requiring sustainable management (Ashworth et al., 2020; Ritz and Merka, 2013).

http://dx.doi.org/10.19103/AS.2025.0166.07

PL consists of a friable bedding-manure mix and a compacted, high-moisture 'caked' fraction, the latter often prioritized for removal and treatment (Grubinger, 1999). Beyond its nutrient value, PL provides substantial on-farm economic benefits by reducing reliance on synthetic fertilizers, decreasing on-site waste accumulation, and minimizing environmental pollution, thereby easing the burden of disposal (Chen and Jiang, 2017a). Harmel et al. (2008) showed that substituting commercial fertilizer with PL lowered input costs while maintaining yields, underscoring its role as a cost-effective alternative. These financial incentives contribute to the widespread land application of raw PL.

Land application of PL plays an important role in agricultural nutrient recycling. Long-term use has been shown to increase soil concentrations of nitrogen, phosphorus, potassium, and micronutrients, thereby enhancing plant nutrient uptake and promoting nutrient cycling through plant–soil interactions (Sauer et al., 2015). Consequently, PL is widely used as a soil amendment because it improves soil fertility and structure, which in turn supports higher crop productivity (Solaiman et al., 2020; Jarosz et al., 2022). Following land application, nitrogen is mineralized into ammonium and subsequently oxidized to nitrate through nitrification, both of which are readily absorbed by plants (Sainju et al., 2010). Similarly, phosphorus is converted into orthophosphate, the primary form available for plant uptake, while residual carbon contributes to microbial biomass and soil organic matter formation. Through these processes, nutrients contained in PL are recycled within agroecosystems as plants absorb nutrients from amended soils and subsequently return them to the system through animal feed and manure production (Ayantokun et al., 2025). On the contrary, PL amendments can influence soil microbial communities and ecosystem processes by altering microbiome structure and introducing antimicrobial residues and resistance genes into soil environments (Lopes et al., 2025).

Untreated PL poses microbiological risks, including the introduction of zoonotic pathogens, antibiotic resistance genes (ARGs), and microbial imbalances that may disrupt soil ecosystems and lead to environmental contamination. Studies have shown the presence of specific pathogens such as *Salmonella*, *Escherichia coli*, *Campylobacter*, *Clostridium perfringens*, and so on in PL, although the diversity of microbial communities and resistant markers remains poorly understood (Li et al., 2007). According to the surveys, *Salmonella* prevalence in PL can vary widely from less than 1% to over 50%–depending on management and storage conditions. Populations can reach levels exceeding 10^5 MPN/g (Dunn et al., 2022), underscoring the critical need for effective treatment prior to use. In addition to high microbial population, PL also contains heavy metals, antibiotic residues, growth and sex hormones such as estrogen, estradiol, and testosterone, as well as pesticides including dioxins, furans, polychlorinated biphenyls, and polycyclic aromatic hydrocarbons (Lu et al., 2003). Moreover, nutrients, particularly nitrogen and phosphorus from

PL, can be released into surrounding ecosystems through runoff and leaching, leading to water pollution and eutrophication (Shakya and Agarwal, 2017). Repeated surface application of PL may also result in nutrient accumulation in soils, particularly phosphorus, which increases the risk of runoff into nearby water bodies if not properly managed (Tabler et al., 2024). In addition, improper treatment or disposal of PL has the potential to pollute air, groundwater, and surface water and may increase the risk of disease transmission in humans and animals (Hubbard et al., 2020)

To reduce these risks, several treatment strategies have been developed to stabilize PL before land application. These strategies are broadly classified into physical, chemical, and biological treatment methods aimed at reducing nitrogen content, acidity, and biological risks (Heinonen-Tanski et al., 2006). Physical treatments include heating processes such as incineration, pasteurization, and radiation to reduce microbial contamination, while chemical treatments use additives such as lime to inhibit the growth of pathogenic microorganisms, including *Salmonella* species (Heinonen-Tanski et al., 2006). Composting and anaerobic digestion are also widely used biological treatment methods. Among the available treatment technologies, composting offers an effective approach for PL management by enabling waste stabilization, renewable energy production, and nutrient recovery (Ayantokun et al., 2025). Hydrolysis, acidogenesis, acetogenesis, and methanogenesis during anaerobic digestion facilitate the removal of excess organic carbon from poultry manure while stabilizing the waste (Ayantokun et al., 2025). However, improper composting conditions, including inadequate temperature, aeration, pH, or moisture levels, may allow pathogens to persist or regrow in the finished product, potentially contaminating crops grown on PL-amended soils (Chen and Jiang, 2017a).

Collectively, repeated PL application returns essential nutrients such as nitrogen, phosphorus, and potassium to soils, thereby stimulating microbial activity and supporting plant productivity. However, if not properly managed, nutrient buildup and the presence of human pathogens and harmful chemicals may result in runoff or leaching, ultimately contributing to environmental pollution. These considerations highlight the importance of developing sustainable PL management and treatment strategies that balance agronomic benefits with environmental protection.

This chapter focuses on how PL treatment methods, such as composting, heat drying, pelletization, and chemical amendments, can be optimized to mitigate microbial hazards while enhancing soil health. We also evaluate the effectiveness of these treatments in reducing pathogen transmission and their impact on soil microbial diversity and functions following soil application, as influenced by various environmental factors. The role of microbial inoculants, particularly competitive exclusion microorganisms (CEMs), is also discussed as a strategy to restore or support soil microbiota post-application.

2 Microbial risks of raw or inadequately managed poultry litter

2.1 Human pathogens

While rich in nutrients and organic matter, PL presents significant microbial risks if applied to soil without adequate treatment. Raw PL contains a wide range of zoonotic and enteric pathogens, posing serious risks to food safety, public health, and environmental quality. Pathogen prevalence and persistence are influenced by litter composition, moisture content, temperature, and treatment history. A recent systematic review and meta-analysis found that PL accounts for roughly 25% of pre-harvest *Salmonella* colonization risk in broilers, making it one of the most important contributors after hatchery environments (Wang et al., 2023). Untreated PL can harbor diverse human pathogens, including antibiotic-resistant strains, posing potential risks to environmental and public health (Fatoba et al., 2021; Kelley et al., 1994; Pope and Cherry, 2000). These risks make treating PL before land application imperative, especially in agricultural systems producing fresh produce or crops consumed raw.

Table 1 summarizes recent findings on the prevalence of major human pathogens and antimicrobial resistance risks associated with PL and following soil application. Raw broiler litter often contains a wide range of pathogens, with reported prevalence varying from 0% to 100%. Bacterial concentrations frequently exceed 10^6–10^8 CFU/g, and spore-forming organisms and viruses can persist for up to approximately 77 days. Pathogen regrowth has also been observed, including *Salmonella* regrowth within 48 h in pelletized litter under high moisture (>30%) (Shah et al., 2019) and 1.5~2 log increases in compost with >40% moisture stored at 20°C for 3 days (Reynnells et al., 2014). Longitudinal field studies showed *L. monocytogenes* (up to 49%) and *Salmonella* (≤51%) were common in amendments, with detection in soils and water influenced by environmental conditions (Ferguson et al., 2023). In addition, long-term manure use increased antibiotic resistance gene (ARG) abundance and the prevalence of multidrug-resistant (MDR) *Enterococcus*.

Many studies have confirmed the presence of *Salmonella* spp., *Listeria monocytogenes, E. coli* O157:H7, *Campylobacter* spp., *C. perfringens*, and other pathogens in untreated PL from broiler operations (Kelley et al., 1994; Pope and Cherry, 2000; Terzich et al., 2000). For instance, a national survey across 12 US regions found *Staphylococcus* spp. and *E. coli* at levels exceeding 10^8 CFU/g and noted high microbial loads in fresh litter, with regional variation linked to moisture content and pH (Terzich, et al., 2000). Kelley et al. (1994) reported that *Salmonella, Listeria*, and *Campylobacter* could persist in PL fractions for up to 16 weeks, with reductions affected by storage time and pile depth. In addition to bacterial pathogens, raw PL can also carry viruses (e.g. avian reovirus), protozoa (e.g. *Cryptosporidium*), and MDR organisms such as

Table 1 Prevalence of major human pathogens and antimicrobial resistance (AMR) risks associated with PL and its soil application

PL type/context	Pathogen or target	Key findings	Reference
Raw broiler litter	*Salmonella* spp., *L. monocytogenes*, *Campylobacter* spp., *E. coli*, *Staphylococcus* spp., *C. perfringens*	Multiple pathogens frequently isolated from untreated litter. Prevalence ranges 0–100% depending on flock cycle, reuse, farm, and environmental conditions. *E. coli* and *Staphylococcus* spp. often reach >10^6–10^8 CFU/g. Spore-formers (e.g. *C. perfringens*) and viral agents may persist up to ~77 days.	(Kelley et al., 1994; Pope and Cherry, 2000; Terzich et al., 2000)
PL compost (storage and use)	*Salmonella* spp., *E. coli* O157:H7	1–2 log regrowth under moderate temperature (20°C) and high moisture (>40%) within 3 days.	(Reynnells et al., 2014)
PL (field-applied, long-term)	*Salmonella enterica*	Persisted up to 210 days in clay-rich soil under less frequent (weekly) irrigation.	(Bardsley et al., 2021)
PL and dairy manure (field-applied, USA)	*Salmonella* spp., *L. monocytogenes*, *Campylobacter* spp., Shiga toxin–producing *E. coli* (STEC), *Arcobacter*	Longitudinal monitoring: *L. monocytogenes* 49% in amendments, *Arcobacter* 32%, *Salmonella* ≤51%. Soil samples: *L. monocytogenes* 13%, *Arcobacter* 7%. Prevalence influenced by weather, irrigation sources, and soil fertility.	(Ferguson et al., 2023)
Long-term manured soils (China)	ARGs; DNA signatures of human pathogenic bacteria (HPB): *E. coli*, *Salmonella* spp., *Klebsiella*, *Pseudomonas*, etc.	Higher ARG abundance and diversity; metagenomic detection of multiple HPB in manured soils; stronger ARG–HPB associations with longer-term manure use.	(Fang et al., 2015)
Chicken manure-amended soils (China)	ARGs, virulence factor genes (VFGs), HPB	Metagenomics analysis identified 157 HPB (~1.3% of sequences); high ARG/VFG abundance from chicken manure; lower contamination risk relative to swine manure and moderate compared to other manures.	(Zhu et al., 2022)
PL-fertilized agricultural soils	ARGs (*tet*, *erm*, *sul*, *intI1*)	ARGs persisted in fertilized soils for up to 120 days post-application; resistome composition shaped by history of PL application.	(Lopes et al., 2025)
Chicken litter-amended soil (South Africa)	*Enterococcus* spp. (MDR)	Among MDR *Enterococcus* isolates (n=130), 67.7% were from litter-amended soil. Common resistance to tetracycline, erythromycin, and trimethoprim-sulfamethoxazole.	(Fatoba et al., 2021)

Enterococcus spp. (Fatoba et al., 2021). They also showed that over 55% of *Enterococcus* isolates from chicken litter and amended soil were resistant to at least one antibiotic, and nearly 28% were MDR. Notably, MDR isolates were significantly more prevalent in PL-amended soil (67.7%) than in unamended soil (10%). Importantly, these microbes are resilient in the environment and can persist or even multiply under favorable conditions.

Once secreted from animal bodies, some human pathogens could survive for an extended time in the fecal materials. Studies have also demonstrated that under favorable storage conditions, particularly when moisture and nutrients are sufficient, enteric pathogens such as *Salmonella* can regrow in manure- and compost-based amendments (Reynnells et al., 2014). Kim et al. (2009) reported that as few as 10 cells of *E. coli* O157:H7, adapted to low-nutrient conditions, were able to regrow to approximately 6.54 log CFU/g in autoclaved dairy compost when the moisture content was at least 20%. PL compost stored under suboptimal conditions, particularly when it retains excess moisture and residual nutrients, and lacks effective microbial competition, may also support regrowth of *E. coli* and *Salmonella* by 1.5~2 log CFU/g in some cases (Sharma and Reynnells, 2016). While these studies demonstrate microbial survival and regrowth, little is known about the role of alternative survival strategies, such as the viable but nonculturable (VBNC) or persister states in PL or other biological soil amendments. This knowledge gap is critical because such subpopulations can escape detection by culture-based assays and potentially resuscitate under favorable conditions, which further increases the risk.

2.2 Antibiotic resistance

The widespread use of antimicrobials in poultry production poses significant antimicrobial resistance challenges that extend beyond farm boundaries. Poultry operations account for approximately 80% of agricultural antimicrobial use, with southern Asian countries administering as much as 52-276 mg per kg of chicken annually (Zalewska et al., 2021). Although overall antimicrobial use in US poultry declined by 85% in 2021, the use of specific compounds such as avilamycin and bambermycin increased by 0.6 and 0.1 mg/kg live weight, respectively (Singer et al., 2023). Commonly used antibiotics, including tetracyclines, beta-lactams, sulfonamides, and bacitracin, continue to exert strong selective pressure and remain among the highest-selling antimicrobials for food animals (Atanasova et al., 2025).

When PL is applied to agricultural soils, it introduces not only antibiotic residues but also antimicrobial-resistant bacteria carrying extensive resistance gene repertoires. Microorganisms harbor critical resistance genes, including *emrD, tetA,* and *tetX* (encoding efflux pumps), *lnuf* genes (producing antibiotic-inactivating enzymes), and *aadA* (conferring aminoglycoside resistance)

(Atanasova et al., 2025). Sulfonamide resistance genes *sul1* and *sul2,* which encode enzymes that substitute antibiotic targets, frequently associate with mobile genetic elements, facilitating horizontal gene transfer throughout environmental microbial communities (Atanasova et al., 2025). Research has identified 374 distinct antibiotic resistance genes (ARGs) in chicken manure and anaerobic digestate, with fresh manure showing higher resistance loads than processed materials, though both retain substantial ARG diversity (Atanasova et al., 2025). Critically, while anaerobic digestion reduces viable resistant organisms, it fails to eliminate the resistance genes themselves. Field studies demonstrate that PL application immediately increases populations of erythromycin-resistant and tetracycline-resistant enterococci in soil, with resistance primarily linked to *ermB* and *tetM* genes, respectively (Agga et al., 2025). The persistence of antimicrobial residues impacts these problems significantly. Active compounds like sulfachloropyridazine, oxytetracycline, and enrofloxacin remain detectable for 61, 244, and 514 days, respectively, maintaining selective pressure that promotes resistant *E. coli* survival. Furthermore, this environmental contamination triggers measurable shifts in soil resistome composition and microbiome structure, altering diversity patterns and resistance gene abundance while reshaping indigenous microbial communities (Chen and Jiang, 2017a). Leaching and runoff transport mobile genetic elements and resistant bacteria into food systems and water bodies, creating additional opportunities for horizontal gene transfer. These environmental and microbial challenges associated with PL accumulation highlight the need for effective management and treatment strategies.

Current treatment strategies, including composting and chemical treatment, target the reduction of viable microbial loads and weeds. Understanding how these treatment processes influence nutrient availability, microbial activity, and soil ecosystem responses following land application is essential for developing sustainable PL management practices. Critical research gaps remain regarding resistant gene transfer mechanisms under varying soil conditions, strategies to minimize environmental dissemination, and the complex interactions among nutrient enrichment, antimicrobial residues, and resistance selection pressure in agricultural systems.

2.3 Transmission of microbial hazards in poultry litter

Microbial risks associated with the application of untreated PL extend beyond environmental contamination and may contribute to the spread of infectious agents between animals and potentially to humans. Pathogen dissemination can occur through multiple pathways, including vector-mediated transmission, wind-driven aerosolization, surface runoff, and mechanical transfer between poultry farms (Nguyen et al., 2023). Microorganisms associated with PL can become aerosolized by attaching to airborne dust particles, which may migrate within and between poultry houses through ventilation systems or natural airflow (Diao et al., 2015; Pal et al., 2021). Pathogens such as *E. coli* and *Salmonella* have been shown to associate with dust particles originating from PL and can be transported within poultry production environments and potentially beyond farm boundaries (Pal et al., 2021). In addition to airborne transmission, rodents, wild birds, insects, and human activities may act as biological or mechanical vectors capable of transporting microorganisms from litter-amended fields or contaminated environments to nearby poultry facilities (Nguyen et al., 2023). This risk is further exacerbated by the ability of many pathogens to persist for extended periods in bedding materials and organic substrates when PL is not adequately treated or stabilized prior to land application (Hauck et al., 2017).

Collectively, these findings highlight the importance of effective PL management practices, including appropriate litter treatment prior to application, minimizing dust generation during land spreading, maintaining adequate buffer distances between litter-amended fields and poultry houses, and limiting wildlife or vector access to litter-treated areas. Such measures are essential for reducing the risk of pathogen dissemination and protecting both poultry health and broader agricultural biosecurity.

To reduce microbial risks associated with biological soil amendments of animal origin (BSAAOs), regulatory frameworks such as the FDA's Food Safety Modernization Act (FSMA) Produce Safety Rule have established specific microbiological thresholds. Treated BSAAOs must meet limits of <0.3 MPN/g for *E. coli* O157:H7, <3 MPN/4 g for *Salmonella* spp., and <1 CFU/5 g for *L. monocytogenes*. While not mandatory for PL, compost producers and researchers often reference the EPA's 40 CFR Part 503 Class A biosolids standard (particularly the <3 MPN/4 g threshold for *Salmonella*) as a benchmark for compost microbial safety. These thresholds highlight the importance of implementing validated treatment strategies before PL application.

3 Treatment methods to control microbial risks in poultry litter

Several treatment strategies have been developed and validated to mitigate the microbial hazards associated with raw PL to ensure its safe use as a soil amendment. These treatments aim to inactivate or significantly reduce enteric pathogens and antibiotic resistance reservoirs while preserving the nutrient value that makes PL a valuable soil amendment/biofertilizer. Effective treatment also supports compliance with microbial safety standards set forth by the FSMA Produce Safety Rule and benchmarked EPA biosolids guidelines as described previously. A summary of validated treatment methods, their quantitative pathogen reduction levels, and agronomic benefits is presented in Table 2.

Table 2 Summary of validated PL treatments: microbial and agronomic impacts

Treatment method	Pathogen reduction	Nutrient retention/soil benefit	Key references
Raw/untreated PL (baseline)	Serves as the highest risk benchmark: frequently harbors zoonotic pathogens and antimicrobial resistance determinants; also prone to nutrient losses via NH_3 volatilization and P runoff.	Nutrients highly available but unstable; significant food safety and environmental risks without treatment.	(Kelley et al., 1994; Pope and Cherry, 2000; Terzich et al., 2000)
Thermophilic composting	>5-log reduction of *Salmonella, Listeria, E. coli* O157:H7 when piles exceed 55°C for several days. Proper turning + optimal moisture essential. Regrowth possible under 20°C, >40% MC. Compost history influences resistome in amended fields (ARGs persisted up to 120 d).	Stabilizes organic matter; increases microbial diversity and activity; improves soil structure and water retention.	(Sharma and Reynnells, 2016; Esperón et al., 2020; Song et al., 2020; Lopes et al., 2025)
Pelletization	Validated >5-log reduction of *Salmonella*.	Produces stable, transportable product; slow nutrient release.	(Wang et al., 2021)
Heat drying	Consistent inactivation of bacteria at >80°C. Drying at 100–120°C provides near-complete kill of *Salmonella* and *E. coli*.	Conserves N and P; reduces odor and pathogens; increases handling safety. Operational design influences energy efficiency.	(Wilkinson et al., 2011; Ghaly and Alhattab, 2013)
Two-step heat process (e.g. pasteurization + drying)	Enhanced pathogen inactivation, meeting FSMA microbial reduction standards across multiple batches; validated control against *Salmonella*.	Maintains nutrient value with minimal N loss; consistent product safety across lots.	(Chen et al., 2015a)
Alum amendment	~50% reduction in *Salmonella* prevalence; inhibits urease activity, limiting microbial growth.	Increases P retention in soil, decreases soluble P runoff by >70%, suppresses NH_3 volatilization by 30–60%. Also reduces odor.	(Cook et al., 2008; Chen and Jiang, 2017a)
Field application (PL-amended soils)	*Salmonella enterica* persisted up to 210 d in clay-rich soils with limited irrigation. Soil type and moisture strongly influenced persistence.	Provides nutrients to crops; risk of pathogen and ARG carryover if not managed with waiting periods before harvest.	(Bardsley et al., 2021)
AMR considerations (PL and compost)	67.7% of MDR *Enterococcus* isolates originated from PL-amended soils; resistance common to tetracycline and erythromycin. ARGs (*tet*, *erm*, *sul*, *intI1*) persisted in soil and crops for up to 120 d post-application.	Adds organic matter but carries risk of ARG dissemination, requiring cautious integration with crop production.	(Fatoba et al., 2021; Lopes et al., 2025)

3.1 Composting

Composting is one of the most widely adopted treatment methods for PL. Composting is a managed biological process in which animal manure and other organic materials are decomposed through microbial activity under aerobic or anaerobic conditions. Typically, the composting process proceeds through four phases: mesophilic, thermophilic, cooling, and maturation. During each phase, microbial populations change substantially in response to shifts in environmental conditions. The heat generated by microbial activities during the thermophilic phase (approximately 50~70°C) is sufficient to inactivate many microorganisms, including human and plant pathogens, as well as weed seeds and fly larvae. According to the USDA National Organic Program (NOP) guidelines, composting operations should maintain a temperature in the range of 55–70°C for a minimum of 3 days for static aerated pile or in-vessel systems and 15 days with 5 turns for windrow systems, provided the carbon to nitrogen ratio (C:N) is in the range of 25:1 to 40:1 (NOSB 2002). A number of studies have demonstrated that properly managed thermophilic composting–characterized by frequent turning, maintenance of optimal C:N ratios and moisture levels, and elevated temperatures–effectively inactivates non-spore-forming bacterial pathogens such as *Salmonella*, *L. monocytogenes*, and *E. coli* O157:H7, achieving greater than a 5-log reduction in microbial loads (Chen et al., 2013; Esperón et al., 2020; Shepherd et al., 2007; Singh et al., 2012; Song et al., 2020; Wang et al., 2025). However, both Fremaux et al. (2007) and Shepherd et al. (2007) have shown the uneven pathogen inactivation throughout the compost heaps with longer survival on the surface or around peripheral of the heaps due to low temperature there. Beyond microbial safety, compost improves soil health by enhancing structure, nutrient cycling, and microbial activity, making it an important practice for sustainable crop production (Ghaly and Alhattab, 2013; Gurtler et al., 2018). Composting also promotes organic matter stabilization; however, improper conditions or rewetting events may allow pathogen regrowth (Sharma and Reynnells, 2016). These further underscore composting as a reliable strategy for pathogen inactivation and soil quality improvement, playing a vital role in ensuring food safety and supporting sustainable crop production.

3.2 Pelletization and heat drying

Physical heat-based treatments, including pelletization and drying, are increasingly used in commercial operations. These processes apply high heat (often >65–80°C) through steam conditioning and compression to kill pathogens and reduce moisture content. Under laboratory conditions, the thermal inactivation of *Salmonella* in PL compost has revealed that *Salmonella* cells were killed at 70~150°C within 30 min to 6 h, depending on the bacterial physiological status (desiccation-adapted or not), moisture level, and experimental design (Chen et al., 2013). Chen et al. (2015a) optimized a two-step heat treatment (wet heat followed by dry heat) to improve pathogen inactivation efficacy without compromising nutrient content. However, changes in some chemical, physical, and microbiological properties during PL storage could possibly influence the effectiveness of further heat treatment (Chen et al., 2015b). When heat-treated at 75°C, 80°C, 85°C, and 150°C, the inactivation rate of *Salmonella* in chicken litter slowed as the storage time increased from 0 to 6 months, likely due to increased dryness and reduced ammonia levels. As suggested by the researchers, the chicken litter processing industry should heat-process the chicken litter immediately after removal from the chicken house. Although laboratory studies showed complete inactivation of *Salmonella* in PL under thermal treatment, these results may not fully translate to industrial settings due to scale-up and material heterogeneity. Wang et al. (2021) conducted a validation study of the thermal processes used at two PL processing facilities under a range of operating conditions, employing *Enterococcus faecium* and presumptive indigenous enterococci as surrogate and indicator microorganisms, respectively. Despite substantial variation in processing conditions between those two facilities, the results demonstrated that *Salmonella*, if present, could be reduced by at least 5 log units, as inferred from the observed reductions in the surrogate and indicator organisms. Although physical heat-based treatments such as pelletization and drying can substantially reduce pathogen loads, subsequent storage conditions remain critical. Studies of raw PL and the composts have shown that if sufficient moisture and nutrients remain, pathogens like *Salmonella* and *E. coli* may regrow during storage; this is especially true for heat-treated PL, as the indigenous microbiota has been reduced significantly by heat (Kim and Jiang, 2010; Reynnells et al., 2014). Therefore, careful monitoring and controlling of post-treatment storage conditions is essential to ensure the long-term microbiological safety of heat-processed PL.

3.3 *Chemical amendments*

Chemical amendments such as alum (aluminum sulfate), lime, and acidifiers provide additional safety benefits by modifying pH and inhibiting microbial activity. Alum treatment has been shown to reduce *Salmonella* by 50%, suppress ammonia volatilization, and increase phosphorus retention (Cook et al., 2008). On-farm studies in the Delmarva region found that alum-treated PL reduced dissolved reactive phosphorus runoff by 60~70%, minimized soil phosphorus accumulation, and decreased the PL pH, thus reducing ammonia emissions (Sims and Luka-McCafferty, 2002). Further field studies demonstrated that alum-amended PL significantly reduces water-soluble phosphorus, resulting in up to 75% lower phosphorus runoff compared to untreated litter (Warren et al., 2006). In addition to improving nitrogen use efficiency and crop yield, alum also reduces the mobility of arsenic and heavy metals such as copper and zinc (Warren et al., 2006). These properties help reduce environmental losses and improve nutrient stability while minimizing odor and runoff potential.

4 Soil application of poultry litter

4.1 Pathogen persistence in soil

To support sustainable agricultural practices, raw PL is frequently applied directly to soil because it is inexpensive, easy to use, and readily available. However, pathogens originating from PL can persist in soil for several weeks to over 200 days, depending on environmental conditions (Shah et al., 2019; Lopes et al., 2025). Therefore, the application of untreated PL poses a risk of pathogen dissemination to nearby farms and agricultural land through windborne dust, runoff, and vectors such as birds or rodents, potentially contributing to disease in neighboring flocks. When PL is composted under thermophilic conditions, its application not only reduces microbial risks but also enhances soil enzyme activity, increases microbial biomass, and improves soil properties. Acosta-Martínez and Harmel (2006) found that dehydrogenase and phosphatase activities were elevated in soils amended with composted PL, indicating enhanced microbial activity and nutrient mineralization. However, this practice may pose some microbial risks, as even with heat-treated PL, *Salmonella* Newport survived in amended soil for 40~62 days (Shah et al., 2019).

Environmental factors strongly modulate the fate of pathogens and microbial communities following PL soil application, and multiple studies highlight the interplay of moisture, temperature, soil type, and management practices. Moisture is a critical factor influencing microbial dynamics in PL compost, and studies have shown that elevated moisture levels can facilitate the regrowth of *E. coli* O157:H7 and *Salmonella*, even after these pathogens have been previously inactivated (Reynnells et al., 2014; Sharma and Reynnells, 2016). Also, rewetting of treated PL by rainfall and irrigation can potentially encourage the regrowth of pathogens even after heat or compost treatment. Soil type significantly influences pathogen persistence: clay-rich soils can extend *Salmonella enterica* survival up to 210 days, compared with 84 days in sandy soils (Bardsley et al., 2021). Additionally, limited irrigation has been shown to further prolong *Salmonella* viability (Shah et al., 2019). Chemical amendments can also alter microbial ecology; for example, alum amendment has been reported to reduce *Salmonella* prevalence in soils (Cook et al., 2008). Seasonal effects have been observed: pathogens applied with PL survive longer in fall and winter compared with spring, and *Salmonella* has been recovered from produce surfaces such as cantaloupe rinds (Natvig et al., 2002). Similarly, the controlled studies show longer survival at lower temperatures and higher moisture (Wilkinson et al., 2011). In addition to pathogens, resistance determinants may also be mobilized. Lopes et al. (2025) showed that antibiotic resistance genes (*tet, erm, sul, intI1*) persisted in PL-amended agricultural soils for up to 120 days, with soil resistomes strongly influenced by prior amendment history. These environmental drivers of pathogen persistence and regrowth are consolidated in Table 3 to provide a comparative summary of key environmental factors and their documented impacts on microbial risks, facilitating reference for applied decision-making.

Collectively, effective control requires more than just initial pathogen reduction–it demands careful management of the post-treatment environment. Factors such as moisture content, soil type, temperature, and microbial interactions are critical, as insufficient treatment and improper storage can allow PL to contribute to direct crop contamination and indirect environmental transmission through soil and water.

Table 3 Environmental factors affecting pathogen or ARG genes in PL-amended soil

Environmental factor	Condition/range	Effect on pathogens/ARGs	References
Moisture	>40% in compost	Regrowth of *E. coli* O157:H7, *Salmonella*	Sharma and Reynnells (2016)
Temperature	>55°C (thermophilic)	>5-log pathogen reduction	Song et al. (2020); Esperón et al.(2020)
	Ambient (<40°C)	Incomplete inactivation, possible regrowth	Song et al. (2020)
Soil type	Clay-rich	Extended *Salmonella* survival up to 210 days	Bardsley et al. (2021)
	Sandy	Reduced persistence to 84 days	Bardsley et al. (2021)
Irrigation	Limited irrigation	Prolonged *Salmonella* viability	Shah et al. (2019)
Soil amendments	Alum-treated	~50% reduction in *Salmonella*, ammonia suppression	Cook et al. (2008)
Season/crop timing	Fall/winter	Longer pathogen survival on crops (e.g. cantaloupe)	Natvig et al. (2002); Wilkinson et al. (2011)
ARG persistence	120 days post-PL application	*tet, erm, sul, intI1* genes persisted; influenced by past amendment history	Lopes et al. (2025)

4.2 Produce contamination

There are very limited studies on produce contamination following PL application. A few field studies have suggested that land application of raw or inadequately treated PL poses direct risks of pathogen transfer to fresh produce. Islam et al. (2004a,b) conducted several field studies to determine the fates of *E. coli* O157:H7 and *Salmonella* in soil and on vegetables grown in fields treated with contaminated manure composts (poultry manure compost, dairy manure compost, or alkaline pH–stabilized dairy cattle manure compost) or irrigation water. *E. coli* O157:H7 survived for 154 and 217 days in soil amended with contaminated manure compost during lettuce and parsley cultivation, with survival not differing among compost types (Islam et al., 2004a). *Salmonella* survived greatest in soil amended with poultry compost as compared with dairy cattle manure compost and alkaline-pH-stabilized dairy cattle manure compost (Islam et al., 2004b). Importantly, a single application of contaminated compost led to pathogen contamination of radishes and carrots well beyond their growing cycle. Bucher et al. (2020) and Parente et al. (2021) reported the transfer of *Salmonella* from PL-amended soil to lettuce up to 21 days post-application. Natvig et al. (2002) demonstrated that *Salmonella* persisted in soil for up to 119 days following raw litter application and was subsequently recovered from cantaloupe rinds up to 78 days, underscoring the potential for pre-harvest contamination. Composting PL reduces microbial risks but does not eliminate them. Sharma and Reynnells (2016) reported that *Salmonella* and *E. coli* O157:H7 can regrow in PL compost under moderate temperatures and high moisture conditions, creating potential pathways for crop contamination.

Collectively, those limited studies highlight how both raw and insufficiently treated PL can serve as reservoirs of human pathogens in soil-plant systems, emphasizing the need for validated treatment and careful management of PL before field application.

4.3 Impact on soil microbiota

Beyond pathogen survival, PL and the compost also drive distinct shifts in the broader soil microbiome. Neher et al. (2020) reported that soils amended with PL compost harbored different bacterial communities compared with dairy compost or unamended soils, with higher abundances of Bacteroidetes (e.g. Flavobacteria, Pedobacter) and altered ecoenzyme activities, highlighting how compost origin can shape both community structure and functional potential. Song et al. (2020) demonstrated that composted chicken manure used in anaerobic soil disinfestation promoted beneficial microbial functions and suppressed pathogen viability, while also enhancing strawberry growth and yield. These beneficial functions were achieved by creating anaerobic soil conditions using composted manure amendments covered by impermeable film. This approach enriched microbial groups involved in organic matter degradation and biocontrol (e.g. *Trichoderma* spp.), while simultaneously generating antagonistic conditions that suppressed plant pathogens such as *Fusarium* and *Phytophthora*. However, these benefits can be decreased with excessive application rates associated with reduced microbial diversity and potential pathogen regrowth. Long-term field studies have shown that repeated application of PL compost can reduce soil microbial diversity and alter microbiota structure (Wang et al., 2024). Long-term compost inputs have been shown to enrich ammonia-oxidizing and plant growth-associated bacteria, although such shifts can reduce overall microbial diversity (Liu et al., 2022). The reduced microbial diversity has been associated with diminished suppression of enteric pathogens, whereas highly diverse soil communities generally exhibit stronger pathogen suppression (Major et al., 2020; Schierstaedt et al., 2020). Kharel et al. (2025) reported that soils amended with heat-treated poultry pellets supported longer *E. coli* survival than soils amended with composted PL throughout a 140-day field study period in the southeastern US. This effect may be attributable to heat treatment during pelletization reducing or eliminating indigenous microbial populations, thereby diminishing microbial competition and increasing pathogen persistence in agricultural soils. Schierstaedt et al. (2020) also reported that imbalanced and reduced soil microbial diversity correlated with increased *Salmonella* abundance, while more diverse microbial communities tended to inhibit *Salmonella* proliferation.

Emerging evidence also highlights that treatment methods vary in their impacts on soil microbial communities. Poultry compost, for example, tends to support greater microbial diversity and enzyme activity than raw or pelleted litter, while over-application or poor moisture control can lead to reduced microbiome stability and elevated ARG persistence. Pelletized and heat-dried PL products provide additional advantages by offering uniform nutrient release and enhanced microbial safety. Although these treatments mitigate microbial risks, they also reduce indigenous microbiota substantially, and their impact on soil microbial communities can vary depending on post-application environmental conditions. Studies have shown shifts in non-pathogenic microbial communities, including reduced fungal abundance and altered ratios of beneficial soil bacteria such as *Actinobacteria* and *Proteobacteria* (Bao et al., 2021). These findings underscore the need for co-optimization strategies that target both microbial safety and ecological sustainability.

A summary of validated studies comparing composted, heat-treated, pelletized, chemical amendments, and raw PL applications is presented in Table 4, which illustrates how treatment choice and application rate influence pathogen reduction, ARG burden, and microbiome stability in soil systems. Collectively, these PL treatments not only reduce microbial hazards but also

Table 4 Treated vs. untreated and application rate effects on soil microbiota

Treatment	Key findings	Reference
Composted vs. raw PL	Composted PL caused milder microbiota shifts and lower ARG burden than raw or pelleted litter; >5-log pathogen reductions under thermophilic conditions.	(Sharma and Reynnells, 2016; Esperón et al., 2020; Song et al., 2020; Lopes et al., 2025)
Over-application	High PL rates reduced microbial richness, suppressed enzyme activity, and prolonged pathogen survival in soil	(Acosta-Martínez and Harmel, 2006; Shah et al., 2019)
Residual poultry gut microbes	Soil amended with raw PL retained poultry-associated taxa (e.g. *Enterococcus, Clostridium*) for weeks to months. Field studies demonstrated transfer of *Salmonella* from PL-amended soil to lettuce up to 21 days post-application.	(Bucher et al., 2020; Parente et al., 2021)
Heat/thermal treatments vs. raw	Heat drying >80°C consistently inactivated *Salmonella* and viruses; reduced odor, improved handling, reduced risks of pathogen carry over, and minimized shifts in soil microbiota. Lower moisture content decreases pathogen survival risk during storage.	(Wilkinson et al., 2011; Ghaly and Alhattab, 2013)

influence PL's physical, chemical, and biological characteristics, making it easier to transport, store, and apply in field settings. Properly treated PL, such as composted or pelletized forms, reduces the risks of introducing zoonotic pathogens, enhances microbial activity, stimulates nutrient turnover, and supports crop growth more sustainably than untreated or overly applied raw litter. In contrast, raw or inadequately treated PL may prolong pathogen persistence, disrupt microbial balance, and facilitate the accumulation of ARGs, underscoring the critical importance of treatment prior to land application to safeguard both soil health and food safety. Integrated approaches that combine effective treatment with responsible storage and application practices are critical for maximizing both microbial safety and agronomic value. With validated efficacy, many of these methods support compliance with FSMA Produce Safety Rule standards and are increasingly adopted in commercial compost and pellet production systems (Chen and Jiang, 2017b; Wang et al., 2021).

Key findings on microbial shifts, ARG persistence, and enzyme activities in soils amended with PL highlight both the benefits of composting (e.g. enhanced enzyme activities and reduced ARG abundance) and pelletized PL, as well as the risks associated with raw or over-applied litter, including the reduced microbial diversity and persistence of MDR *Enterococcus*. These observations emphasize that environmental factors–such as moisture, temperature, soil physicochemical properties, amendment history, crop interactions, and management practices–play a more decisive role in shaping pathogen survival and soil microbial dynamics than treatment strategies alone. Ultimately, selecting an appropriate treatment method is not only critical for microbial safety but also presents an opportunity to improve soil structure, support microbial diversity, and enhance nutrient efficiency. When properly applied, treated PL contributes positively to soil fertility and long-term sustainability while mitigating the risks of pathogen dissemination and ARG accumulation.

4.4 Strategies to improve microbial safety and restore soil balance post-poultry litter application

Although PL treatments effectively reduce pathogen load, its application can still interfere with native soil microbial communities and enable the regrowth of surviving pathogens under favorable conditions. As shown in studies of PL-amended soils (Table 3), over-application of raw PL reduces microbial diversity and enriches resistance reservoirs. Given the microbial risks, there is growing interest in biological strategies such as microbial inoculants that can reinforce soil microbial defenses and counteract the long-term shifts in microbial diversity caused by repeated PL applications. Researchers are increasingly exploring microbial inoculants, particularly CEMs, to enhance the biological safety and resilience of PL-amended soils, as a complementary strategy to suppress pathogens and restore microbial balance. Unlike composting or chemical stabilization, microbial inoculants are applied after PL reaches the soil. However, they represent an essential biological strategy to mitigate microbiome disturbance and reduce the risk of pathogen regrowth or ARG persistence. CEMs function by outcompeting or directly inhibiting harmful organisms through niche exclusion, antimicrobial production, or metabolic competition. Jiang and Wang (2022) isolated CEMs from compost products and then used them as the biological control agents against *L. monocytogenes*. The addition of CEM cultures (n=15), including effective taxa of *Bacillus*, *Paenibacillus*, and *Actinobacteria* strains, resulted in up to 2 log reduction of *L. monocytogenes* population in both dairy and poultry composts after 168 h incubation. In another study, Jiang and Wang (2024) reported that over 99% of *E. coli* O157:H7 populations introduced via cross-contaminated compost with ≥30% moisture was effectively inactivated by CEMs within 2 days during colder seasons. Similarly, Bucher et al. (2020) reported that reused PL fosters microbial communities capable of suppressing *Salmonella* invasion by over 50%. The enrichment of beneficial taxa such as *Lactobacillus*, *Faecalibacterium*, and *Ruminococcaceae* suggests that natural microbial succession may be harnessed to enhance soil health and safety. However, the broader soil microbiota context is also critical. Peng et al. (2021) conducted greenhouse/field studies and showed *Bacillus subtilis* and *Pseudomonas fluorescens* inoculants partially suppressed *E. coli* and *Salmonella,* improved phosphorus solubilization, and reduced native soil pathogen loads. Furthermore, field trials with compost amendments showed partial restoration of soil microbial diversity within 21~30 days and reductions in ARG markers (*sul1*, *tetM*) by ~1 log with outcomes depending on amendment history (Lopes et al., 2025; Major et al., 2020). These findings highlight the importance of microbial inoculants–such as compost-derived consortia and plant-beneficial probiotics–and maintaining soil biodiversity to mitigate microbial risks in PL–amended systems. However, commercial deployment of these inoculants faces challenges, including strain compatibility, storage stability, field efficacy, and regulatory approval. In the United States, products intended for pathogen suppression must be registered under the EPA's Federal Insecticide, Fungicide, and Rodenticide Act (FIFRA), which requires rigorous testing for identity, genetic stability, safety, and performance (Chen and Jiang, 2017a,b). These regulatory demands have limited the availability of pathogen-targeting soil probiotics, despite strong research support for their role in enhancing crop safety and nutrient cycling.

Ensuring microbial safety post-application also depends on sound management practices and regulatory compliance. Under FIFRA, commercial microbial products must meet strict standards that can slow innovation and limit market access (Lehman et al., 2015). Compliance with the FSMA is uneven, particularly among smaller producers who face barriers such as high testing costs and limited access to validated treatment technologies. While the FSMA Produce Safety Rule requires documentation of treatment processes and microbial benchmarks (Clements et al., 2019), these gaps can compromise food safety (Adalja and Lichtenberg, 2018).

Furthermore, post-treatment monitoring is essential for managing microbial risks during the storage and application of properly treated PL. Routine moisture control, microbial testing, and soil health indicators can support risk-based decision-making. This highlights the importance of treatment validation and tailoring application practices to site-specific conditions like soil type and moisture. To enhance pathogen monitoring–typically conducted using indicators like *E. coli* or molecular tools such as qPCR targeting *Salmonella*–the inclusion of antibiotic resistance genes (ARGs) and soil health indicators, including enzyme activities (e.g. dehydrogenase, phosphatase) and microbial biomass, has been proposed as a practical strategy for assessing both treatment efficacy and the long-term ecological impacts of PL application (Acosta-Martínez and Harmel, 2006; Peng et al., 2021). These findings underscore the need for integrated monitoring frameworks that address microbial safety and soil health, while balancing regulatory requirements with the operational realities faced by producers.

Future research should focus on field validation of CEMs, improving formulation stability, and developing affordable, standardized monitoring tools. These innovations are key to unlocking the full potential of treated PL as a safe, effective, and sustainable soil amendment/biofertilizer.

5 Conclusion and future trends in research

The optimization of PL treatments for safe and effective land application is inherently complex. There is no one-size-fits-all solution, as the ideal treatment strategy depends on a balance between pathogen inactivation, nutrient retention, and long-term soil health outcomes. While physical heat treatments such as thermophilic composting and pelletization have demonstrated strong efficacy in reducing pathogens (>5-log reduction under ideal conditions), their performance is highly contingent on post-treatment conditions such as moisture content, storage practices, application rate, and soil characteristics at the application site. Integrating microbial inoculants with real-time monitoring systems into PL management offers a proactive strategy for sustainably enhancing soil fertility while ensuring environmental and crop safety.

Future research should prioritize the development and application of standardized microbial indicators to evaluate treatment effectiveness beyond traditional culture-based approaches. In addition, intelligent monitoring tools, including real-time sensors for moisture and temperature, and DNA-based microbial diagnostics, offer promising avenues for adaptive management. Expanding the toolbox of validated biocontrol agents, supported by regulatory pathways that balance safety and innovation, will also be critical. Additionally, future studies are needed to evaluate whether pathogens that survive PL treatment through composting, physical heat treatment, or chemical methods can regrow following rewetting from irrigation or rainfall, and to identify strategies to mitigate such potential microbial risk. Ultimately, sustainable use of PL requires an integrated approach that aligns treatment technologies, site-specific management, regulatory compliance, and ecological monitoring. Continued investment in interdisciplinary research and stakeholder engagement will be key to translating scientific advances into scalable, field-ready solutions.

6 Where to look for further information

6.1 Further reading

- Chen and Jiang (2014): An extensive review of microbiological safety on the prevalence of foodborne pathogens in poultry litter and poultry litter-based organic fertilizers.

- Clements, D. et al. (2019): Overview of documentation requirements for commercial soil amendment suppliers under the FSMA Produce Safety Rule.
- Gurtler, J.B. et al. (2018): A comprehensive review on composting strategies aimed at inactivating foodborne pathogens for safe crop soil application.
- Harmel, D. et al. (2008): A study evaluating the agro-economic impacts of using poultry litter as fertilizer in cropland systems.
- Sharma, M. and Reynnells, R. (2016): Importance of Soil Amendments: Survival of Bacterial Pathogens in Manure and Compost Used as Organic Fertilizers.
- Wang, J. et al. (2023): A systematic review and meta-analysis of pre-harvest sources of *Salmonella* in poultry production and their relative contributions to contamination risks in poultry meat.

6.2 Key journals/conferences

- Applied and Environmental Microbiology: Publishes high-impact, novel research on applied and environmental microbiology, including microbial safety and agricultural applications.
- Bioresource Technology: Covers original research on biomass, organic waste treatment, and bioconversion technologies.
- US Composting Council (USCC) Annual Conference: The world's largest conference dedicated to composting and organics recycling, offering valuable insights for both industry and research sectors.
- ASM Microbe Annual Meeting: the premier global event for the microbiology community covering disciplines from basic science to clinical applications, environmental microbiology, antimicrobial resistance, and biotechnology
- International Association for Food Protection (IAFP) Annual Meeting: Provides up-to-date information on emerging food safety issues, including pre-harvest risks and environmental pathogen control.

7 References

Acosta-Martínez, V. and Harmel, R.D. (2006) "Soil microbial communities and enzyme activities under various poultry litter application rates." *Journal of Environmental Quality*, 35(4): 1309–1318. Available at: https://doi.org/10.2134/jeq2005.0470.

Adalja, A. and Lichtenberg, E. (2018) "Implementation challenges of the food safety modernization act: evidence from a national survey of produce growers." *Food Control*, 89: 62–71. Available at: https://doi.org/10.1016/j.foodcont.2018.01.024.

Agga, G.E. et al. (2025) "Poultry litter as a soil amendment enriches antibiotic resistant enterococci." *Environmental Pollution*, 380: 126531. Available at: https://doi.org/10.1016/j.envpol.2025.126531.

Ashworth, A.J. et al. (2020) Nutrient characteristics of poultry manure and litter. In: Waldrip, H. et al. (Eds.), *Animal manure: Production, characteristics, environmental concerns and management* (pp. 63–87). John Wiley and Sons, Ltd.

Atanasova, A. et al. (2025) "Temporal dynamics of antimicrobial resistance gene abundances in chicken manure and anaerobic digestate." *Frontiers in Antibiotics*, 4: 1612886. Available at: https://doi.org/10.3389/frabi.2025.1612886.

Ayantokun, A.S. et al. (2025) "A critical review of food waste and poultry manure anaerobic co-digestion: an eco-friendly valorization for sustainable waste management and biogas production." *Frontiers in Sustainable Food Systems*, 9: 1695945. Available at: https://doi.org/10.3389/fsufs.2025.1695945.

Bao, Y. et al. (2021) "Important ecophysiological roles of non-dominant Actinobacteria in plant residue decomposition, especially in less fertile soils." *Microbiome*, 9(1): 84. Available at: https://doi.org/10.1186/s40168-021-01032-x.

Bardsley, C.A. et al. (2021) "Strain, soil-type, irrigation regimen, and poultry litter influence Salmonella survival and die-off in agricultural soils." *Frontiers in Microbiology*, 12: 590303. Available at: https://doi.org/10.3389/fmicb.2021.590303.

Bucher, M.G. et al. (2020) "Reused poultry litter microbiome with competitive exclusion potential against Salmonella Heidelberg." *Journal of Environmental Quality*, 49(4): 869–881. Available at: https://doi.org/10.1002/jeq2.20081.

Chen, Z. and Jiang, X. (2017a) "Microbiological safety of animal wastes processed by physical heat treatment: an alternative to eliminate human pathogens in biological soil amendments as recommended by the food safety modernization act." *Journal of Food Protection*, 80(3): 392–405. Available at: https://doi.org/10.4315/0362-028X.JFP-16-181.

Chen, Z. and Jiang, X. (2017b) "Selection of indigenous indicator micro-organisms for validating desiccation-adapted Salmonella reduction in physically heat-treated poultry litter." *Journal of Applied Microbiology*, 122(6): 1558–1569. Available at: https://doi.org/10.1111/jam.13464.

Chen, Z. and Jiang, X. (2014) "Microbiological safety of chicken litter or chicken litter-based organic fertilizers: a review." *Agriculture*, 4: 1–29. Available at: https://doi.org/10.3390/agriculture4010001.

Chen Z. et al. (2013) "Thermal inactivation of desiccation-adapted *Salmonella spp.* in aged chicken litter". *Applied and Environmental Microbiology*, 79: 7013–7020. Available at: https://doi.org/10.1128/AEM.01969-13.

Chen, Z. et al. (2015a) "Developing a two-step heat treatment for inactivating desiccation-adapted Salmonella spp. in aged chicken litter." *Foodborne Pathogens and Disease*, 12(2): 104–109. Available at: https://doi.org/10.1089/fpd.2014.1822.

Chen Z. et al. (2015b) "Effects of chicken litter storage time and ammonia content on thermal resistance of desiccation-adapted *Salmonella spp.*" *Applied and Environmental Microbiology*, 81: 6883–6889. Available at: https://doi.org/10.1128/AEM.01876-15.

Clements, D. et al. (2019) "FSMA produce safety rule: documentation requirements for commercial soil amendment suppliers." *Produce Safety Alliance*. Available at: https://resources.producesafetyalliance.cornell.edu/documents/FSMA-PSR-Documentation-Requirements-for-Commercial-Soil-Amendment-Suppliers.pdf (Accessed: 20 April 2026).

Cook, K.L. et al. (2008) "Effect of alum treatment on the concentration of total and ureolytic microorganisms in poultry litter." *Journal of Environmental Quality*, 37(6): 2360–2367. Available at: https://doi.org/10.2134/jeq2008.0024.

Diao, J. et al. (2015) "Factors affecting pathogen survival in finished dairy compost with different particle sizes under greenhouse conditions." *Foodborne Pathogens & Disease*. 12: 749–758. Available at: https://doi.org/10.1089/fpd.2014.1912.

Dunn, L.L. et al. (2022) "The prevalence and concentration of Salmonella enterica in poultry litter in the southern United States." *PLoS One*, 17: e0268231. Available at: https://doi.org/10.1371/journal.pone.0268231.

Esperón, F. et al. (2020) "Assessing the benefits of composting poultry manure in reducing antimicrobial residues, pathogenic bacteria, and antimicrobial resistance genes: a field-scale study." *Environmental Science and Pollution Research*, 27(22): 27738–27749. Available at: https://doi.org/10.1007/s11356-020-09097-1.

Fang et al. (2015) "Prevalence of antibiotic resistance genes and bacterial pathogens in long-term manured greenhouse soil as revealed by metagenomic survey." *Environmental Science & Technology*. 49(2): 1095–1104.

Fatoba, D.O. et al. (2021) "Rethinking manure application: increase in multidrug-resistant Enterococcus spp. in agricultural soil following chicken litter application." *Microorganisms*, 9(5): 885. Available at: https://doi.org/10.3390/microorganisms9050885.

Ferguson, M. et al. (2023) "A longitudinal study to examine the influence of farming practices and environmental factors on pathogen prevalence using structural equation modeling." *Frontiers in Microbiology*, 14: 1141043. Available at: https://doi.org/10.3389/fmicb.2023.1141043.

Fremaux, B. et al. (2007) "Growth and survival of non-O157:H7 Shiga-toxin-producing *Escherichia coli* in cow manure." *Journal of Applied Microbiology*, 102: 89–99. Available at: https://doi.org/10.1038/ismej.2010.80.

Ghaly, A.E. and Alhattab, M. (2013) "Drying poultry manure for pollution potential reduction and production of organic fertilizer." *American Journal of Environmental Sciences*, 9(2): 88–102. Available at: https://doi.org/10.3844/ajessp.2013.88.102.

Grubinger, V.P. (1999) *Sustainable vegetable production from start-up to market (NRAES 104 - Front Matter Only)*. Available at: https://hdl.handle.net/1813/69441 (Accessed: 16 September 2025).

Gurtler, J.B. et al. (2018) "Composting to inactivate foodborne pathogens for crop soil application: a review." *Journal of Food Protection*, 81(11): 1821–1837. Available at: https://doi.org/10.4315/0362-028X.JFP-18-217.

Harmel, D. et al. (2008) "The MANAGE database: nutrient load and site characteristic updates and runoff concentration data." *Journal of Environmental Quality*, 37(6): 2403–2406. Available at: https://doi.org/10.2134/jeq2008.0079.

Hauck, R. et al. (2017) "Persistence of highly pathogenic and low pathogenic avian influenza viruses in footbaths and poultry manure." *Avian Diseases*, 61(1): 64–69. Available at: https://doi.org/10.1637/11495-091916-Reg.

Heinonen-Tanski, H. et al. (2006) "Methods to reduce pathogen microorganisms in manure." *Livestock Science*, 102(3): 248–255. Available at: https://doi.org/10.1016/j.livsci.2006.03.024.

Hubbard, L.E. et al. (2020) "Poultry litter as potential source of pathogens and other contaminants in groundwater and surface water proximal to large-scale confined poultry feeding operations." *Science of The Total Environment*, 735: 139459. Available at: https://doi.org/10.1016/j.scitotenv.2020.139459.

Islam, M. et al. (2004a) "Persistence of enterohemorrhagic *E. coli* O157:H7 in soil and on leaf lettuce and parsley grown in fields treated with contaminated manure composts or irrigation water." *Journal of Food Protection*, 67: 1365–1370. Available at: https://doi.org/10.4315/0362-028x-67.7.1365.

Islam, M. et al. (2004b) "Fate of *Salmonella enterica* serovar Typhimurium on carrots and radishes grown in fields treated with contaminated manure composts or irrigation water." *Applied and Environmental Microbiology*, 70: 2497–2502.

Jarosz, R. et al. (2022) "Changes in quantity and quality of organic matter in soil after application of poultry litter and poultry litter biochar–5-year field experiment." *Biomass Conversion and Biorefinery*, 12(7): 2925–2934. Available at: https://doi.org/10.1007/s13399-020-01005-4.

Jiang, X.P. and Wang, J.X. (2024) "Biological control of *Escherichia coli* O157:H7 in dairy manure-based compost using competitive exclusion microorganisms." *Pathogens*, 13(5): 361. Available at: https://doi.org/10.3390/pathogens13050361.

Kelley, T.R. et al. (1994) "Fate of selected bacterial pathogens and indicators in fractionated poultry litter during storage." *Journal of Applied Poultry Research*, 3(3): 279–288. Available at: https://doi.org/10.1093/japr/3.3.279.

Kharel, K. et al. (2025) "The effect of heat-treated poultry pellets and composted poultry litter on *E. coli* survival in Southeastern US soils: Florida and Georgia." *Journal of Food Protection*, 88: 100439. Available at: https://doi.org/10.1016/j.jfp.2024.100439.

Kim, J. et al. (2009) "Factors impacting the regrowth of *Escherichia coli* O157:H7 in dairy manure compost." *Journal of Food Protection*, 72(7): 1576–1584. Available at: https://doi.org/10.4315/0362-028X-72.7.1576.

Kim, J. and Jiang, X. (2010) "The growth potential of *Escherichia coli* O157: H7, *Salmonella* spp. and *Listeria monocytogenes* in dairy manure-based compost in a greenhouse setting under different seasons." *Journal of Applied Microbiology*, 109(6): 2095–2104. Available at: https://doi.org/10.1111/j.1365-2672.2010.04841.x.

Lehman, R.M. et al. (2015) "Soil biology for resilient, healthy soil." *Journal of Soil and Water Conservation*, 70: 12A–18A. Available at: https://doi.org/10.2489/jswc.70.1.12A.

Li, X. et al. (2007) "Salmonella populations and prevalence in layer feces from commercial high-rise houses and characterization of the Salmonella isolates by serotyping, antibiotic resistance analysis, and pulsed field gel electrophoresis." *Poultry Science*, 86(3): 591–597. Available at: https://doi.org/10.1093/ps/86.3.591.

Liu, X. et al. (2022) "Long-term application of bio-compost increased soil microbial community diversity and altered its composition and network." *Microorganisms*, *10*(2): 462. Available at: https://doi.org/10.3390/microorganisms10020462.

Lopes, E.D.S. et al. (2025) "Microbiome and resistome in poultry litter-fertilized and unfertilized agricultural soils." *Antibiotics*, 14(4): 355. Available at: https://doi.org/10.3390/antibiotics14040355.

Lu, J. et al. (2003) "Evaluation of broiler litter with reference to the microbial composition as assessed by using 16S rRNA and functional gene markers." *Applied and Environmental Microbiology*, 69(2): 901–908. Available at: https://doi.org/10.1128/AEM.69.2.901-908.2003.

Major, N. et al. (2020) "Composted sewage sludge influences the microbiome and persistence of human pathogens in soil." *Microorganisms*, 8: 1020. Available at: https://doi.org/10.3390/microorganisms8071020.

Natvig, E.E. et al. (2002) "Salmonella enterica serovar Typhimurium and Escherichia coli contamination of root and leaf vegetables grown in soils with incorporated bovine manure." *Applied and Environmental Microbiology*, 68(6): 2737–2744. Available at: https://doi.org/10.1128/AEM.68.6.2737-2744.2002.

Neher, D.A. et al. (2020) "Bacterial community dynamics distinguish poultry compost from dairy compost and non-amended soils planted with spinach." *Microorganisms*, 8(10): 1601. Available at: https://doi.org/10.3390/microorganisms8101601.

Nguyen, X.D. et al. (2023) "Modeling long-distance airborne transmission of highly pathogenic avian influenza carried by dust particles." *Scientific Reports*, 13(1): 16255. Available at: https://doi.org/10.1038/s41598-023-42897-2.

National Organic Standards Board (NOSB). (2002) *Compost task force recommendations, as amended by the NOP*. Available at: https://www.ams.usda.gov/sites/default/files/media/NOP%20Final%20Rec%20Guidance%20use%20of%20Compost.pdf (Accessed: 18 April 2026).

Pal, A. et al. (2021) "Impact of poultry litter Salmonella levels and moisture on transfer of Salmonella through associated in vitro generated dust." *Poultry Science*, 100(8): 101236. Available at: https://doi.org/10.1016/j.psj.2021.101236.

Parente, C.E.T. et al. (2021) "Bacterial diversity changes in agricultural soils influenced by poultry litter fertilization." *Brazilian Journal of Microbiology*, 52(2): 675–686. Available at: https://doi.org/10.1007/s42770-021-00437-y.

Peng, S. et al. (2021) "Chicken manure and mushroom residues affect soil bacterial community structure but not the bacterial resistome when applied at the same rate of nitrogen for 3 years." *Frontiers in Microbiology*, 12. Available at: https://doi.org/10.3389/fmicb.2021.618693.

Pope, M.J. and Cherry, T.E. (2000) "An evaluation of the presence of pathogens on broilers raised on poultry litter treatment-treated litter." *Poultry Science*, 79(9): 1351–1355. Available at: https://doi.org/10.1093/ps/79.9.1351.

Reynnells, R. et al. (2014) "Comparison of U.S. Environmental Protection Agency and U.S. Composting Council microbial detection methods in finished compost and regrowth potential of Salmonella spp. and Escherichia coli O157:H7 in finished compost." *Foodborne Pathogens and Disease*, 11(7): 555–567. Available at: https://doi.org/10.1089/fpd.2013.1698.

Ritz, C.W. and Merka, W.C. (2013) "Maximizing poultry manure use through nutrient management planning." Available at: https://api.semanticscholar.org/CorpusID:127545310.

Sainju, U.M. et al. (2010) "Poultry litter application increases nitrogen cycling compared with inorganic nitrogen fertilization." *Agronomy Journal*, 102(3): 917–925. Available at: https://doi.org/10.2134/agronj2009.0482.

Sauer, T.J. et al. (2015) "Nutrient cycling in an agroforestry alley cropping system receiving poultry litter or nitrogen fertilizer." *Nutrient Cycling in Agroecosystems*, 101(2): 167–179. Available at: https://doi.org/10.1007/s10705-014-9667-0.

Shah, M.K. et al. (2019) "Salmonella enterica in soils amended with heat-treated poultry pellets survived longer than bacteria in unamended soils and more readily transferred to and persisted on spinach." *Applied and Environmental Microbiology*, 85(10): e00334-19. Available at: https://doi.org/10.1128/AEM.00334-19.

Shakya, A. and Agarwal, T. (2017) "Poultry litter biochar: an approach towards poultry litter management – a review." *International Journal of Current Microbiology and Applied Sciences*, 6(10): 2657–2668. Available at: https://doi.org/10.20546/ijcmas.2017.610.314.

Sharma, M. and Reynnells, R. (2016) "Importance of soil amendments: survival of bacterial pathogens in manure and compost used as organic fertilizers." *Microbiology Spectrum*, 4(4): PFS-0010-2015. Available at: https://doi.org/10.1128/microbiolspec.PFS-0010-2015.

Schierstaedt, J. et al. (2020) "Salmonella persistence in soil depends on reciprocal interactions with indigenous microorganisms." *Environmental Microbiology*, *22*: 2639–2652. Available at: https://doi.org/10.1111/1462-2920.14972.

Shepherd, M.W. et al. (2007) "Fate of *Escherichia coli* O157: 117 during on-farm dairy manure-based composting". *Journal of Food Protection*, 70(12): 2708–2716. Available at: https://doi.org/10.4315/0362-028x-70.12.2708.

Sims, J.T. and Luka-McCafferty, N.J. (2002) "On-farm evaluation of aluminum sulfate (alum) as a poultry litter amendment." *Journal of Environmental Quality*, 31(6): 2066–2073. Available at: https://doi.org/10.2134/jeq2002.2066.

Singer, R.S. et al. (2023) "Antimicrobial usage in broiler chicken production in the United States, 2013–2021." *Frontiers in Veterinary Science*, 10: 1139908. Available at: https://doi.org/10.3389/fvets.2023.1139908.

Singh, R. et al. (2012) "Heat inactivation of *Salmonella* spp. in fresh poultry compost by simulating early phase of composting process." *Journal of Applied Microbiology*, 112: 927–938. Available at: https://doi.org/10.1111/j.1365-2672.2012.05268.x.

Solaiman, Z.M. et al. (2020) "Poultry litter biochar increases mycorrhizal colonisation, soil fertility and cucumber yield in a fertigation system on sandy soil." *Agriculture*, 10(10): 480. Available at: https://doi.org/10.3390/agriculture10100480.

Song, Z. et al. (2020) "Composted chicken manure for anaerobic soil disinfestation increased the strawberry yield and shifted the soil microbial communities." *Sustainability*, 12(16): 6313. Available at: https://doi.org/10.3390/su12166313.

Tabler, T. et al. (2024) *Poultry litter management (inside and outside the house)*. University of Tennessee Institute of Agriculture.

Terzich, M. et al. (2000) "Survey of pathogens in poultry litter in the United States." *Journal of Applied Poultry Research*, 9(3): 287–291. Available at: https://doi.org/10.1093/japr/9.3.287.

USDA. (2025) *"Chicken meat" USDA foreign agricultural service*. Available at: https://www.fas.usda.gov/data/production/commodity/0115000 (Accessed: 16 September 2025).

USDA NASS. (2025) "Poultry – production and value 2024 summary." Available at: https://www.nass.usda.gov/Publications/Todays_Reports/reports/plva0425.pdf (Accessed: 16 April 2026).

Wang, D. et al. (2024) "Effects of compost application of green waste on soil properties: a meta-analysis." *Sustainability*, 16(20): 8877. Available at: https://doi.org/10.3390/su16208877.

Wang, H. et al. (2021) "Plant-scale validation of physical heat treatment of turkey and laying hen litter composts using surrogate and indicator microorganisms for

Salmonella." *Applied and Environmental Microbiology*, 87(5): e02234-20. Available at: https://doi.org/10.1128/AEM.02234-20.

Wang, H. et al. (2025) Managing poultry litter to improve safety and reduce environmental impact. In S. C. Ricke (Ed.), *Improving poultry meat safety and sustainability*. Burleigh Dodds Science Publishing.

Wang, J. et al. (2023) "A systematic review and meta-analysis of pre-harvest sources of *Salmonella* in poultry production and their relative contributions to contamination risks in poultry meat." *Poultry Science*, 102(5): 102566. Available at: https://doi.org/10.1016/j.psj.2023.102566.

Warren, J.G. et al. (2006) "Environmental and production consequences of using alum-amended poultry litter as a nutrient source for corn." *Journal of Environmental Quality*, 35(1): 172–182. Available at: https://doi.org/10.2134/jeq2004.0418.

Wilkinson, K.G. et al. (2011) "Effect of heating and aging of poultry litter on the persistence of enteric bacteria." *Poultry Science*, 90(1): 10–18. Available at: https://doi.org/10.3382/ps.2010-01023.

Zalewska, M. et al. (2021) "Antibiotics and antibiotic resistance genes in animal manure – consequences of its application in agriculture." *Frontiers in Microbiology*, 12: 610656. Available at: https://doi.org/10.3389/fmicb.2021.610656.

Zhu et al. (2022) "The profile and persistence of clinically critical antibiotic resistance genes and human pathogenic bacteria in manure-amend farm land soils". *Frontiers in Cellular and Infection Microbiology*, 12: 1073118.

Chapter 7

Optimising on-farm management and use of slurry to minimise its environmental impact

Lisa Norton, UK Centre for Ecology and Hydrology, UK; and Emma Cardwell and Claire Waterton, Lancaster University, UK

1 Introduction

Slurry is a significant by-product of livestock agriculture in the United Kingdom (as with other countries with significant livestock sectors). It consists of either exclusively liquid or semi-liquid excreta or excreta mixed with livestock bedding, rainwater and washings from yards, which can be pumped or discharged by gravity (SSAFO, 2010). In 2016, an estimated 83 million tonnes fresh weight of livestock manure was produced during the housing period in the UK. Of this, 67 Mt (80%) came from cattle: 47% of this was estimated to be undiluted slurry and 53% solid, mainly straw-based farmyard manure (FYM) (Smith and Williams, 2016). Regional differences in cattle manure management mean that there is a predominance of slurry in South West Scotland, Northern Ireland and Wales, compared to the relative importance of straw-based FYM systems in England

http://dx.doi.org/10.19103/AS.2025.0166.20

(ADAS, 2008). Within England, slatted floor housing systems producing cattle slurry are more often found in the North and North-West.

Slurry management incorporates its handling, storage and use. Slurry can be a valuable source of nutrients (phosphorus, nitrogen, sulphur, potassium and trace elements) if managed efficiently within the agricultural sector in the UK. However, poor management results in the loss of these nutrients to water and air and the consequent pollution of wider ecosystems through both point and diffuse pollution. Ensuring that slurry is used as an effective resource is essential to avoiding pollution of land, water and air and to creating a more sustainable circular bioeconomy (Rothwell et al., 2020).

Farmers have to make decisions about what types of cattle they farm, how many and whether and how to house their cattle, as well as how to manage slurry storage (management and investment), application (rate, timing and method) and use all year round. In the UK, they are required to do this within new legal regulations brought in 2024 to protect the environment and animal health and welfare, as well as within Nitrate Vulnerable Zones, where they apply.

It is well recognised that the way in which government and industry support good practice for handling and utilising slurry across the UK is important for business efficiencies as well as for farm, catchment and wider environmental sustainability. Much recent work in this area continues to focus on technological and natural science solutions to slurry management including familiar areas like slurry separation, slurry covers and application methods alongside new areas such as the role of diet in reducing methane outputs from livestock and biological means of waste processing (e.g. use of black soldier fly larvae) (see Symeon et al., 2025).

However, compared to the vast quantity of science and technological research on slurry management (close to 15 000 papers between January 2022 and December 2025), there has been relatively little on how well this research has been translated into practical guidance and tools that farmers are willing to engage with and use (Rose et al., 2016; Rose et al., 2018).

2 Exploring farmer engagement with best practice in slurry management: the SLURRY-MAX Project

Funded by over 100 000 producers across the dairy, beef, lamb, pork and cereal sectors, the Agriculture and Horticulture Development Board (AHDB) is one of the leading providers of independent advice to British farmers. The Sustainable Agriculture Research and Innovation Club (SARIC) is an initiative funded by the Biotechnology and Biological Sciences Research Council (BBSRC) and the Natural Environment Research Council (NERC), two of the leading funders of science research in the UK, to fund innovative research programmes promoting more sustainable practices in UK farming.

In 2016, AHDB approached SARIC to help in 'understanding more about the provision of decision-support on organic slurry storage and treatment techniques.' The project was partly framed around the idea (prevalent among advisers and the agricultural industry at the time) that, despite the existence of many decision support tools, a large proportion of livestock farmers were still not engaging with them or managing slurries and manures effectively.

This led to the SLURRY-MAX Research Project funded from 2016 to 2018 and designed, in part, to understand this 'information behaviour gap' (Waterton et al., 2018). In contrast to most work around slurry which focuses on dairy systems, the SLURRY-MAX Project focused on beef slurry which constituted over 40% of all slurry by weight at the time of the project (Smith and Williams, 2016). Slurry made up the majority of all handled manures for beef production in Wales, Scotland and Northern Ireland (NI) (86%, 56% and 83% respectively; see Smith and Williams, 2016).

The project did not consider farmer decision-making around types or numbers of animals, although these have an obvious bearing on slurry volumes for beef production, with hardier cattle breeds often able to stay outside on upland pastures for more of the year and even through winter. Instead, the project focused on storage and management challenges for beef slurry which were under-represented in advice provision at the time of the project. However, these farms, often in agriculturally Less Favoured Areas of the UK, may have important environmental impacts by virtue of being located in sensitive habitats including headwater catchments, see (Norton et al., 2012). Beef farmers, in contrast to dairy farmers, tend to manage smaller areas and have reduced infrastructure and financial turnover compared to dairy farms. This often leads to them being less likely to access professional farm advisors given the costs of doing so (Defra, 2024). Key challenges for beef farmers include how best to manage their slurry given the sensitivity of the environments in which they farm and where to access relevant free advice, including online resources, for slurry management.

The project included both social and natural scientists working together with farmers and advisors specifically around beef farming systems (though many results were applicable to dairy) to:

- enhance the accessibility of decision support tools around farmers' practical questions about slurry management through working alongside them to better understand their needs; and
- consider how other support mechanisms (besides decision support tools) may help to improve nutrient management practices on farms.

The project looked at how beef farmers used decision support tools in slurry management and sought to understand the social practices of farmers around slurry management and use. Subsequently, it focused on finding solutions to enhance decision support around slurry that would both support the resilience of agricultural production and minimise environmental impacts.

Broadly speaking, research on farmer behaviour has steadily moved beyond simple models based solely on individual attitudes or economic rationality, towards more integrated and multi-faceted perspectives that consider social, institutional and contextual influences. Reviews of related literature, such as those by Baumgart-Getz et al. (2012), Dessart et al. (2019) and Prokopy et al. (2019), highlight how farmer decision-making is shaped by a combination of factors, including economic incentives, environmental values, social norms, policy frameworks, knowledge networks and structural constraints. These studies demonstrate that the adoption of agricultural practices is typically influenced by a confluence of psychological, social and institutional drivers, rather than any single determinant. Such insights have encouraged researchers to adopt a more holistic, analytical framework when examining how and why farmers adopt or reject particular management practices.

With this in mind, the research also took a social practices (Chatterton and Wilson, 2014) approach to understanding farmers management of slurry, recognising 'that patterns of behaviour are tied up in shared meanings and skills within interpersonal networks, communities and other social institutions.' Rather than considering the behaviours and attitudes of individuals, the work investigated how patterns of behaviour are dependent on wider networks and routines which are themselves constrained by physical infrastructures, bodies and technologies. This approach was deliberately taken to step away from traditional assumptions that farmers' decision-making is solely dependent on individual attitudes or simple financial calculation.

Other work focusing on the adoption of sustainable agricultural practices more broadly has looked at factors affecting uptake of a wide range of practices. Han and Niles (2023) propose an 'adoption spectrum' framework that conceptualises farmer uptake of practices as a dynamic and multidimensional process. Their framework highlights a number of elements responsible for shaping adoption, emphasising that adoption is rarely a simple, one-dimensional choice but involves varying levels of implementation and persistence over time. The framework also identifies different types of adopters including non-adopters, early adopters, conservative and innovative new adopters. By highlighting the variability in adoption pathways and farmer engagement, this study shows that farmer decision-making can vary over time and in different contexts.

While this chapter focuses primarily on the UK and local contexts and on beef farms, slurry decision-making internationally faces similar challenges. A review by Niles et al. (2019) on determinants of dairy farmers' manure management decisions in high-income countries highlights that farmers' choices are influenced by a combination of economic incentives, environmental regulations and social networks. Studies like this can demonstrate that challenges – such as balancing productivity with environmental compliance, managing labour and equipment costs and integrating scientific advice into routine farm practice – are common across diverse contexts and locations.

It is, of course, no surprise that the adoption of high-tech manure management systems is strongly influenced by economic and technical considerations as well as social factors, even in relatively intensive dairy systems. Research indicates that establishing and maintaining effective slurry storage, separation and spreading infrastructure can entail substantial upfront costs, including investments in tanks, pumps and spreading equipment (Bittman et al., 2026). However, these investments can also offer long-term financial benefits through improved nutrient use efficiency, reduced fertiliser purchases and potential avoidance of environmental fines. Understanding the balance of these costs and benefits is crucial, as farmers frequently cite financial barriers as a disincentive to adopting best-practice slurry management.

Sutton (2009) highlighted how manure and compost had been reframed from waste to economic inputs; the pricing of manure was shown to be contingent on factors such as nutrient content, soil benefits and application context, while also being constrained by practical costs like transportation and handling – meaning its economic viability was highly situational. This instance reflects a broader principle in agricultural economics, namely the cost-effectiveness of sustainable farming techniques hinging not only on theoretical benefits but on logistical, spatial and market factors that shape real-world adoption (Sutton, 2009). As an example, capital-intensive technologies such as anaerobic digesters are often financially unviable without subsidies and are limited to large-scale operations (Anderson et al., 2013), while lower-cost practices (e.g. in managing slurry) offer more scalable but still incentive-dependent pathways to achieve environmental and economic gains (De Laporte et al., 2022).

Taken together, this range of examples suggests that, while sustainable practices can deliver long-term cost benefits and environmental efficiencies, their uptake is ultimately determined by farmers wider networks and routines which are themselves constrained by factors such as upfront costs for technologies, physical infrastructures and the presence of policy or market support mechanisms.

3 SLURRY-MAX Project findings: decision support tools and their use

We conducted focus groups (5 with between 15 and 28 students), questionnaires (84), in-depth interviews (>40) and email consultations (~50)

Table 1 Tools investigated and score of user-friendliness

Tool	Score on user-friendly criteriaLow scores = more use friendly
PLANET	34
The Farm Crap App	26
MANNER_NPK	31
DAERA Nutrient Calculator (Northern Ireland)	26
Scottish Government N Max Calculator (Scotland)	26
DairyCo – Slurry Wizard (housing)	29

NB, the lowest score for each of the 15 questions was 1 - hence a very user friendly tool would score 15 and a very difficult to use tool would score 45.

with farmers, agricultural advisors, contractors, regulators and decision support tool developers to better understand tool use and slurry management issues. Participants were recruited on a voluntary basis at agricultural colleges, cattle markets and agricultural shows and all responses were anonymised. Researchers also attended four tool training sessions (PLANET, Slurry Wizard and Farm Crap App).

The 'tools' that were assessed included software, leaflets and online resources for advice on some aspect of slurry management which were available at the time of the study. The software-based tools chosen for further investigation are included in Table 1. An example of a decision support tool developed to aid manure management is the Farm Crap App, created by Rothamsted Research. It is a mobile application designed to help farmers optimise the use of livestock manure, including slurry, farmyard manure and poultry litter. The App allows farmers to input data (e.g. on soil and crop type and type of manure) which is then used to calculate nutrient values of manure (which can then be compared e.g. to conventional fertiliser nutrient values and potential savings from using manure). A calculator estimates the amount of crop available key nutrients (nitrogen, potassium and phosphorus) in the manure at different spreading rates, enabling farmers to determine how much to apply to meet crop requirements.

All tools draw on the scientific knowledge contained in RB209 (Research Book 209) historically produced by government in collaboration with research and industry bodies. Interviews with advisors as part of this research highlighted the fact that RB209 is now managed by an industry body (the AHDB), rather than government. The AHDB carries out annual reviews to establish areas that require updating in conjunction with partners and a Steering Group who co-ordinate research and knowledge exchange to generate independent and robust research on crop nutrient management[1] including grass and forage.

[1] https://ahdb.org.uk/rb209.

One issue, both for the software tools investigated and for the development of any new tools, is the extent to which the tools are able to take advantage of these (no longer public) RB209 updates. This is likely to be highly dependent on how they are supported, with those supported by national governments potentially more likely to receive funds for updates. This issue could lead to future inconsistency in nutrient management information across tools.

The SLURRY-MAX research benefitted from the work by Rose et al. (2016) which took place shortly before the project began. This work pulled together and assessed a number of Decision Support Tools for Agriculture. In our research, the usability of the slurry management software tools was considered by experts against a list of 15 criteria which had been shown to influence decision support tool uptake (and which reflect the broad range of economic, social and other factors involved in decision-making) (Rose et al., 2016):

Core factors:

- performance
- ease of use
- peer recommendation
- trust
- cost
- habit
- relevance to user
- farmer advisor compatibility

Enabling factor:

- facilitating conditions

Driving factors:

- compliance (legislation)
- level of marketing

Modifying factors:

- age
- scale of business
- farming type
- IT education

Experts carrying out the assessments had previously interviewed farmers about their use of tools for slurry management and had carried out focus group workshops with agricultural students at four colleges in England, Scotland,

Wales and Northern Ireland during which students tested and evaluated the tools and discussed their findings with members of the research team.

Many of those interviewed did not use slurry management tools, with just 12 of 84 farmers surveyed using either paper or software tools. Most of those using them did not report positively on their usability. Of 11 farmers interviewed in depth, 3 had not heard of any of the tools we considered and 8 had tried them but had chosen not to use them. The reasons given were repeated in tool tests with students, where tools were assessed as being difficult to use, too time consuming and providing too little information of value.

In general, the expert scores for the tools investigated were low to medium in terms of how user-friendly the tools are for farmers. Rose et al. (2016) and Rose and Bruce (2018) showed that tools were much more useful and usable when they were co-developed with users. While a lack of co-development is likely to have influenced the usability of the tools investigated in Table 1, it is also clear that some farmers may never engage with tools (or their co-development). This may be because they have other priorities relating to slurry to consider, such as infrastructure or they may not like using IT within their farm management. Even in areas with heavy regulation in place, such as in Northern Ireland, evidence from our interviews with advisors revealed that many farmers still do not use the DAERA Nutrient Calculators.

Interviews with tool developers revealed that these results were not surprising, and responses tended to focus on factors beyond the tool developers' control, such as lack of computer literacy or record keeping by farmers. Such arguments had led to the development of paper resources. However, farmer interviews revealed that many farmers (or family members) did use digital systems for other farming applications, perhaps pointing to tool design around slurry management being a limiting factor. This is explored further in the following section.

4 SLURRY-MAX Project findings: farmers' social practices around slurry

Slurry storage and management is carried out by farmers, farm workers and contractors; they play a vital role in its beneficial and/or detrimental impacts. We investigated how their patterns of behaviour relating to slurry are tied up in, and influenced by, wider networks and routines, and the extent to which they may be shaped and constrained by physical infrastructures, bodies and technologies.

In-depth interviews were semi-structured with defined sets of questions based on the role of the interviewee. Interviews were subsequently transcribed and coded for analysis based on the themes emerging from them. Two categories were created based specifically on:

- potential improvements to decision support tools; and
- wider slurry management contexts.

The social science data collected provided a long list of recommendations for tools (a) which were ranked according to:

- the priority/need and potential benefits (as expressed by farmers);
- the extent and availability of relevant knowledge; and
- the feasibility and ease of implementing that knowledge in tools for farmers.

The results from this part of the social science research are presented in Section 5. Information on non-tool specific ways in which the slurry management landscape could be improved in the beef sector (b) is presented as follows.

Slurry infrastructures and wet weather were seen as the most important determining factors for the effective use (and avoided misuse) of slurry from farmer survey data. For farmers with animals inside for 7 months of the year, having sufficient storage of the right type is essential. Farmers, including tenants, may have inherited infrastructures from pre-existing systems and have limited capacity (particularly money) to change systems. One farmer summarised this well in the following interview response:

> Most farmers have got a grasp that you should use nutrients to their best possible effect, but… if your tank is full, your tank is full. What do you do? At the end of the day it's the cost of getting a new pit. As nearly every farmer would say, 'well, I'd like a new one, but where am I going to get forty grand from?'
>
> *(Beef farmer, Wales)*

Even when funds are available (which appeared to be very uneven across the UK), they cannot always be taken up because the funds/infrastructures available don't always fit with the infrastructures present on farm. Such 'lock-ins' can constrain efforts to move to better or more efficient systems, leaving farmers with no choice but to use what they have (Shove, 2010).

Slurry as a substance is not benign and can be very harmful to work with or to use on land, as evidenced by much of the data collected from farmers, including the following quote from one of the interviewees:

> We can't mix the slurry when the cows are on [the slats]. It kills them. Which is a downside, because there are times when you could spread the slurry, eh?
>
> *(Beef farmer, England)*

The interviews uncovered, in preference, the merits of working with solid manures with reduced risk to land from damaging soil organisms or polluting waterways and reduced risk to humans managing and storing it from toxic fumes or even death from falling into stores.

The importance of face-to-face advice has frequently been shown in research with those in the farming sector, and this also relates to the delivery of plans or tools which are designed to assist farmers in their management (Rose and Bruce, 2018). Likewise, our research showed that both farmers and advisors recognised the need for advice to support software that was not easily accessible for farmers. Cost was cited as a common issue restricting advice, with the agendas of advisors (particularly those representing companies with products to sell) being a further risk to farm finances. In contrast, the Catchment Sensitive Farming programme was offered up as an example of a source of impartial, farmer-oriented, understandable and trustworthy advice, but this advice is limited to those within targeted catchments.

Agricultural colleges may be a key educational opportunity as evidenced by the difficulties of getting farmers to attend tool training (as reported by tool developers) and the current training approaches in most colleges (where publicly available tools are not used). Northern Ireland was the exception where students were familiar with the DAERA nutrient Calculator. Potentially providing certification to students trained in such tools could help ensure their wider use.

Governance and support for nutrient planning was shown to be a key area currently under the aegis of the AHDB levy body. Our research revealed the need for maintaining consistency across nutrient management advice given the proliferation of public tools and for ensuring that all levy payers (regardless of farm size) have equal access to nutrient management support.

5 SLURRY-MAX Project findings: finding solutions for enhancing decision support tools

This part of the research focused on simplifying tools for clarity and ease of use and reducing data input requirements by ensuring the use of existing data. Several key issues were identified. The first was that the tool must answer a question that farmers are asking. If infrastructure, or lack of it, is the key issue affecting how the farmer manages their slurry, then a tool is unlikely to provide anything useful. If, however, they have to comply with Nitrate Vulnerable Zone requirements or consider which fertiliser and how much of it they may need to buy to supplement the nutrients provided by their slurry, a tool may be useful.

Our results indicated that the Tool interfaces, which are typically spreadsheet interfaces, were not intuitive for users and prevented them from using the tools effectively. Users suggested that a narrative interface where the users are taken through step-by-step, signposting all the information needed and in what order, would be preferable, enabling them to focus on the plan rather than the tool itself.

Users found the information required complex and detailed; work to simplify this level of detail is essential, e.g. built-in calculations that average out cattle ages across the year. In addition, the ability to transfer standard information from other tools would significantly reduce the time and effort involved in using the tool. Such information could include farm maps, as these are available for all farms registered with the Rural Payments Agency. Also publicly available are cattle tracing information, environmental status of waterbodies, NVZ and Stewardship areas.

It is clear that decision support tools should be part of a wider landscape of decision support, potentially with different but compatible interfaces for farmers and advisors, as well as ongoing human-based technical support. As soil testing is integral to the effective use of nutrients including slurry/manure, it is important that advice around tool use should include how to soil test, and results should be straightforward to incorporate into the tools. Incorporating cost-benefit analysis around slurry storage and use – including often neglected areas like costs of covering slurry – would also help to encourage use by farmers.

Following from the initial research, the SLURRY-MAX project endeavoured to create a prototype tool which addressed the multiple issues raised in the research and provided recommendations for enhancing decision support for all farmers and future farmers for effective slurry management. The tool and report on which this chapter is based can be accessed online.[2]

6 Future trends in research

This different approach to research on slurry, contrasting with the overwhelming majority of work in this area, highlighted the potential contribution of this type of research to improving the use of slurry, both to enhance production and minimise its negative impacts. Since SLURRY-MAX was completed, some of the areas covered by the research have been addressed in new decision support tools, such as the integration of automatic weather data and pre-mapped nutrient zones. Guidance has been improved for tools like Slurry Wizard, now managed by the AHDB, with short films and better signposting now available.

Slurry Infrastructure Grants have also been available in recent years in England and Wales and in Scotland through its own scheme. These are positive signs that industry is responding to research findings. However, the balance of research around slurry remains heavily biased towards the natural sciences, and there is a continuing need for ensuring that this is balanced by a better understanding of the factors that constrain farmers actions relating to slurry storage and management and how they are influenced by physical infrastructures, agricultural management bodies and available technologies.

[2] https://wp.lancs.ac.uk/slurry-max/.

Since the SLURRY-MAX research project finished, some progress has been made in thinking about the applicability of tools. The need for recognition of the social aspects of slurry management, including improving shared learning across countries (Higgins et al., 2023) and acknowledging and managing the burdens of extra administration and record keeping on farmers of using such tools (Velthof et al., 2024) have been highlighted. However, despite this, there has been very little to follow the work of SLURRY-MAX and the work of Rose and colleagues (Rose et al., 2016; Rose et al., 2018) on the practical effectiveness of tools for users. The need to ensure that development is grounded by effective engagement with end users and to understand how that engagement influences uptake remain vital areas for applied research in agricultural tool development (Fielke et al., 2022; Schillings et al., 2024).

7 Conclusion

Scientific responses to slurry primarily focus on the substance itself and on its potential use and impacts. Slurry managers and their relationships with slurry, as both a resource and a problem, are rarely considered. This work emphasises the need to work with farmers and better understand the factors that constrain their actions relating to slurry storage and management. It is only by addressing farmers' needs around physical infrastructures, advice and improving the accessibility of management technologies, that progress on ensuring best practice around slurry will be resolved.

8 Where to look for further Information

To access the various tools investigated, please go to the tool websites or websites listed below.

PLANET: https://www.planet4farmers.co.uk/Content.aspx?name=PLANET.

MANNER NPK: https://www.planet4farmers.co.uk/manner.aspx.

AHDB Slurry Wizard: https://ahdb.org.uk/slurry-wizard.

The Farm Crap App: https://apps.apple.com/gb/app/farm-crap-app-pro/id1218868576.

DAERA nutrient calculator: https://www.daera-ni.gov.uk/services/daera-online-services.

Scotland's N Max calculator: https://www.gov.scot/publications/nitrate-vulnerable-zones-nmax-calculator/.

9 References

ADAS. (2008). *The national inventory and map of livestock manure loadings to agricultural land: MANURES-GIS*. Final Report for Defra Project WQ0103.

Anderson, R.C. et al. (2013). An economic and functional tool for assessing the financial feasibility of farm-based anaerobic digesters. *Renewable Energy*, 51: 85–92. https://doi.org/10.1016/j.renene.2012.08.081.

Baumgart-Getz, A. et al. (2012). Why farmers adopt best management practice in the United States: a meta-analysis of the adoption literature. *Journal of Environmental Management*, 96(1): 17–25.

Bittman, S. et al. (2026). Optimising manure application on intensive dairy farming operations. In J. Webster (Ed.), *Achieving net zero dairy farming*. Burleigh Dodds Science Publishing.

Chatterton, T. and Wilson, C. (2014). The 'Four Dimensions of Behaviour' framework: a tool for characterising behaviours to help design better interventions. *Transportation Planning and Technology*, 37: 38–61.

Defra. (2024). *Farm business management practices in England 2022/23 – statistics notice*. https://www.gov.uk/government/statistics/farm-business-management-practices/farm-business-management-practices-in-england-202223-statistics-notice#:~:text=Over%20half%20of%20farms%20(57,accessing%20it%20at%20a%20charge.

De Laporte, A. et al. (2022). *Costs and benefits of on-farm beneficial management practices that reduce net GHG emissions*. Farmers for Climate Solutions. (https://static1.squarespace.com/static/5dc5869672cac01e07a8d14d/t/62a9d65619bdd5782852a5d3/1655297622603/FCS+APF+Recommendation-Technical+Economics+Report_June+2022.pdf).

Dessart, F.J. et al. (2019). Behavioural factors affecting the adoption of sustainable farming practices: a policy-oriented review. *European Review of Agricultural Economics*, 46(3): 417–471.

Fielke, S. et al. (2022). A call to expand disciplinary boundaries so that social scientific imagination and practice are central to quests for 'responsible' digital agri-food innovation. *Sociologia Ruralis*, 62: 151–161.

Han, G. and Niles, M.T. (2023). An adoption spectrum for sustainable agriculture practices: a new framework applied to cover crop adoption. *Agricultural Systems*, 212: 103771.

Higgins, S. et al. (2023). Stocktake study of current fertilisation recommendations across Europe and discussion towards a more harmonised approach. *European Journal of Soil Science*, 74: e13422.

Niles, M.T. et al. (2019). A review of determinants for dairy farmer decision making on manure management strategies in high-income countries. *Environmental Research Letters*, 14(5): 053004.

Norton, L. et al. (2012). Using models to bridge the gap between land use and algal blooms: an example from the Loweswater catchment, UK. *Environmental Modelling & Software*, 36: 64–75.

Prokopy, L.S. et al. (2019). Adoption of agricultural conservation practices in the United States: evidence from 35 years of quantitative literature. *Journal of Soil and Water Conservation*, 74(5): 520–534.

Rose, D.C. and Bruce, T.J.A. (2018). Finding the right connection: what makes a successful decision support system? *Food and Energy Security*, 7: e00123.

Rose, D.C. et al. (2018). Exploring the spatialities of technological and user re-scripting: the case of decision support tools in UK agriculture. *Geoforum*, 89: 11–18.

Rose, D.C. et al. (2016). Decision support tools for agriculture: Towards effective design and delivery. *Agricultural Systems*, 149: 165–174.

Rothwell, S.A. et al. (2020). Phosphorus stocks and flows in an intensive livestock dominated food system. *Resources Conservation and Recycling*, 163: 105065.
Schillings, J. et al. (2024). Managing end-user participation for the adoption of digital livestock technologies: Expectations, performance, relationships, and support. *Journal of Agricultural Education & Extension*, 30: 277–295.
Shove, E. (2010). Beyond the ABC: Climate change policy and theories of social change. *Environment and Planning A: Economy and Space*, 42: 1273–1285.
Smith, K.A. and Williams, A.G. (2016). Production and management of cattle manure in the UK and implications for land application practice. *Soil Use and Management*, 32: 73–82.
SSAFO. (2010). *Water resources* (Control of Pollution) (Silage, Slurry and Agricultural Fuel Oil) (England) Regulations 2010.
Sutton, S. (2009). *Putting a price tag on manure and compost.* Country Guide. https://www.country-guide.ca/guide-business/putting-a-price-tag-on-manure-and-compost/.
Symeon, G.K. et al. (2025). Manure management as a potential mitigation tool to eliminate greenhouse gas emissions in livestock systems. *Sustainability*, 17: 112.
Velthof, G.L. et al. (2024). Managing organic resources in agriculture: Future challenges from a scientific perspective. *Frontiers in Sustainable Food Systems*, 8. https://doi.org/10.3389/fsufs.2024.1369874.
Waterton, C. et al. (2018). *SLURRY-MAX 2018: Holistic decision support for cattle slurry storage and treatment*. Lancaster University. (http://wp.lancs.ac.uk/slurry-max/files/2018/10/Slurry-Max-Report.pdf).

Chapter 8

Optimising use of frass as a soil amendment

David Houben and Anne-Maïmiti Dulaurent, UniLaSalle, France

1 Introduction

Global population is projected to exceed 9.7 billion by 2050, which will necessitate a 70% increase in food production (Crist et al., 2017). To feed the world without environmental degradation, traditional agricultural practices reliant on synthetic fertilisers are unsustainable due to their environmental impacts, including greenhouse gas emissions, soil degradation, and nutrient imbalances (Penuelas et al., 2023). One emerging solution to address these challenges is insect farming and the utilisation of its by-products, particularly frass. Frass, a mixture of insect excreta, exuviae (shed exoskeletons), and uneaten feed, is increasingly considered a promising organic amendment for soil (Poveda, 2021). This is especially related to its nutrient-rich composition and potential to stimulate soil health, while its production aligns with the principles of the circular economy (Mannaa et al., 2024; Watson et al., 2022), namely minimising and recycling waste, using biomass as a source of energy and feed, and regenerating natural systems while preventing the depletion of natural resources (Faucon et al., 2023). Its use might thus offer a dual benefit by improving soil fertility while recycling waste from insect farming, thereby reducing reliance on finite resources (Hénault-Ethier et al., 2024; Houben et al., 2020; Lopes et al., 2022).

Overall, insect farming aims at converting organic waste into valuable products (Berggren et al., 2019; Madau et al., 2020). Beyond providing an alternative source of proteins for humans and animals, insect farming, especially

http://dx.doi.org/10.19103/AS.2025.0166.10

black soldier fly (BSF; *Hermetia illucens*) and mealworms (*Tenebrio molitor*), has also been considered to produce biofuel, some rare lipids, and chitin, which may have interesting applications in cosmetics, pharmaceuticals, textile, etc. (Gortari and Hours, 2013). Moreover, for every ton of edible insect biomass produced, up to four tons of frass are generated, making it a key by-product of the industry (Hénault-Ethier et al., 2024). If frass is then used as a soil amendment, this bioconversion process closes the nutrient loop by returning essential elements to agricultural systems (Abd Manan et al., 2024; Houben et al., 2021; Klammsteiner et al., 2020).

According to Beesigamukama et al. (2023), the largest fraction of publications dealing with the use of frass as a soil amendment has focused on the BSF frass (67%) followed by mealworm frass (13%) and cricket frass (4.8%). Irrespective of the insect species, most of studies have suggested that frass might be particularly valued for its nutrient content, which includes high levels of nitrogen (N), phosphorus (P), and potassium (K), as well as secondary nutrients like calcium (Ca) and micronutrients such as zinc (Zn), copper (Cu), and iron (Fe) (Gärttling and Schulz, 2022). For instance, a comparative analysis of the nutrient composition of mealworm frass and poultry litter demonstrates higher N and P but lower K concentrations in frass, highlighting the importance of understanding its composition when designing soil amendment strategies (Amorim et al., 2024). In addition, studies have shown that replacing synthetic fertilisers with frass could reduce reliance on fossil fuel-derived fertilisers, lowering the global warming potential of agricultural systems while also contributing to soil carbon (C) sequestration through the addition of organic matter by frass (Lopes et al., 2022; Poveda, 2021). In addition to providing nutrients, frass also contains bioactive compounds like chitin, which enhance soil microbial activity and promote plant resilience against pathogens (Poveda, 2020; Quilliam et al., 2020). All in all, this might make frass a valuable resource not only for improving soil fertility but also for supporting plant health in the face of biotic and abiotic stresses (Kisaakye et al., 2024; Poveda et al., 2019).

Despite its benefits, optimising frass use presents several challenges. Its composition can vary significantly depending on the insect species and rearing substrates, potentially leading to inconsistent nutrient delivery. Furthermore, safety concerns such as the presence of pathogens or potentially toxic elements (PTEs) in some frass samples must be addressed through proper processing and application guidelines (Safitri et al., 2024). The absence of standardised regulations governing frass use further complicates its integration into mainstream agricultural practices. Nonetheless, ongoing research continues to demonstrate its potential as a transformative tool in soil management, making frass an essential component of sustainable agriculture (Houben et al., 2020; Poveda, 2021).

This chapter explores the potential of frass as a soil amendment, with a focus on its composition, nutrient dynamics, and impact on soil and plant health. The discussion begins with an analysis of frass as a source of organic C, its role in nutrient availability, and its potential as a fertiliser. The chapter then examines microbial interactions, its influence on soil biota, and the broader environmental benefits. Finally, future research directions and regulatory considerations are outlined.

2 Frass as a soil amendment

2.1 Frass and soil carbon dynamics

Frass contains diverse C fractions that significantly influence its decomposition dynamics and soil interactions. For instance, mealworm frass has been shown to contain 49.3% soluble C, 31.0% hemicellulose-like compounds, 15.2% cellulose-like compounds, and only 4.4% lignin-like compounds (Houben et al., 2020). Similarly, BSF frass was reported to have a low proportion of lignin (11.3%) compared to more easily degradable compounds such as hemicellulose (17.0%) and cellulose (19.2%) (Papa et al., 2022), even though proportions may vary depending on the insect diet (Dal Magro et al., 2024; Gebremikael et al., 2022). Overall, the high proportion of easily degradable compounds has direct implications for decomposition rates and nutrient release. As relatively simple polymers, hemicellulose and cellulose are rapidly degraded by soil microorganisms which promote C and nutrient mineralisation (Beesigamukama et al., 2021a; Houben et al., 2021). Consequently, frass supports rapid nutrient cycling and microbial activity, making it an effective organic fertiliser for short-term agricultural applications. On the other hand, its low lignin content may limit its interest for enhancing soil organic C sequestration, potentially positioning it as a sustainable amendment that prioritises immediate soil fertility over long-term C storage. However, more research needs to be carried out to better quantify the impact of frass on soil C storage since lignin content is not the only driver of C sequestration in soil and compounds of microbial origin have a longer lifespan in soil (Basile-Doelsch et al., 2020). While frass contributes to immediate soil fertility through rapid decomposition and nutrient release, its potential for long-term C sequestration remains an area of active research. Understanding these dual roles is essential for optimising frass application strategies.

As an organic amendment, frass is logically expected to enhance soil organic matter. This was shown by Antoniadis et al. (2023) who demonstrated that frass application at 1% increased soil organic matter from 2.7% to 3.2% in spinach cultivation. Similarly, Malheiro et al. (2024) observed improvements in soil organic matter and nutrient availability when applying BSF frass, particularly at moderate rates of 0.64% to 1.28% of soil volume. The study by Borkent and Hodge (2021) supports these findings, showing that HexaFrass™, a BSF-derived product, significantly improved soil organic matter and provided sustained nutrient release. However, Fuertes-Mendizábal et al. (2023) found that mealworm frass, while effective in enriching organic C, had variable outcomes depending on the application rate. The frass contribution to soil organic matter may also depend on the composition of the feedstock and post-processing techniques which influence C stability (Lopes et al., 2022). For instance, Gebremikael et al. (2022) reported that the chemical composition of the feedstock, including cellulose, hemicellulose, and lignin fractions, influences C mineralisation and humification of frass as well as its stable C fraction. Even though it has not been deeply studied, the stimulation of soil microbial activity by frass due to its high concentration in the easily degradable C fraction (Houben et al., 2020) could also be indirectly involved in C storage since it is now recognised that microbial by-products play a significant role in C sequestration in soil (Kallenbach et al., 2016). Further research should however be carried out to quantify the relative importance of this process on the additional C storage induced by frass application. In addition, except for the study by Ashworth et al. (2025), which demonstrated in field that frass application could double soil C storage compared to synthetic fertilisers, most of the experiments have been carried out under short-term lab or greenhouse conditions which therefore restrict the quantification of the impact of frass on the long-term *in situ* C storage. Using simulations with the AMG model (Levavasseur et al., 2020), it has however been shown that applying frass may significantly increase soil organic C storage, compared to conventional fertiliser (Fig. 1; unpublished data).

2.2 Fertiliser potential of frass

2.2.1 Frass nutrient composition

Frass may be a substantial source of macronutrients, primarily N, P, and K, which are essential for plant growth. Comparative analyses show that, on average, mealworm frass contains 4.53 ± 2.19% N, 1.43 ± 0.53% P, and 1.89 ± 0.54 K, while BSF frass has slightly lower N (2.95 ± 1.22%) and P (1.28 ± 0.97%) but similar K content (1.97 ± 1.29%) (Safitri et al., 2024). The usually low carbon-to-nitrogen (C:N) ratio of frass, e.g. equals on average to 13.8 and 10.8 as reported by for BSF and mealworm frass, respectively (Safitri et al., 2024), enhances N mineralisation, making nutrients readily available for plant uptake (Watson et al., 2021). In addition to N, P, and K, frass provides other macronutrients such as calcium (Ca) and magnesium (Mg), as well as essential micronutrients like zinc (Zn), manganese (Mn), and copper (Cu). For example, by analysing the composition of BSF frass from various studies, Lopes et al. (2022) revealed

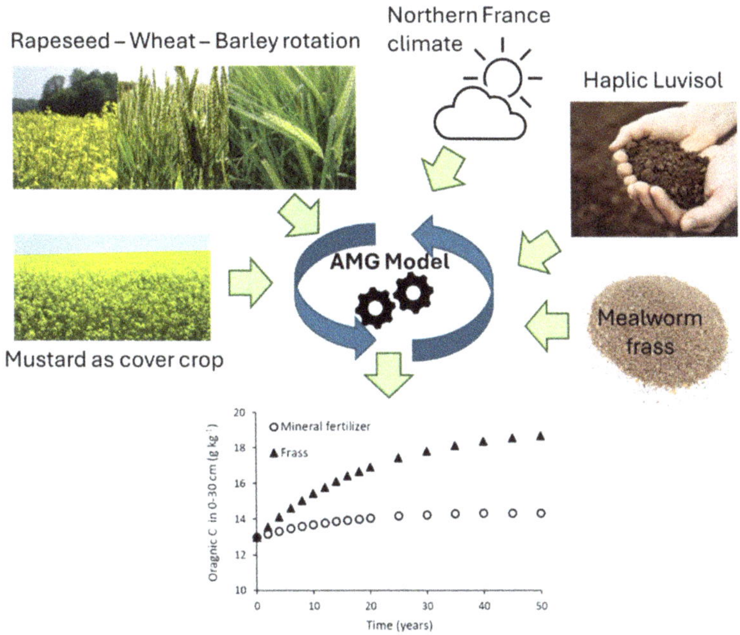

Figure 1 Impact of mineral fertiliser substitution by frass on soil organic carbon content in a representative cropping system of Northern France. Simulations are made using the AMG model using a rapeseed – wheat – barley rotation, mustard as cover crop, climate data of Northern France (oceanic climate), Haplic luvisol characteristics, and mealworm frass I_{ROC} properties as input data (unpublished data).

that BSF frass contains on average 9.9 g kg^{-1} of Ca and 1.0 g kg^{-1} of Mg while mealworm frass was shown to contain 5 g kg^{-1} of Ca and 4 g kg^{-1} of Mg (Houben et al., 2020). The nutrient composition of frass is however influenced by insect diet, with carbohydrate-rich diets yielding frass relatively rich in P, K, Mg, and Mn, whereas protein-rich diets yield frass with higher Ca concentrations (Poveda et al., 2019). Micronutrient concentrations are less documented, but for instance, Romano et al. (2023) reported 9.2 mg kg^{-1} to 51.8 mg kg^{-1} of Zn and 6.1 mg kg^{-1} to 8.9 mg kg^{-1} of Cu for BSF frass depending on the larvae diet while a recent literature review indicated a mean of 85.3 mg kg^{-1} and 16.3 mg kg^{-1} of Zn and Cu, respectively (Gärttling and Schulz, 2022). For mealworm frass, reported concentrations ranged from 94.2 mg kg^{-1} to 102.7 mg kg^{-1} of Zn and 10.0 mg kg^{-1} to 12.6 mg kg^{-1} of Cu (Beesigamukama et al., 2022; Houben et al., 2020). Interestingly, most of studies report that nutrients in frass are under a readily soluble form (Amorim et al., 2024; Poveda, 2021; Praeg and Klammsteiner, 2024; Watson et al., 2021), supplying quickly nutrients to plants (Houben et al., 2021). This may notably be explained by the uniform distribution of nutrients (P, K, and Ca) within the frass, suggesting the absence

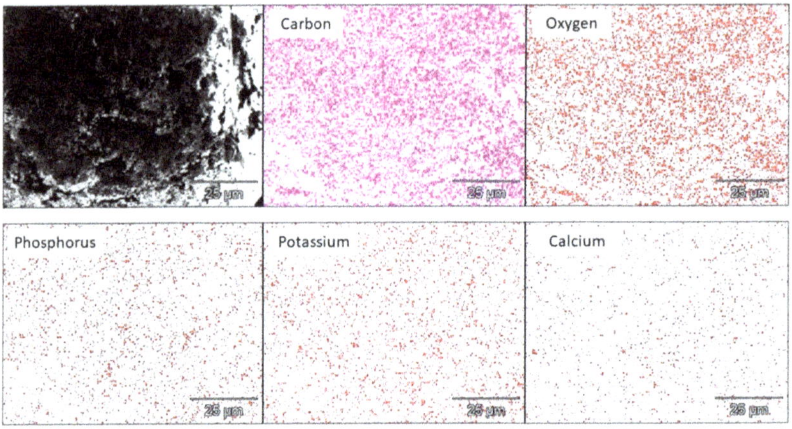

Figure 2 Element distribution in frass determined using a scanning electron microscope (SEM) equipped with an energy-dispersive X-ray spectrometer (EDS) (Houben et al., 2020).

of isolated mineral phases and ensuring an homogenous release of nutrients by frass after its incorporation into the soil (Fig. 2).

2.2.2 How frass influences nutrient availability

Frass is increasingly recognised for its contributions to soil fertility, encompassing chemical and biological parameters that enhance nutrient availability and soil function. Several studies highlight the multifaceted roles played by frass in influencing soil fertility, not only by supplying macronutrients such as N, P, and K but also by modifying soil microbial dynamics and enhancing chemical properties essential for nutrient retention and cycling.

Many studies have explored the potential of BSF larvae frass as a fertiliser due to its nutrient composition. Choi et al. (2009) were among the first to evaluate food waste-derived BSF frass. The authors found that frass was virtually as efficient as commercial fertiliser to supply nutrients and sustain the growth of cabbage. Similarly, Boudabbous et al. (2023) demonstrated that BSF frass significantly improved N and P cycling in durum wheat fields. This effect was attributed to the slow-release nature of N in frass, which helps sustain soil fertility over extended periods. Beesigamukama et al. (2021b) examined the mineralisation of nutrients in composted frass over 125 days, reporting net immobilisation initially (30–60 days) but a subsequent release of N (mostly ammonium), P, and Mg. The gradual release of N not only reduces the risk of leaching but also ensures its availability throughout the crop growth cycle, thereby enhancing N uptake by plants. In contrast, mealworm frass was found to be rapidly mineralised in soil (Fig. 3), releasing N quickly into the

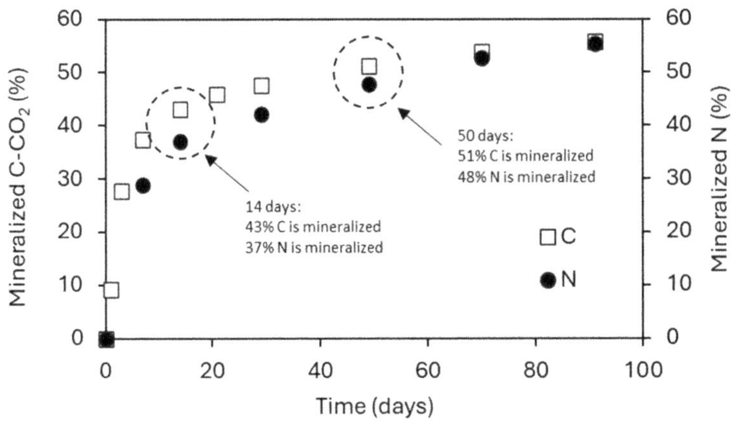

Figure 3 Carbon and nitrogen mineralisation dynamics of frass (adapted from Houben et al., 2020).

soil solution (Houben et al., 2020). Similarly, Fuertes-Mendizábal et al. (2023) found that the high N content of mealworm frass (3.64%) supported rapid N mineralisation, directly translating into enhanced N uptake by plants. While this immediate availability may support early plant growth, it requires careful management to prevent nutrient imbalances. For instance, Houben et al. (2021) warned that excessive application rates of frass could lead to N leaching, emphasising the importance of site-specific calibration. In addition, frass application may also impact N emissions. For instance, Rummel et al. (2021) investigated the impact of BSF frass, derived from carbohydrate- or protein-based diets, on soil greenhouse gas emissions, mineralisation dynamics, and microbial properties through a 40-day incubation experiment. They found that carbohydrate-based frass led to high initial CO_2 and N_2O emissions due to its high C and N availability, whereas protein-based frass showed lower emissions despite higher soil nitrate levels. The authors concluded that insect diet had a significant influence on frass composition, which in turn affected N cycling and greenhouse gas emissions, with overall emissions being high enough to undermine the presumed environmental benefits of insect-based protein production.

Phosphorus and K availability may also be positively affected by frass. Elissen et al. (2023) reported that BSF frass provided P in plant-available forms, while Fuertes-Mendizábal et al. (2023) indicated that improved P availability was further amplified when frass was combined with biostimulants. This synergistic effect increased P solubilisation, improving nutrient use efficiency. The potential of mealworm frass as a P and K fertiliser has also been shown by many studies (Poveda et al., 2019). For instance, Houben et al. (2020) found that while the available P and K concentrations in frass-amended soil were lower than those observed with mineral fertilisers, frass acts as a slow-release fertiliser and effectively supplied P and K for barley growth, leading to nutrient uptake comparable to mineral fertiliser. The authors concluded that frass could serve as a partial or complete substitute for mineral fertilisers, preventing P and K losses through sorption and enhancing soil microbial activity, thereby promoting long-term nutrient availability. Beyond macronutrients, studies have also revealed that frass provides essential micronutrients such as Zn, Cu, and B, which are critical for enzymatic processes and plant stress tolerance (Watson et al., 2021).

In addition to directly releasing nutrients, frass may also influence soil nutrient availability through its effects on soil biota. Green (2023) observed that BSF frass introduced N_2-fixing bacteria, enhancing N availability in soils. Similarly, Fuertes-Mendizábal et al. (2023) reported that mealworm frass combined with biostimulants further boosted microbial activity, leading to improved nutrient uptake and crop performance. The findings by Menino et al. (2021) highlighted the significant role of BSF frass in improving microbial biomass and enzymatic activity in ryegrass fields. In particular, their study showed that increasing doses of frass led to a progressive rise in soil dehydrogenase activity, which doubled at the highest application rate (equivalent to 210 kg N ha^{-1}) compared to the control, suggesting a dose-dependent stimulation of microbial activity. Watson et al. (2021) further emphasised that frass application increased enzymatic activities such as phosphatases and dehydrogenases, which are crucial for P and C cycling. Houben et al. (2021) also observed higher microbial activity related to amine catabolism in frass-treated soils compared to NPK treatments, resulting in higher N availability for plants.

2.2.3 Nutrient uptake and plant biomass

The beneficial effects of frass on nutrient availability in soils are reflected in increased plant biomass production and enhanced nutrient uptake (Table 1). Houben et al. (2020) demonstrated that barley grown with mealworm frass had similar yields and nutrient uptake as conventional NPK fertiliser applied at an equivalent rate. This finding aligns with Boudabbous et al. (2023), who reported that BSF frass application in durum wheat fields led to improved N and P uptake, enhancing both crop quality and yield. Similarly, Beesigamukama et al. (2020) observed significant biomass increases in maize grown in tropical agroecosystems, attributing the improvement to the slow-release N properties of BSF frass, which minimised N leaching while sustaining nutrient availability over time. In perennial ryegrass, Menino et al. (2021) and Houben et al. (2021) noted an increase in dry matter yield and plant vigour, which they attributed to both the nutrient supply and the microbial activity stimulation by frass.

Table 1 Effects of frass on nutrient uptake and biomass production by plants

Frass source	Application rate	Plant	Main effects	Reference
Hermetia illucens	5%, 10%, and 15% of soil weight	*Brassica rapa*	Improved biomass yield, increased P solubilising bacteria and enzyme activity.	Agustiyani et al. (2021)
Hermetia illucens	2.5 Mg ha^{-1}, 5 Mg ha^{-1}, 7.5 Mg ha^{-1}	*Zea mays*	Improved maize yield and nitrogen efficiency.	Beesigamukama et al. (2020)
Hermetia illucens	7.5 Mg ha^{-1}, 15 Mg ha^{-1}, 30 Mg ha^{-1}	*Triticum durum*	Improved soil health, higher productivity, best results at 7.5 Mg ha−1.	Boudabbous et al. (2023)
Hermetia illucens	15, 30, and 45 kg N ha^{-1}	*Phaseolus vulgaris*	Boosted biological nitrogen fixation, improved yield and profitability.	Chepkorir et al. (2024)
Hermetia illucens	Variable depending on feedstock	*Zea mays*	Increased microbial activity and carbon sequestration; some frass types immobilised N temporarily.	Gebremikael et al. (2022)
Hermetia illucens	Trace amounts used, nitrogen dynamics studied	*Triticum aestivum*	11% increase in shoot length and biomass attributed to rhizobacteria activity.	Green (2023)
Hermetia illucens	Various, adjusted to crop needs	Broad agriculture, tested on major grains and leafy greens	Varied per crop: nutrient uptake, pathogen resilience bettered.	Lomonaco et al. (2024)
Hermetia illucens	0.64–1.28% of soil volume	*Allium cepa*, *Brassica rapa*, *Solanum lycopersicum*	Promoted plant growth at intermediate rates; potential phytotoxicity at high rates.	Malheiro et al. (2024)
Hermetia illucens	Variable rates based on feedstock	*Lolium perenne*	Improved soil properties, steady ryegrass yield increase.	Menino et al. (2021)
HexaFrass™	0.5% to 3.5% NPK equivalent	Varied vegetables and herbs	Improved dry weight of shoots and roots; performance similar to chicken manure.	Borkent and Hodge (2021)

(Continued)

Table 1 *Continued*

Frass source	Application rate	Plant	Main effects	Reference
Tenebrio molitor	10 Mg ha^{-1}	*Hordeum vulgare*	Similar effectiveness to mineral NPK fertiliser; prevented P loss; stimulated microbial activity.	Houben et al. (2020)
Tenebrio molitor	1% of soil weight	*Spinacia oleracea*	Boosted soil carbon and nitrogen, enhanced spinach growth.	Antoniadis et al. (2023)
Tenebrio molitor	2% of soil weight	*Solanum lycopersicum*, *Helianthus annuus*, *Arabidopsis thaliana*	Increased growth in tomato; activated systemic defences in Arabidopsis.	Blakstad et al. (2023)
Tenebrio molitor	120 kg N, 43 kg P, and 47 kg K per ha	*Helianthus annuus*	Similar biomass as NPK fertiliser; reduced germination rates for some crops.	Foscari et al. (2024)
Tenebrio molitor	1%, 2.5%, and 5% of substrate	*Lactuca sativa*	Enhanced biomass and quality with microbial synergy; high rates led to root stress.	Fuertes-Mendizábal et al. (2023)
Tenebrio molitor	0.5% v/v	Urban horticulture crops (e.g., beets, carrots, flowers)	16-fold biomass boost, flower numbers increased.	Hénault-Ethier et al. (2023)
Tenebrio molitor	5 Mg ha–1, 10 Mg ha–1	*Lolium multiflorum*	At 10 Mg ha–1, similar N, P, and K uptake as mineral fertilisers; rapid N mineralisation.	Houben et al. (2021)
Tenebrio molitor	10 Mg ha^{-1} (with earthworms)	*Hordeum vulgare*	Increased N, P, K, Ca uptake; synergistic effect with earthworms.	Dulaurent et al. (2020)
Tenebrio molitor, *Acheta domesticus*, *Hermetia illucens*	Multiple rates tested	*Brassica oleracea*	Improved plant growth and rhizosphere bacterial diversity.	van de Zande et al. (2024)

For vegetables, frass has also shown strong potential as an organic fertiliser. Hénault-Ethier et al. (2023) found that BSF frass was as efficient as hen manure to sustain vegetable, herbs, and flowers production, with a total biomass production 16 times higher compared to the control without fertiliser. Antoniadis et al. (2023) reported a six-fold increase in spinach biomass following frass application at 1% of soil volume, highlighting its efficacy in supporting leafy vegetable growth. The improvement was attributed to the high organic matter and N content of frass, which enhanced soil fertility and nutrient availability. Similarly, Fuertes-Mendizábal et al. (2023) found that mealworm frass improved lettuce aerial biomass, particularly when combined with plant growth-promoting biostimulants. These biostimulants helped mitigate root stress observed at higher frass application rates (5%), suggesting a synergistic effect between frass and microbial amendments. Comparisons across different crops reveal that leafy vegetables such as spinach and lettuce consistently benefit from frass application due to their high nutrient demands and rapid growth cycles (Antoniadis et al., 2023; Foscari et al., 2024). In contrast, root crops like onions and turnips appeared more sensitive to high frass rates, with potential adverse effects on germination and early growth stages (Malheiro et al., 2024).

2.2.4 Effects of frass on soil water retention and drought resilience

As an organic amendment, frass is expected to affect not only nutrient dynamics but also the physical properties of soils, notably water retention. This is consistent with broader evidence that soil organic matter plays a critical role in improving water holding capacity, infiltration, and aggregate stability (Abbott et al., 2018). Soil organic matter enhances soil structure and increases porosity, especially in sandy soils, where its contribution to moisture retention is most pronounced (Védère et al., 2022). Thus, frass, being rich in labile and partially humified organic matter, could theoretically exert similar effects on soil water dynamics.

Evidence on this topic remains limited, yet emerging studies provide promising insights. In controlled greenhouse experiments, BSF frass improved basil and lettuce performance under drought, suggesting among others higher water availability in the rhizosphere (Radzikowska-Kujawska et al., 2023; Sawinska et al., 2024). Moreover, the combined application of BSF frass and biochar significantly enhanced bell pepper growth and enriched the abundance of bacteria with putative stress-resilience functions under water-limited conditions, indicating potential improvements in soil water retention and drought buffering capacity (Gurung et al., 2025). Overall, the recent review by Zunzunegui et al. (2025) confirms that frass could enhance plant resilience to hydric stresses such as drought. The authors highlight that the underlying mechanism remains unclear, but they propose that frass-derived chitin could improve drought tolerance in plants due to its ability to retain water.

While these findings remain preliminary, they point toward an additional functional benefit of frass as a soil amendment, supporting both nutrient and moisture management and warranting further field-based and soil-physical investigations.

2.2.5 Potential risks and considerations

Overall, frass application also presents challenges, particularly regarding application rates and potential nutrient imbalances. In particular, field and greenhouse experiments indicate that excessive frass application can have negative impact on plant growth. Alattar et al. (2016) reported that BSF frass negatively impacted maize growth, likely due to high ammonia concentrations. Similarly, Gärttling et al. (2020) found that BSF frass application resulted in poor maize growth. Unlike other studies showing that frass may act as a N fertiliser, the authors found in their study that frass had low N-fertilisation effects, which was attributed to volatilisation losses, mainly driven by high pH and ammonium contents. Kawasaki et al. (2020) recommended frass application rates of 1/20–1/30 relative to soil volume for optimal plant growth, as higher rates (i.e. 1/10) led to chlorotic symptoms. Chiam et al. (2021) tested okara-derived BSF frass on lettuce and observed poor growth at high frass concentrations (20–30%) in growing media. The authors attributed this adverse effect to the low C/N ratio (7.2) of frass, which induces rapid nutrient mineralisation and an excess nutrient supply, thereby disrupting the balance of elements within the plant, inhibiting growth, and increasing susceptibility to disease. Similarly, Borkent and Hodge (2021) found that moderate applications of HexaFrass™ significantly enhanced the dry shoot weight of vegetables and herbs, but excessive applications led to nutrient imbalances and inhibited growth. Likewise, Malheiro et al. (2024) noted that frass application rates above 5% could disrupt microbial balance and lead to phytotoxic effects, particularly in tomato and onion cultivation, where increased salinity and nutrient imbalances were observed. Salomon et al. (2025) also showed that excessive frass applications could lead to ammonia toxicity for tomato and almost complete inhibition of arbuscular mycorrhizal root colonisation when frass was applied at 100 and 250 kg N ha^{-1}. The discrepancy between studies highlights the need for crop-specific frass application strategies, emphasising the importance of optimising application rates based on specific crop requirements and soil conditions to maximise benefits while avoiding potential negative effects.

Moreover, frass composition, influenced by insect species and feedstock, introduces variability in its effects. For instance, although Foscari et al. (2024)

found that sunflower biomass responded positively to mealworm frass, the authors showed that its effectiveness varied greatly depending on the rearing substrate of the insects, underlining the need for standardised frass composition. In addition, Malheiro et al. (2024) emphasised the need for standardisation in frass production and application to ensure predictable agronomic outcomes because even though most studies report consistent improvements in soil fertility and plant growth, others highlight inconsistencies that necessitate further research.

Together, these studies highlight the potential of frass as a valuable organic fertiliser but also underscore the complexity of its interactions with soil and crops. While frass can significantly enhance plant growth, nutrient uptake, and soil microbial activity, its efficiency depends on factors such as C/N ratio, feedstock source, application rate, and interactions with soil biota. Further research is needed to optimise frass application strategies and mitigate risks such as N losses, salinity issues, and nutrient imbalances, ensuring that frass can be fully harnessed as a sustainable alternative to synthetic fertilisers.

2.3 How frass enhances soil and plant health

2.3.1 Impact of frass on soil biota

Frass plays a significant role in enhancing soil health by influencing microbial communities, soil fauna, and enzymatic activities critical for nutrient cycling. Its unique composition, including organic matter, nutrients, and chitin, contributes to improving soil biodiversity, fostering beneficial microbial populations, and supporting key soil organisms such as earthworms and enchytraeids (Poveda, 2021; Dulaurent et al., 2020; Malheiro et al., 2024). Mechanistically, the impact of frass on soil biota is linked to its diverse composition. Untreated frass can introduce beneficial microorganisms directly into the soil, enhancing its microbial biodiversity. Additionally, frass serves as a C source and contains chitin, which stimulates the activity of chitinase-producing microbes (Mirabello et al., 2025; Watson et al., 2025). These microbes degrade chitin into smaller compounds that not only act as nutrient sources but also trigger plant defence responses. The organic matter in frass further supports microbial respiration and growth, creating a cascade of biological activity that enriches the soil ecosystem (Lopes et al., 2022).

A distinguishing feature of frass is its chitin content, derived from insect exoskeletons and present in quantities ranging from 10.2 to 11.9% in BSF frass, as reported by Anedo et al. (2024). Chitin, a biopolymer composed of N-acetylglucosamine units, serves multiple functions in soil ecosystems. It acts as a substrate for chitinolytic microorganisms, thereby enhancing microbial activity and nutrient cycling (Hallmann et al., 1999; Sharp, 2013). Additionally, chitin has been shown to elicit plant immune responses and suppress plant-parasitic nematodes, contributing to integrated pest management strategies (Poveda, 2020; Quilliam et al., 2020). The presence of pupal exuviae and other insect-derived organic residues further enriches the biological complexity of frass, providing a slow-release C source that supports microbial respiration and soil stability (Ilijin et al., 2024; Machado et al., 2024).

Frass has been widely recognised for its ability to foster beneficial microbial communities. The microbial composition of frass itself is influenced by the rearing feedstock of the insects, which affects its capacity to stimulate microbial diversity and activity in the soil (Beesigamukama et al., 2020). Green (2023) observed that BSF frass introduced N-fixing bacteria and P-solubilising microorganisms, enhancing nutrient availability and cycling. Similarly, Fuertes-Mendizábal et al. (2023) found that mealworm frass, when combined with biostimulants, boosted microbial activity, leading to improved nutrient uptake and crop performance. Schmitt and de Vries (2020) further highlighted that the chitin and exuviae components of frass play a key role in stimulating soil biodiversity, serving as both microbial activators and elicitors of plant and soil organism defences. This interaction promotes agroecosystem resilience, particularly by enhancing enzymatic activities such as phosphatases, dehydrogenases, and chitinases, which are crucial for P mobilisation and organic matter breakdown (Amadou et al., 2021; Fuertes-Mendizábal et al., 2023; Gebremikael et al., 2022; Watson et al., 2021).

The impact of frass on soil health extends beyond microbial dynamics to its effects on soil fauna. Earthworms, enchytraeids, and other invertebrates play essential roles in soil structure and nutrient cycling, and their populations can be significantly influenced by frass amendments. Dulaurent et al. (2020) observed that earthworms in frass-amended soils exhibited increased biomass and activity. The authors attributed the positive effect of frass on earthworms to the high organic C and nutrient content of frass, which creates favourable conditions for their growth. Beneficial effects of frass on earthworms are especially relevant as earthworms contribute to soil aeration, organic matter decomposition, and aggregate formation, further amplifying the benefits of frass on soil health. Beyond earthworms, other soil invertebrates such as enchytraeids and collembolans respond differently to frass amendments. Malheiro et al. (2024) investigated the effects of BSF frass on the survival and reproduction of the enchytraeid *Enchytraeus crypticus* and the collembolan *Folsomia candida*, two key soil invertebrate species. Their study found that frass did not negatively affect survival but significantly enhanced juvenile production of *E. crypticus* at moderate application rates, suggesting that frass might support mesofauna populations by improving soil organic matter content and providing a nutrient-rich environment. However, excessive frass application (above 5%) led to increased soil salinity and altered pH, which could pose risks to sensitive soil fauna. The neutral response observed in *F. candida* populations indicates that not all soil fauna benefit equally from frass amendments, highlighting the need for further research into species-specific responses.

2.3.2 Impact of frass on plant health

Insect frass plays a crucial role in enhancing plant health by inducing resistance against biotic stresses and improving tolerance to abiotic challenges (Table 2). One of its key protective properties is its chitin content, which serves as an elicitor of plant immune responses. Chitin fragments in frass can trigger plant defence pathways, leading to increased resistance against fungal pathogens. In cowpea, for instance, frass biofertiliser significantly reduced plant losses due to *Fusarium* wilt, which was attributed to chitin-induced disease resistance (Quilliam et al., 2020). Additionally, frass can influence plant-insect interactions by modulating plant defence responses. Studies have shown that frass from various insect species, including the fall armyworm (*Spodoptera frugiperda*) and European corn borer (*Ostrinia nubilalis*), alters the balance between herbivore-induced and pathogen-induced defences in crops such as maize and rice (Ray et al., 2016b, 2016a). For example, frass from *S. frugiperda* steadily induced herbivore defences in rice while simultaneously suppressing pathogen defences, potentially affecting the plant resilience to stress (Ray et al., 2016b). Similarly, maize treated with frass from *O. nubilalis* exhibited sustained herbivore-induced defences while experiencing a suppression of pathogen-induced defences (Ray et al., 2016b).

Beyond pathogen resistance, frass has been found to suppress nematodes and insect pests through multiple mechanisms. Soil amendment with BSF frass significantly reduced potato cyst nematode (*Globodera pallida*) infestations, decreasing egg and reproduction rate by up to 98%, with suppression effects increasing as the chitin concentration in the frass increased (Anedo et al., 2024). Similarly, BSF frass was effective in reducing *Meloidogyne incognita*-induced root galling in soybean, lowering the number of galls from 104 in untreated plants to just 3 in the highest frass treatment (Kisaakye et al., 2024). Frass also exhibits deterrent properties against insect pests. In tomato, *Helicoverpa zea* frass induced sustained herbivore defences in leaves while showing an initial suppression followed by an induction of defences in fruit tissues (Ray et al., 2016b). Frass may boost plant defence after not only soil but also foliar application. For instance, Ray et al. (2020) showed that fall armyworm frass deposits on maize leaves suppressed the herbivory performance of aphids.

In addition to these protective functions, frass contributes to plant health by enhancing beneficial microbial communities in the soil. Studies have shown that frass harbours bacteria with biocontrol potential, such as *Bacillus velezensis*, which survived the BSF larval gut passage and exhibited antifungal and

Table 2 Effects of frass on pests and pathogens in the soil-plant system

Frass source	Plant	Pest or pathogen	Main effects	Reference
Acheta domesticus	*Brassica oleracea*	*Delia radicum*	No effect	Wantulla et al. (2023)
Agrotis ipsilon	*Solanum tuberosum*	*Phthorimaea operculella*	Oviposition deterrent effect	Ahmed et al. (2013)
Anoplophora glabripennis	*Salix babylonica*	*Anoplophora glabripennis*	Emission of volatile organic compounds capable of attracting parasitoid (*Dastarcus helophoroides*)	Wei et al. (2013)
Apriona swainsoni	*Sophora japonica*	*Apriona swainsoni*	Emission of volatile organic compounds capable of attracting parasitoid (*Dastarcus helophoroides*)	Wei et al. (2013)
Helicoverpa zea	*Solanum lycopersicon*	*Helicoverpa zea*	Initial suppression followed by induction of herbivore defences	Ray et al. (2016b)
Hermetia illucens	*Vigna unguiculata*	*Fusarium oxysporum*	Disease resistance induced by fragments of chitin	Quilliam et al. (2020)
Hermetia illucens	*Lepidium sativum*	*Pythium ultimum*	Disease suppression	Elissen et al. (2019)
Hermetia illucens	*Brassica oleracea*	*Delia radicum*	Cabbage root fly survival and biomass reduction	Wantulla et al. (2023)
Hermetia illucens	*Glycine max*	*Meloidogyne incognita*	Nematode reduction by chitin	Kigeso et al. (2024)
Hermetia illucens	n/a	*Phytophthora capsici, Botrytis cinerea, Alternaria solani, Fusarium oxysporum, Sclerotinia sclerotiorum, Rhizoctonia solani, Pythium ultimum*	Inhibition of pathogens by *Bacillus velezensis*	Arabzadeh et al. (2023)

Hermetia illucens (5% chitin fortified)	*Solanum tuberosum*	*Globodera rostochiensis* + *Globodera pallida*	Nematode reduction by chitin	Anedo et al. (2024)
Hermetia illucens (5% chitin fortified)	*Solanum tuberosum*	*Globodera rostochiensis* + *Globodera pallida*	Nematode reduction and plant yield increase by chitin	Anedo et al. (2024)
Hermetia illucens (chitin fortified)	*Spinacia oleracea*	*Meloidogyne incognita*	Nematode reduction by chitin	Kisaakye et al. (2024)
Leptinotarsa decemlineata	*Solanum tuberosum*	*Leptinotarsa decemlineata*	Suppression of plant defences; larval growth increase by bacterial symbionts	Gao et al. (2022)
Ostrinia nubilalis	*Zea mays*	*Ostrinia nubilalis*	Herbivore defences induction; pathogen defences suppression	Ray et al. (2016a)
Phthorimaea operculella	*Solanum tuberosum*	*Phthorimaea operculella*	Oviposition deterrent effect	Zhang et al. (2019)
Spodoptera frugiperda	*Oryza sativa*	*Spodoptera frugiperda*	Herbivore defences induction; pathogen defences suppression	Ray et al. (2016b)
Spodoptera frugiperda	*Zea mays*	*Spodoptera frugiperda* + *Cochliobolus heterostrophus*	Herbivore performance increase; pathogen defences induction; pathogen performance reduction by proteins	Ray et al. (2015)
Spodoptera frugiperda	*Zea mays*	*Spodoptera frugiperda* + *Cochliobolus heterostrophus*	Herbivore defences suppression; pathogen defences induction by chitinase enzymes	Ray et al. (2016a)
Spodoptera frugiperda	*Zea mays*	*Rhopalosiphum maidis*	Aphids performance reduction by chitinases	Ray et al. (2020)
Tenebrio molitor	*Arabidopsis thaliana*	*Botrytis cinerea*	Activation of systemic defences genes	Blakstad et al. (2023)

(*Continued*)

Table 2 *Continued*

Frass source	Plant	Pest or pathogen	Main effects	Reference
Tenebrio molitor (bio-oil)	n/a	*Plodia interpunctella*	Insecticidal activity by fumigant and contact exposure of frass bio-oils	Urrutia et al. (2023)
	n/a	*Tribolium castaneum*	Insecticidal activity by fumigant frass bio-oils	
	n/a	*Culex pipiens pipiens*	Larvicidal activity	
Trichoplusia ni	*Brassica oleracea*	*Trichoplusia ni*	Oscillating pattern of herbivore defences	Ray et al. (2016b)

anti-oomycete activity in the resulting frass (Arabzadeh et al., 2023). Moreover, mealworm frass was found to induce systemic resistance in *Arabidopsis thaliana*, activating defence genes even in the absence of a pathogen, further enhancing the plant resilience (Blakstad et al., 2023). Additionally, proteins from fall armyworm frass were found to mediate suppression of herbivore-induced defences while simultaneously triggering pathogen defence pathways, reducing fungal pathogen growth on maize leaves (Ray et al., 2015). These findings highlight the potential of frass not only as a source of nutrients but also as a bioactive amendment that strengthens plant defences against multiple stress factors.

However, the impact of frass on plant health is highly dependent on its composition, which varies with the insect species, diet, and processing methods (Ray et al., 2016a). Further research is needed to optimise frass applications and understand its interactions with different plant species and soil microbiomes. Nevertheless, its ability to suppress diseases, deter pests, and enhance plant resilience might make frass a valuable tool in sustainable agriculture.

3 Future trends in frass research

The increasing demand for sustainable agricultural practices has highlighted the potential of insect frass, particularly from BSF, mealworm, and other insect species used in large-scale production systems, as a soil amendment. Frass, a by-product of insect rearing, contains essential nutrients, organic matter, and beneficial microbial communities that enhance soil fertility and plant growth. However, despite the promising applications, several research gaps remain. Future investigations should address frass standardisation, its long-term soil interactions, biostimulant properties, including underlying mechanisms, pathogen suppression potential, and regulatory frameworks.

3.1 Standardisation of frass composition and quality control

Frass composition varies significantly based on insect species, feed substrates, and processing conditions. Mealworm frass, for example, has different nutrient compositions compared to BSF frass due to variations in diet and metabolic processing. To maximise its agronomic potential, future research should focus on developing standardised protocols for frass production, ensuring consistent nutrient content, and minimising contamination risks (Safitri et al., 2024). Investigations should also explore post-treatment methods such as composting or heat treatment to stabilise frass and reduce variability (Lopes et al., 2022). Additionally, comparative studies across different insect-derived frass types will be essential to establish their respective benefits and optimal application methods.

3.2 Long-term effects of frass on soil and microbial communities

Although studies indicate that frass can improve soil fertility and microbial diversity, its long-term effects remain unclear as most of the findings derive from short-term lab or greenhouse experiments while *in situ* studies are very scarce (Ashworth et al., 2025). Future research should explore how frass interacts with different soil types over multiple growing seasons. Additionally, studies should assess whether repeated applications lead to nutrient accumulation, microbial shifts, or potential negative impacts such as soil salinisation (Schmitt and de Vries, 2020). Research should also compare how different insect frass types influence soil microbial ecology, given the variation in microbial load and chitin content between species (Barragán-Fonseca et al., 2022). To date, most studies have primarily examined specific enzymatic functions or isolated soil organisms, while the broader ecological interactions among soil organisms remain largely unexplored. There is a critical need for research that investigates how these interactions influence overall soil functioning and health, providing a more comprehensive understanding of frass application in agricultural ecosystems.

3.3 Biostimulant and disease-suppressive properties

Frass contains chitin, a component known for its role in inducing plant defence responses against pathogens (Poveda, 2021). While preliminary studies suggest that frass may act as a biostimulant, promoting plant growth and stress resistance, more research is needed to identify the specific mechanisms involved (Chavez and Uchanski, 2021). Future work should investigate whether frass-derived biostimulants can be optimised for use in commercial biofertiliser formulations. Further, comparative trials between BSF, mealworm, and other frass sources should be conducted to determine which insect frass provides the most significant biostimulant benefits depending on the conditions in which it is used (soil type, climate, targeted crops, etc.).

3.4 Integration with other soil amendments and fertilisers

Frass is often compared to conventional fertilisers, composts, and manure, but little research has explored its synergistic effects when combined with other organic or synthetic fertilisers (van Huis, 2020). Woo et al. (2024) demonstrate that co-composting frass with fish meal, bone meal, boiler ash, and sawdust effectively improves its nutrient content, providing a sustainable alternative for organic waste management and nutrient recycling. Jasso et al. (2024) showed that integrating BSF frass and sheddings into composting was a viable option and maintained key nutrient parameters, but stressed that further research was needed to optimise inclusion rates for improved compost quality. Overall, future studies should evaluate optimal application rates and combinations to enhance nutrient availability and crop yield while minimising environmental risks such as nutrient leaching or volatilisation (Hénault-Ethier et al., 2024). Additionally, even though preliminary investigations have shown promising results, research should further explore how different insect frass types interact with biochar, compost, and microbial inoculants to optimise soil health and nutrient uptake.

3.5 Environmental and economic benefits of frass

While frass recycling aligns with circular economy principles, more research is needed to quantify its environmental footprint relative to other organic fertilisers (Beesigamukama et al., 2023). Life cycle assessments (LCAs) should analyse frass production's energy use, greenhouse gas emissions, and potential benefits in C sequestration (Schmitt and de Vries, 2020). Economic studies should evaluate the feasibility of large-scale frass commercialisation, including market demand, cost-benefit analyses, and incentives for adoption by farmers. Understanding the market potential for different frass types and their price competitiveness relative to synthetic and other organic fertilisers is also crucial for industry growth.

3.6 Regulatory and safety considerations

As frass use expands, clear regulatory guidelines must be established to ensure its safety and efficacy (van Huis, 2020). For instance, in Europe, the placing on the market of insect frass as fertiliser is harmonised by Commission Regulation (EU) 2021/1925 and Regulation (EU) No 142/2011, provided that the frass is subject to heat treatment of at least 70°C for at least 60 min, and representative samples must be taken to monitor compliance with some microbiological criteria. However, future research should support policy development by addressing concerns related to heavy metal accumulation, pathogen risks, and residue limits in agricultural soils (Lopes et al., 2022; Safitri et al., 2024). Comparative studies between different regulatory frameworks in the EU, North America, and other regions will help harmonise global frass standards (Schmitt and de Vries, 2020). Additionally, research should focus on food safety regulations regarding frass applications in organic agriculture and its potential integration into established organic certification programs.

4 Conclusion

Frass, as a by-product of insect farming, presents a promising avenue for enhancing soil fertility, promoting plant growth, and supporting sustainable agricultural practices (Fig. 4). Its nutrient-rich composition, including essential macronutrients (N, P, K), secondary elements, and micronutrients, makes it a valuable organic fertiliser with the potential to reduce reliance on synthetic fertilisers. Additionally, its organic C content supports soil C storage, microbial activity, influencing soil health and nutrient cycling. Beyond its fertilisation potential, frass also exhibits biostimulant properties, fostering beneficial soil microbes, improving plant resilience against pathogens, and even suppressing nematodes and pests through its chitin content.

Despite these benefits, optimising the use of frass as a soil amendment requires addressing several challenges. Variability in its composition, influenced by insect species, diet, and processing methods, affects its agronomic performance. Application rates must be carefully managed to avoid nutrient imbalances, salinity issues, or excessive N losses. Furthermore, while initial studies highlight its capacity to improve soil organic matter, its long-term impacts on C sequestration and microbial communities remain poorly understood. Regulatory frameworks also need to be refined to ensure safe and standardised use in agricultural systems.

Future research should focus on standardising frass composition, developing guidelines for its integration with other organic or mineral amendments, and assessing its long-term effects on soil and plant health under field conditions. Moreover, life cycle assessments are needed to quantify its environmental footprint and economic feasibility on a larger scale. By addressing these knowledge gaps, frass could emerge as a key component of circular and sustainable agriculture, closing nutrient loops while enhancing soil productivity and resilience.

5 Where to look for further information

For readers seeking a broad introduction to the use of insect frass in agriculture, the review by Poveda (2021) provides one of the most comprehensive overviews of frass as a biofertilizer and biostimulant, covering nutrient dynamics, plant health and soil biological interactions. More recent syntheses, such as Abd Manan et al. (2024), offer updated perspectives specifically focused on black soldier fly frass and its agronomic potential in sustainable agriculture. Readers interested in the broader context of insect farming and circular bioeconomy approaches may also consult van Huis (2020), which discusses the integration of insect-derived products into sustainable food and agricultural systems. A useful collection of papers dedicated specifically to frass valorisation is the

Figure 4 Expected results and challenges related to the use of frass as a soil amendment.

Frontiers Research Topic 'Frass: The Legacy of Larvae – Benefits and Risks of Residues from Insect Production', which gathers contributions on nutrient cycling, soil fertility, plant health, microbial interactions, and environmental risks associated with insect frass (https://www.frontiersin.org/research-topics/16226/frass-the-legacy-of-larvae---benefits-and-risks-of-residues-from-insect-production/magazine). Key organisations and research infrastructures working on insect farming and frass valorisation include the International Platform of Insects for Food and Feed (IPIFF) (https://ipiff.org/), which provides updates on regulations, industrial developments, and sustainability initiatives in Europe, as well as the Wageningen University & Research ACRRES insect larvae and compost worms facility, dedicated to research on insect production, organic waste valorisation, and circular agriculture (https://www.wur.nl/en/research/facilities/acrres-insect-larvae-and-compost-worms-facility). Readers may also follow the activities of leading research groups working on insect frass and soil health, including:- UniLaSalle (France), particularly on frass impacts on soil functioning and nutrient cycling;- International Centre of Insect Physiology and Ecology (ICIPE, Kenya), especially on black soldier fly valorisation and circular agriculture;- Wageningen University & Research (Netherlands), notably on insect production systems and circular bioeconomy;- Rhine-Waal University of Applied Sciences (Germany), particularly on insect-derived fertilisers and organic residue valorisation; University of Udine (Italy), notably on frass agronomic performance, soil ecotoxicology, and application strategies. For readers interested in the regulatory framework surrounding insect frass in agriculture, the European Commission Regulation (EU) 2021/1925 provides important guidance regarding the placing on the market of insect frass as fertilising products in Europe, including hygienisation requirements and microbiological safety criteria (https://eur-lex.europa.eu/legal-content/EN/TXT/HTML/?uri=CELEX:32021R1925).

As the field is evolving rapidly, readers are encouraged to monitor emerging research on frass standardisation, long-term field applications, interactions with soil microbiomes, and the integration of frass into regenerative and circular agricultural systems.

6 References

Abbott, L.K., et al., 2018. Potential roles of biological amendments for profitable grain production – A review. *Agric. Ecosyst. Environ.* 256, 34–50. https://doi.org/10.1016/j.agee.2017.12.021.

Abd Manan, F., et al., 2024. Unlocking the potential of black soldier fly frass as a sustainable organic fertilizer: a review of recent studies. *J. Environ. Manage.* 367, 121997. https://doi.org/10.1016/j.jenvman.2024.121997.

Agustiyani D, et al. 2021. The effect of application of compost and frass from Black Soldier Fly Larvae (*Hermetia illucens* L.) on growth of Pakchoi (*Brassica rapa* L.). *IOP*

Conference Series: Earth and Environmental Science 762(1). *IOP Publishing:* 012036. https://10.1088/1755-1315/762/1/012036.
Ahmed, A.A.I., et al., 2013. Protection of potato crop against Phthorimaea operculella (Zeller) infestation using frass extract of two noctuid insect pests under laboratory and storage simulation conditions. *Arch. Phytopathol. Plant Prot.* 46, 2409–2419.
Alattar, M.A., et al., 2016. Effects of microaerobic fermentation and black soldier fly larvae food scrap processing residues on the growth of corn plants (Zea mays). *Plant Sci. Today* 3, 57–62. https://doi.org/10.14719/pst.2016.3.1.179.
Amadou, I., et al., 2021. Unravelling the role of rhizosphere microbiome and root traits in organic phosphorus mobilization for sustainable phosphorus fertilization: a review. *Agronomy* 11, 2267. https://doi.org/10.3390/agronomy11112267.
Amorim, H.C.S., et al., 2024. Insect frass composition and potential use as an organic fertilizer in circular economies. *J. Econ. Entomol.* 117, 1261–1268. https://doi.org/10.1093/jee/toad234.
Anedo, E.O., et al., 2024. Evolving dynamics of insect frass fertilizer for sustainable nematode management and potato production. *Front. Plant Sci.* 15. https://doi.org/10.3389/fpls.2024.1343038.
Antoniadis, V., et al., 2023. Insect frass as a novel organic soil fertilizer for the cultivation of spinach (Spinacia oleracea): Effects on soil properties, plant physiological parameters, and nutrient status. *J. Soil Sci. Plant Nutr.* 23, 5935–5944. https://doi.org/10.1007/s42729-023-01451-9.
Arabzadeh, G., et al., 2023. Assessment of antifungal/anti-oomycete activity of frass derived from black soldier fly larvae to control plant pathogens in horticulture: involvement of Bacillus velezensis. *Sustainability* 15, 10957. https://doi.org/10.3390/su151410957.
Ashworth, A.J., et al., 2025. Insect frass fertilizer as soil amendment for improved forage and soil health in circular systems. *Sci. Rep.* 15, 3024. https://doi.org/10.1038/s41598-025-87075-8.
Barragán-Fonseca, K.Y., et al., 2022. Insect frass and exuviae to promote plant growth and health. *Trends Plant Sci.* 27, 646–654. https://doi.org/10.1016/j.tplants.2022.01.007.
Basile-Doelsch, I., et al., 2020. Reviews and syntheses: the mechanisms underlying carbon storage in soil. *Biogeosciences* 17, 5223–5242. https://doi.org/10.5194/bg-17-5223-2020.
Beesigamukama, D., et al., 2020. Exploring black soldier fly frass as novel fertilizer for improved growth, yield, and nitrogen use efficiency of maize under field conditions. *Front. Plant Sci.* 11. https://doi.org/10.3389/fpls.2020.574592.
Beesigamukama, D., et al., 2021a. In situ nitrogen mineralization and nutrient release by soil amended with black soldier fly frass fertilizer. *Sci. Rep.* 11, 14799. https://doi.org/10.1038/s41598-021-94269-3.
Beesigamukama, D., et al., 2021b. Low-cost technology for recycling agro-industrial waste into nutrient-rich organic fertilizer using black soldier fly. *Waste Manag.* 119, 183–194. https://doi.org/10.1016/j.wasman.2020.09.043.
Beesigamukama, D., et al., 2022. Nutrient quality and maturity status of frass fertilizer from nine edible insects. *Sci. Rep.* 12, 7182. https://doi.org/10.1038/s41598-022-11336-z.
Beesigamukama, D., et al., 2023. Waste to value: Global perspective on the impact of entomocomposting on environmental health, greenhouse gas mitigation and soil bioremediation. *Sci. Total Environ.* 902, 166067. https://doi.org/10.1016/j.scitotenv.2023.166067.

Berggren, Å., et al., 2019. Approaching ecological sustainability in the emerging insects-as-food industry. *Trends Ecol. Evol.* 34, 132–138. https://doi.org/10.1016/j.tree.2018.11.005.
Blakstad, J.I., et al., 2023. Frass from yellow mealworm (*Tenebrio molitor*) as plant fertilizer and defense priming agent. *Biocatal. Agric. Biotechnol.* 53, 102862. https://doi.org/10.1016/j.bcab.2023.102862.
Borkent, S. and Hodge, S., 2021. Glasshouse evaluation of the black soldier fly waste product HexaFrass™ as an organic fertilizer. *Insects* 12, 977. https://doi.org/10.3390/insects12110977.
Boudabbous, K., et al., 2023. Black soldier fly (Hermetia illucens) larvae frass organic fertilizer improves soil quality and the productivity of durum wheat. *Commun. Soil Sci. Plant Anal.* 54, 2491-2507. https://doi.org/10.1080/00103624.2023.2227208.
Chavez, M. and Uchanski, M., 2021. Insect left-over substrate as plant fertiliser. *J. Insects Food Feed.* 7, 1612. https://doi.org/10.3920/JIFF2020.0063.
Chiam, Z., et al., 2021. Evaluating the potential of okara-derived black soldier fly larval frass as a soil amendment. *J. Environ. Manage.* 286, 112163. https://doi.org/10.1016/j.jenvman.2021.112163.
Choi, Y., et al., 2009. Potential usage of food waste as a natural fertilizer after digestion by Hermetia illucens (Diptera: Stratiomyidae). *Int. J. Ind. Entomol. Biomater.* 19, 171–174.
Crist, E., et al., 2017. The interaction of human population, food production, and biodiversity protection. *Science* 356, 260–264. https://doi.org/10.1126/science.aal2011.
Dal Magro, A., et al., 2024. The potential of insect frass for sustainable biogas and biomethane production: a review. *Bioresour. Technol.* 412, 131384. https://doi.org/10.1016/j.biortech.2024.131384.
Dulaurent, A.-M., et al., 2020. Earthworms (Lumbricus terrestris L.) mediate the fertilizing effect of frass. *Agronomy* 10, 783. https://doi.org/10.3390/agronomy10060783.
Elissen, H., et al., 2019. Disease suppression in cress and sugarbeet seedlings with frass of the Black Soldier Fly (Hermetia illucens).Wageningen Research, Report WPR-816. https://doi.org/10.18174/515047.
Elissen, H., et al., 2023. Effects of black soldier fly frass on plant and soil characteristics : a literature overview. https://doi.org/10.18174/587213.
Faucon, M.-P., et al., 2023. Combining agroecology and bioeconomy to meet the societal challenges of agriculture. *Plant Soil.* https://doi.org/10.1007/s11104-023-06294-y.
Foscari, A., et al., 2024. Frass from *Tenebrio molitor* as alternative to NPK-mineral fertilization: results from a germination test and pot experiment on sunflower. *Ital. J. Agron.* 19, 100010. https://doi.org/10.1016/j.ijagro.2024.100010.
Fuertes-Mendizábal, T., et al., 2023. Mealworm frass as a potential organic fertilizer in synergy with PGP-based biostimulant for lettuce plants. *Agronomy* 13, 1258. https://doi.org/10.3390/agronomy13051258.
Gao, Z., et al., 2022. Colorado potato beetle exploits frass-associated bacteria to suppress defense responses in potato plants. Pest Manag. *Sci.* 78, 3778–3787. https://doi.org/10.1002/ps.6823.
Gärttling, D., et al., 2020. Assessment of the N- and P-fertilization effect of black soldier fly (Diptera: Stratiomyidae) By-Products on Maize. *J. Insect Sci.* 20, 8. https://doi.org/10.1093/jisesa/ieaa089.

Gärttling, D. and Schulz, H., 2022. Compilation of black soldier fly frass analyses. *J. Soil Sci. Plant Nutr.* 22, 937–943. https://doi.org/10.1007/s42729-021-00703-w.
Gebremikael, M.T., et al., 2022. The impacts of black soldier fly frass on nitrogen availability, microbial activities, C sequestration, and plant growth. *Front. Sustain. Food Syst.* 6. https://doi.org/10.3389/fsufs.2022.795950.
Gortari, M.C. and Hours, R.A., 2013. Biotechnological processes for chitin recovery out of crustacean waste: a mini-review. *Electron. J. Biotechnol.* 16, 14–14. https://doi.org/10.2225/vol16-issue3-fulltext-10.
Green, T., 2023. A biochemical analysis of Black Soldier fly (Hermetia illucens) larval frass plant growth promoting activity. *PLoS One* 18, e0288913. https://doi.org/10.1371/journal.pone.0288913.
Gurung, S.K., et al., 2025. Turning waste into resilience: the effect of black soldier fly larvae frass and wheat straw biochar on bell pepper growth, soil fertility and rhizosphere bacteria under water stress. *Plant Soil.* https://doi.org/10.1007/s11104-025-07672-4.
Hallmann, J., et al., 1999. Chitin-mediated changes in bacterial communities of the soil, rhizosphere and within roots of cotton in relation to nematode control. *Soil Biol. Biochem.* 31, 551–560. https://doi.org/10.1016/S0038-0717(98)00146-1.
Hénault-Ethier, L., et al., 2024. Opportunities and challenges in upcycling agri-food byproducts to generate insect manure (frass): a literature review. *Waste Manag.* 176, 169–191. https://doi.org/10.1016/j.wasman.2023.12.033.
Hénault-Ethier, L., et al., 2023. Growth trials on vegetables, herbs, and flowers using mealworm frass, chicken manure, and municipal compost. *ACS Agric. Sci. Technol.* 3, 249–259. https://doi.org/10.1021/acsagscitech.2c00217.
Houben, D., et al., 2021. Assessment of the short-term fertilizer potential of mealworm frass using a pot experiment. *Front. Sustain. Food Syst.* 5, 714596. https://doi.org/10.3389/fsufs.2021.714596.
Houben, D., et al., 2020. Potential use of mealworm frass as a fertilizer: Impact on crop growth and soil properties. *Sci. Rep.* 10, 1–9. https://doi.org/10.1038/s41598-020-61765-x.
Ilijin, L., et al., 2024. Sourcing chitin from exoskeleton of *Tenebrio molitor* fed with polystyrene or plastic kitchen wrap. *Int. J. Biol. Macromol.* 268, 131731. https://doi.org/10.1016/j.ijbiomac.2024.131731.
Jasso, B., et al., 2024. Black soldier fly larvae (Hermetia illucens) frass and sheddings as a compost ingredient. *Front. Sustain. Food Syst.* 7. https://doi.org/10.3389/fsufs.2023.1297858.
Kallenbach, C.M., et al., 2016. Direct evidence for microbial-derived soil organic matter formation and its ecophysiological controls. *Nat. Commun.* 7, 13630. https://doi.org/10.1038/ncomms13630.
Kawasaki, K., et al., 2020. Evaluation of fertilizer value of residues obtained after processing household organic waste with black soldier fly larvae (Hermetia illucens). *Sustainability* 12, 4920. https://doi.org/10.3390/su12124920.
Kigeso, A.M., et al., 2024. Management of root knot nematode Meloidogyne incognita using black soldier fly frass in soybean [Glycine max (L.)]. *J. Current Opinion Crop Sci.* 5, 103–112. https://doi.org/10.62773/jcocs.v5i2.247.
Kisaakye, J., et al., 2024. Chitin-enriched insect frass fertilizer as a biorational alternative for root-knot nematode (Meloidogyne incognita) management. *Front. Plant Sci.* 15. https://doi.org/10.3389/fpls.2024.1361739.

Klammsteiner, T., et al., 2020. Suitability of black soldier fly frass as soil amendment and implication for organic waste hygienization. *Agronomy* 10, 1578. https://doi.org/10.3390/agronomy10101578.

Levavasseur, F., et al., 2020. The simple AMG model accurately simulates organic carbon storage in soils after repeated application of exogenous organic matter. *Nutr. Cycl. Agroecosys.* 117, 215–229. https://doi.org/10.1007/s10705-020-10065-x.

Lomonaco, G., et al., 2024. Larval frass of Hermetiaillucens as organic fertilizer: composition and beneficial effects on differentcrops. *Insects 15, 293*. https://doi.org/10.3390/insects15040293.

Lopes, I.G., et al., 2022. Frass derived from black soldier fly larvae treatment of biodegradable wastes. A critical review and future perspectives. *Waste Manag*. 142, 65–76. https://doi.org/10.1016/j.wasman.2022.02.007.

Machado, S.S.N., et al., 2024. Insect residues as an alternative and promising source for the extraction of chitin and chitosan. *Int. J. Biol. Macromol*. 254, 127773. https://doi.org/10.1016/j.ijbiomac.2023.127773.

Madau, F.A., et al., 2020. Insect farming for feed and food production from a circular business model perspective. *Sustainability* 12, 5418. https://doi.org/10.3390/su12135418.

Malheiro, C., et al., 2024. Insect frass from upcycling vegetable by-products with cereals: effects on the soil properties, plant development and soil invertebrate fitness. *J. Environ. Manage*. 372, 123234. https://doi.org/10.1016/j.jenvman.2024.123234.

Mannaa, M., et al., 2024. Insect-based agri-food waste valorization: agricultural applications and roles of insect gut microbiota. *Environ. Sci. Ecotechnol.* 17, 100287. https://doi.org/10.1016/j.ese.2023.100287.

Menino, R., et al., 2021. Agricultural value of Black Soldier Fly larvae frass as organic fertilizer on ryegrass. *Heliyon* 7, e05855. https://doi.org/10.1016/j.heliyon.2020.e05855.

Mirabello, L., et al., 2025. Comparative effects of black soldier fly frass, poultry litter, and NPK fertilizer on plant growth, soil microbial activity and earthworm biomass. *J. Soil Sci. Plant Nutr.* https://doi.org/10.1007/s42729-025-02668-6.

Papa, G., et al., 2022. Valorizing the organic fraction of municipal solid waste by producing black soldier fly larvae and biomethane in a biorefinery approach. *J. Clean. Prod.* 379, 134422. https://doi.org/10.1016/j.jclepro.2022.134422.

Penuelas, J., et al., 2023. A better use of fertilizers is needed for global food security and environmental sustainability. *Agric. Food Secur*. 12, 5. https://doi.org/10.1186/s40066-023-00409-5.

Poveda, J., 2020. Use of plant-defense hormones against pathogen-diseases of postharvest fresh produce. *Physiol. Mol. Plant Pathol.* 111, 101521. https://doi.org/10.1016/j.pmpp.2020.101521.

Poveda, J., 2021. Insect frass in the development of sustainable agriculture. A review. *Agron. Sustain. Dev.* 41, 5. https://doi.org/10.1007/s13593-020-00656-x.

Poveda, J., et al., 2019. Mealworm frass as a potential biofertilizer and abiotic stress tolerance-inductor in plants. *Appl. Soil Ecol.* 142, 110–122. https://doi.org/10.1016/j.apsoil.2019.04.016.

Praeg, N. and Klammsteiner, T., 2024. Primary study on frass fertilizers from mass-reared insects: Species variation, heat treatment effects, and implications for soil application at laboratory scale. *J. Environ. Manage.* 356, 120622. https://doi.org/10.1016/j.jenvman.2024.120622.

Quilliam, R.S., et al., 2020. Integrating insect frass biofertilisers into sustainable peri-urban agro-food systems. https://doi.org/10.3920/JIFF2019.0049.

Radzikowska-Kujawska, D., et al., 2023. Hermetia illucens frass improves the physiological state of basil (Ocimum basilicum L.) and its nutritional value under drought. *PLoS One* 18, e0280037. https://doi.org/10.1371/journal.pone.0280037.

Ray, S., et al., 2015. Maize plants recognize herbivore-associated cues from caterpillar frass. *J. Chem. Ecol.* 41, 781–792. https://doi.org/10.1007/s10886-015-0619-1.

Ray, S., et al., 2016a. Turnabout is fair play: herbivory-induced plant chitinases excreted in fall armyworm frass suppress herbivore defenses in maize. *Plant Physiol.* 171, 694–706. https://doi.org/10.1104/pp.15.01854.

Ray, S., et al., 2016b. Lessons from the far end: Caterpillar frass-induced defenses in maize, rice, cabbage, and tomato. *J. Chem. Ecol.* 42, 1130–1141. https://doi.org/10.1007/s10886-016-0776-x.

Ray, S., et al., 2020. Asymmetry in herbivore effector responses: Caterpillar frass effectors reduce performance of a subsequent herbivore. *J. Chem. Ecol.* 46, 76–83. https://doi.org/10.1007/s10886-019-01131-y.

Romano, N., et al., 2023. Mineral composition in black soldier fly (Hermetia illucens) larvae and resulting frass from fruit and their peels. https://doi.org/10.3920/JIFF2022.0019.

Rummel, P.S., et al., 2021. Black soldier fly diet impacts soil greenhouse gas emissions from frass applied as fertilizer. *Front. Sustain. Food Syst.* 5. https://doi.org/10.3389/fsufs.2021.709993.

Safitri, R.A., et al., 2024. Exploring potential uses of insect frass for agricultural production considering its nutrients, and chemical and microbiological safety. *J. Insects Food Feed.* https://doi.org/10.1163/23524588-00001224.

Salomon, M.J., et al., 2025. Potential of black soldier fly larvae frass (BSFL) as a novel fertilizer: impacts on tomato growth, nutrient uptake, and mycorrhizal formation. *Plant Soil.* https://doi.org/10.1007/s11104-024-07187-4.

Sawinska, Z., et al., 2024. Hermetia illucens frass fertilization: a novel approach for enhancing lettuce resilience and photosynthetic efficiency under drought stress conditions. *Appl. Sci.* 14, 2386. https://doi.org/10.3390/app14062386.

Schmitt, E. and de Vries, W., 2020. Potential benefits of using *Hermetia illucens* frass as a soil amendment on food production and for environmental impact reduction. *Curr. Opin. Green Sustain. Chem.* 25, 100335. https://doi.org/10.1016/j.cogsc.2020.03.005.

Sharp, R.G., 2013. A review of the applications of chitin and its derivatives in agriculture to modify plant-microbial interactions and improve crop yields. *Agronomy* 3, 757–793. https://doi.org/10.3390/agronomy3040757.

Urrutia, R.I., et al., 2023. From waste to food and bioinsecticides: An innovative system integrating *Tenebrio molitor* bioconversion and pyrolysis bio-oil production. *Chemosphere* 340, 139847. https://doi.org/10.1016/j.chemosphere.2023.139847.

van de Zande, E.M., et al., 2024. Soil amendment with insectfrass and exuviae affects rhizosphere bacterial community, shoot growth andcarbon/nitrogen ratio of a brassicaceous plant. Plant Soil 495, 631–648.https://doi.org/10.1007/s11104-023-06351-6.

van Huis, A., 2020. Insects as food and feed, a new emerging agricultural sector: a review. *J. Insects Food Feed* 6, 27–44. https://doi.org/10.3920/JIFF2019.0017.

Védère, C., et al., 2022. How does soil water status influence the fate of soil organic matter? A review of processes across scales. *Earth-Sci. Rev.* 234, 104214. https://doi.org/10.1016/j.earscirev.2022.104214.

Wantulla, M., et al., 2023. The potential of soil amendment with insect exuviae and frass to control the cabbage root fly. *Journal of Applied Entomology* 147, 181–191. https://doi.org/10.1111/jen.13097.

Watson, C., et al., 2022. Frass: the legacy of larvae–benefits and risks of residues from insect production. *Front. Sustain. Food Syst.* 6, 889004.

Watson, C., et al., 2025. Urease inhibition stimulates fungal degradation of chitin in frass-amended soil. *J. Plant Nutr. Soil Sci.* 188, 45–51. https://doi.org/10.1002/jpln.202300349.

Watson, C., et al., 2021. Excellent excrement? Frass impacts on a soil's microbial community, processes and metal bioavailability. *Appl. Soil Ecol.* 168, 104110. https://doi.org/10.1016/j.apsoil.2021.104110.

Wei, J.-R., et al., 2013. Monoterpenes from larval frass of two Cerambycids as chemical cues for a parasitoid, Dastarcus helophoroides. *J. Insect Sci.* 13, 59. https://doi.org/10.1673/031.013.5901.

Woo, K.T., et al., 2024. Co-composting of Black Soldier Fly Frass (BSFF) with various organics additives for nutrient enhancement. *Bioresour. Technol.* Rep. 27, 101944. https://doi.org/10.1016/j.biteb.2024.101944.

Zhang, X.G., et al., 2019. Oviposition deterrents in larval frass of potato tuberworm moth, *Phthorimaea operculella* (Lepidoptera: Gelechiidae). *Neotrop. Entomol.* 48, 496–502. https://doi.org/10.1007/s13744-018-0655-y.

Zunzunegui, I., et al., 2025. Insect frass as an agricultural resource against abiotic and biotic crop stresses: Mechanisms of action and possible negative effects. *Appl. Sci.* 15, 3606. https://doi.org/10.3390/app15073606.

Part 2

Using biostimulants and promoting beneficial microorganisms in soil

Chapter 9

Leveraging plant growth-promoting bacteria for enhanced crop nutrient use efficiency and growth

A. S. S. Sousa, P. M. L. Castro and S. I. A. Pereira, Centre for Biotechnology and Fine Chemistry (CBQF) – Catholic University of Portugal, Portugal

http://dx.doi.org/10.19103/AS.2025.0166.11

1 Introduction

The agriculture sector is under growing pressure to meet a projected 50% increase in global food demand by 2050, driven by population growth (Kumar et al., 2022; Yuan et al., 2024). At the same time, soil and freshwater systems are nearing their productive thresholds (FAO, 2021). Conventional agriculture depends heavily on synthetic fertilizers and pesticides, but their long-term sustainability is uncertain due to major environmental impacts such as soil degradation, water pollution, biodiversity loss, and increased greenhouse gas emissions (Nuruzzaman et al., 2025). Moreover, excessive nitrogen (N) and phosphorus (P) inputs do not reliably boost yields (Cai et al., 2025). This inefficiency paired with escalating costs and resource depletion underscores the urgent need for strategies that improve nutrient use efficiency (NUE) and support sustainable crop production.

Microbial biostimulants (also known as biofertilizers, microbial inoculants, and bioinoculants) are increasingly recognized as a sustainable and eco-friendly approach in modern agriculture. By promoting greater NUE, they reduce reliance on environmentally damaging synthetic agrochemicals and enhance plant resilience to diverse abiotic stresses (Fadiji et al., 2022). These formulations contain living beneficial microorganisms, such as plant growth-promoting bacteria (PGPB) and arbuscular mycorrhizal fungi (AMF), that enhance nutrient uptake (Severo et al., 2025) and improve NUE (Thioub et al., 2019).

PGPB offer a multifaceted approach to enhancing soil and plant health. They can fix atmospheric N, solubilize P, and produce key phytohormones such as auxins (which promote plant growth and improve response to biotic and abiotic stresses), along with the secretion of siderophores (with the ability to scavenge iron as a plant nutrient). PGPB also suppress soil-borne pathogens by producing antibiotics and hydrolytic enzymes and by inducing systemic resistance in plants (Castiglione et al., 2021; Chaudhary et al., 2022; Singh et al., 2023). Their beneficial traits and interactions with host plants make them valuable components of integrated nutrient management systems aimed at improving overall crop growth and health.

This chapter explores the key mechanisms by which PGPB-based biostimulants enhance crop growth and NUE, while reviewing the main benefits that plants derive from associations with PGPB and examining the key factors influencing the effectiveness of PGPB-based biostimulants. It provides a critical evaluation of the current literature, identifying major knowledge gaps, technical challenges, and emerging approaches, including the selection of high-performing strains, the development of effective formulations using suitable carriers to enhance microbial viability and shelf-life, and the integration of PGPB-based biostimulants into sustainable agricultural systems. Together, these components are essential for their successful development and practical field application.

2 Types of plant growth-promoting bacteria and mechanisms of action: overview

PGPB are a group of bacteria that can establish mutualistic relations with host plants and stimulate their growth (Aloo et al., 2022). A bacteria can only qualify as a PGPB if it produces a positive effect on the host plant (Grover et al., 2021). Understanding of PGPB and their potential role as an alternative and sustainable solution to conventional crop fertilization developed in the 1980s with concepts such as 'biofertilizers' (Aloo et al., 2022).

A key aspect of the relationship between PGPB and plants is that both derive benefits from their interaction in a mutualistic relationship. Plants benefit from bacteria through a wide range of direct mechanisms, including nutrient solubilization and mobilization as well as the production of plant growth-stimulating hormones, among others (Singh et al., 2022). In addition, bacteria indirectly support plants by enhancing tolerance to abiotic stresses and increasing resistance against pathogens (Grover et al., 2021). In return, bacteria take advantage of plant exudates as a nutrient source, resulting in a microbial community in the rhizosphere that is 100 to 1000 times more abundant and diverse than in surrounding bulk soil (Goswami et al., 2016).

PGPB can be classified into two main types based on their location within the soil/root system (Hardoim et al., 2008):

- Those that colonize the area surrounding the root (rhizosphere) and
- Endophytic bacteria, which, through various mechanisms, can establish themselves within the root endosphere.

Any bacteria can potentially become an endophyte if present in the rhizosphere, for instance due to passive penetration via root cracks or tips. However, certain bacteria, such as nodulating species, have evolved specific mechanisms, including the presence of *nod* genes, that enable bacteria to actively penetrate and colonize the root system. Besides this, the production of cell-wall degrading enzymes and twitching motility are also identified as active colonization mechanisms (Compant et al., 2010). Although both types of bacteria can enter plant roots, those with specific adaptations for active entry are generally more effective in promoting plant growth and development. PGPB include key genera such as *Rhizobium*, *Azospirillum*, *Azotobacter*, *Bacillus*, and *Pseudomonas* (Glick, 2012; Gómez-Godínez et al., 2023).

There are several mechanisms through which PGPB can enhance plant growth, differing among species and strains. They can support the plant either directly or indirectly, depending on the mode of action of each mechanism, whether it targets the plant itself by providing essential resources or modifies the surrounding environment to indirectly enhance plant performance (Jeyanthi and Kanimozhi, 2018). The difference between the two categories is not always evident, but by grouping the different mechanisms, it is possible to have a clearer picture of what distinguishes them (Goswami et al., 2016).

Direct mechanisms include (Goswami et al., 2016; Jeyanthi and Kanimozhi, 2018):

- acquisition of nutrients, for instance through biological nitrogen fixation (BNF) or the solubilization and mobilization of nutrients through the production of phosphatases and siderophores and
- production of phytohormones and exopolysaccharides.

Indirect mechanisms include:

- biocontrol of phytopathogens;
- induction of plant systemic resistance; and
- mitigation of abiotic stresses.

The following sections focus on some of these key mechanisms.

3 Nutrient acquisition: nitrogen and phosphorus

3.1 Nitrogen fixation

Nitrogen is an essential macronutrient for plants, playing a key role in several physiological processes like photosynthesis and protein synthesis (Alori et al., 2017). Although N_2 constitutes c.a. 79% of the Earth's atmosphere, its low reactivity makes it unavailable for direct uptake by plants, which can only utilize N as ammonia or nitrate (Barrett and Burke, 2000). To compensate, mineral fertilizers are widely applied in agriculture to ensure adequate N availability for crop production. However, their overuse and consequent soil saturation disrupt the natural balance of the N cycle and lead to environmental problems, including soil degradation and water pollution, ultimately reducing agricultural sustainability and crop performance (Shah et al., 2021).

In this context, N_2-fixing bacterial strains have gained special attention for their ability to convert atmospheric N into plant-available forms (primarily NH_3), thereby enhancing N availability in agroecosystems. These bacteria are broadly classified as symbiotic and free-living. Symbiotic bacteria, like *Rhizobium*, form specialized root nodules, while free-living N fixers, like *Azospirillum* and *Azotobacter*, can live in the rhizosphere without infecting plant roots. Over time, other bacterial genus have gained importance for their ability to fix N_2, such as *Bacillus* and *Paenibacillus*. These genera have been reported to carry the *nif* gene cluster, which encodes for the nitrogenase enzyme, a key enzyme in BNF (Goswami et al., 2016).

By naturally supplying N to plants, bacteria involved in BNF contribute directly to improved NUE and reduce N losses through leaching, volatilization, and runoff. Increased NUE leads to more efficient N assimilation by crops which, in turn, promotes plant growth, productivity, and yield stability. Globally, it is estimated that 50–70 teragrams (Tg) of N are fixed biologically in agricultural systems every year, and 52–130 Tg in all terrestrial ecosystems (Zeng et al., 2022), highlighting the significant role of BNF in the global N cycle. The use of N_2-fixing bacteria represents a sustainable alternative to mineral fertilizers and a promising strategy to enhance crop growth while minimizing environmental impact.

3.2 Phosphorus solubilization

Phosphorus is an essential macronutrient for the proper functioning of plant metabolism, playing a central role in energy transfer, membrane structure, and nucleic acid synthesis. Despite its abundance in soil, most P occurs in insoluble inorganic and organic forms, which greatly limits its availability to plants.

Some PGPB are recognized as capable of solubilizing unavailable forms of P and converting them into plant-accessible orthophosphate ions, like monophosphate (HPO_4^-) and dibasic phosphate ($H_2PO_4^-$) (Riaz et al., 2020). In addition, certain PGPB produce enzymes such as phytases and phosphomonoesterases (acid or alkaline phosphatases) that mineralize organic P into inorganic form available to plants (Rawat et al., 2021; Ali et al., 2022). In agricultural soils, inorganic P is frequently immobilized through precipitation and adsorption to Ca^{2+} and Fe^{2+}, which drastically reduces fertilizer efficiency (Pii et al., 2015). Indeed, only a small fraction, c.a. 10–20% of applied P fertilizers are efficiently used by crops, highlighting the low phosphorus use efficiency (PUE) of conventional fertilization practices.

PGPB enhance PUE by releasing organic acids such as gluconic, citric, oxalic, and tartaric acids, which help to solubilize P. These acids chelate P-bound cations like Fe^{2+}, Al^{3+}, and Ca^{2+}, and lower the rhizosphere pH, thus increasing P solubility and plant uptake. Rawat et al. (2021) reported several bacterial strains capable of solubilizing P and promoting plant growth. For instance, *Bacillus megaterium* CS22 enhanced both biomass and P content in *Brassica napus,* while *Pseudomonas putida* PSE3 was reported to increase shoot biomass and P accumulation in *Pisum sativum*. Likewise, bacterial genera such as *Azospirillum, Azotobacter, Burkholderia, Rhizobium,* and *Serratia* have been widely identified as effective P-solubilizers (Pii et al., 2015). As global demand for sustainable and efficient management strategies intensifies, the use of P-solubilizing bacteria offers a promising strategy to improve PUE. Consequently, these microorganisms are strong candidates for the development of biofertilizers that support environmentally responsible and high-yield agricultural systems (Rizvi et al., 2021).

4 Nutrient acquisition: production of siderophores

Iron is essential for both plant and microbial metabolism, yet its bioavailable form is often limited in soils. Siderophores are low molecular weight compounds (e.g. hydroxamates, phenol catecholates, carboxylates, and pyoverdines) that have a dual mechanism of action: they not only facilitate Fe acquisition by plants while inhibiting its absorption by pathogenic microorganisms (Zhou and Bai, 2025), but also aid in inorganic P solubilization by strongly chelating Fe^{3+} in Fe-P complexes (Rawat et al., 2021). This dual action of the siderophore-producing bacteria enhances overall NUE and promotes better nutrient balance in plants.

Several genera have been identified as siderophore producers, including *Bacillus*, *Azotobacter*, *Enterobacter*, *Pseudomonas*, and *Rhodococcus*. Among these, *Bacillus subtilis* is among the most extensively studied and widely recognized siderophore-producing species (Singh et al. 2022; Ali et al. 2022). Singh et al. (2022) have reported several studies reinforcing the ability of siderophore-producing bacteria to enhance the growth and development of numerous crops such as tomato, potato, and wheat, some of them under salinity and drought stress conditions. Also, in terms of biocontrol, beneficial PGPB outcompete phytopathogens for Fe in the soil, effectively limiting pathogen access to this nutrient and suppressing disease development. Consistent with this, *Bacillus niabensis* PT-32-1 and *B. subtilis* SWI16b presented antagonistic activity against banana wilt pathogens (Kesaulya et al., 2018).

5 Nutrient acquisition: production of phytohormones

Chemical compounds such as auxins, cytokinins, gibberellins, abscisic acid, and ethylene are key players in the physiological processes of plants, intervening in their growth (e.g. cell division and elongation), development (e.g. shoot differentiation, flower and fruit appearance, number of root tips), and adaptation to different environments (Poveda and González-Andrés, 2021; Ali et al., 2022). These compounds are classified as phytohormones, and PGPB can synthesize them, thereby influencing the regulation of the plant hormonal system and helping maintain balanced hormone levels. Through this hormonal modulation, PGPB enhance root system architecture, including root length, branching, and root hair development, which directly improves nutrient acquisition from the soil (Vacheron et al., 2013). As a result, NUE is increased, leading to improved plant growth, biomass accumulation, and overall crop performance.

5.1 Auxins

Auxins are a low-molecular-weight group of phytohormones that play a fundamental role in the development and growth of plants. They are involved in several different biological processes (Pantoja-Guerra et al., 2023). Among the various mechanisms influenced by phytohormones, notable examples include the regulation of the cell cycle, gravitropism, and phototropism of roots and shoots, determination of shoot and root meristems' size, flower morphogenesis, and shadow avoidance responses (Tsukanova et al., 2017; Gomes and Scortecci, 2021).

The most common auxins are indole-3-acetic acid (IAA), indole-3-butyric acid (IBA), 4-chloroindole-3-acetic acid (4-Cl-IAA), and phenylacetic acid (PAA), with different morphogenic activities. IAA is the most common and extensively studied auxin produced by PGPB, playing a key role as a signaling molecule in plant–microbe interactions (Gomes and Scortecci, 2021). The production of IAA is notably prominent in several bacterial genera, such as *Acinetobacter*, *Azospirillum*, *Bacillus*, *Pseudomonas*, *Rhizobium*, *Rhodococcus*, and *Azotobacter*, which makes it a valuable and recognized plant growth precursor. IAA uptake regulation by the plants is related to the internal and surrounding levels of IAA, including those attributed to bacterial production. To maintain hormonal balance, plants regulate IAA levels through both synthesis and degradation pathways. In addition, the formation of IAA conjugates is essential for storing the active hormone (Etesami and Glick, 2024).

IAA production by bacteria occurs through several mechanisms, with the primary IAA synthesis pathway based on L-tryptophan exuded by plants' roots (Ilangumaran and Smith, 2017) or provided by any external means. The indole-3-acetamide (IAM) and indole-3-pyruvic acid (IPA) pathways are the predominant biosynthetic routes for IAA in microorganisms, both dependent on the precursor, L-tryptophan (Patten et al., 2013; Etesami and Glick, 2024). However, L-tryptophan is an amino acid synthesized at a very high energetic cost by bacterial cells, which makes the production of IAA from the endogenous L-tryptophan relatively low. Moreover, available L-tryptophan is primarily allocated to protein synthesis to avoid compromising essential cellular functions (Patten et al., 2013). It is also noteworthy that the different IAA production pathways exhibited by bacterial strains can intersect and complement each other, which makes the process resilient when biological alterations occur (e.g. genetic mutation) (Duca and Glick, 2020).

5.2 Cytokinins

PGPB are also known for producing cytokinins, which are plant hormones involved in various stages of development, from seed germination to leaf expansion and senescence. With participation in cell division, shoot initiation, and mechanisms such as root development and root hair formation, bacterial cytokinins can directly affect the growth of plants (Jeyanthi and Kanimozhi, 2018).

Goswami et al. (2016) have demonstrated that bacterial genera like *Pseudomonas*, *Azospirillum*, *Bacillus*, *Escherichia*, and *Xanthomonas* are notable producers of cytokinins. Bacterial strains contain specific enzymes known as cytokinin synthases, which are vital in converting precursor molecules into active cytokinins. Different types of cytokinins can be produced by PGPB, including isopentenyladenine (iP), cis-zeatin (cZ), and trans-zeatin (tZ) (Palberg et al., 2022; Kejela, 2024). In recent studies Palberg et al. (2022) analyzed the phytohormone-production potential of 46 different strains of the *Methylobacterium* genus, most of them collected from plant tissues or soil samples. The findings highlighted the characterization of a less common type of cytokinin (tZ), which was produced in higher quantities compared to other PGPB. This cytokinin is known to influence plant growth, alleviate abiotic stress, and regulate the immune system, playing a role in plant defense responses against bacterial, fungal, and insect pathogens (Akhtar et al., 2020). Some bacterial strains synthesize cytokinins through *de novo* pathways, while others modify existing cytokinins from their environment (Kejela, 2024).

5.3 Gibberellins

Gibberellins are another type of signaling and bioactive compound involved in processes such as seed germination, leaf expansion, flowering, and fruit development. Physiologically active gibberellins have been detected in different bacteria, isolated from the rhizosphere or surrounding soil. These bacteria include organisms from the genera *Acinetobacter, Bacillus*, *Burkholderia*, *Flavobacterium*, *Enterococcus*, *Leifsonia,* *Rhizobium*, and *Xanthomonas* (Tsukanova et al., 2017).

5.4 Abscisic acid

Phytohormones play a central role in regulating plant growth and mediating responses to abiotic stress. Abscisic acid (ABA) synthesized by plants is involved in several physiological processes, including seed dormancy and response to abiotic stresses such as water scarcity, drought, and salinity (Poveda and González-Andrés, 2021). During osmotic stress, ABA is rapidly produced in the roots and transported to the leaves via the vascular system, where it regulates stomatal closure and opening, thereby helping the plant to withstand drought conditions (Sati et al., 2023).

PGPB are also known for their ability to produce ABA, although the mechanisms by which this happens are still not well understood (Kejela, 2024). Bacteria can influence plant ABA levels by modifying the surrounding biotic environment, thereby modulating stress responses. Numerous studies have shown that PGPB-mediated regulation of ABA triggers a cascade of physiological responses that help to alleviate abiotic stress (Cohen et al., 2009, 2015; Sheteiwy et al., 2021; Jha et al., 2023). These studies have involved different crop species (e.g. pigeon pea, maize, soybean), as well as the model plant *Arabidopsis thaliana*, underscoring the consistent and robust influence of PGPB on ABA regulation across different plant systems.

6 Nutrient acquisition: production of exopolysaccharides

Exopolysaccharides (EPS) are large biopolymers composed primarily of galactose, glucose, and mannose monomers. When combined with other components such as amino sugars and uronic acids, they form a protective matrix around bacterial cells, which is commonly called a biofilm. This is typically produced in response to adverse environmental conditions, enhancing bacterial survival and resilience.

EPS enhance soil structure by promoting aggregation, either through direct binding or adhesion to soil particles, resulting in a more stable and cohesive soil matrix. This structural improvement reduces soil dispersion and improves overall soil health. EPS-producing bacteria play a crucial role in supporting plant development by (Bhat et al., 2020; Grover et al., 2021):

(i) Creating a resilient root environment capable of withstanding environmental fluctuations; and
(ii) Enhancing soil cation exchange capacity and water retention, thereby improving water availability under drought and saline conditions.

EPS also exhibit high water retention capacity, which further facilitates nutrient uptake and helps maintain a hydrated and supportive environment around the root (Fig. 1).

Dar et al. (2021) investigated the potential of EPS-producing bacteria to enhance maize growth and yield under greenhouse conditions. *Pseudomonas bathysetes* and *Rhizobium phaseoli* Mn-6 showed superior performance in promoting soil aggregation. Among them, *R. phaseoli* demonstrated particularly strong effects on plant development, significantly improving plant

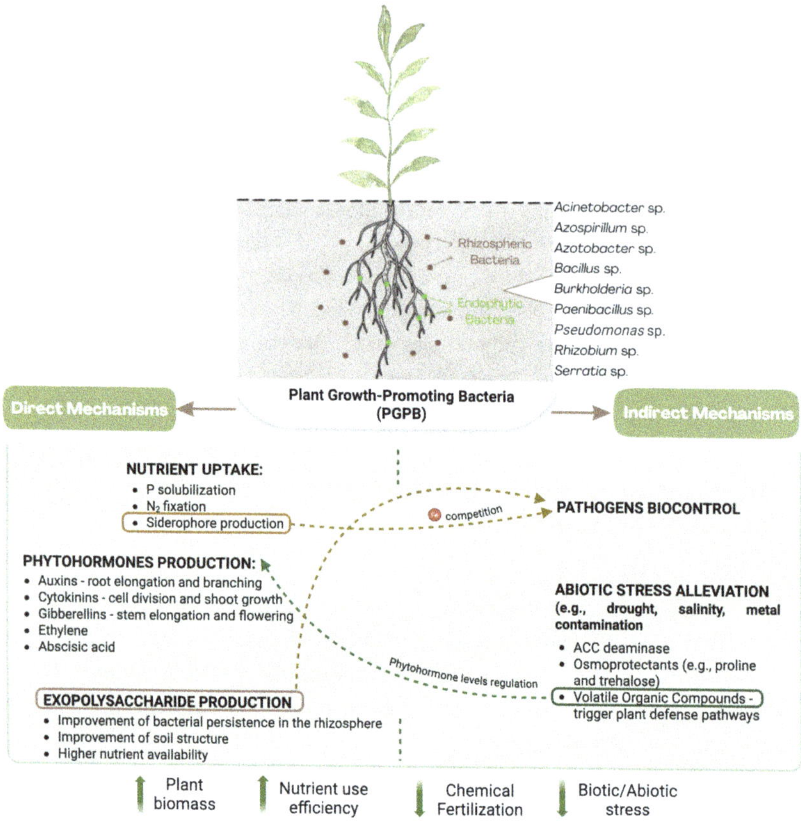

Figure 1 Mechanisms of PGPB-mediated plant growth promotion.

height, root length, fresh and dry biomass, root length density, SPAD values, leaf area index, and photosynthetic rate. Inoculating sunflower (*Helianthus annuus* L.) seeds and soil with *Rhizobium* sp. YAS34 also led to an increase in root-adhering soil per unit of root dry mass, along with a significant rise in soil macropore volume. Notably, shoot and root dry mass increased by up to 50% and 70%, respectively, under both normal and water stress conditions (Alami et al., 2000).

7 Abiotic stress alleviation

Plant growth depends on specific environmental conditions, and any deviation from this can disrupt plant physiological processes, ultimately reducing crop productivity and/or quality. Key abiotic stresses include drought, salinity, extreme temperatures, and metal contamination (Shah et al., 2021; Sati et al., 2023; Karnwal et al., 2024).

Drought affects multiple stages of plant development, causing delayed seed germination, stunted shoot growth, limited leaf expansion, and decreased photosynthetic activity, ultimately leading to lower crop yields. Moreover, nutrient uptake is impaired due to reduced soil moisture, since the mass flow and diffusion of water-soluble nutrients such as Ca^{2+}, Mg^{2+}, SO_4^-, and NO_3^- are restricted (Steiner et al., 2017; Li et al., 2020).

Salinity is another major constraint on agricultural productivity, often intensified by anthropogenic practices such as improper irrigation and fertilizer management, which impair soil quality and disrupt plant homeostasis, and development. Salinity first induces osmotic stress, followed by ion toxicity due to the excess of Na^+ and Cl^- accumulation in plant tissues. This leads to disrupted photosynthesis, hormone imbalance, reduced nutrient translocation, and metabolic dysfunction, ultimately compromising plant health and development (Ilangumaran and Smith, 2017). Under both drought and salinity stress, nutrient imbalances arise not only from limited nutrient mobility in soil but also from impaired root function, leading to reduced NUE and suboptimal plant growth.

To counteract the negative effects of abiotic stress, PGPB deploy multiple complementary beneficial mechanisms that directly contribute to stress alleviation and improved nutrient acquisition. These include the modulation of phytohormone levels to stimulate root development, regulation of stomatal conductance to improve water-use efficiency, and production of EPS that improve soil aggregation and moisture retention (Sati et al., 2023). A review of Karnwal et al. (2024) has highlighted the effectiveness of key microbial genera, including *Bacillus Azospirillum*, and *Pseudomonas,* in mitigating water stress, salinity, extreme temperatures, and metal pollution across various experimental systems.

Several studies support these findings. Bouremani et al. (2024) demonstrated that osmotolerant PGPB (*Providencia vermicola* ME1, *Pantoea agglomerans* Pa, *Pseudomonas knackmussi* MR6, and *Bacillus* sp. D13) significantly improved wheat growth and chlorophyll content under moderate (50% field capacity; FC) and severe (25% FC) drought conditions through ACC-deaminase activity and biofilm formation, which lowered stress ethylene levels, enhanced root colonization, and reduced oxidative damage under harsh conditions.

Dauala et al. (2025) also reported that *B. subtilis* IPACC29 and *Paenibacillus* sp. IPACC38 substantially improved drought tolerance in forage cactus pear, with the strongest effects observed under severe water stress and without N fertilization. Although drought reduced growth in all treatments, these strains enhanced root development and plant performance by promoting N accumulation and optimizing photosynthetic pigment content, indicating improved NUE and metabolic adjustment under stress. This study also showed that drought tolerance in cactus pear is primarily driven by metabolic reprogramming rather than improved plant water content, with drought conditions inducing intrinsic defense responses such as increased N accumulation in cladodes and elevated levels of photosynthetic pigments and carotenoids to sustain photosynthesis and limit oxidative damage.

The potential of ACC deaminase–producing PGPB to alleviate ethylene stress under salinity was clearly highlighted by Win et al. (2023), who showed that dual inoculation with *Bradyrhizobium ottawaense* SG09 and *Pseudomonas* OFT2 or OFT5 significantly enhanced soybean salt tolerance by reducing stress-induced ethylene levels, improving nutrient uptake, and restoring nodulation and BNF. Collectively, these studies reinforce the role of PGPB as versatile, sustainable tools for enhancing crop tolerance to multiple abiotic stresses.

Bacterial adaptability to harsh environments plays a key role in their effectiveness. Therefore, selecting PGPB strains isolated from stress-prone habitats is considered a promising strategy for maximizing their potential in stress mitigation and improving crop performance under adverse conditions.

8 Pathogen suppression and biocontrol

Soil bacteria that promote healthy plant growth can also contribute to plant protection by suppressing pathogens and reducing disease incidence, thereby acting as effective biocontrol agents. In addition, pathogen attack often damages the root system and its architecture (Xue et al., 2020). By limiting pathogen pressure, PGPB help preserve root integrity and physiological function, which are essential for efficient nutrient and water uptake. PGPB thus represent a sustainable and environmentally friendly alternative to chemical pesticides in modern agriculture.

PGPB employ several general mechanisms of biocontrol that support plant health and suppress pathogen development (Fig. 2). These include (Boro et al., 2022):

- competition for nutrients and space, which restricts access to essential resources such as C and Fe and limits pathogen establishment in the rhizosphere;
- production of antibiotics and lytic enzymes (e.g. chitinases, proteases, and glucanases) that degrade pathogen cell walls and inhibit their growth;
- siderophore secretion, which chelates Fe^{3+}, thereby depriving pathogens of this essential element while improving its availability to plants; and
- induction of plant immune responses through systemic resistance, which enhances host defense mechanisms and increases resistance against a broad spectrum of pathogens.

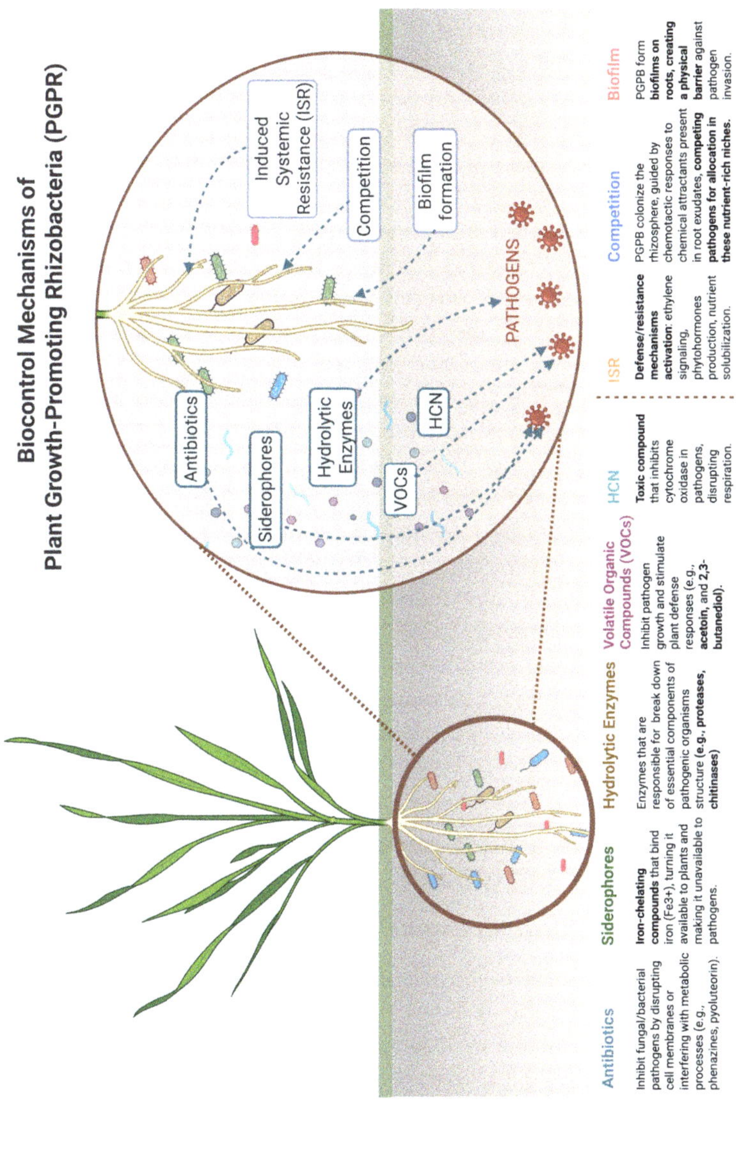

Figure 2 Biocontrol mechanisms of PGPR.

These mechanisms have been validated by multiple experimental studies. Triantafyllou et al. (2023) evaluated the biocontrol potential of 49 PGPB strains against fungal pathogens and demonstrated that several strains were able to induce plant immune response. In particular, *Azotobacter chroococcum* SAESo-E4 significantly suppressed *Sclerotinia sclerotiorum* infection in lettuce by triggering host-defense pathways. In addition to its biocontrol activity, this strain enhanced photosynthesis and promoted plant growth, highlighting its dual function as both a disease suppressor and growth promoter.

Further evidence supporting the role of enzymatic activity and siderophore production has been reported by Lau et al. (2020), who showed that *B. subtilis, Brevibacillus gelatini, Burkholderia territorii, Pseudomonas geniculate*, and *Pseudomonas beteli* produced antifungal siderophores (against soil-borne *Fusarium solani)*, exhibited chitinase activity (with a chitinolytic index ranging from 1.19 to 1.76), solubilized P, and produced ammonia. These strains also promoted plant growth through the synthesis of IAA at concentrations ranging from 10.5 to 42.6 $\mu g\ mL^{-1}$. Indeed, siderophore production is crucial under field conditions due to limited Fe availability and intense microbial competition.

Gu et al. (2020) also demonstrated that siderophore-mediated interactions play a central role in shaping microbiome function in natural field environments. As a representative group of PGPB, *Bacillus* sp. strains are frequently used as models to evaluate biocontrol potential. For example, *Bacillus thuringiensis* has been shown to interfere with bacterial quorum sensing, a key regulatory system often associated with pathogen virulence, thereby reducing disease development (Ramírez-Pool et al., 2024). Under field conditions, the ability of bacterial strains to interfere with quorum sensing is particularly important, as agricultural soils often harbor multispecies pathogen communities in which disruption of cell-to-cell signaling can suppress biofilm formation and reduce virulence, thereby limiting disease development (Zhu et al., 2023).

9 Benefits of plant growth-promoting bacteria for crop nutrient uptake and growth

9.1 Improvement of nutrient acquisition and use

Efficient nutrient use is critical for sustainable crop production. However, despite the abundance of macronutrients in soil, many nutrients remain unavailable to plants. PGPB play a crucial role in making essential nutrients more accessible to plants through BNF, P solubilization, and by stimulating nutrient uptake processes.

Naher et al. (2021) demonstrated the effectiveness of a PGPB-based bio-organic fertilizer in improving NUE and reducing reliance on inorganic fertilizers in rice. Their formulation contained several *Bacillus* sp. and *Paenibacillus* spp. strains known for their ability to fix N_2, solubilize P, and produce IAA. The formulation was tested across 16 field experiments and 18 on-farm trials over 3 years. Results showed significant increase in agronomic use efficiency of N and P, allowing reductions of 30% in N fertilization and up to 100% in P fertilization without yield loss. These large-scale field studies provide strong evidence that multispecies consortia can significantly reduce fertilizer inputs while maintaining productivity through improved nutrient availability. Similarly, Akintokun et al. (2019) also demonstrated the efficiency of *B. cereus* to solubilize essential nutrients and confirmed its plant growth-promoting effects through pot experiments in maize, where inoculation significantly increased root and shoot length and biomass.

The integration of PGPB, mineral fertilizers, and soil amendments seems to further enhance their effectiveness. Phares et al. (2022) evaluated the combined application of NPK fertilizers, strains from *Bacillus* and *Pseudomonas* genera, and biochar in maize field trials. This combined approach increased soil organic carbon, enhanced soil dehydrogenase activity, and improved N, P and K use efficiency. Treatments combining microbial inoculants with biochar outperformed those with NPK alone or NPK plus biochar, demonstrating the value of integrated nutrient management for maximizing crop productivity and soil health.

9.2 Effects on root architecture and morphology

Beyond improved nutrient mobilization, PGPB influence root system architecture, which plays a critical role in water and nutrient acquisition. Root traits such as branching intensity, lateral root development, and primary root length determine the plant's capacity to explore soil resources (Grover et al., 2021). PGPB can modify these traits through hormonal regulation and enhanced rhizosphere activity (Mantelin et al., 2006). de Sousa et al. (2021) studied the effect of tropical *Bacillus* strains on maize root morphology in a hydroponic system and observed increases in root surface area, nutrient content, and seedling biomass. When the most effective strains were later tested in field trials under different P fertilization regimes, inoculated plants exhibited higher yields and greater P accumulation in the grain, confirming the functional link between root morphology modulation and improved nutrient uptake.

9.3 Yield improvements in different crops

Several studies have demonstrated that PGPB application leads to increased yield across economically important crops. In maize, *B. megaterium*, *B. subtilis*, and *A. chroococcum* significantly boosted yield compared with non-inoculated controls in field studies (Efthimiadou et al., 2020). Notably, *B. subtilis* also increased post-harvest seed quality by increasing total solid content by 92%.

In wheat, Chandra et al. (2019) reported that seed inoculation with *Variovorax paradoxus* RAA3, *Pseudomonas palleroniana* DPB16, and *Pseudomonas* sp. UW4 strains significantly improved yields in both irrigated and rain-fed trials. Under rain-fed conditions, *V. paradoxus* RAA3 increased grain yield by 47.4%, straw yield by 26%, and harvest index by 15.3%, highlighting the role of PGPB in improving crop performance under water-limited environments.

Overall, PGPB represents a promising strategy to enhance NUE, increase crop yields, and reduce reliance on chemical fertilizers, while decreasing the risk of nutrient losses through leaching, runoff, and volatilization. This reduction in fertilizer use contributes to lower greenhouse gas emissions associated with fertilizer manufacturing and application, mitigates soil and water contamination, and promotes healthier soil microbial communities. Together, these effects position PGPB as an important tool for advancing environmentally sustainable crop production and improving the long-term resilience of agroecosystems.

To fully realize the potential of PGPB, it is crucial to bridge the gap between promising laboratory and greenhouse findings and consistent and reliable performance under real-world agricultural conditions. The case studies summarized in Table 1 highlight the importance of robust field-based evidence for validating both the efficacy and stability of PGPB-based biostimulants across diverse cropping systems and environmental contexts.

However, results obtained under field conditions revealed that PGPB application is constrained by several recurrent limitations. Crop responses to inoculation are often non-uniform and are strongly shaped by interactions with native soil microbiota, plant species or genotype, and prevailing environmental conditions, resulting in variable outcomes across locations and seasons. Moreover, single-strain inoculations often show limited or no impact on key agronomic parameters, including biomass accumulation, yield, 1000-grain weight, nutrient uptake, and grain protein content. Under controlled greenhouse conditions, inoculant effects are often amplified due to reduced abiotic stress and limited microbial competition. In contrast, field environments are inherently heterogeneous, with fluctuating climate conditions, variable soil chemistry, and established native microbial communities that can constrain PGPB activity. Consequently, field-level efficacy remains inconsistent, underscoring the need for improved formulations, careful strain selection, and environment-specific validation strategies.

Table 1 Effects of PGPB (strains or commercial PGPB-based formulations) use on crop growth and nutrient uptake under field conditions. CGI: crop growth improvement; NUE: Nutrient use efficiency; ASA: abiotic stress alleviation; BSA: Biotic stress alleviation; SS: Soil stabilization; SAPT: sustainable agricultural practices transition; n.a: not available.

Crop	Field trial design	PGPB	Application mode	Benefits	Main results	Limitations/Future research	References
Common bean (*Phaseolus vulgaris* L.)	2 years; pre-cleaned field **Treatments:** × Strains RPN5, BPR7, PPR8 single inoculated × Consortium 1: RPN5+ BPR7 × Consortium 2: RPN5+PPR8 × Consortium 3: RPN5+ BPR7+ PPR8	*Rhizobium leguminosarum* RPN5 *Bacillus* sp. BPR7 *Pseudomonas* sp. PPR8	Seed coating	CGI	**All bacterial treatments:** + Seed germination + Shoot and root length, fresh and dry weight + Grain yield Consortium 3 outperformed the others' inoculation treatments	**Future Research:** × Strains monitoring to assure bacteria survival in the soil × Proper farming practices management (fertility, irrigation) to maximize benefits	Kumar et al. (2016)
	No-tillage system in the previous 8 years **Treatments:** × *Rhizobium or Azospirillum* seed inoculation × *Rhizobium* + *Azospirillum* seed inoculation × *Azospirillum* foliar application × *Rhizobium* seed inoculation + *Azospirillum* foliar application × *Rhizobium* + *Azospirillum* seed inoculation + *Azospirillum* foliar application	Starfix feijão®: *Rhizobium tropici*Azos®: *Azospirillum brasilense*	Seed inoculation Foliar application	NUE; CGI	*Rhizobium* + *Azospirillum* seed inoculation + *Azospirillum* foliar application treatment presented the best results: + Shoot biomass + N accumulation + Grain yield (95.3% compared to control)	**Limitations:** × Non-uniform response to inoculation, due to native microbiota, plant species/variety, and environmental conditions parameters × Single inoculation with *Rhizobium* or *Azospirillum* did not significantly increase biomass, 1000-grain weight or yield **Future research:** × Field trials in different environmental conditions × Understand the underlying mechanisms of plant-bacteria and bacteria-bacteria interactions	Filipini et al. (2021)

(*Continued*)

Table 1 *Continued*

Crop	Field trial design	PGPB	Application mode	Benefits	Main results	Limitations/Future research	References
Maize (*Zea mays* L.)	1 year of experiment with maize 2 field trials: hybrid AG1051 and a transgenic AG8088PRO Low fertility soil 1 year experiment with maize (hybrid AG 2040) High fertility soil **Treatments:** Application of humic acids + inoculation	*Herbaspirillum seropedicae* Z67	Foliar application + humic acids	CYI; SS	In field trials with low fertility soil, inoculation + humic acids: + Shoot fresh weight (28-30%) + Shoot dry weight (only for hybrid AG1051) + Root system surface + Root dry mass (160-245%) In field trial with high fertility soil: - Grain production was not consistently affected by inoculation	**Limitations:** - Bioinoculation was less effective under optimal fertility conditions than in low fertility soils	Canellas et al. (2015)
	No-tillage; weed control **Treatments:** P_2O_5 rates (0, 17.5, 35, 70 e 105 kg/ha): × non inoculated × single inoculated with strains Ab-V5, Ab-V6, CCTB04, CCTB03	AzoTotal®: *Azospirillum brasilense* Ab-V5, Ab-V6Vult ®: *Bacillus subtilis* CCTB04 Audax ®: *Pseudomonas fluorescens* CCTB03	Seed inoculation	CYI; NUE	+ P use efficiency when inoculated with *B. subtilis* at 17.5 kg/ha P_2O_5 + Grain yield increased with *B. subtilis* and *A. brasilense* inoculation	**Limitations:** × *P. fluorescens* had little impact on P uptake, plant development, and grain yield × Number of rows per ear without response to P_2O_5 rates or PGPB inoculation **Future research:** × More studies are needed to better understand the role of PGPB, in single or combined inoculation, under biotic or/and abiotic conditions	Pereira et al. (2020a)
	3-years experiment (2015 – 2018) 2017-2018- drought period **Treatments:** × Two applications of atrazine and glyphosate: with or without foliar inoculation × One application of atrazine and two applications of glyphosate: with or without foliar inoculation	*Azospirillum brasilense* Az39	Foliar application	ASA; CYI	Strain Az39: + Ears perimeter and length + Number and weight of grains per ear and per surface Although yield in the drought affected season was significanctly decreased, inoculation showed improvements in the evaluated parameters compared to the control without inoculation	n.a.	Cardozo et al. (2022)

1-year experiment (2024-growing season) **Treatments:** Four inoculant doses (0, 100, 130, 160 mL/da), applied in four different commercial maize hybrids	Consortium: *Azotobacter* spp., *Bacillus* sp., *Paenibacillus* sp., *Pantoea* sp., and *Pseudomonas*sp.	Soil application via drip irrigation at 4 phenological stages: sowing, V4 (4 leaves), V8 (8 leaves), and VT (tasseling)	CGI, NUE, SS	Effects on soil (mainly for VT stage): + Soil organic matter + P availability + Cationic exchange capacity Effects on maize (more prevalent with 130 mL/da): + Grain productivity + Plant height and ear weight	**Limitations:** × Inoculation showed inconsistent effects on plant parameters, including height, ear length and diameter and grain yield × Inoculation enhanced the solubilization of micronutrients (Zn and Cu); however, this may pose a potential risk of nutrient imbalance **Future research:** × Microbial persistence and colonization assessment × Multi-season, multi-location field trials across diverse agroecological contexts × Economic feasibility analyses to assess scalability	Kurt (2025)

(Continued)

Table 1 *Continued*

Crop	Field trial design	PGPB	Application mode	Benefits	Main results	Limitations/Future research	References
Wheat (*Triticum aestivum* L.)	2-years experiment (2019 to 2021) **Treatments:** × Control - full irrigation during the growth period + PGPB × Moderate stress - full irrigation until the pollination stage and no irrigation up to 60% of available moisture until the end of the growth period + PGPB × Severe stress - full irrigation until the pollination stage and no irrigation until the end of the growth period + PGPB	*Azotobacter* sp. *Azospirillum* sp. *Microbacterium* sp. *Pseudomonas* sp.	Seed inoculation	ASA; CGI	All inoculated treatments: + Grain yield (about 3.6 grains per spike) + Root system traits (dry weight (48%), volume (31%), surface area (18.5%) with inoculation Although increased drought affected root development and crop yield, inoculation significantly relieved these negative impacts	**Future research:** × Optimization of microbial consortia for crops in real field conditions	Rostamian et al. (2023)
	1-year experiment (2022- 2023) **Treatments:** × Full N dose (180 kg/ha) – with and without bacterial foliar application at end tillering × Reduced N dose (130 kg/ha) - with and without bacterial foliar application at end tillering	*Methylobacterium symbioticum*	Foliar application	SAPT; NUE	PGPB inoculation: + SPAD readings and normalized difference vegetation index reflecting improved N status and extended leaf greenness Major benefits achieved under reduced N fertilization	**Limitations:** × Inoculation had no significant effect on yield or grain protein content **Future Research:** × Testing across multiple growing seasons and soil types × Improving understanding of *M. symbioticum* mechanisms and optimizing its use in agricultural systems	Valente et al. (2024)
	4-years experiment (2002–2006) **Treatments:** × Seed inoculation: liquid formulation containing the *A. brasilense* - 10 mL kg^{-1} of seed, about 6 h before sowing × Crop management: according to the best locally recommended practices for high yields achievement	*Azospirillum brasilense* INTA Az-39	Seed inoculation	CGI (only in 70% of the fields)	PGPB inoculation: + Shoot and root dry matter + Grains (increase in 6.1%), with a yield of ca. 260kg ha^{-1})	**Limitations:** × PGPB effectiveness was strongly dependent on environmental conditions, showing no significant effects in severely limiting environments such as shallow soils, semi-arid regions and low productivity systems	Díaz-Zorita et al. (2009)

Treatments: × DAP+FYM: Chemical fertilizer (150 kg/ha DAP) + organic fertilizer (10 t/ha farmyard manure) × SA: Seed inoculated with strains from Samahni (*Bacillus subtilis, Stenotrophomonas humi, Azospirillum brasilense*) × BH: Seed inoculated with strains from Bhimber (*Azospirillum oryzae, Azotobacter armeniacus, Rhizobium pusense*) × BA: Seed inoculated with strains from Barnala (*Azospirillum zeae, Rhizobium pusense, Pseudomonas fragi*) × DAP+FYM+SA: Combination of chemical, organic fertilizers, and SA strains × DAP+FYM+BH: Combination of chemical, organic fertilizers and BH strains × DAP+FYM+BA: Combination of chemical, organic fertilizers, and BA strains	*Bacillus subtilis, Stenotrophomonas humi, Azospirillum brasilense, Azospirillum oryzae, Azotobacter armeniacus, Rhizobium pusense, Azospirillum zeae, Pseudomonas fragi*	Seed inoculation	CGI; NUE	DAP+FYM+SA/BH/BA: + Plant height + Grains per spike + Shoot and root dry weight DAP+FYM+BA: + Grain quality (protein, carbohydrates, lipids)	**Limitations:** × PGPB alone showed no significant effect on grain quality or nutrient mobilization in the plant **Future research:** × Optimization of biofertilizers to increase shelf life and bacteria viability	Asghar et al. (2023)

(Continued)

Table 1 *Continued*

Crop	Field trial design	PGPB	Application mode	Benefits	Main results	Limitations/Future research	References
Tomato (*Solanum lycopersicum* L.)	Transplantation of greenhouse seedlings to the field **Greenhouse Treatments**: × C – Control - commercial substrate × VC - Soil + vermicompost × VCi - Soil + vermicompost + humates + inoculation **Field Treatments:** Control, VC and VCi with and without foliar application of humates + PGPB	*Herbaspirillum seropedicae* HRC54	Foliar application	SAPT	Foliar application: + Root biomass (31% and 60% in plants grown previously in the greenhouse in VC and C) + Root area (159 and 435% in plants previously grown in VC and VCi, respectively) + Total protein and nitrate content higher than in VC and VCi treatments. Maximum fruit production in plants treated with VCi + humates + PGPB (increased yield by 41–48% compared to non-inoculated plants)	**Limitations:** × None of the treatments affected the Brix index (3.04–3.51), which remained below the industrial threshold (4.2) × VC treatment did not produce measurable effects on plant performance	Olivares et al. (2015)
	Two consecutive trials in the same field (2009-2010) **Treatments:** × Control – no inoculation or chemical control × Chemical control - thiram and fytolan application × Single inoculation of Bacillus sp. strains (9×10^8 cfu/ml mixed with sterile talc powder formulation)	*Bacillus* sp. isolates BETS11, BETS14, BETR11, BETR13, BETS5, BETL10	Seed inoculation	BSA	Most of the isolates significantly controlled the incidences of plant diseases BETS11 isolate demonstrated superior efficiency in reducing the incidence of major phytopatogenic diseases (damping off, root rot, leaf spot, wilt), performing equal to or better than chemical controls BETS11 and BETR11 isolates increased the fruit yield comparable to the chemical control (11.73 t/ha, 11.24 t/ha and 11.81 t/ha, respectively)	**Limitations:** × Results were strain-dependent **Future research:** × Formulating microbial isolates to reduce or replace the use of harmful pesticides and chemical fertilizers	Amaresan et al. (2019)

Chickpea (*Cicer arietinum* L.)	Trials conducted under two consecutive dry seasons (2018 and 2019) **Treatments:** × Full irrigation regimes (100% water requirements (WR), 50%, 25%) – without inoculation, inoculation with PGPB consortium and inoculation with PGPB consortium + AMF consortium. × Irrigation only during the reproductive stages (100% WR) - without inoculation, inoculation with PGPB consortium and inoculation with PGPB consortium + AMF consortium. × Rainfed irrigation - without inoculation, inoculation with PGPB consortium and inoculation with PGPB consortium + AMF consortium.	PGPB: *Mesorhizobium* sp. UTADM31 *Burkholderia* sp. UTADB34 *Pseudomonas* sp UTAD11.3 AMF - Symbiom Ltd: *Rhizophagus irregularis* BEG140, *Funneliformis geosporum* BEG199, *Claroideoglomus claroideum* BEG210	Seed inoculation (PGPB); Manually added at sowing (AMF)	CGI	PGPB+AMF inoculation: + 25%, 23%, 23%, 26% and 24% in shoot dry weight, number of pods, pod weight, number of seeds, and seed weight, respectively, compared to non- inoculated plants (C). + Grain yield was 19% higher in plants irrigated only during the reproductive stages than in fully irrigated plants (100% WR); In all irrigation regimes, co-inoculation with PGPB+AMF resulted in higher grain yield when compared to the single inoculation (PGPB) Irrigation regimes: Applying 100% WR during the reproductive stage resulted in increased grain yield across all inoculated and control treatments + Shoot dry weight under PGPB+AMF treatment when 100% WR was applied during reproductive stage	**Limitations:** × Under 100% WR, PGPB alone resulted in substantially lower yield than co-inoculation treatments (PGPB+AMF) under optimized irrigation strategies	Laranjeira et al. (2021)

10 Factors affecting efficacy of plant growth-promoting bacteria-based biostimulants: microbial competition, host specificity and environmental conditions

10.1 Microbial competition

The effectiveness of PGPB-based biostimulants is often limited by competition with native soil microbial communities, which represents one of the main barriers to successful establishment and plant growth promotion (Rilling et al., 2019; Massa et al., 2022). This limitation is particularly relevant in the context of NUE, as failure of introduced strains to successfully colonize and persist in the rhizosphere directly constrains their ability to contribute to key nutrient-related processes.

Studies have reported enhanced performance of bacterial inoculants in sterilized soils, where native microbial communities are absent (Rojas-Tapias et al., 2012; Pereira et al., 2015; Moreira et al., 2016, 2019). However, such conditions fail to reflect the ecological complexity of natural soil ecosystems. In real field settings, soils harbor diverse and highly adapted microbial communities that resist colonization by introduced strains (Mallon et al., 2018). For this reason, evaluating inoculants in non-sterile soil is strongly recommended to provide a more realistic assessment of performance (de-Bashan and Nannipieri, 2024).

Successful colonization in such competitive environments depends on several ecological traits, including persistence in the rhizosphere, efficient resource utilization, and the formation of stable root associations. These traits are often supported by community-forming traits, such as biofilm formation and exopolysaccharide production, which enhance survival and functionality within the rhizosphere (Rilling et al., 2019; Hossain et al., 2023; Pathak et al., 2024). Some strains also produce antimicrobial compounds that suppress competitors, thereby improving their ecological fitness (Benizri et al., 2001).

10.2 Host specificity

The host plant also plays a significant role in shaping the composition and activity of the root-associated microbiome, thereby influencing the success of PGPB colonization (Pascale et al., 2020; Zhang et al., 2023). This host effect is a major determinant of NUE because microbial-mediated nutrient transformation is only effective when compatible plant–microbe partnerships are established. Host plants can selectively recruit microbial taxa through root exudates, creating a rhizosphere environment that favors specific functional groups involved in nutrient uptake and assimilation (Pascale et al., 2020).

Tovi et al. (2019) demonstrated that three *Pseudomonas* strains isolated from wheat preferentially colonized wheat roots compared with those of cucumber, highlighting strong host specificity. Similar preferences were observed for barley and maize over soybean, reinforcing the concept that host identity governs microbial recruitment patterns and, consequently, nutrient provisioning routes. Similarly, Thokchom et al. (2017) reported that bacterial strains isolated from the rhizoplane mucilage-soil complex of mandarin orange in citrus orchards were capable of successfully colonizing citrus roots, suggesting that locally adapted plant-microbe systems often outperform generic inoculation strategies. Host specificity also operates at the genotype level.

Schultz et al. (2022) showed genotype-specific interactions between bacterial endophytes and different maize germplasms, resulting in variable and inconsistent growth-promotion outcomes. Likewise, inoculation of different wheat genotypes with three strains of *A. brasilenses* resulted in markedly different outcomes, underscoring that microbial performance is tightly dependent on host genotype. These observations highlight that failure to account for host compatibility can lead to suboptimal nutrient acquisition and unreliable performance under field conditions. Thus, targeted inoculant-cultivar matching represents a critical step toward improving NUE through biological strategies.

10.3 Environmental variability and abiotic stress

Environmental variability across field sites further affects the effectiveness of PGPB applications. Factors such as soil composition, pH, temperature, moisture, and agronomic practices influence plant-microbe interactions and often make field outcomes highly site dependent (Ganesh et al., 2023).

Research indicates that many rhizobacteria thrive optimally within a temperature range of 25°C to 30°C (Askari-Khorasgani et al., 2019), and high soil temperatures can profoundly impair the efficiency of microbial strains by hindering their survival, metabolic activity, and the dynamics of plant-microbe interactions within the rhizosphere (Landa et al., 2013). For example, Landa et al. (2004) reported that while chickpea inoculation with four PGPB strains enhanced plant growth, the magnitude of this effect declined as temperature increased from 20°C to 30°C.

Water availability also plays a critical role. Pereira et al. (2020b) reported that PGPB inoculation increased maize shoot biomass by up to 89% under moderate drought conditions, but beneficial effects were absent under severe drought stress, likely due to reduced microbial survival or activity.

Microbial inoculants frequently exhibit reduced effectiveness under abiotic stress conditions such as drought, salinity, extreme temperatures, and nutrient-poor soils, primarily due to impaired survival, colonization, and metabolic activity (Landa et al., 2013; Pereira et al., 2020b; Ganesh et al., 2023).

Environmental stresses severely limit microbial growth and persistence in the rhizosphere, thereby constraining key plant growth-promoting functions including BNF, P solubilization, and phytohormone production (Dimkpa et al., 2009; Nadeem et al., 2014).

To overcome these limitations, increasing attention has been given to the isolation and application of stress-tolerant PGPB, capable of functioning under adverse conditions. For example, a heat-tolerant strain of *B. cereus* (SA1) enhanced soybean tolerance to heat stress by promoting growth and activating the antioxidant defense system (Khan et al., 2020). The use of stress-adapted PGPB strains can therefore substantially improve inoculant performance under abiotic stress, as these microorganisms are naturally equipped to persist and remain metabolically active in hostile environments. Such strains exhibit greater root colonization capacity and sustained expression of key stress-alleviating traits, including ACC-deaminase activity, EPS production, osmolyte accumulation, and induction of antioxidant enzymes, across different stresses (Timmusk et al., 2014; Nadeem et al., 2014).

The use of microbial consortia, combining bacteria, fungi, or multiple microbial taxa has been highlighted as a powerful approach to enhance functional complementarity, ecological fitness, and resilience, leading to improved nutrient mobilization and overall crop performance (Negi et al., 2024; Nunes et al., 2024). In addition, advanced bioformulations and carrier-based inoculant technologies (e.g. encapsulation, optimized carrier matrices, polymer-based formulations) have been shown to significantly improve microbial survival, stress tolerance, and rhizosphere establishment under variable field conditions (Khan et al., 2023; Fadiji et al., 2024). In parallel, emerging approaches to microbiome-assisted selection and biofertilizer development, which integrate metagenomics, multi-omics screening and tailored strain selection, offer promising avenues to improve host-microbe compatibility, nutrient cycling efficiency, and long-term stability of microbial communities contributing to NUE and sustainable crop productivity (Shahzad et al., 2025).

11 Enhancing efficacy of plant growth-promoting bacteria-based biostimulants: strain selection and screening

The development of effective microbial formulations is a multi-step process that begins with the isolation and accurate identification of non-pathogenic bacterial strains. Once isolated, strains must be rigorously screened for plant growth-promoting traits, such as nutrient solubilization, phytohormone production, or antagonistic activity against phytopathogens. Traditionally, these traits have been evaluated under optimal, non-stressed *in vitro* conditions (Pereira et al.,

2015). However, such conditions do not reflect the complexity and variability under agricultural environments. The efficiency and functional performance of bacterial strains can vary considerably under abiotic stress conditions, including salinity, drought, and extreme temperatures.

For applications in stress-prone environments, strain selection should be conducted under simulated stress conditions to assess whether growth-promoting traits are maintained, reduced, or enhanced. For example, Moreira et al. (2016) demonstrated that bacterial growth-promoting traits changed in response to increasing metal concentrations. Similarly, Rojas-Tapias et al. (2012) showed that two *A. chroococcum* strains retained their N fixation and P solubilizing abilities under saline conditions, with one strain even exhibiting enhanced IAA production. Pereira et al. (2016) further reported that salt-tolerant PGPB maintained the production of key metabolites, including IAA, hydrogen cyanide, ammonia, and siderophores, across a salinity gradient. These findings highlight the importance of stress-based screening to identify robust and resilient strains. In addition to functional properties, the selection of strains with strong community-forming and colonizing abilities is essential, as these traits improve long-term survival and persistence within native soil microbial populations (de Souza et al., 2015; Khan et al., 2020; Ajijah et al., 2023; Pathak et al., 2024).

12 Enhancing development and use of plant growth-promoting bacteria-based biostimulants: microbial consortia design, *in vitro* and *in vivo* testing

The use of microbial consortia, rather than single-strain inoculations, has been recognized as a promising strategy to enhance the efficacy of biostimulants (Molina-Romero et al., 2021; Santoyo et al., 2021; Liu et al., 2023; Kumar et al., 2024). By combining multiple microorganisms with complementary functional traits, consortia can simultaneously target different plant growth-promoting mechanisms, thereby increasing functional redundancy and resilience under variable environmental conditions (Liu et al., 2023).

However, the successful development of microbial consortia requires a strong understanding of the individual functions of each strain, as well as their potential synergistic or antagonistic interactions (Kumar et al., 2024; Nunes et al., 2024). Inappropriately assembled consortia may result in competitive exclusion, reduced metabolic efficiency, or suppression of key functions, ultimately compromising their performance in the rhizosphere. In contrast, well-designed consortia can enhance mutual survival and colonization by promoting niche differentiation, metabolic cross-feeding, and cooperative biofilm formation, leading to more stable establishment within native soil microbial communities.

From an applied perspective, microbial consortia increase biological complexity, which poses challenges for formulation, quality control, and regulatory approval, emphasizing the need for standardized approaches to consortium design and validation. Overall, advancing consortium-based biostimulants will require integrative strategies that combine microbial ecology, functional genomics, and agronomic validation to ensure consistent performance under field conditions.

12.1 From in vitro *to* in vivo *validation*

Although *in vitro* studies are indispensable for efficiently screening large numbers of bacterial strains, they do not always predict performance under greenhouse or field conditions (Bizjak-Johansson et al., 2025). Consequently, *in vivo* assays are crucial to validate strain efficacy in the presence of native soil microbiota and complex plant-microbe interactions. Combining *in vitro* screening with greenhouse and field validation represents a robust strategy for selecting high-performing strains, capable of delivering consistent benefits under real agricultural conditions.

13 Enhancing development and use of plant growth-promoting bacteria-based biostimulants: formulation, application and integration with existing practices

13.1 Formulation: carriers and shelf-life

Enhancing the efficacy of PGPB-based biostimulants also depends on optimizing the entire formulation process (Vassileva et al., 2021; Hossain et al., 2023). Choosing an appropriate carrier is critical to the success of a formulation. The carrier must preserve bacterial viability, extend shelf life, ensure effective delivery under field conditions, and remain cost-effective (Rojas-Sánchez et al., 2022). Additionally, carriers must be non-toxic to both inoculated bacterial strains and plants, to avoid compromising microbial function or crop health (Sakpirom et al., 2021), possess good moisture retention capacity, and facilitate strong adhesion to microbial cells, ensuring uniform distribution and effective colonization in the soil or seeds. Balancing all these factors is essential for developing reliable and commercially viable PGPB formulations.

Novel formulations based on organic matrices, polymers, and agri-food byproducts have been shown to improve microbial viability, shelf life, and controlled release under field conditions. In particular, agri-food byproducts represent a promising class of carriers, as they are widely available, scalable, and align with circular economy principles by valorizing waste into valuable agricultural inputs.

Bioencapsulation in alginate beads also represents a promising strategy for the efficient delivery of microbial inoculants in agriculture. Balla et al. (2025) showed that alginate capsules produced by ionic gelation and supplemented with proline efficiently encapsulated *Bacillus thuringiensis* B25 and *Pantoea agglomerans* Pa, achieving encapsulation efficiencies above 99% and enabling the controlled release of viable cells after long-term storage at room temperature. Encapsulation ensured sustained bacterial viability, with individually encapsulated strains displaying greater stability than co-inoculated formulations, while maintaining plant growth-promoting activity, effectively colonizing wheat roots, and significantly improving plant growth. Similarly, Hamed et al. (2024) reported the successful development of a chitosan-alginate-humic acid nano-encapsulation system for the co-delivery of essential nutrients (NPK) and beneficial microorganisms, which exhibited high encapsulation efficiency, enhanced water retention capacity, and sustained nutrient release over a 30-day period.

13.2 Optimal dose and application

Despite its importance, the optimal inoculum dose for PGPB application remains poorly understood, with limited studies systematically comparing dose-dependent effects. Several studies have shown that higher microbial doses do not necessarily lead to improved rhizosphere colonization or enhanced plant growth compared to lower doses (Pereira et al., 2015, 2020b; Bueno et al., 2022). The lack of a clear positive dose-response relationship challenges the assumption that 'more is better' and highlights the need for precise inoculum optimization.

Applying PGPB at high concentrations can rapidly exhaust available nutrients, thereby increasing competitive pressure on native microbial communities and reshaping the composition of the rhizosphere microbiome (Samonty et al., 2025). From an agronomic perspective, applying unnecessarily high doses increases production and application costs, thereby limiting large-scale adoption of PGPB-based products. These findings underscore the importance of defining minimum effective doses that balance biological efficacy, economic feasibility, and environmental safety, while highlighting the need for a deeper understanding of dose-dependent effects to maximize biofertilizer efficacy and ensure consistent plant responses.

13.3 Integration with sustainable agricultural practices

Numerous studies have also demonstrated that integrating sustainable farming practices, such as the use of organic amendments and intercropping, with microbial biostimulants can further enhance plant performance. For instance, Ahmed et al. (2025) reported that combining organic amendments with phosphate-solubilizing bacteria significantly improved wheat growth under saline calcareous soil conditions, outperforming the use of individual amendments alone. Similarly, Dragičević et al. (2024) reported that the positive effects of field pea intercropping on maize biomass, chlorophyll content, and yield were further amplified by biofertilizer application, highlighting the value of combining strategies.

Together, these targeted and integrated approaches increase the likelihood of developing PGPB-based biostimulants that are effective, reliable, and adaptable to diverse field conditions. Overall, recent technological advances in microbial formulation, including the development of functionally complementary microbial consortia, the optimization of carrier materials, and the incorporation of circular economy approaches through the use of agri-food byproducts, represent important steps toward improving the effectiveness and reliability of PGPB-based biostimulants. By enhancing microbial survival, colonization, and functional persistence in the rhizosphere, these innovations can improve nutrient mobilization and uptake efficiency, thereby reducing reliance on synthetic fertilizers. When strategically integrated into sustainable farming systems, such advances facilitate a transition toward more resource-efficient and environmentally responsible agriculture, aligning biostimulant technologies with global goals related to soil health, climate resilience, and long-term food security. Nevertheless, ensuring the reliability and effectiveness of biofertilizers requires the urgent development of standardized methodologies for their production, application, and evaluation, supported by robust quality control measures and appropriate regulatory frameworks (Shahzad et al., 2025).

14 Conclusion and future trends

The adoption of PGPB-based biostimulants holds significant promise for sustainable agriculture, particularly as a strategy to improve NUE and reduce dependence on chemical fertilizers. However, several barriers continue to hinder their widespread application. Among the most significant challenges are inconsistent field performance, limited microbial viability under variable environmental conditions, and the complexity of plant–microbe interactions. PGPB strains that perform well under controlled laboratory or greenhouse conditions often show reduced efficacy in diverse field environments, due to fluctuations in soil properties, climate, crop management practices, and competition with native microbial communities. Such variability directly affects their ability to consistently mobilize nutrients such as N and P, thereby limiting predictable gains in NUE.

Beyond technical limitations, non-technical barriers also play a critical role. The lack of harmonized regulatory frameworks and standardized methodologies to produce microbial-based formulations, quality control, application guidelines, and efficacy assessment remains major constraints. Regulatory uncertainty delays product registration and market entry, while insufficient standards for microbial concentration, viability, and functional consistency undermine farmer confidence. In addition, market acceptance among farmers is often limited by variable performance relative to conventional fertilizers, perceived agronomic risk, and limited guidance on how biostimulants should be integrated with existing fertilization regimes to achieve measurable improvements in nutrient uptake and crop growth.

Advances in technology and research are progressively addressing technical challenges and strengthening the contribution of PGPB to nutrient-efficient agriculture. The development of microbial consortia, combining multiple beneficial microbes, including arbuscular mycorrhizal fungi, offers multifunctional benefits and improved resilience under varying environmental conditions. These consortia can provide synergistic effects, enhancing nutrient uptake, stress tolerance, and disease resistance in plants. In addition, emerging genomic and metagenomic tools are providing deeper insights into the functional potential and interactions of rhizosphere microbial communities. The application of next-generation sequencing and other omics approaches enables the identification of highly effective strains and consortia, as well as the optimization of microbial traits related to nutrient solubilization and plant growth promotion (Shahzad et al., 2025). By tailoring applications based on specific crop needs, soil conditions, and environmental factors, farmers can optimize the benefits of PGPB. In addition, the integration into precision agriculture and sustainable farming practices further enhances the potential of PGPB-based biostimulants. These targeted approaches contribute to increased crop productivity, reduced reliance on chemical inputs, and improved environmental sustainability.

In conclusion, while challenges remain in the adoption of PGPB-based biostimulants, ongoing research and technological advancements are paving the way for their effective integration into modern agriculture. Future efforts should focus on bridging the gap between laboratory innovation and field application by integrating ecological principles, standardized testing protocols, multi-site field evaluations, and agronomic validation of NUE gains under reduced fertilizer regimes. At the same time, clearer regulatory pathways, improved product standardization, and stronger engagement with farmers and stakeholders will be essential to increase market acceptance. By aligning technological innovation with agronomic practicality and policy frameworks, PGPB-based biostimulants can become a reliable tool for supporting sustainable and resilient agricultural systems, contributing to soil health, climate resilience, and long-term food security.

15 Where to look for further information

Readers interested in further exploring the role of PGPB in improving nutrient use efficiency, crop productivity, and stress tolerance may consult the following scientific resources, organisations, and research projects.

Key Scientific Reviews, Articles, and Practical Field-Based Examples Plant biostimulants: Definition, Concept, Main Categories and Regulation. https://www.sciencedirect.com/science/article/pii/S0304423815301850?via%3Dihub.

Microbial Biostimulants as Response to Modern Agriculture Needs: Composition, Role and Application of These Innovative Products. https://www.mdpi.com/2223-7747/10/8/1533.

Field Applications of Plant Growth Promoting Rhizobacteria for Crop Protection and Nutrient Enhancement. https://www.frontiersin.org/research-topics/76687/field-applications-of-plant-growth-promoting-rhizobacteria-for-crop-protection-and-nutrient-enhancementy.

Useful information on microbial plant biostimulants and regulatory developments can be accessed through the European Biostimulants Industry Council (EBIC). https://biostimulants.eu/plant-biostimulants/; https://biostimulants.eu/advocacy/microbials/.

Food and Agriculture Organization (FAO) AGRIS database. https://agris.fao.org/search/en/providers/122535/records/65dfe1f44c5aef494fe62eba.

CGIAR https://www.cgiar.org/publications.

The American Phytopathological Society https://www.apsnet.org/ International.

Union of Soil Sciences https://www.iuss.org.

There are also several research projects focused on the application of PGPB to improve crop nutrient use efficiency, plant productivity, and stress tolerance, while identifying current agricultural challenges and sustainable strategies to address them, including:

SoildiverAgro: Soil biodiversity enhancement in European agro ecosystems to promote their stability and resilience by external inputs reduction and crop performance increase. https://cordis.europa.eu/project/id/817819.

BENEFIT-Med Project: Boosting technologies of orphan legumes towards resilient farming systems in the Greater Mediterranean Region from bench to open field. https://www.era-learn.eu/network-information/networks/prima/section-2-2013-multi-topic-2021/boosting-technologies-of-orphan-legumes-towards-resilient-farming-systems-in-the-greater-mediterranean-region-from-bench-to-open-field.

ReCROP: Bioinocula and CROPping systems: an integrated biotechnological approach for improving crop yield, biodiversity and REsilience of Mediterranean agro-ecosystemshttps://recrop.netlify.app/.

16 Acknowledgments

This work was funded by national funds from Fundação para a Ciência e Tecnologia (FCT) and by FEDER under the Thematic Program Innovation and Digital Transition, within the framework of the project MICRODROP – MICRObial formulations to improve crop growth and resilience under DROught stress and Phosphorus deficiency (COMPETE2030-FEDER-00771700). This work was also carried out within the scope of contract 2023.15056. TENURE.047, funded by national funds through FCT and by the Recovery and Resilience Plan (RRP) – Portuguese Republic, through the Recover Portugal Mission Structure. The authors also thank CBQF scientific collaboration under the FCT projects UID/50016/2025 and LA/P/0076/2020 (https://doi.org/10.54499/LA/P/0076/2020). Ana S.S. Sousa thank FCT for the doctoral studentship grant 2024.15499.PRT.

17 References

Ahmed, A. F., et al. (2025), Integration of organic amendments and phosphate-solubilizing bacteria improves wheat growth and yield by modulating phosphorus availability and physiological reponses, *J. Plant Nutr.* 48(7): 1144–1165. https://doi.org/10.1080/01904167.2024.2422586.

Ajijah, N., et al. (2023), Plant Growth-Promoting Bacteria (PGPB) with biofilm-forming ability: a multifaceted agent for sustainable agriculture, *Diversity*. 15(1): 112. https://doi.org/10.3390/d15010112.

Akhtar, S. S., et al. (2020), Role of cytokinins for interactions of plants with microbial pathogens and pest insects, *Front. Plant Sci.* 10: 1777. https://doi.org/10.3389/fpls.2019.01777.

Akintokun, A. K., et al. (2019), Isolation, screening and response of maize to plant growth promoting rhizobacteria inoculants, *Sci. Agric. Bohem.* 50: 181–190. https://doi.org/10.2478/sab-2019-0025.

Alami, Y., et al. (2000), Rhizosphere soil aggregation and plant growth promotion of sunflowers by an exopolysaccharide-producing *Rhizobium* sp. strain isolated from sunflower roots, *Appl. Environ. Microbiol.* 66(8): 3393–3398. https://doi.org/10.1128/AEM.66.8.3393-3398.2000.

Ali, A., et al. (2022), Antagonistic potential of bacterial species against fungal plant pathogens (FPP) and their role in plant growth promotion (PGP): a Review, *Phyton-Inter. J. Exp. Bot.* 91(9): 1859–1877. https://doi.org/10.32604/phyton.2022.021734.

Aloo, B. N., et al. (2022), Plant growth-promoting rhizobacterial biofertilizers for crop production: the past, present, and future, *Front. Plant Sci.* 13: 1002448. https://doi.org/10.3389/fpls.2022.1002448.

Alori, E. T., et al. (2017), Microbial inoculants for soil quality and plant health, in E. Lichtfouse (ed), *Sustainable Agriculture Reviews*, 281–307. Cham: Springer International Publishing. https://doi.org/10.1007/978-3-319-48006-0_9.

Asghar, I., et al. (2023), Characterizing indigenous plant growth promoting bacteria and their synergistic effects with organic and chemical fertilizers on wheat (*Triticum aestivum*), *Front. Plant Sci.* 14: 1232271 https://doi.org/10.3389/fpls.2023.1232271.

Askari-Khorasgani, O., et al. (2019), Managing plant-environment-symbiont interactions to promote plant performance under low temperature stress, *J. Plant Nutr.* 42(16): 2010–2027. https://doi.org/10.1080/01904167.2019.1648682.

Balla, A., et al. (2025), Root colonization dynamics of alginate encapsulated rhizobacteria: implications for *Arabidopsis thaliana* root growth and durum wheat performance, *AIMS Microbiol.* 11(1): 87–125. https://doi.org/10.3934/microbiol.2025006.

Barrett, J. E. and Burke, I. C. (2000), Potential nitrogen immobilization in grassland soils across a soil organic matter gradient, *Soil Biol. Biochem.* 32(11–12): 1707–1716. https://doi.org/10.1016/S0038-0717(00)00089-4.

Benizri, E., et al. (2001), Root colonization by inoculated plant growth-promoting rhizobacteria, *Bio. Sci. Technol.* 11(5): 557–574. https://doi.org/10.1080/09583150120076120.

Bhat, M. A., et al. (2020), Mechanistic insights of the interaction of Plant Growth-Promoting Rhizobacteria (PGPR) with plant roots toward enhancing plant productivity by alleviating salinity stress, *Front. Microbiol.* 11:1952. https://doi.org/10.3389/fmicb.2020.01952.

Bizjak-Johansson, T., et al. (2025), Inoculation with in vitro promising plant growth-promoting bacteria isolated from nitrogen-limited boreal forest did not translate to in vivo growth promotion of agricultural plants, *Biol. Fert. Soils.* 61: 925–940. https://doi.org/10.1007/s00374-025-01910-8.

Boro, M., et al. (2022), Microorganisms in biological control strategies to manage microbial plant pathogens: a review, *Arch. Microbiol.* 204: 666. https://doi.org/10.1007/s00203-022-03279-w.

Bouremani, N., et al. (2024), Osmotolerant plant growth promoting bacteria mitigate adverse effects of drought stress on wheat growth. *AIMS Microbiol.* 10(3): 507–541. https://doi.org/10.3934/microbiol.2024025.

Bueno, C. B., et al. (2022), Effects of chemical fertilization and microbial inoculum on bacillus subtilis colonization in soybean and maize plants, *Front. Microbiol.* 13: 901157. https://doi.org/10.3389/fmicb.2022.901157.

Cai, S., et al. (2025), Towards precise nitrogen fertilizer management for sustainable agriculture, *Earth Crit. Zone*. 2: 100026. https://doi.org/10.1016/j.ecz.2025.100026.

Castiglione, A. M., et al. (2021), Microbial biostimulants as response to modern agriculture needs: Composition, role and application of these innovative products, *Plants* 10(8): 1533. https://doi.org/10.3390/plants10081533.

Chandra, D., et al. (2019), Field performance of bacterial inoculants to alleviate water stress effects in wheat (*Triticum aestivum* L.), *Plant Soil* 441(1–2): 261–281. https://doi.org/10.1007/s11104-019-04115-9.

Chaudhary, P., et al. (2022), Overview of biofertilizers in crop production and stress management for sustainable agriculture, *Front. Plant Sci.* 13: 930340. https://doi.org/10.3389/fpls.2022.930340.

Cohen, A. C., et al. (2009), Participation of abscisic acid and gibberellins produced by endophytic *Azospirillum* in the alleviation of drought effects in maize, *Botany* 87(5): 455–462. https://doi.org/10.1139/B09-023.

Cohen, A. C., et al. (2015), *Azospirillum brasilense* ameliorates the response of *Arabidopsis thaliana* to drought mainly via enhancement of ABA levels, *Physiol. Plant.* 153(1): 79–90. https://doi.org/10.1111/ppl.12221.

Compant, S., et al. (2010), Plant growth-promoting bacteria in the rhizo- and endosphere of plants: their role, colonization, mechanisms involved and prospects for utilization, *Soil Biol. Biochem.* 42(5): 669–678. https://doi.org/10.1016/j.soilbio.2009.11.024.

Dar, A., et al. (2021), Efficacy of rhizobacterial exopolysaccharides in improving plant growth, physiology, and soil properties, *Environ. Monit. Assessment*. 193: 515. https://doi.org/10.1007/s10661-021-09286-6.

Dauala, G. A., et al. (2025), Plant growth-promoting bacteria modulate metabolism and nitrogen accumulation to counteract drought damage in cactus pear plants, *Sci. Rep.* 15: 36432. https://doi.org/10.1038/s41598-025-20509-5.

de-Bashan, L. and Nannipieri, P. (2024), Recommendations for plant growth-promoting bacteria inoculation studies, *Biol. Fert. Soils*. 60(3): 259–261. https://doi.org/10.1007/s00374-024-01798-w.

de Sousa, S. M., et al. (2021), Tropical *Bacillus* strains inoculation enhances maize root surface area, dry weight, nutrient uptake and grain yield, *J. Plant Growth Regul.* 40(2): 867–877. https://doi.org/10.1007/s00344-020-10146-9.

de Souza, R., et al. (2015), Plant growth-promoting bacteria as inoculants in agricultural soils, *Genet. Mol. Biol.* 38(4): 401–419. https://doi.org/10.1590/S1415-475738420150053.

Díaz-Zorita, M. and Fernández-Canigia, M. V. (2009), Field performance of a liquid formulation of *Azospirillum brasilense* on dryland wheat productivity, *Eur. J. Soil Biol.* 45(1): 3–11. https://doi.org/10.1016/j.ejsobi.2008.07.001.

Dimkpa, C., et al. (2009), Plant-rhizobacteria interactions alleviate abiotic stress conditions, *Plant Cell Environ.* 32(12): 1682-94. https://doi.org./10.1111/j.1365-3040.2009.02028.x.

Dragičević, V., et al. (2024), Combined effect of cover crops and bio-fertilizer on sustainable popcorn maize production, *Front. Plant Sci.* 14: 1250903. https://doi.org/10.3389/fpls.2023.1250903.

Duca, D. R. and Glick, B. R. (2020), Indole-3-acetic acid biosynthesis and its regulation in plant-associated bacteria, *Appl. Microbiol. Biotechnol.* 104(20): 8607–8619. https://doi.org/10.1007/s00253-020-10869-5.

Efthimiadou, A., et al. (2020), Effect of foliar and soil application of plant growth promoting bacteria on growth, physiology, yield and seed quality of maize under Mediterranean conditions, *Sci Rep.* 10(1): 1–11. https://doi.org/10.1038/s41598-020-78034-6.

Etesami, H. and Glick, B. R. (2024), Bacterial indole-3-acetic acid: a key regulator for plant growth, plant-microbe interactions, and agricultural adaptive resilience, *Microbiol. Res.* 281: 127602. https://doi.org/10.1016/j.micres.2024.127602.

Fadiji, A. E., et al. (2022), The Potential role of microbial biostimulants in the amelioration of climate change-associated abiotic stresses on crops, *Front. Microbiol.* 12: 829099. https://doi.org/10.3389/fmicb.2021.829099.

Fadiji, A. E., et al. (2024), Formulation challenges associated with microbial inoculants and prospective solutions, *J. Sustain. Agric. Environ.* 3: e70006. https://doi.org/10.1002/sae2.70006.

FAO. (2021), *The State of the World's Land and Water Resources for Food and Agriculture – Systems at Breaking Point Systems at Breaking Point*. Synthesis report 2021. Rome: FAO. https://doi.org/10.4060/cb7654en.

Ganesh, R. S. V., et al. (2023), Constrains in biofertilizer development and environmental factors influencing its survival in soil: a review, *Agric. Rev.* 44(4): 530–535. https://doi.org/10.18805/ag.r-2348.

Glick B. R. (2012), Plant growth-promoting bacteria: mechanisms and applications. *Scientifica* (Cairo): Article No. 963401. https://doi.org/10.6064/2012/963401.

Gomes, G. L. B. and Scortecci, K. C. (2021), Auxin and its role in plant development: structure, signalling, regulation and response mechanisms, *Plant Biol.* 23(6): 894–904. https://doi.org/10.1111/plb.13303.

Gómez-Godínez L. J., et al. 2023. A look at plant-growth-promoting bacteria. *Plants (Basel)* 12(8): Article 1668. https://doi.org/10.3390/plants12081668.

Goswami, D., et al. (2016), Portraying mechanics of plant growth promoting rhizobacteria (PGPR): a review, *Cogent Food and Agricul.* 2(1): 1127500. https://doi.org/10.1080/23311932.2015.1127500.

Grover, M., et al. (2021), PGPR mediated alterations in root traits: way toward sustainable crop production, *Front. Sust. Food Syst.* 4: 618230. https://doi.org/10.3389/fsufs.2020.618230.

Gu, S., et al. (2020), Competition for iron drives phytopathogen control by natural rhizosphere microbiomes, *Nature Microbiol.* 5: 1002–1010. https://10.1038/s41564-020-0719-8.

Hamed, R., et al. (2024), Nano bio fertilizer capsules for sustainable agriculture, *Sci. Rep.* 14: 13646. https://doi.org/10.1038/s41598-024-62973-5.

Hardoim, P. R., et al. (2008), Properties of bacterial endophytes and their proposed role in plant growth, *Trends Microbiol.* 16(10): 463–471. https://doi.org/10.1016/j.tim.2008.07.008.

Hossain, M. A., et al. (2023), Challenges faced by plant growth-promoting bacteria in field-level applications and suggestions to overcome the barriers, *Physiol. and Mol. Plant Pathol.* 126: 102029. https://doi.org/10.1016/j.pmpp.2023.102029.

Ilangumaran, G. and Smith, D. L. (2017), Plant growth promoting rhizobacteria in amelioration of salinity stress: a systems biology perspective, *Front. Plant Sci.* 8: 1768. https://doi.org/10.3389/fpls.2017.01768.

Jeyanthi, V. and Kanimozhi, S. (2018), Plant Growth Promoting Rhizobacteria (PGPR)-prospective and mechanisms: a review, *J. Pure Appl. Microbiol.* 12(2): 733–749. https://doi.org/10.22207/JPAM.12.2.34.

Jha, Y., et al. (2023), Plant growth-promoting bacteria and exogenous phytohormones alleviate the adverse effects of drought stress in pigeon pea plants, *Plant Soil.* 505(1): 163–183. https://doi.org/10.1007/s11104-023-06155-8.

Karnwal, A., et al. (2024), PGPR-mediated breakthroughs in plant stress tolerance for sustainable farming, *J. Plant Growth Regul.* 43(9): 2955–2971. https://doi.org/10.1007/s00344-023-11013-z.

Kejela, T. (2024), Phytohormone-producing rhizobacteria and their role in plant growth, in B. Ali and J. Iqbal (eds), *New Insights Into Phytohormones*, 23. https://doi.org/10.5772/intechopen.1002823.

Kesaulya, H., et al. (2018), Potential of *Bacillus* spp produces siderophores insuppressing the wilt disease of banana plants, *IOP Conference Series: Earth Environ. Sci.*, 102(1): 012016. https://doi.org/10.1088/1755-1315/102/1/01201.

Khan, M. A., et al. (2020), Thermotolerance effect of plant growth-promoting *Bacillus cereus* SA1 on soybean during heat stress, *BMC Microbiol.* 20(1): 175. https://doi.org/10.1186/s12866-020-01822-7.

Khan, A., et al. (2023), Microbial bioformulation: a microbial assisted biostimulating fertilization technique for sustainable agriculture. *Front. Plant Sci.* 14: 1270039. https://doi.org/10.3389/fpls.2023.1270039.

Kumar, P., et al. (2024), Formulation and effectivity of consortia of potential plant growth promoting endophytic bacteria from chickpea (*Cicer arietinum*), *J. Appl. Biol. Biotechnol.* 12(4): 241–247. https://doi.org/10.7324/JABB.2024.163641.

Kumar, R., et al. (2022), Reclamation of salt-affected soils in India: progress, emerging challenges, and future strategies, *Land Degrad. Dev.* 33(13): 2169–2180. https://doi.org/10.1002/ldr.4320.

Kurt, P. O. (2025), Field-based evaluation of multi-strain PGPR to improve zea mays yield and soil nutrient dynamics in semi-arid of Türkiye, *Sci. Rep.* 15: 22553. https://doi.org/10.1038/s41598-025-07643-w.

Landa, B. B., et al. (2004), Influence of temperature on plant-rhizobacteria interactions related to biocontrol potential for suppression of fusarium wilt of chickpea, *Plant Pathol.* 53(3): 341–352. https://doi.org/10.1111/j.0032-0862.2004.01008.x.

Landa, B. B., et al. (2013), Use of PGPR for controlling soilborne fungal pathogens: assessing the factors influencing its efficacacy, in D. K. Maheshwari (ed), *Bacteria in Agrobiology: Disease Management*, 259–297. Berlin: Springer-Verlag. https://doi.org/10.1007/978-3-642-33639-3.

Lau, E. T., et al. (2020), Plant growth-promoting bacteria as potential bio-inoculants and biocontrol agents to promote black pepper plant cultivation, *Microbiol. Res.* 240: 126549. https://doi.org/10.1016/j.micres.2020.126549.

Li, X., et al. (2020), Impacts of growth temperature, water deficit and heatwaves on carbon assimilation and growth of cotton plants (*Gossypium hirsutum* L.), *Environ. Exp. Bot.* 179: 104204. https://doi.org/10.1016/j.envexpbot.2020.104204.

Liu, X., et al. (2023), Inoculated microbial consortia perform better than single strains in living soil: a meta-analysis, *Appl. Soil Ecol.* 190: 105011 https://doi.org/10.1016/j.apsoil.2023.105011.

Mallon, C. A., et al. (2018). The impact of failure: Unsuccessful bacterial invasions steer the soil microbial community away from the invader's niche, *ISME J.* 12(3): 728–741. https://doi.org/10.1038/s41396-017-0003-y.

Mantelin, S., et al. (2006), Nitrate-dependent control of root architecture and N nutrition are altered by a plant growth-promoting *Phyllobacterium* sp, *Planta.* 223(3): 591–603. https://doi.org/10.1007/s00425-005-0106-y.

Massa, F., et al. (2022), Exploitation of plant growth promoting bacteria for sustainable agriculture: hierarchical approach to link laboratory and field experiments, *Microorganisms.* 10(5): 865. https://doi.org/10.3390/microorganisms10050865.

Molina-Romero, D., et al. (2021), A Bacterial consortium interacts with different varieties of maize, promotes the plant growth, and reduces the application of chemical fertilizer under field conditions, *Front. Sust. Food Syst.* 4: 616757. https://doi.org/10.3389/fsufs.2020.616757.

Moreira, H., et al. (2016), Selection of metal resistant plant growth promoting rhizobacteria for the growth and metal accumulation of energy maize in a mine soil - Effect of the

inoculum size, *Geoderma*, 278: 1–11. https://doi.org/10.1016/j.geoderma.2016.05.003.
Moreira, H., et al. (2019), Effects of soil sterilization and metal spiking in plant growth promoting rhizobacteria selection for phytotechnology purposes, *Geoderma*. 334: 72–81. https://doi.org/10.1016/j.geoderma.2018.07.025.
Nadeem, S. M., et al. (2014), The role of mycorrhizae and Plant Growth Promoting Rhizobacteria (PGPR) in improving crop productivity under stressful environments. *Biotechnol. Adv.* 32: 429–448. http://dx.doi.org/10.1016/j.biotechadv.2013.12.005.
Naher, U. A., et al. (2021), Bio-organic fertilizer: a green technology to reduce synthetic N and P fertilizer for rice production, *Front. Plant Sci.* 12: 602052. https://doi.org/10.3389/fpls.2021.602052.
Negi, R., et al. (2024), Microbial consortia: promising tool as plant bioinoculants for agricultural sustainability. *Curr. Microbiol.* 81: 222. https://doi.org/10.1007/s00284-024-03755-0.
Nunes, P. S. O., et al. (2024), Microbial consortia of biological products: do they have a future?, *Biol. Control.* 188: 105439. https://doi.org/10.1016/j.biocontrol.2024.105439.
Nuruzzaman, M., et al. (2025), Diffuse soil pollution from agriculture: impacts and remediation, *Sci. Total Environ.* 962: 178398. https://doi.org/10.1016/j.scitotenv.2025.178398.
Palberg, D., et al. (2022), A survey of Methylobacterium species and strains reveals widespread production and varying profiles of cytokinin phytohormones, *BMC Microbiol.* 22(1): 22–49. https://doi.org/10.1186/s12866-022-02454-9.
Pantoja-Guerra, M., et al. (2023), Total auxin level in the soil–plant system as a modulating factor for the effectiveness of PGPR inocula: a review, *Chem. Biol. Technol. Agric.* 10(1): 1–17. https://doi.org/10.1186/s40538-022-00370-8.
Pascale, A., et al. (2020), Modulation of the root microbiome by plant molecules: the basis for targeted disease suppression and plant growth promotion, *Front. Plant Sci.* 10: 1741. https://doi.org/10.3389/fpls.2019.01741.
Pathak, D., et al. (2024), Community-forming traits play role in effective colonization of plant-growth-promoting bacteria and improved plant growth, *Front. Plant Sci.* 15: 1332745. https://doi.org/10.3389/fpls.2024.1332745.
Patten, C. L., et al. (2013), Activity, distribution and function of indole-3-acetic acid biosynthetic pathways in bacteria, *Crit. Rev. Microbiol.* 39(4): 395–415. https://doi.org/10.3109/1040841X.2012.716819.
Pereira, S. I. A., et al. (2015), Rhizobacteria isolated from a metal-polluted area enhance plant growth in zinc and cadmium-contaminated soil, *Int. J. Environ. Sci. Technol.* 12(7): 2127–2142. https://doi.org/10.1007/s13762-014-0614-z.
Pereira, N. C. M., et al. (2020a), Corn yield and phosphorus use efficiency response to phosphorus rates associated with plant growth promoting bacteria. *Front. Environ. Sci.* 8: 40. https://doi.org/10.3389/fenvs.2020.00040.
Pereira, S. I. A., et al. (2020b), Plant Growth Promoting Rhizobacteria (PGPR) improve the growth and nutrient use efficiency in maize (*Zea mays* L.) under water deficit conditions, *Heliyon*. 6(10): e05106. https://doi.org/10.1016/j.heliyon.2020.e05106.
Pereira, S. I. A., et al. (2016), Promotion of sunflower growth under saline water irrigation by the inoculation of beneficial microorganisms, *Appl. Soil Ecol.* 105: 36–47. https://doi.org/10.1016/j.apsoil.2016.03.015.

Phares, C. A., et al. (2022), Biochar and NPK fertilizer co-applied with plant growth promoting bacteria (PGPB) enhanced maize grain yield and nutrient use efficiency of inorganic fertilizer, *J. Agric. Food Res.* 10: 100434. https://doi.org/10.1016/j.jafr.2022.100434.

Pii, Y., et al. (2015), Microbial interactions in the rhizosphere: beneficial influences of plant growth-promoting rhizobacteria on nutrient acquisition process: a review, *Biol. Fert. Soils*. 51(4): 403–415. https://doi.org/10.1007/s00374-015-0996-1.

Poveda, J. and González-Andrés, F. (2021), *Bacillus* as a source of phytohormones for use in agriculture, *Appl. Microbiol. Biotechnol.* 105(23): 8629–8645. https://doi.org/10.1007/s00253-021-11492-8.

Ramírez-Pool, J. A., et al. (2024), *Bacillus* strains as effective biocontrol agents against phytopathogenic bacteria and promoters of plant growth, *Microb. Ecol.* 87: 1. https://doi.org/10.1007/s00248-024-02384-1.

Rawat, P., et al. (2021), Phosphate-Solubilizing microorganisms: mechanism and their role in phosphate solubilization and uptake, *J. Soil Sci. Plant Nutr.* 21(1): 49–68. https://doi.org/10.1007/s42729-020-00342-7.

Riaz, U., et al. (2020), Plant Growth-Promoting Rhizobacteria (PGPR) as biofertilizers and biopesticides, in Hakeem, K. R., et al. (eds) *Microbiota and Biofertilizers*, 181–196. Springer, Cham. https://doi.org/10.1007/978-3-030-48771-3.

Rilling, J. I., et al. (2019), Current opinion and perspectives on the methods for tracking and monitoring plant growth–promoting bacteria, *Soil Biol. Biochem.* 130: 205–219. https://doi.org/10.1016/j.soilbio.2018.12.012.

Rizvi, A., et al. (2021), Sorghum-phosphate solubilizers interactions: crop nutrition, biotic stress alleviation, and yield optimization, *Front. Plant Sci.* 12: 746780. https://doi.org/10.3389/fpls.2021.746780.

Rojas-Sánchez, B., et al. (2022), Bioencapsulation of microbial inoculants: mechanisms, formulation types and application techniques, *Appl. Biosci.* 1(2): 198–220. https://doi.org/10.3390/applbiosci1020013.

Rojas-Tapias, D., et al. (2012), Effect of inoculation with Plant Growth romoting Bacteria (PGPB) on amelioration of saline stress in maize (*Zea mays*), *Appl. Soil Ecol.* 61: 264–272. https://doi.org/10.1016/j.apsoil.2012.01.006.

Sakpirom, J., et al. (2021), Optimization of carriers and packaging for effective biofertilizers to enhance *Oryza sativa* L. growth in paddy soil, *Rhizosphere* 19: 100383. https://doi.org/10.1016/j.rhisph.2021.100383.

Samonty, I., et al. (2025), The PGPB paradox: a critical review of field performance and practical constraints in agriculture *World J. Microbiol. Biotechnol.* 41: 321 https://doi.org/10.1007/s11274-025-04552-y.

Santoyo, G., et al. (2021), Plant growth stimulation by microbial consortia, *Agronomy* 11(2): 219. https://doi.org/10.3390/agronomy11020219.

Sati, D., et al. (2023), Recent Advances in PGPR and molecular mechanisms involved in drought stress resistance, *J. Soil Sci. Plant Nutr.* 23(1): 106–124. https://doi.org/10.1007/s42729-021-00724-5.

Severo, H. C. A. R., et al. (2025), Co-inoculation of arbuscular mycorrhizal fungi and *Bacillus subtilis* enhances morphological traits, growth, and nutrient uptake in maize under limited phosphorus availability, *Sci. Rep.* 15: 25448. https://doi.org/10.1038/s41598-025-10038-6.

Schultz, C. R., et al. (2022), The role of genetic variation in *Zea mays* response to beneficial endophytes, *Plant Growth Regul.* 98(1): 167–177. https://doi.org/10.1007/s10725-022-00842-9.

Shah, A., et al. (2021), PGPR in Agriculture: a sustainable approach to increasing climate change resilience, *Front. Sust. Food Sys.* 5: 667546. https://doi.org/10.3389/fsufs.2021.667546.

Shahzad, M., et al. (2025), Biofertilizers in sustainable agriculture: mechanisms and recent advancements, *Discov. Agric.* 3: 224. https://doi.org/10.1007/s44279-025-00318-0.

Sheteiwy, M. S., et al. (2021), Inoculation with *Bacillus amyloliquefaciens* and mycorrhiza confers tolerance to drought stress and improve seed yield and quality of soybean plant, *Physiol. Plant.* 172(4): 2153–2169. https://doi.org/10.1111/ppl.13454.

Singh, P., et al. (2022), Mechanistic Insights and potential use of siderophores producing microbes in rhizosphere for mitigation of stress in plants grown in degraded land, *Front. Microbiol.* 13: 1–16. https://doi.org/10.3389/fmicb.2022.898979.

Singh, R., et al. (2023), Bacterial biostimulants for climate smart agriculture practices: Mode of action, effect on plant growth and roadmap for commercial products, *J. Sustain. Agric. Environ.* 3(1): 1–20. https://doi.org/10.1002/sae2.12085.

Steiner, F., et al. (2017), Drought tolerance of wheat and black oat crops at early stages of seedling growth, *Revista Ciências Agrárias*, 40(3): 576–586. http://dx.doi.org/10.19084/RCA16118.

Thioub, M., et al. (2019), Arbuscular mycorrhizal fungi inoculation enhances phosphorus use efficiency and soybean productivity on a Haplic Acrisol, *Soil Till. Res.* 192: 174–186. https://doi.org/10.1016/j.still.2019.05.001.

Thokchom, E., et al. (2017), Root colonization by host-specific rhizobacteria alters indigenous root endophyte and rhizosphere soil bacterial communities and promotes the growth of mandarin orange, *Eur. J. Soil Biol.* 79, 48–56. https://doi.org/10.1016/j.ejsobi.2017.02.003.

Timmusk, S., et al. (2014), Drought-tolerance of wheat improved by rhizosphere bacteria from harsh environments: enhanced biomass production and reduced emissions of stress volatiles, *PLoS One* 9(5): e96086. https://doi.org/10.1371/journal.pone.0096086.

Tovi, N., et al. (2019), Host specificity and spatial distribution preference of three *Pseudomonas* isolates, *Front. Microbiol.* 10: 3263. https://doi.org/10.3389/fmicb.2018.03263.

Triantafyllou, A., et al. (2023), Evaluation of the biocontrol potential of PGPB strains isolated from drought-tolerant tomatoes against fungal pathogens, *J. Plant Pathol.* 105(3): 1013–1029. https://doi.org/10.1007/s42161-023-01422-0.

Tsukanova, K. A., et al. (2017), Effect of plant growth-promoting rhizobacteria on plant hormone homeostasis, *South Afr. J. Bot.* 113: 91–102. https://doi.org/10.1016/j.sajb.2017.07.007.

Vacheron, J., et al. (2013), Plant growth-promoting rhizobacteria and root system functioning, *Front. Plant Sci.* 4, 356. https://doi.org/10.3389/fpls.2013.00356.

Vassileva, M., et al. (2021), Fermentation strategies to improve soil bio-inoculant production and quality, *Microorganisms* 9: 1254. https://doi.org/10.3390/microorganisms9061254.

Win, K. T., et al. (2023), Synergistic N_2-fixation and salt stress mitigation in soybean through dual inoculation of ACC deaminase-producing *Pseudomonas* and *Bradyrhizobium*. *Sci. Rep.* 13: 17050. https://doi.org/10.1038/s41598-023-43891-4.

Xue, H., et al. (2020), Insights into the root invasion by the plant pathogenic bacterium *Ralstonia solanacearum*. *Plants* 9(4): 516. https://doi.org/10.3390/plants9040516.

Yuan, X., et al. (2024), Impacts of global climate change on agricultural production: a comprehensive review, *Agronomy* 14(7): 1360. https://doi.org/10.3390/agronomy14071360.

Zeng, Q., et al. (2022), Insight into soil nitrogen and phosphorus availability and agricultural sustainability by plant growth-promoting rhizobacteria, *Envir. Sci. Poll. Res.* 29(30): 45089–45106. https://doi.org/10.1007/s11356-022-20399-4.

Zhang, H., et al. (2023), A New Algivorous Heterolobosean Amoeba, *Euplaesiobystra perlucida* sp. nov. (Tetramitia, Discoba), isolated from pilot-scale cultures of *Phaeodactylum* tricornutum, *Microbiol. Spec.* 11(4). https://doi.org/10.1128/spectrum.00817-23.

Zhou, Y. and Bai, Y. (2025), The functional mechanisms, screening methods, and agricultural application potential of plant growth-promoting bacteria, *Geogr. Res. Bull.* 4: 158–174. https://doi.org/10.50908/grb.4.0_158.

Zhu, X., et al. (2023), Innovative microbial disease biocontrol strategies mediated by quorum quenching and their multifaceted applications: a review. *Front. Plant Sci.* 13: 1063393. https://doi.org/10.3389/fpls.2022.1063393.

Chapter 10

Understanding and promoting existing communities of plant growth-promoting rhizobacteria (PGPR) in agricultural soils

Sajeevee S. Sarathchandra and Zakaria M. Solaiman, The University of Western Australia, Australia

http://dx.doi.org/10.19103/AS.2025.0166.21

1 Introduction

Plant growth-promoting rhizobacteria (PGPR) are a functionally defined group of root-associated bacteria inhabiting the rhizosphere and, in some cases, the root interior that enhance plant growth and health through multiple direct and indirect mechanisms (Compant et al., 2023; Vurukonda et al., 2023). These mechanisms include nutrient mobilisation through biological nitrogen fixation, phosphate and potassium solubilisation, and siderophore-mediated iron acquisition; production of phytohormones such as indole-3-acetic acid, cytokinins, and gibberellins; alleviation of abiotic stress via modulation of ethylene levels and osmolyte production; and suppression of soil-borne pathogens through antibiosis, niche competition, and induction of systemic resistance (Backer et al., 2018; Liu et al., 2024). Owing to this multifunctionality, PGPR are widely recognised as critical contributors to crop productivity, nutrient-use efficiency, and stress resilience in agroecosystems (Ahemad and Kibret, 2013; Mendes et al., 2022).

PGPR are taxonomically diverse and do not form a discrete phylogenetic group. Commonly reported genera include *Bacillus*, *Pseudomonas*, *Azotobacter*, *Azospirillum*, *Paenibacillus*, and *Rhizobium*, alongside representatives of *Burkholderia*, *Enterobacter*, and *Streptomyces* (Santoyo et al., 2021; Vurukonda et al., 2023). Importantly, PGPR traits are phylogenetically widespread and frequently strain-specific, reflecting functional convergence rather than taxonomic relatedness (Yang et al., 2024; Compant et al., 2023). This implies that most agricultural soils harbour a substantial reservoir of organisms with latent or conditionally expressed plant growth-promoting potential, even in the absence of external microbial inputs.

Plants play a central role in structuring rhizosphere microbiomes through active recruitment and filtering processes (Fig. 1). Root exudates – complex mixtures of sugars, amino acids, organic acids, phenolics, and secondary metabolites – serve as both nutrient sources and signalling molecules that selectively stimulate microbial taxa and functions (Kuzyakov and Razavi, 2024). Recent studies demonstrate that plants can dynamically alter exudation patterns in response to nutrient limitation or abiotic stress, thereby enriching for PGPR capable of nitrogen fixation, phosphorus mobilisation, or stress mitigation (Liu et al., 2024). In addition, plant immune systems act as selective filters, favouring beneficial or commensal microbes while restricting potential pathogens (Trivedi et al., 2020).

These insights suggest a complementary paradigm: rather than relying solely on introduced strains, we can manage soils and cropping systems to promote existing PGPR communities and their networks with other beneficial microbes. This chapter focuses on this paradigm.

Increasing attention has been directed towards the use of PGPR consortia, where multiple compatible microbial strains are combined to exploit

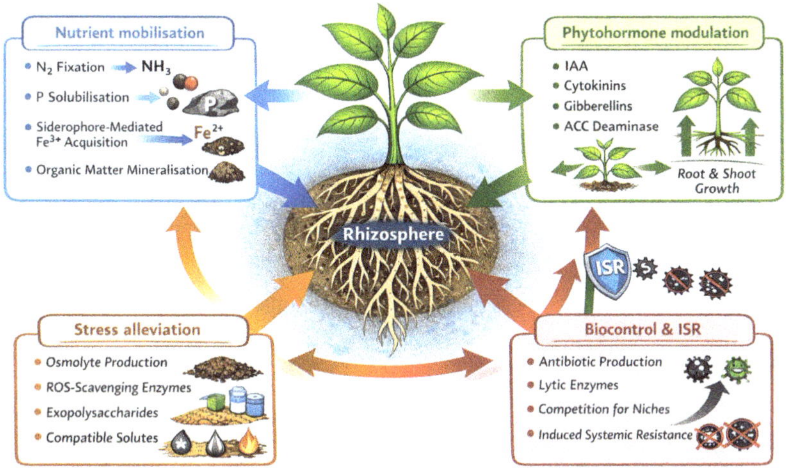

Figure 1 Mechanistic pathways through which indigenous PGPR influence plant performance. There are mechanisms outlined in boxes: (1) 'Nutrient mobilisation' including sub-bullets or icons for N fixation, P solubilisation, siderophore-mediated Fe acquisition and organic matter mineralisation, with arrows from soil pools (N_2, mineral/organic P, Fe^{3+}) into the root; (2) 'Phytohormone modulation' with bullets for IAA, cytokinins, gibberellins and ACC deaminase, with arrows back to the root indicating stimulation of root branching and elongation, and a small icon showing balanced shoot–root growth (Basu et al., 2024; Kumar et al., 2020); (3) 'stress alleviation' including osmolyte production, ROS-scavenging enzymes, exopolysaccharides and compatible solutes, with arrows from stress icons (salt, drought, metals, temperature) towards PGPR, then from PGPR to the plant root indicating enhanced stress tolerance (Kumar et al., 2020); and (4) 'Biocontrol and induced systemic resistance (ISR)' including antibiotic production, lytic enzymes, competition for niches and ISR signalling to the shoot, represented by a long arrow from the rhizosphere up the stem to leaves labelled 'ISR' and small pathogen icons being blocked (Song et al., 2022; Moore et al., 2022). Interconnections among the four boxes can be shown with double-headed arrows to indicate that mechanisms are not independent (e.g. nutrient mobilisation supports growth, which in turn conditions stress tolerance and ISR).

synergistic interactions and multifunctional traits. Single-strain inoculants often show inconsistent performance under field conditions due to environmental variability and microbial competition. In contrast, microbial consortia provide complementary mechanisms such as enhanced nutrient solubilisation, phytohormone production, and pathogen suppression, resulting in more stable plant growth promotion across diverse agroecosystems (Yu et al., 2025). Studies have demonstrated that PGPR consortia can significantly improve crop biomass, nutrient uptake efficiency, and stress tolerance compared with individual strains, highlighting the importance of cooperative microbial interactions in the rhizosphere (Joshi et al., 2025). For example, synthetic consortia composed of *Pseudomonas* and *Azotobacter* species have been shown to influence soil microbial communities and enhance crop growth under field conditions (Dobrzyński et al., 2025). Similarly, experimental studies reported increased plant biomass and improved physiological performance when lettuce plants were inoculated with multi-strain PGPR consortia under nutrient-limited conditions (Abou Jaoudé et al., 2025). These findings demonstrate that microbial interactions within consortia can enhance functional stability and improve colonisation efficiency in the rhizosphere.

Despite promising advances, challenges remain in understanding the ecological complexity of PGPR interactions in natural soils. Variability in soil physicochemical properties, plant genotype specificity, environmental stress conditions, and competition with native microbiota can influence the effectiveness of inoculated strains. Furthermore, the mechanisms regulating microbial community assembly, metabolic cooperation, and signalling networks in multi-species consortia remain poorly understood (Maciel-Rodríguez et al., 2025). These limitations highlight the need for improved experimental approaches that integrate microbiome engineering, synthetic biology, and systems-level modelling to predict functional outcomes of microbial inoculation strategies.

Future research is expected to focus on designing tailored microbial consortia with improved functional compatibility, resilience, and adaptability to specific agroecosystems. Integration of artificial intelligence, precision agriculture tools, and bioinformatics will enable better prediction of plant–microbe interactions and optimisation of biofertiliser formulations. Additionally, advances in phytobiome research are facilitating a systems-level understanding of plant–soil–microbe communication networks, providing new opportunities to develop targeted microbial inoculants for climate-resilient agriculture (Ahmad et al., 2022;Iqbal et al., 2024). Overall, PGPR and their consortia represent an essential strategy for reducing dependence on chemical fertilisers and improving agricultural sustainability. Continued research into microbial ecology, functional genomics, and rhizosphere signalling mechanisms will support the development of next-generation bioinoculants that enhance crop productivity while maintaining soil health. Addressing current challenges in formulation stability, field performance, and microbial compatibility will be critical to translating PGPR-based technologies into practical agricultural applications that contribute to global food security and environmental sustainability.

2 Plant growth-promoting rhizobacteria within the soil and rhizosphere microbiome

2.1 Functional definition and traits

Plant growth-promoting rhizobacteria are defined primarily by their functional attributes rather than their taxonomic placement and thus encompass a broad spectrum of microbial groups that enhance plant performance through nutrient mobilisation, hormonal regulation, biocontrol and stress mitigation (Basu et al., 2024; Kumar et al., 2020; see Table 1). Among the core functional categories, nutrient-mobilising PGPR include diazotrophic bacteria capable of biological nitrogen fixation, such as *Azotobacter*, *Azospirillum*, and free-living *Rhizobium*, as well as phosphate- and potassium-solubilising genera such as *Pseudomonas*, *Bacillus* and *Enterobacter*, which release organic acids, phosphatases and chelators that improve nutrient availability in the rhizosphere (Song et al., 2022; S. Zhang et al., 2025). Many of these taxa also produce siderophores that facilitate iron acquisition under limiting conditions, further supporting plant nutrition (Kumar et al., 2020). A second major group comprises phytohormone-modulating PGPR, which synthesise key signalling molecules, including indole-3-acetic acid (IAA), gibberellins and cytokinins while also expressing ACC deaminase, an enzyme that lowers stress-induced ethylene concentrations and promotes sustained root growth under adverse conditions (Khoso et al., 2024; Basu et al., 2024). Biocontrol and induced-resistance PGPR represent another essential functional domain, characterised by the production of antibiotics, lytic enzymes, and hydrogen cyanide (HCN), along with strong competitive abilities that suppress pathogens and activate induced systemic resistance (ISR) via jasmonic acid and ethylene pathways, thereby enhancing plant immune capacity (Moore et al., 2022; Mmotla et al., 2025). Stress-alleviating PGPR form a further functional category, contributing to plant tolerance to salinity, drought, heavy metals, and temperature extremes through mechanisms involving osmoprotectants, exopolysaccharides, antioxidant enzymes, and modulation of ion homeostasis (Ahmad et al., 2022; Kumar et al., 2020). Importantly, indigenous soil communities typically harbour multiple PGPR functional types simultaneously, and these groups often interact synergistically or exhibit functional redundancy, creating resilient microbial networks that stabilise nutrient cycling, suppress pathogens, and buffer plants against fluctuating environmental conditions (Maciel-Rodríguez et al., 2025; X. Zhang et al., 2025). Indigenous communities usually contain multiple functional types, often interacting synergistically or redundantly.

2.2 PGPR as part of a microbiome network

PGPR rarely function in isolation; instead, they operate as integral components of complex, multi-kingdom microbiome networks that include fungi – particularly arbuscular mycorrhizal fungi (AMF) – as well as diverse bacterial taxa, archaea and micro-eukaryotes that collectively inhabit the rhizosphere (Maciel-Rodríguez et al., 2025; Song et al., 2022). These networks orchestrate the movement and transformation of nutrients and carbon by regulating key processes such as nitrogen mineralisation, phosphorus mobilisation, organic

Table 1 Functional groups of PGPR, key traits, representative genera, and supporting references

PGPR functional group	Key traits/Mechanisms (with references)	Representative genera/ Examples (with references)	Functional contribution in soil-plant systems (with references)
Nutrient-mobilising PGPR	Biological nitrogen fixation (Basu et al., 2024; Kumar et al., 2020) Phosphate and potassium solubilisation (Song et al., 2022) Siderophore-mediated Fe acquisition (Basu et al., 2024)	*Azotobacter*, *Azospirillum*, free-living *Rhizobium* (Kumar et al., 2020) *Pseudomonas*, *Bacillus*, *Enterobacter* (Song et al., 2022)	Enhanced N, P, K uptake; reduced fertiliser dependency; improved plant vigour (Basu et al., 2024; S. Zhang et al., 2025)
Phytohormone-modulating PGPR	IAA, gibberellin & cytokinin production (Song et al., 2022) ACC deaminase reducing ethylene (Khoso et al., 2024)	*Pseudomonas*, *Bacillus*, *Enterobacter* (Song et al., 2022) *Azospirillum* spp. (Kumar et al., 2020)	Enhanced root branching; improved seedling establishment; balanced shoot–root ratio; improved stress resilience (Basu et al., 2024; Ferrante et al., 2024)
Biocontrol & induced systemic resistance (ISR) PGPR	Antibiotics, lytic enzymes, HCN (Song et al., 2022) Competition for niches and nutrients (Moore et al., 2022) ISR via JA/ET pathways (Mmotla et al., 2025)	*Pseudomonas*, *Bacillus*, *Streptomyces*, *Paenibacillus* (Song et al., 2022; Moore et al., 2022)	Suppression of soil-borne pathogens; activation of systemic resistance; improved yield stability (Moore et al., 2022; X. Zhang et al., 2025)
Stress-alleviating PGPR	Osmolytes, exopolysaccharides, antioxidant enzymes (Kumar et al., 2020) Salinity and drought tolerance mechanisms, Metal and temperature-stress mitigation (Kumar et al., 2020)	*Bacillus*, *Pseudomonas*, *Rhizobium* spp. (Kumar et al., 2020) *Enterobacter* spp.	Improved tolerance to salinity, drought, metals and temperature extremes; stabilised photosynthesis; enhanced survival under climate stress (Ferrante et al., 2024; Kumar et al., 2020)
Interactions within Indigenous PGPR communities	Functional redundancy & synergy (Maciel-Rodríguez et al., 2025) Biofilms, quorum sensing, microbial network interactions (Mmotla et al., 2025)	Mixed native consortia including *Pseudomonas*, *Bacillus*, actinobacteria (Maciel-Rodríguez et al., 2025)	Enhanced rhizosphere stability; improved nutrient cycling; buffering against environmental fluctuations; higher ecosystem resilience (S. Zhang et al., 2025; Moore et al., 2022)

matter decomposition, and rhizodeposition-mediated carbon turnover, with PGPR often acting as biochemical catalysts that enhance plant–microbe and microbe–microbe interactions (Basu et al., 2024; Kumar et al., 2020). Co-operative interactions between PGPR and AMF are particularly well documented; PGPR can stimulate fungal spore germination and hyphal growth, while AMF increase the absorptive surface area of plant roots, creating a synergistic nutrient acquisition system that improves nitrogen, phosphorus and micronutrient uptake beyond the capacity of either group alone (Savastano and Bais, 2024; Yan et al., 2025; X. Zhang et al., 2025).

Such microbial consortia also contribute to soil structure and aggregation through the production of extracellular polymeric substances (EPS), glomalin from AMF and root-associated mucilage, all of which stabilise soil aggregates and enhance aeration, porosity, and water retention (Maciel-Rodríguez et al., 2025; Kumar et al., 2020). Importantly, these networks provide functional redundancy – multiple microbial taxa performing similar ecological roles – thereby buffering plants against environmental disturbances such as drought, nutrient fluctuations, and pathogen invasion, and supporting greater resilience of the rhizosphere ecosystem (Moore et al., 2022; Mmotla et al., 2025). Because PGPR influence and are influenced by the broader soil microbiome, effective management cannot target single strains alone but must instead promote entire beneficial consortia and the environmental conditions – such as organic matter input, reduced disturbance, and biologically diverse cropping systems – that allow PGPR guilds to thrive, interact, and sustain key soil functions over time (Basu et al., 2024; Maciel-Rodríguez et al., 2025).

3 Ecology and functioning of indigenous plant growth-promoting rhizobacteria communities

3.1 Soil properties shaping PGPR communities

Soil physicochemical properties are major determinants of the diversity, abundance, and functional potential of PGPR communities, influencing microbial survival, metabolism, competition, and interactions within the rhizosphere. Soil pH exerts the strongest single control over bacterial community structure, with slightly acidic to neutral soils typically supporting higher abundances of *Pseudomonas*, *Bacillus*, and free-living diazotrophs, whereas extreme pH conditions tend to reduce PGPR diversity and favour more stress-tolerant taxa (Maciel-Rodríguez et al., 2025; Basu et al., 2024). Soil organic matter (SOM) provides carbon and energy sources critical for heterotrophic PGPR, and its quality – labile versus recalcitrant fractions – modulates microbial growth dynamics, nutrient turnover, and the expression of PGPR functional traits such as siderophore production and phosphate solubilisation (Kumar et al., 2020;

Song et al., 2022). Texture and mineralogy further shape PGPR communities by affecting water-holding capacity, aeration, and microsite availability; clay-rich soils often create protective microhabitats conducive to microbial persistence, while sandy soils may favour rapid turnover but lower biomass (S. Zhang et al., 2025). Nutrient availability, particularly of nitrogen and phosphorus, influences PGPR function by altering plant nutrient demand and, consequently, root exudation patterns that regulate microbial recruitment. Excess mineral N inputs frequently decrease the competitive advantage of nitrogen-fixing PGPR, whereas moderate nutrient limitation stimulates their activity and plant reliance on microbial nutrient mobilisation (Basu et al., 2024). Collectively, these abiotic factors interact to form the 'soil habitat template' that selects for specific PGPR guilds and influences their stability, function and responsiveness to management interventions (see Table 2).

3.2 Plant genotype and domestication effects on PGPR assembly

Plant genotype plays a pivotal role in shaping rhizosphere microbiome assembly, with distinct cultivars and species differing markedly in root architecture, exudation chemistry, and immune signalling – all of which influence PGPR recruitment and activity (Duret et al., 2024; Tang et al., 2023; Song et al., 2022). Modern crop domestication and breeding, often conducted under high-input conditions, have unintentionally altered plant–microbiome interactions, frequently reducing the ability of elite cultivars to attract or sustain beneficial PGPR compared with their wild relatives or landraces (Hemelda and Noutoshi, 2025; S. Zhang et al., 2025). Root traits such as exudate composition, mucilage production, suberisation, and fine-root density control the quantity and diversity of carbon resources available to PGPR and other rhizosphere microbes (Basu et al., 2024). Meanwhile, plant immune receptors mediate selective microbial filtering via pattern recognition, thereby allowing beneficial bacteria to colonise while excluding pathogens (Kumar et al., 2020). Domestication-related reductions in root chemical diversity have been linked to narrower and less resilient microbiomes, with implications for nutrient use efficiency and stress tolerance (Tang et al., 2023). Consequently, there is growing interest in microbiome-aware breeding strategies that intentionally incorporate PGPR compatibility traits – such as stable exudation profiles, enhanced root surface area, or synergy with AMF – to develop crop ideotypes capable of forming strong and consistent partnerships with indigenous microorganisms (X. Zhang et al., 2025). Such approaches signal a shift towards co-designing plant genotypes and soil microbiomes in future sustainable agroecosystems.

Table 2 Understanding and promoting existing communities of PGPR in soil

Aspect	Current findings	Implications for management
Ecological role of indigenous PGPR	Indigenous PGPR communities already present in soil contribute substantially to nutrient mobilisation, phytohormone balance, stress tolerance, and disease suppression, often matching or exceeding the contribution of introduced strains when conditions are favourable (Basu et al., 2024; Song et al., 2022).	Management should prioritise creating conditions that enhance the activity of existing PGPR rather than relying solely on external inoculants.
Drivers of PGPR community structure	Soil pH, texture, organic matter, moisture, and nutrient status interact with plant genotype and root exudation profiles to determine the composition and function of PGPR communities, with domestication sometimes having reduced modern cultivars' capacity to recruit beneficial microbes (Tang et al., 2023).	Site- and crop-specific strategies are needed; breeding and variety selection should consider microbiome–plant compatibility alongside conventional traits.
Root exudates and plant control	Plants exert strong control over the rhizosphere via dynamic exudation of sugars, amino acids, organic acids, and secondary metabolites that selectively stimulate or inhibit particular microbial guilds, including PGPR and pathogens (Mmotla et al., 2025).	Manipulating plant nutrition, stress status and developmental stage, as well as selecting for favourable exudation traits, can be used to indirectly steer indigenous PGPR communities.
Community and network interactions	PGPR act within complex microbial networks involving other bacteria, fungi and micro-eukaryotes; beneficial effects on plants often emerge from community-level properties such as functional redundancy, keystone taxa, and biofilm formation rather than from single strains alone (Moore et al., 2022).	Interventions should aim to support whole beneficial consortia (e.g. via organic inputs and reduced disturbance) rather than focusing narrowly on one or two 'hero' strains.
Effects of tillage and disturbance	Reduced and no-till systems generally increase microbial biomass, diversity, and stability in the rhizosphere, with consistent enrichment of PGPR-associated genera compared with intensive tillage (Maciel-Rodríguez et al., 2025).	Moving towards conservation tillage is a robust lever for promoting indigenous PGPR and improving their habitat quality.
Organic amendments and residues	Compost, manure and crop residues increase labile carbon and microhabitats, stimulating PGPR abundance, soil enzyme activity, and nutrient cycling, although responses are context-dependent and excessive or poorly processed amendments can favour undesirable microbes (Maciel-Rodríguez et al., 2025).	Use well-characterised, appropriately dosed organic amendments tailored to soil constraints; combine with residue retention to build long-term PGPR-supportive soil organic matter.

(*Continued*)

Table 2 *Continued*

Aspect	**Current findings**	**Implications for management**
Crop rotation, cover crops and diversification	Diverse rotations and cover crops generate heterogeneous root exudate inputs and rooting patterns, thereby enhancing PGPR diversity and stability and often reducing disease pressure and input requirements relative to monocultures (Basu et al., 2024).	Implement diversified rotations and multifunctional cover crop mixtures as core tools for sustaining beneficial PGPR communities.
Fertilisation and pesticide regimes	Moderate fertiliser inputs, slow-release sources, and organic nutrient forms tend to preserve or enhance PGPR-mediated nutrient functions, whereas high mineral N and some pesticides can suppress key PGPR groups or disrupt community balance (S. Zhang et al., 2025).	Design 'microbiome-aware' input regimes that meet crop needs while avoiding over-fertilisation and unnecessary pesticide applications.
Performance of PGPR inoculants in the field	Meta-analyses show that PGPR inoculants yield, on average, modest positive effects but with large variability among sites and seasons; many inoculants act transiently yet still shift the indigenous community towards more beneficial states (Ferrante et al., 2024; Moore et al., 2022; X. Zhang et al., 2025).	Treat inoculants as tools to *nudge* indigenous communities within a supportive management framework, rather than as stand-alone solutions; expect context-dependent outcomes.
PGPR in degraded and stressed soils	In salinised, drought-prone, or contaminated soils, PGPR can facilitate plant establishment, improve tolerance, and accelerate soil restoration, especially when combined with organic amendments and appropriate vegetation (Kumar et al., 2020; Maciel-Rodríguez et al., 2025).	Restoration programmes should explicitly integrate PGPR-supportive practices (pioneer plants, organic inputs, reduced disturbance) to rebuild beneficial microbial networks.
Evidence gaps and variability	Causal links between specific PGPR taxa or traits and field-scale agronomic outcomes remain incomplete; outcomes are strongly context-dependent, and long-term ecological effects of microbiome manipulation are not yet fully known (Moore et al., 2022).	Future work must combine mechanistic experiments, long-term field trials, and risk assessment to refine recommendations and avoid overgeneralisation.
Emerging tools and future directions	Multi-omics, synthetic consortia, microbiome-aware breeding, and precision microbiome management are rapidly advancing and offer pathways to more predictable PGPR promotion, but their translation into routine agronomy and decision-support tools remains in its early stages (Basu et al., 2024; Mmotla et al., 2025; S. Zhang et al., 2025).	Investment in diagnostics, breeding, modelling, and advisory tools is required to mainstream PGPR-centred soil management and to integrate it into broader soil health and climate-resilient agriculture strategies.

3.3 Root exudates and rhizodeposition as drivers of PGPR dynamics

Root exudates serve as the primary biochemical interface between plants and rhizosphere microbes, shaping PGPR community composition and behaviour by providing carbon substrates, signalling molecules and chemical cues that regulate microbial gene expression (Mmotla et al., 2025). Exudates consist of sugars, amino acids, organic acids, fatty acids, phenolics, and secondary metabolites, each with distinct ecological roles. Labile compounds, such as simple sugars and amino acids, often promote fast-growing copiotroph PGPR, including *Pseudomonas* and *Enterobacter*, whereas organic acids and phenolic compounds can selectively enrich for functional groups involved in phosphate solubilisation or pathogen suppression (Basu et al., 2024). Root exudation patterns change dynamically with plant developmental stage, nutrient status, and environmental stress, leading to temporally variable PGPR recruitment patterns (Song et al., 2022). For instance, during early root establishment, plants often exude high levels of IAA precursors and signalling molecules that attract IAA-producing PGPR, promoting root elongation, while under nutrient limitation, exudates shift towards organic acids that encourage P-solubilising bacteria (Kumar et al., 2020). Rhizodeposition – including sloughed cells, mucilage, and decaying lateral roots – provides longer-term carbon inputs that sustain stable microbial networks and biofilm structures essential for PGPR persistence (Maciel-Rodríguez et al., 2025). Overall, the root secretome functions not only as a nutrient resource but also as a selective filter and communication platform that determines which PGPR guilds thrive in the rhizosphere.

3.4 Temporal dynamics and management influences on PGPR communities

PGPR communities exhibit strong temporal dynamics influenced by crop growth stage, seasonal variation, environmental fluctuations, and agricultural management practices. Early developmental stages typically favour fast-colonising PGPR such as *Pseudomonas*, while later stages – characterised by increased root biomass and exudate complexity – support more diverse communities including *Bacillus*, actinobacteria, and mycorrhizal-associated bacteria (Roesti et al., 2006; S. Zhang et al., 2025). Seasonal changes in temperature, moisture, and nutrient cycling shift microbial competitiveness and functional expression, resulting in pronounced temporal turnover in PGPR abundance and activity (Maciel-Rodríguez et al., 2025). Agricultural practices, including tillage, residue retention, fertilisation, pesticides, and irrigation, exert some of the most powerful influences on PGPR community trajectories. Reduced or no-till systems enhance aggregate stability and habitat continuity, supporting more abundant and stable PGPR populations, whereas intensive tillage disrupts microbial niches and reduces beneficial consortia (Basu et al., 2024). Organic amendments such as compost or manure increase labile carbon inputs, often stimulating beneficial PGPR such as *Bacillus* and *Paenibacillus*, while promoting synergistic interactions with AMF and other beneficial microorganisms (Maciel-Rodríguez et al., 2025). Conversely, excessive mineral nitrogen fertilisation can depress nitrogen-fixing PGPR by lowering plant dependence on microbial nutrient inputs (Kumar et al., 2020). Crop rotations and cover crops introduce diverse root exudation patterns across the season, broadening microbial functional diversity and enhancing the resilience of PGPR networks (Song et al., 2022). Together, these temporal and management drivers interact to shape dynamic, context-specific PGPR communities whose structure and function evolve over the cropping cycle.

4 Approaches to characterising existing plant growth-promoting rhizobacteria communities

Historically, research and commercial application of PGPR have been dominated by an inoculant-centric paradigm, centred on isolating, mass-producing, and applying individual strains or simplified consortia as biofertilisers or biostimulants (Bashan et al., 2014; Santoyo et al., 2021). While this strategy has generated valuable mechanistic understanding and several commercial products, its agronomic performance has often been inconsistent across sites, seasons, and cropping systems (Mitter et al., 2021; Trivedi et al., 2020). High-throughput sequencing and metagenomic studies have since revealed that agricultural soils and rhizospheres already harbour highly diverse indigenous microbial communities containing numerous PGPR taxa, which are increasingly recognised as the primary contributors to microbially mediated plant benefits under field conditions (Trivedi et al., 2020; Compant et al., 2023). The composition, structure, and functional capacity of these resident communities are shaped by long-term interactions among soil physicochemical properties, climate, land-use history, and crop management practices (Table 1; Hartmann and Six, 2023).

Meta-analyses and long-term field experiments consistently demonstrate that the effectiveness of PGPR inoculants is strongly context-dependent, with outcomes influenced by soil characteristics, climate, crop species, and management regimes (Mitter et al., 2021; Liu et al., 2024). A key limitation of inoculant-based approaches is the need for introduced strains to compete with complex, well-adapted native microbiomes that often exhibit high diversity and functional redundancy, reducing the likelihood of inoculant persistence or expression of beneficial traits at agronomically relevant levels (Mendes et al., 2022). Agricultural practices further exert strong selective pressures on soil and rhizosphere microbial communities. Intensive tillage disrupts soil structure and microbial habitats, while high mineral fertiliser inputs can suppress microbial nutrient-mobilisation functions such as biological nitrogen fixation and phosphorus solubilisation (Mendes et al., 2022; Hartmann and Six, 2023). In contrast, reduced tillage, organic amendments, and conservation agriculture tend to enhance microbial diversity, stability, and PGPR functional potential, while crop diversification strategies promote more complex and resilient microbial networks, although responses remain highly context-specific (Trivedi et al., 2020; Liu et al., 2024).

Recent evidence also suggests that microbial inoculants often function indirectly by transiently modifying resident microbiomes rather than acting as persistent dominant colonisers, for example, by altering resource availability, microbial signalling, or competitive interactions (Mitter et al., 2021; Mendes et al., 2022). This has led to a reconceptualisation of inoculants as tools for steering the microbiome rather than as stand-alone solutions. Collectively, these findings support a paradigm shift toward PGPR-centred microbiome stewardship, which prioritises soil health, crop diversity, and management practices that enhance the recruitment and activity of indigenous beneficial microbes (Compant et al., 2023). Despite this progress, important knowledge gaps remain, including limited causal links between specific PGPR community changes and yield stability, long-term ecosystem services, and the potential ecological risks associated with deliberate microbiome manipulation (Liu et al., 2024).

4.1 Culture-dependent methods

Culture-dependent methods have long underpinned research on PGPR and continue to play an important role in soil and rhizosphere microbiology. Classical approaches typically involve the isolation of microorganisms on selective or semi-selective media, followed by trait-based screening for functions associated with plant growth promotion, such as phosphate solubilisation, indole-3-acetic acid (IAA) production, siderophore release, ACC deaminase activity, and biological nitrogen fixation (Bashan et al., 2014; Santoyo et al., 2021). These assays provide direct, experimentally tractable evidence of functional potential and remain widely used in applied PGPR research.

One key strength of culture-dependent methods is their ability to identify cultivable PGPR members within indigenous microbial communities. Isolated strains can be taxonomically characterised, functionally validated, and tested in planta, thereby establishing clear links between microbial traits and plant responses (Mendes et al., 2022). In addition, classical isolation enables the development of local strain collections that reflect site-specific soil properties, climate, and cropping history. Such collections are valuable resources for developing locally adapted inoculants, which may perform more reliably than non-native commercial strains when reintroduced into their original environments (Compant et al., 2023). Consequently, culture-based PGPR isolation remains common in applied studies, and inoculants frequently modulate the resident microbiome rather than act in isolation (Darriaut et al., 2025).

Despite these advantages, culture-dependent approaches have well-recognised limitations. Only a small fraction of soil and rhizosphere microorganisms can be cultivated under standard laboratory conditions, thereby substantially underestimating microbial diversity and functional breadth (Trivedi et al., 2021). Many ecologically important taxa are slow-growing, obligately symbiotic, or require complex nutrient or signalling conditions that are not replicated in routine media. Furthermore, trait-based screening focuses on individual functions in isolation and fails to capture microbial interactions, functional redundancy, or emergent properties arising within complex community networks (Hartmann and Six, 2023).

As a result, functional performance observed in vitro often fails to translate directly to consistent field outcomes (Mitter et al., 2021). While culture-dependent methods remain indispensable for mechanistic studies and inoculant development, they are increasingly most effective when combined with culture-independent approaches to better contextualise PGPR function within whole microbial communities.

4.2 Culture-independent methods

Culture-independent, DNA- and RNA-based methods have fundamentally transformed the study of PGPR by enabling comprehensive characterisation of microbial communities without the biases inherent to cultivation-based assays. High-throughput sequencing of 16S rRNA gene amplicons remains the most widely used approach for profiling PGPR communities, as it provides detailed taxonomic resolution of bacterial assemblages in bulk soil, rhizosphere and endosphere compartments, and enables researchers to quantify how soil management strategies, environmental gradients, or microbial inoculation influence community composition at fine spatial and temporal scales (Song et al., 2022). While amplicon sequencing offers robust insights into community structure, shotgun metagenomics provides a deeper layer of understanding by simultaneously recovering taxonomic and functional gene information from the entire microbial community, thereby allowing identification of genes associated with key PGPR traits such as nitrogen fixation (*nif* operons), phosphate solubilisation enzymes, siderophore biosynthesis pathways, and phytohormone production (Basu et al., 2024). These functional datasets also permit inference of putative PGPR guilds and the reconstruction of metabolic networks that underpin nutrient mobilisation, biocontrol capacity, and stress mitigation in situ. Complementing these DNA-based tools, metatranscriptomics and metaproteomics reveal the subset of microbial functions that are actively expressed under field conditions, offering a dynamic view of PGPR activity that captures real-time responses to environmental cues, plant developmental stages, and agronomic practices (Mmotla et al., 2025). Despite their power, these approaches remain less commonly applied due to high costs, complex bioinformatic pipelines, and challenges in preserving RNA and proteins from soil samples; nonetheless, emerging methodological advances and decreasing sequencing costs are rapidly expanding their feasibility for routine PGPR research and soil-health monitoring. Together, culture-independent methods provide an integrated toolbox for elucidating the structure, function and ecological dynamics of PGPR communities, enabling more precise and evidence-based strategies for leveraging these microbes in sustainable agriculture.

4.3 Stable isotope probing (SIP)

Stable isotope probing (SIP) provides a powerful, culture-independent approach for identifying active PGPR within complex soil microbiomes by linking microbial identity to specific metabolic processes. SIP works by supplying plants or soils with isotopically labelled substrates – most commonly ^{13}C, ^{15}N or ^{18}O – which are incorporated into microbial DNA, RNA, or lipids when particular microbial taxa metabolise these compounds (Jameson et al., 2023; Basu et al., 2024). In rhizosphere studies, ^{13}C-labelled CO_2 or ^{13}C-glucose introduced through plant photosynthesis or root exudates allows researchers to trace carbon assimilation into actively growing PGPR populations, distinguishing functionally active members from dormant or transient cells (Wang et al., 2021). Similarly, ^{15}N-labelled nitrogen sources can be used to pinpoint diazotrophic PGPR involved in biological nitrogen fixation, providing clear insights into which taxa are responsible for N input under field conditions (Kumar et al., 2020). DNA-SIP has been particularly effective in identifying PGPR guilds associated with carbon turnover, siderophore production, and root-induced nutrient mobilisation. At the same time, RNA-SIP offers higher sensitivity for detecting rapid responses of PGPR to exudation pulses or environmental stress (Wahab et al., 2024). Lipid-SIP is increasingly used to track membrane lipid biomarkers of bacteria and fungi, helping resolve interactions between PGPR and arbuscular mycorrhizal fungi within the rhizosphere network (Maciel-Rodríguez et al., 2025). Although SIP is technically demanding – requiring density-gradient ultracentrifugation, high-quality isotopic substrates, and rigorous controls – it remains one of the most direct approaches for linking PGPR identity to real-time ecological function. As sequencing costs decline and SIP protocols become more accessible, SIP is poised to play an increasingly central role in eco-physiological studies of PGPR and soil nutrient cycling.

4.4 Multi-omics integration for understanding PGPR ecology

Multi-omics approaches integrate complementary datasets – metagenomics, metatranscriptomics, metaproteomics, metabolomics, and soil physicochemical profiles – to provide a holistic understanding of PGPR ecology and function in situ. Metagenomics provides a foundational layer by cataloguing taxonomic diversity and functional gene repertoires within PGPR communities, enabling the identification of genes encoding key traits such as nitrogen fixation, ACC deaminase activity, siderophore biosynthesis, and hormone-regulation pathways (Jain et al., 2024; Song et al., 2022; Zhao et al., 2025). Metatranscriptomics builds on this by revealing which of these genes are actively expressed under specific environmental or management conditions, providing a dynamic picture of PGPR responses to nutrient pulses, moisture fluctuations, or plant developmental stages (Mellidou et al., 2021; Mmotla et al., 2025). Metaproteomics further validates functional activity by quantifying microbial proteins involved in nutrient mobilisation, detoxification, signalling, and biocontrol processes, offering biochemical confirmation of PGPR activity in the rhizosphere (Miller et al., 2023). Meanwhile, metabolomics captures the chemical interactions between plants and microbial communities by measuring root exudates, signalling molecules, and microbial metabolites that shape PGPR recruitment and function (Kellogg and Kang, 2020; Basu et al., 2024). Integrating these datasets into spatially explicit soil models or co-occurrence network analyses enables researchers to identify keystone PGPR taxa, infer trophic interactions, map metabolic pathways, and predict how PGPR functions shift under varying management regimes (Maciel-Rodríguez et al., 2025; X. Zhang et al., 2025). Although multi-omics integration remains technically complex – requiring high computational power, advanced bioinformatics, and careful experimental design – it is rapidly becoming the gold standard for mechanistic PGPR research. By linking microbial identity, gene expression, biochemical activity, and soil environmental context, multi-omics approaches enable unprecedented insights into how PGPR contribute to nutrient cycling, plant health, and soil ecosystem resilience.

5 Management strategies for promoting indigenous plant growth-promoting rhizobacteria communities

5.1 Principles underpinning PGPR-centred soil management

Management strategies that effectively enhance indigenous PGPR communities share several core principles: minimising physical disturbance to preserve microbial habitat continuity; increasing soil organic matter to supply energy and substrates; diversifying plant inputs to stimulate a broader range of microbial guilds; balancing fertiliser use to avoid suppressing microbial nutrient mobilisation pathways; and integrating practices that strengthen beneficial microbial networks, including synergistic interactions between PGPR, AMF, and other rhizosphere organisms (see Fig. 2 and Table 3; Basu et al., 2024; Maciel-Rodríguez et al., 2025). These strategies collectively shift soils towards biologically active, structurally stable, and functionally resilient states – conditions under which PGPR naturally thrive.

5.2 Tillage and soil disturbance

Physical disturbance is a major driver of microbial turnover, as it disrupts soil aggregates, oxygen gradients, microbial hotspots, and hyphal networks. Conventional ploughing reduces microbial biomass, destabilises biofilms, and impairs PGPR populations, particularly those associated with aggregate

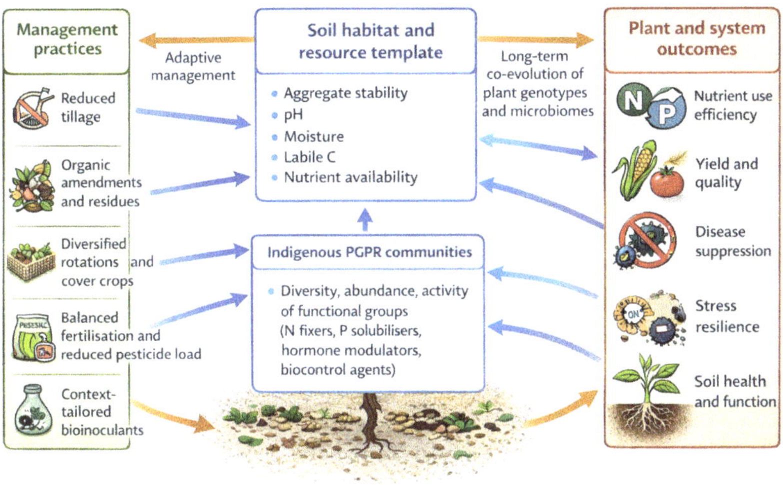

Figure 2 Soil and crop management drivers for promoting indigenous PGPR. On the left, a column of boxes representing management practices: 'Reduced tillage,' 'Organic amendments and residues', 'Diversified rotations and cover crops', 'Balanced fertilisation and reduced pesticide load', and 'Context-tailored bioinoculants'; arrows flow from these columns into a central box labelled 'Soil habitat and resource template' that lists key properties such as aggregate stability, pH, moisture, labile C, and nutrient availability (Maciel-Rodríguez et al., 2025). An arrow leads from a second central box labelled 'Indigenous PGPR communities', which includes sub-labels for diversity, abundance, and activity of functional groups (N fixers, P solubilisers, hormone modulators, biocontrol agents) towards the first central box. From this PGPR box, arrows extend to a right-hand column labelled 'Plant and system outcomes' with items such as 'Nutrient use efficiency', 'Yield and quality', 'Disease suppression', 'Stress resilience', and 'Soil health and function' (Zhang et al., 2025a).

Table 3 PGPR mechanisms, scientific outcomes, and representative evidence

Mechanism/ functional domain	Key PGPR traits and processes	Main scientific outcomes in plants and soil systems	References
Nutrient mobilisation (N, P, K, micronutrients)	Biological N fixation by free-living and associative diazotrophs; solubilisation of mineral and organic P via organic acids and phosphatases; siderophore production for Fe acquisition; mobilisation of K and micronutrients	Increased N and P uptake; higher nutrient-use efficiency; reduced requirement for mineral fertilisers; enhanced root and shoot biomass; improved grain yield and quality in cereals, legumes and horticultural crops	Basu et al. (2024); Kumar et al. (2020); S. Zhang et al. (2025).
Phytohormone modulation and root architecture	Production of IAA, cytokinins and gibberellins; ACC deaminase activity lowering stress ethylene; modulation of ABA and jasmonate signalling; stimulation of lateral roots and root hairs	Enhanced root surface area and soil exploration; improved water and nutrient foraging; better seedling establishment; more balanced shoot–root allocation and improved biomass partitioning under both optimal and stress conditions	Basu et al. (2024); Song et al. (2022); Ferrante et al. (2024)
Stress alleviation (abiotic stress tolerance)	Production of osmoprotectants, exopolysaccharides and antioxidants; induction of ROS-scavenging enzymes; modulation of ion homeostasis under salinity; sequestration or transformation of toxic metals	Improved tolerance to drought, salinity, temperature extremes and heavy metal contamination; maintenance of photosynthetic capacity and membrane integrity; better survival and productivity under climate and pollution stress scenarios	Kumar et al. (2020); Ferrante et al. (2024); X. Zhang et al. (2025).
Biocontrol of soil-borne pathogens and induced systemic resistance (ISR)	Production of antibiotics, lytic enzymes, siderophores and volatile organic compounds; niche and nutrient competition with pathogens; elicitation of systemic resistance via jasmonic acid and ethylene pathways	Reduced incidence and severity of root and foliar diseases; decreased dependence on chemical pesticides; improved plant vigour and yield stability; enhanced resilience of microbiome networks against pathogen invasion	Song et al. (2022); Moore et al. (2022)
Soil structure, aggregation and carbon dynamics	Exopolysaccharide production; contribution to aggregate formation and stabilisation; stimulation of rhizodeposition; synergy with fungi and other microbes in forming micro-aggregates	Improved soil aggregation and porosity; increased water infiltration and retention; enhanced protection of organic carbon in aggregates; positive feedback between soil physical health and PGPR habitat quality	Maciel-Rodríguez et al. (2025); Basu et al. (2024)
Community-level and network effects (microbiome scale)	Interactions with other bacteria, fungi and archaea; quorum sensing and biofilm formation; keystone roles in co-occurrence networks; modulation of diversity and evenness	Greater functional redundancy and resilience of the soil microbiome; stabilisation of nutrient cycling and disease suppression functions; emergent properties such as buffering against environmental fluctuations	Mmotla et al. (2025); Moore et al. (2022); X. Zhang et al. (2025).

surfaces or root-adhered microhabitats (Roesti et al., 2006; S. Zhang et al., 2025). In contrast, reduced tillage and no-till systems enhance spatial stability, promote stronger biofilm formation, and maintain higher abundances of PGPR such as *Pseudomonas*, *Bacillus*, and free-living nitrogen fixers (Basu et al., 2024). Conservation tillage also supports synergistic interactions between PGPR and AMF by maintaining intact hyphal networks and reducing physical disruption of root–microbe interfaces (Maciel-Rodríguez et al., 2025). Over time, these systems develop more robust microbial networks that exhibit greater functional redundancy and higher capacity for nutrient cycling and stress resilience.

5.3 Fertilisation practices and nutrient management

Fertiliser regimes exert strong top-down effects on PGPR communities. Excessive mineral nitrogen – particularly nitrate and ammonium – can suppress nitrogen-fixing PGPR by diminishing plant dependence on biological N acquisition and altering exudation patterns that normally recruit diazotrophs (Kumar et al., 2020; Basu et al., 2024). Similarly, over-application of phosphorus fertiliser reduces the ecological niche for phosphate-solubilising bacteria, lowering their abundance and activity (Song et al., 2022). In contrast, moderate fertilisation, slow-release formulations, and organic nutrient sources tend to create more balanced nutrient environments that rely partly on microbial nutrient mobilisation processes. Integrated nutrient management strategies – combining composts, manure, biofertilisers and reduced mineral inputs – have consistently been shown to enhance PGPR richness and functional trait expression (Maciel-Rodríguez et al., 2025). Such systems also support stronger PGPR–AMF complementarities, improving overall nutrient-use efficiency.

5.4 Organic amendments, compost and manure

Organic amendments provide carbon substrates, micronutrients and structural complexity that stimulate PGPR recruitment and activity. Compost and well-stabilised manure increase labile organic carbon, which fuels microbial respiration and elevates populations of copiotroph PGPR, including *Pseudomonas*, *Bacillus*, and *Enterobacter* (S. Zhang et al., 2025). These amendments also enhance soil aggregation through EPS production, creating persistent microhabitats that protect microbial populations and maintain favourable moisture regimes (Maciel-Rodríguez et al., 2025). Importantly, compost-derived microbial consortia often contain beneficial PGPR and complementary fungal communities, which can further enrich the soil microbiome via inoculation effects. However, amendment quality is critical;

immature composts or high-salinity manures can stimulate undesirable microbes or compromise plant health, underscoring the need for quality control and context-specific application rates (Basu et al., 2024).

5.5 Crop rotations and cover crops

Crop diversification is among the most effective ecological levers for promoting the use of indigenous PGPR. Different plant species release distinct exudate chemistries that select for functionally diverse PGPR guilds, thereby increasing microbial diversity, functional redundancy, and resilience across temporal scales (Song et al., 2022; Maciel-Rodríguez et al., 2025). Cover crops such as legumes, brassicas, and grasses support distinct PGPR groups while increasing root biomass, soil organic matter, and biological nitrogen inputs. Leguminous cover crops enrich nitrogen-fixing PGPR, while brassicas stimulate biocontrol bacteria through glucosinolate breakdown products (S. Zhang et al., 2025). Multispecies cover crop mixtures extend these benefits by providing a continuous and chemically diverse rhizosphere environment, encouraging stable PGPR consortia across seasons. Long-term crop rotation systems have consistently been associated with improved nutrient cycling, enhanced disease suppression, and higher PGPR abundance compared with monoculture systems.

5.6 Biochar and soil conditioners

Biochar amendments can positively influence PGPR communities by enhancing soil structure, increasing habitat heterogeneity and improving moisture retention. The porous architecture of biochar provides microsites that protect PGPR from predation, promote biofilm formation and enhance colonisation of root surfaces (Maciel-Rodríguez et al., 2025). Biochar can also adsorb signalling molecules, root exudates, and microbial metabolites, creating nutrient-rich hotspots that stimulate PGPR-linked processes such as siderophore production and organic acid secretion. However, biochar effects are highly context dependent, varying with feedstock type, pyrolysis temperature, soil texture, and co-application with fertilisers or organic amendments (Basu et al., 2024). When used judiciously, biochar functions as a long-term soil conditioner that fosters stable, beneficial microbial networks.

5.7 PGPR inoculants and their interaction with indigenous communities

Although indigenous PGPR communities often play the dominant ecological role, microbial inoculants – single-strain or consortia – remain important tools for enhancing rhizosphere function. Many commercial inoculants, including *Azospirillum*, *Pseudomonas*, *Bacillus*, and *Rhizobium* strains, have demonstrated variable but often positive effects on root growth, nutrient mobilisation, and stress tolerance (Ferrante et al., 2024; Moore et al., 2022). Field studies increasingly show that inoculants rarely persist long-term but can nonetheless shift indigenous PGPR communities by priming functional guilds, altering exudation patterns, or stimulating beneficial feedback loops (S. Zhang et al., 2025). Successful inoculation strategies account for environmental context, soil pH, organic matter level, and co-applications with carriers or carbon substrates. Consortia-based inoculants generally outperform single strains, as they better mimic the natural functional diversity of indigenous communities (Maciel-Rodríguez et al., 2025). Ultimately, inoculants should be seen not as replacements for indigenous PGPR but as catalysts that enhance the performance of existing microbial networks.

5.8 Integrating management practices into microbiome-aware farming systems

Maximising the use of indigenous PGPR requires coordinated management that simultaneously targets multiple soil functions. Microbiome-aware farming integrates reduced disturbance, diversified plant inputs, organic amendments, balanced fertilisation, and targeted inoculation to create soil environments that sustain resilient PGPR communities (Basu et al., 2024; X. Zhang et al., 2025). Such systems recognise that PGPR interactions with AMF, saprotrophic fungi, and other beneficial microbes form the foundation of soil health and plant productivity. Decision-support tools, including soil microbiome diagnostics and multi-omics assessments, are increasingly enabling farmers to tailor management to site-specific PGPR dynamics. As agriculture transitions towards sustainability and climate resilience, PGPR-centred management will become integral to soil stewardship frameworks.

6 Future innovations and research priorities in plant growth-promoting rhizobacteria ecology and application

6.1 Advancing multi-omics technologies for mechanistic insight

Future research on PGPR will increasingly rely on integrated multi-omics approaches to resolve the mechanistic bases of plant–microbe interactions at high spatial and temporal resolution. While metagenomics has revealed the broad functional potential of PGPR communities, the next frontier lies in the fine-scale integration of metatranscriptomics, metaproteomics, and metabolomics to track real-time gene expression, enzymatic activity, and metabolite exchange between roots and microbial guilds (Mmotla et al., 2025). Spatially resolved omics, including nanoSIMS, laser microdissection-RNAseq, and rhizosphere microfluidics, will enable localisation of active PGPR niches along the root axis and within soil aggregates (Zhang et al., 2022). Such advances will allow researchers to distinguish between dormant, potentially active, and functionally dominant PGPR populations under field conditions, providing mechanistic clarity to inform targeted intervention strategies.

6.2 Predictive modelling and machine learning for PGPR-based decision support

As high-resolution microbiome datasets expand, predictive modelling and machine learning (ML) will play a pivotal role in forecasting PGPR behaviour and plant responses under varying environmental conditions. ML-driven models that integrate soil physicochemical properties, climate variables, crop genotype, and microbiome profiles are already demonstrating promise for predicting bacterial guild shifts and functional outputs such as nitrogen fixation or disease suppression (S. Zhang et al., 2025; Basu et al., 2024). Future tools may allow agronomists to prescribe management actions – such as optimal cover crop mixtures, organic amendment rates, or inoculant formulations – based on predicted PGPR responses, effectively enabling 'precision microbiome management'. Coupling ML with field robotics and sensor networks may further enable real-time microbiome monitoring in commercial agriculture.

6.3 Designing synthetic and semi-synthetic PGPR consortia

Although indigenous PGPR communities provide broad functional coverage, engineered or synthetic microbial consortia offer opportunities to create complementary trait combinations tailored to specific agroecosystems. Advances in microbial ecology, metabolic modelling, and CRISPR-based genome editing are expected to facilitate the design of multi-strain PGPR consortia with optimised nutrient mobilisation, hormone regulation, stress tolerance, and biocontrol traits (Moore et al., 2022; Maciel-Rodríguez et al., 2025). Next-generation consortia will likely rely on strains derived from local soils to maximise ecological compatibility and persistence. These consortia may be formulated with synergistic AMF or beneficial fungi, biochar carriers, or encapsulated delivery systems that improve survival and colonisation. However, the ecological risks and long-term stability of synthetic consortia require rigorous assessment through field-scale experiments and ecological modelling (Mmotla et al., 2025).

6.4 Microbiome-aware plant breeding and root phenotyping

Plant breeding represents one of the most promising, yet underexploited frontiers for enhancing PGPR recruitment and activity. Research increasingly indicates that root traits such as exudation chemistry, rhizosphere pH modulation, root hair density, suberisation patterns, and association with AMF strongly influence PGPR assembly (Tang et al., 2023; Song et al., 2022). Emerging tools such as automated rhizotrons, X-ray CT imaging, and phenomics platforms allow high-throughput characterisation of belowground traits relevant to microbiome interactions. Breeding programmes that integrate PGPR recruitment traits alongside yield, disease resistance, and stress tolerance will yield crop ideotypes capable of maintaining beneficial microbiomes even under stress or low-input conditions (Dwivedi et al., 2025; S. Zhang et al., 2025; X. Zhang et al., 2025). Such strategies may become essential as agriculture transitions towards reduced dependence on fertiliser.

6.5 Field-scale manipulation of PGPR through habitat engineering

Rather than relying solely on inoculants, future PGPR-based management will emphasise habitat engineering – creating soil conditions that naturally support functional microbial networks. Key innovations include precision organic amendments tailored to specific microbial guilds; targeted use of biochar–compost blends to enhance microbial refugia; and adaptive crop rotations based on predicted exudation profiles (Maciel-Rodríguez et al., 2025; Basu et al., 2024). Microbial habitat engineering may also involve controlled irrigation strategies, micro-irrigation of PGPR hotspots, and controlled disturbance regimes that selectively benefit beneficial guilds over pathogens. Such approaches require a deeper understanding of spatial heterogeneity, microsite dynamics, and microbe–microbe interactions within soils.

6.6 Development of rapid microbiome diagnostics for on-farm use

One of the most immediate research priorities is the creation of affordable, farmer-friendly diagnostic tools for monitoring PGPR communities. Current laboratory-based sequencing, enzyme assays, and metabolomics are too costly and slow for widespread on-farm adoption (Tripathi et al., 2024). Promising innovations include qPCR-based PGPR functional marker panels, handheld biosensors for detecting siderophores or phytohormones, microfluidic 'lab-on-chip' devices, and rapid soil health kits that infer microbial activity from enzyme proxies or carbon fluxes (Dadzie et al., 2024). As these technologies mature, diagnostic tools may become embedded into digital decision-support platforms, enabling farmers to monitor PGPR status and adjust management in real time.

6.7 Long-term ecological monitoring and risk assessment

As PGPR-based technologies expand, long-term ecological monitoring is essential to ensure ecological safety, functional stability, and beneficial outcomes. Key areas of concern include potential disruption of indigenous microbial networks, unintended selection of opportunistic pathogens, alteration of nutrient cycling pathways, and horizontal gene transfer among introduced and native microbes (Moore et al., 2022). Networks of long-term field observatories – similar to those developed for soil carbon and climate monitoring – are needed to evaluate the persistence, resilience, and ecosystem-level impacts of PGPR interventions across seasons, soil types, and cropping systems (Maciel-Rodríguez et al., 2025). Integrating ecological modelling with empirical monitoring will help guide regulatory frameworks and best-practice guidelines for safe and effective PGPR deployment.

6.8 Policy, incentives and socio-economic drivers

Sustainable implementation of PGPR-centred management strategies also requires policy support, farmer engagement, and economic incentives. Payment-for-ecosystem-services schemes, nutrient management policies, and soil health programmes can stimulate the adoption of microbiome-friendly practices such as reduced tillage, organic amendments, and cover cropping (Basu et al., 2024). Education and extension programmes that demystify microbiome science and provide practical PGPR management guidelines will be essential, especially for smallholder farmers. Integrating PGPR research with socio-economic analysis will help identify barriers to adoption and ensure that microbiome innovations benefit diverse agricultural communities.

7 Critical discussion

7.1 Context dependence and inconsistent outcomes

Despite the widely acknowledged potential of PGPR to enhance crop productivity and soil health, their performance under field conditions remains highly inconsistent across ecosystems and management regimes. Meta-analyses consistently show that PGPR and other microbial inoculants produce positive average yield gains, yet these averages mask substantial variability among crop species, climate zones, soil types, and fertilisation contexts (Biswas et al., 2024; Li et al., 2024). Some studies report negligible benefits, particularly in high-input or microbially saturated soils, while others find that

PGPR effects emerge primarily under abiotic stress such as drought, salinity, or nutrient limitation (Jan et al., 2021). This variability reflects the multifactorial nature of PGPR–plant–soil interactions, in which soil properties, climate, plant genotype, and the resident microbiome interact in complex, often nonlinear ways to shape PGPR establishment and activity. Competition, predation, and priority effects within the soil microbial community can suppress introduced or stimulated PGPR strains, preventing them from securing ecological niches or expressing key traits (Kong et al., 2025; Liu et al., 2026; Wang and Kuzyakov, 2024). Moreover, plant responses to PGPR are often threshold-dependent and regulated by intricate hormonal networks, meaning that small environmental changes can tip the system towards weak, neutral, or even negative outcomes (Jalmi and Sinha, 2022). As such, 'one-size-fits-all' PGPR strategies are unlikely to succeed; promoting indigenous PGPR communities will require site-specific or at least system-specific approaches that consider local soil ecology, plant genetics and existing microbiome composition.

7.2 Complexity of cause-and-effect relationships

A major conceptual and methodological challenge in PGPR research lies in distinguishing causal relationships from correlative patterns within complex soil–plant–microbe systems. PGPR abundance or diversity often increases under management practices that improve soil health – such as higher organic matter inputs or reduced tillage – but these shifts do not necessarily demonstrate that specific PGPR taxa are responsible for subsequent gains in plant performance. Instead, PGPR enrichment may simply co-occur with broader improvements in soil physicochemical conditions or microbial community structure. Establishing causality requires tools that move beyond correlation-based inference. Multi-omics approaches, including metagenomics, metatranscriptomics, and metabolomics, can identify candidate taxa, regulatory pathways, and potential functional guilds associated with plant benefits, but mechanistic validation requires synthetic microbial communities, mutant strains lacking key traits, and replicated field experiments that separate direct effects from emergent community-level interactions (Jing et al., 2024). Without such mechanistic clarity, interventions risk being based on associative patterns that may not translate across soils, climates or crop species, thereby limiting generalisability and practical utility. Addressing this challenge is critical to developing robust PGPR-based management strategies capable of delivering reliable outcomes across diverse agricultural landscapes.

7.3 Potential ecological risks

Although most studies report limited ecological harm associated with PGPR use, it is essential to recognise potential risks, particularly as microbial inoculants become more widely deployed. Introducing large numbers of inoculant strains or strongly manipulating soil conditions to favour specific PGPR groups may inadvertently alter competitive dynamics in the soil microbiome, suppressing non-target beneficial taxa, or modifying nutrient cycling in unintended ways. Shifts in microbial composition may also influence community stability, with long-term inoculation potentially reducing microbial resilience to perturbations such as drought, pathogens, or chemical inputs (Rubin et al., 2017; Cordero et al., 2023). Horizontal gene transfer represents another theoretical concern, particularly when strains producing antibiotics or harbouring resistance genes are applied; gene transfer among soil bacteria could alter ecological balances or compromise management of antimicrobial resistance (Abdullah Ibn Mafiz et al., 2021). Although current evidence indicates that serious negative ecological impacts are rare, the limited number of long-term and landscape-scale studies means that subtle or cumulative effects may not yet be fully understood (Zhong et al., 2023). As PGPR use scales up in commercial agriculture, comprehensive risk assessments and long-duration monitoring efforts will be crucial to ensure environmental safety and ecological compatibility.

7.4 Practical constraints

From a practical standpoint, several constraints limit the widespread adoption of PGPR-focused management strategies in farming systems. One major barrier is the lack of rapid, affordable diagnostics that can characterise PGPR communities and translate microbial data into actionable recommendations for farmers. Existing sequencing-based methods remain too slow and costly for routine on-farm use, while enzyme assays or soil health indicators provide only indirect or coarse approximations of microbial function. Economic considerations also shape adoption; practices that enhance indigenous PGPR – such as reduced tillage, cover cropping, diversified rotations, and organic amendments – may require new equipment, increased labour, or temporary yield trade-offs during transition periods, which can deter farmers operating under tight margins. Knowledge transfer presents an additional challenge. While many farmers and advisers are familiar with commercial 'biofertiliser products', far fewer are equipped to design or manage microbiome-aware systems that integrate soil physics, plant traits, and microbial ecology. Addressing these constraints will require strengthened research–extension partnerships, co-design of PGPR management strategies with farmers, and clear demonstrations of cost–benefit in real-world contexts. Policies, incentives, and educational programmes that support soil health-based farming will also be essential to overcome adoption barriers and accelerate the transition to microbiome-centred agricultural systems.

8 Research and management priorities

8.1 Towards microbiome-informed agronomy

Advancing PGPR-based agricultural sustainability requires a transition from isolated microbial inoculation trials toward truly microbiome-informed agronomy, in which PGPR ecology is embedded within broader soil health frameworks. Future research must prioritise long-term, replicated field experiments that capture the complexity of real farming environments, integrating microbial community profiling, root exudation ('exudome') analyses, and ecosystem-level measurements of nutrient flows and plant performance under realistic management scenarios (see Fig. 3; Molefe et al., 2023). Such long-term studies are essential because PGPR communities, exudation patterns,

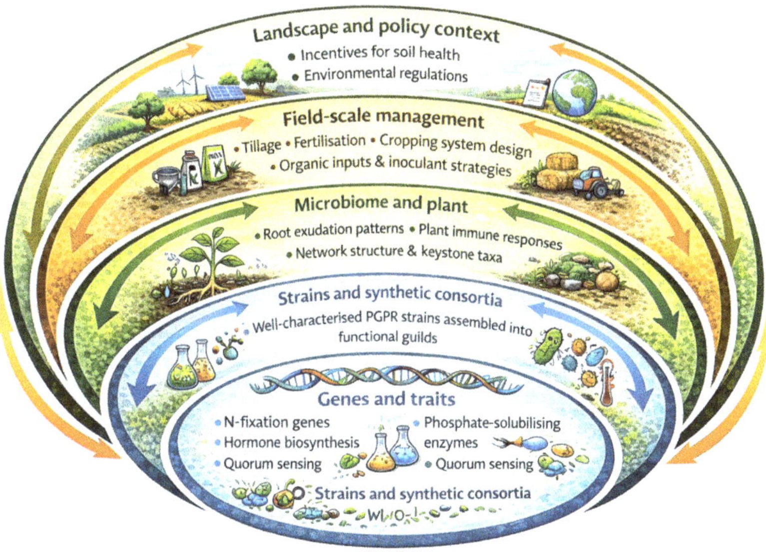

Figure 3 Multi-scale framework for future PGPR research and application (from genes to landscapes: a hierarchical framework for PGPR-based soil management). Figure was arranged as nested layers or concentric rings: the innermost layer is "Genes and traits" (N-fixation genes, phosphate-solubilising enzymes, hormone biosynthesis, quorum sensing, and stress-response determinants), surrounded by "Strains and synthetic consortia" (well-characterised PGPR strains assembled into functional guilds); the next ring is "Microbiome and plant" (root exudation patterns, plant immune responses, network structure, and keystone taxa); the next ring is "Field-scale management" (tillage, fertilisation, cropping system design, organic inputs, and inoculant strategies); the outermost ring is "Landscape and policy context" (incentives for soil health, environmental regulations, climate resilience targets) (Mmotla et al., 2025; Moore et al., 2022; Maciel-Rodríguez et al., 2025). Arrows linking rings highlight bidirectional interactions – for example, policy influences management choices, which feed back on microbiomes and selection for PGPR traits, while discoveries at the gene/trait level inform breeding and inoculant design.

and soil properties co-evolve over seasons, and short-term trials frequently underestimate soil biological inertia and lag effects. Equally important is the development of mechanistic and predictive models that link management practices – such as fertilisation regimes, cover crop mixtures, tillage intensity, and organic amendments – to shifts in PGPR traits and downstream effects on plant growth, while accounting for climate variability, soil heterogeneity and crop genotype interactions (Moukanni et al., 2022). To strengthen causal inference, experimental systems employing synthetic microbial communities, gene-knockout mutant strains, and controlled manipulations of root chemistry will be needed to identify essential PGPR functions and the 'interaction rules' that govern microbial assembly, cooperation, and competition in the rhizosphere (Correa de Souza et al., 2020). Collectively, such approaches will enable a more predictive, quantitative, and context-sensitive integration of PGPR ecology into agronomic decision-making.

8.2 Co-designing plants and PGPR communities

A major opportunity for the future lies in the co-design of plant genotypes and PGPR communities, an approach that aligns plant breeding with microbiome engineering to create more resilient crop–soil systems. Breeding programmes increasingly recognise that belowground traits – including root architecture, exudation chemistry, mucilage composition, rhizosphere pH regulation, and suberisation patterns – play a decisive role in determining which PGPR taxa colonise roots and how effectively they express beneficial functions (Wen et al., 2022). Selecting for genotypes that consistently recruit beneficial PGPR across contrasting soils and climates could reduce dependence on synthetic fertilisers and external inoculants, especially in low-input or stress-prone environments where microbial functions become more critical. Likewise, breeding for traits that enhance compatibility with PGPR – such as stable exudate signatures, increased lateral rooting, or synergistic interactions with arbuscular mycorrhizal fungi – may yield more robust outcomes than strategies that prioritise plant traits in isolation. Integrating microbiome-supportive agronomy – such as diversified rotations, biochar–organic amendment blends, or reduced disturbance – into plant breeding pipelines will create systems in which plants and microbes are optimised together, rather than treated as independent components. This co-design paradigm represents a transformative step toward resilient, biology-centred agricultural systems.

8.3 Decision-support tools and functional indicators

Translating PGPR ecology into on-farm management requires diagnostic tools and indicators that are both scientifically robust and practically accessible. A critical research priority is the development of simplified indicators of PGPR functional potential, including enzymatic assays for nutrient mobilisation (e.g. phosphatases, dehydrogenases), qPCR markers targeting functional genes such as *nifH* (nitrogen fixation), *acdS* (ACC deaminase), or siderophore biosynthesis pathways, and rapid profiling of key taxonomic groups known to correlate with specific soil functions (Renoud et al., 2022). These indicators must be validated against field performance and made affordable for routine agronomic monitoring. Decision-support tools should integrate soil test results, crop histories, weather data, management practices, and microbial indicators to generate context-specific recommendations – for example, identifying which cover crops best promote indigenous PGPR, which residue management strategies maximise PGPR stability, or when an inoculant might provide added value. Such tools would help farmers move beyond generic prescriptions to tailored, microbiome-aware management. As machine learning and multi-omics datasets mature, these decision-support platforms could evolve into predictive systems capable of forecasting PGPR community responses under changing climatic and agronomic conditions (see Table 4).

8.4 Policy and incentive frameworks

Because the benefits of PGPR-friendly management – improved soil structure, enhanced carbon sequestration, reduced fertiliser losses, increased resilience to climate stress – extend beyond individual farms to broader ecosystems and communities, policy frameworks will play a critical role in accelerating adoption. Payments for ecosystem services that reward improvements in soil health, incentives for reduced tillage or diverse cover cropping, subsidies for high-quality composts and organic amendments, and policy support for on-farm microbiome diagnostics could collectively create an enabling environment for widespread PGPR-centred soil management (Mitchell and Bellows, 2020). Extension services and farmer networks should be supported to facilitate knowledge exchange, participatory trials and co-designed management innovations. Policies promoting integrated nutrient management – rather than dependence on chemical fertilisers – would further reinforce the ecological niche for indigenous PGPR. At larger scales, PGPR-informed management aligns with national goals for climate adaptation, nutrient efficiency and biodiversity protection. As microbiome science advances, regulatory frameworks will need to evolve to evaluate PGPR-based interventions, ensure ecological safety and support innovation without imposing unnecessary barriers. Effective policy integration will be essential to embed PGPR into mainstream agricultural systems and achieve long-term sustainability outcomes.

Table 4 Futuristic perspectives and research priorities for promoting indigenous PGPR communities

Future area/concept	Strategic aim	Anticipated benefits for crops and soil systems	Key research and implementation needs
Synthetic yet locally tuned PGPR consortia	Develop functionally complementary consortia constructed from well-characterised indigenous strains adapted to local soils and climates	More reliable and robust field performance than single-strain inoculants; improved multi-trait support (nutrients, hormones, stress, biocontrol) within a single product; reduced risk of ecological disruption by using native genotypes	Deeper understanding of interaction rules within consortia; long-term field trials quantifying stability and impact on resident microbiomes; regulatory frameworks that accommodate strain mixtures; cost-effective production and formulation technologies (Basu et al., 2024; Moore et al., 2022).
Microbiome-aware plant breeding and ideotypes	Select crop genotypes with root traits and exudation profiles that reliably recruit beneficial PGPR and support diverse microbiomes under realistic management	Co-adapted plant–microbiome systems with higher nutrient-use efficiency and resilience; reduced fertiliser and pesticide demand; improved performance in low-input and stress-prone environments	Integration of microbiome metrics into breeding pipelines; high-throughput phenotyping of root traits and exudates; understanding genotype × microbiome × environment interactions; participatory breeding with farmers to match ideotypes to management (Tang et al., 2023; S. Zhang et al., 2025).
Precision microbiome management at field scale	Use data-driven decision tools to match management levers (tillage, fertilisation, cover crops, organic amendments, inoculants) to target PGPR functions in specific fields and seasons	Increased probability of positive PGPR responses; efficient use of inputs; site-specific strategies for soil health enhancement; reduced risk of failure associated with generic recommendations	Development of simple indicators or biomarkers of PGPR status; coupling soil testing with microbiome profiles; modelling tools linking management scenarios to PGPR outcomes; integration with digital agriculture platforms (Maciel-Rodríguez et al., 2025; X. Zhang et al., 2025).

Multi-omics diagnostics and rapid monitoring tools	Translate metagenomics, metatranscriptomics, metabolomics, and soil enzyme assays into practical diagnostics for PGPR activity and function	Early detection of beneficial or degraded microbiome states; capacity to monitor responses to management or inoculation; evidence-based adjustment of practices over time	Standardised sampling and bioinformatic pipelines; identification of key functional markers associated with agronomic outcomes; portable and low-cost assays (e.g. qPCR panels, biosensors) suitable for advisory services (Mmotla et al., 2025; Basu et al., 2024).
Long-term field observatories and restoration programmes	Establish networks of long-term experiments and restored sites to track how PGPR communities respond to sustained changes in management, climate and land use	Understanding of temporal dynamics, legacy effects, and resilience thresholds; evidence for policy and incentive schemes; guidance for scaling PGPR-friendly practices in degraded or marginal lands	Co-ordinated multi-site studies across pedoclimatic gradients; harmonised microbiome and soil-health measurements; integration of socio-economic data; collaboration between researchers, extension services, and land managers (Maciel-Rodríguez et al., 2025).
Risk assessment, governance and policy instruments	Develop frameworks to evaluate ecological risks (e.g. community disruption, resistance gene transfer) and create policies that reward PGPR-supportive management	Safe deployment of inoculants and microbiome engineering; public and farmer trust; alignment of PGPR use with environmental regulations and climate-mitigation goals	Long-term ecological monitoring of inoculant use; assessment of non-target effects; incorporation of microbiome indicators into soil-health certification and payment-for-ecosystem-services schemes; clear guidelines for strain selection and release (Moore et al., 2022; Maciel-Rodríguez et al., 2025).

9 Conclusion

Indigenous PGPR form a foundational component of soil ecosystem functioning, contributing directly to nutrient acquisition, hormonal signalling, pathogen suppression, and stress resilience. Yet, for decades, agricultural research and industry have disproportionately focused on individual microbial inoculant strains rather than on the complex, self-organising microbial communities that naturally inhabit the rhizosphere. Recent advances in microbiome science now make it unequivocally clear that plant–microbe interactions operate as dynamic, context-dependent partnerships shaped by a confluence of root exudation chemistry, soil physicochemical properties, and local management regimes (Korenblum et al., 2022). PGPR are not isolated agents of plant growth but members of an intricate multi-kingdom network whose collective functions underpin soil health and ecosystem resilience.

Importantly, field studies demonstrate that management practices profoundly affect PGPR abundance, diversity, and trait expression. Reduced disturbance systems preserve microbial habitats; organic amendments and residue retention increase carbon availability; crop diversification broadens functional niches; and judicious fertiliser and pesticide use reduce physiological and ecological disruption to beneficial communities (Topa et al., 2025). These practices do not simply increase PGPR numbers; they also modulate microbial network structure, enhance functional redundancy, and stabilise rhizosphere processes that support long-term plant productivity. In this sense, management practices act as ecological levers that shape the selective environment in which indigenous PGPR operate.

Meanwhile, research increasingly shows that microbial inoculants rarely function as stand-alone solutions. Their effects often arise not from persistent colonisation, but from transient shifts they induce in the structure or activity of resident microbial communities (Li et al., 2024). This insight reframes inoculants as potential ecological 'steering agents' rather than replacements for indigenous microbiota, reinforcing the need to promote native PGPR guilds as a central objective in microbiome-aware agroecosystem design. By nurturing the existing microbiome rather than attempting to overwrite it, agronomic strategies can harness the adaptive capacity and ecological stability inherent to indigenous communities.

Yet caution is warranted. PGPR outcomes remain highly variable across soils, climates, and crop genotypes, reflecting the deep context-dependence of root–microbiome interactions. Mechanistic understanding remains incomplete, particularly regarding multi-trophic interactions, priority effects, and the non-linear plant hormonal responses that mediate growth promotion. There is a real danger that PGPR could be oversold as a 'green silver bullet' without adequate consideration of ecological constraints, long-term effects, or economic feasibility for farmers. Overenthusiastic application of microbial products, without clear mechanistic grounding or site-specific tailoring, risks undermining both scientific credibility and farmer trust.

For PGPR-centred strategies to become mainstream and effective, a conceptual shift is required. Rather than treating soils as inert substrates into which microbial products are added, we must recognise them as living, evolving microbiome habitats that self-organise in response to plant inputs and agronomic management. This perspective emphasises systems thinking, designing agronomy that supports the resilience, diversity and functional expression of beneficial indigenous microbial communities, while using inoculants only as complementary tools when contextually justified. Such an approach aligns with broader transitions toward soil health, regenerative agriculture and climate-resilient food systems.

Ultimately, the future of PGPR-based sustainability lies in harmonising scientific innovation with ecological realism, building on indigenous microbial potential, guided by rigorous mechanistic evidence, and implemented through management practices that maintain the integrity and adaptive capacity of soil ecosystems. By foregrounding indigenous PGPR communities within agroecosystem design, we can move towards more productive, resilient, and environmentally sustainable cropping systems capable of meeting global food needs in a changing world.

10 Author contributions

SSS and ZMS contributed to the original draft writing, revision, and preparation for publication.

11 Where to look for further information

- Pandey, A., et al. (eds.) (2025) *Plant-microbe Interactions for Environmental and Agricultural Sustainability*. Springer (e-Book) p. 740.
- Iqbal, M. Z., et al. (2024) Recent advances of plant growth promoting rhizobacteria (PGPR) for eco-restoration of polluted soil. *Cleaner Eng Technol* 23, 100845.

12 Funding

No funding has been secured for this work.

13 Acknowledgements

We thank the editor and anonymous reviewers for their critical reading of this manuscript and for their constructive feedback.

14 Declaration of competing interest

The authors declare that they have no competing financial interests.

15 Declaration of use of OpenAI tools

OpenAI tools were used in this work to generate illustrative figures and to support paraphrasing and refining textual content. All outputs were reviewed and edited by the authors to ensure accuracy, originality, and compliance with academic standards.

16 References

Abdullah Ibn Mafiz, A. I., et al. (2021) Soil bacteria in urban community gardens have the potential to disseminate antimicrobial resistance through horizontal gene transfer. *Front Microbiol* 12, 771707.

Abou Jaoudé, R., et al. (2025) Effect of plant growth-promoting rhizobacteria synthetic consortium on growth, yield, and metabolic profile of lettuce (*Lactuca sativa* L.) grown under suboptimal nutrient regime. *Horticulturae* 11(1), 64. https://doi.org/10.3390/horticulturae11010064.

Ahemad, M., and Kibret, M. (2013) Mechanisms and applications of plant growth-promoting rhizobacteria: current perspective. *J King Saud Univ Sci* 26(1), 1–20.

Ahmad, H. M., et al. (2022) Plant growth-promoting rhizobacteria eliminate the effect of drought stress in plants: a review. *Front Plant Sci* 13, 875774.

Backer, R., et al. (2018) Plant growth-promoting rhizobacteria: Context, mechanisms of action, and roadmap to commercialization of biostimulants for sustainable agriculture. *Front Plant Sci* 9, 1473. https://doi.org/10.3389/fpls.2018.01473.

Bashan, Y., et al. (2014) Advances in plant growth-promoting bacterial inoculant technology. *Plant Soil* 378, 1–33.

Basu, A., et al. (2024) Recent advances of plant growth promoting rhizobacteria (PGPR) towards sustainable agriculture. *J Agric Food Res* 23, 100845.

Biswas, D., et al. (2024) Plant Growth Promoting Rhizobacteria (PGPR): Reports on their colonization, beneficial activities, and use as bioinoculant. *Adv Agric* 20, 8173024. https://doi.org/10.1155/2024/8173024.

Compant, S., et al. (2023) A review on the plant microbiome: ecology, functions, and emerging trends. *Trends Plant Sci* 28, 123–138.

Cordero, I., et al. (2023) High intensity perturbations induce an abrupt shift in soil microbial state. *The ISME Journal* 17, 2190–2199. https://doi.org/10.1038/s41396-023-01512-y.

Correa de Souza, R. S., et al. (2020) From microbiome to traits: designing synthetic microbial communities for improved crop resiliency. *Front Plant Sci* 11, 1179.

Dadzie, F. A. et al. (2024) Inoculating native microorganisms improved soil function and altered the microbial composition of a degraded soil. *Restor Ecol* 32, e14025.

Darriaut, R., et al. (2025) Rhizobacteria from vineyard and commercial arbuscular mycorrhizal fungi induce synergistic microbiome shifts within grapevine root systems. *Sci Rep* 15, 27884.

Dobrzyński, J., Z., et al. (2025) Non-native PGPB consortium consisting of *Pseudomonas* sp. G31 and *Azotobacter* sp. PBC2 promoted winter wheat growth and slightly altered the native bacterial community. *Sci Rep* 15, 3248. https://doi.org/10.1038/s41598-025-86820-3.

Duret, M., et al. (2024) Rhizosphere microbiome assembly, drivers and functions in perennial ligneous plant heath. *Microbiol Res* 287, 127860.

Dwivedi, S.L. et al. (2025) Exploitation of rhizosphere microbiome biodiversity in plant breeding. *Trends Plant Sci* 30, 1033–1045.

Ferrante, R., et al. (2024) Meta-analysis of plant growth-promoting rhizobacteria under drought stress. *Front Plant Sci* 14, 1282553.

Hartmann, M., and Six, J. (2023) Soil structure and microbial functions in agroecosystems. *Soil Biol Biochem* 176, 108892.

Hemelda, N. M., and Noutoshi, Y. (2025) Root-exuded sugars as drivers of rhizosphere microbiome assembly. *Plant Biotechnol* 42, 215–227.

Iqbal M. Z., et al. (2024). Chandra, recent advances of plant growth promoting rhizobacteria (PGPR) for eco-restoration of polluted soil. *Cleaner Engineer Technol* 23, 100845. https://doi.org/10.1016/j.clet.2024.100845.

Jain, A., et al. (2024) Omics approaches in understanding the benefits of plant–microbe interactions. *Front Microbiol* 15, 1391059. https://doi.org/10.3389/fmicb.2024.1391059.

Jalmi, S. K., and Sinha, A. K. (2022) Ambiguities of PGPR-induced plant signaling and stress management. *Front Microbiol* 13, 899563. https://doi.org/10.3389/fmicb.2022.899563.

Jameson, E., et al. (2023) DNA-, RNA-, and protein-based stable-isotope probing for high-throughput biomarker analysis of active microorganisms. In: Streit, W.R., and Daniel, R. (eds.), *Metagenomics: Methods in Molecular Biology*, vol. 2555. Humana, New York, NY. https://doi.org/10.1007/978-1-0716-2795-2_17.

Jan, B., et al. (2021) Plant Growth Promoting Rhizobacteria (PGPR): eco-friendly approach for sustainable agriculture. In: *Plant-Microbes Dynamics* (1st edition, Edited by Pirjadah et al.). CRC Press, p. 16.

Jing, J., et al. (2024) Strategies for tailoring functional microbial synthetic communities, *The ISME Journal*, 18, wrae049. https://doi.org/10.1093/ismejo/wrae049.

Joshi, N., et al. (2025) The art of amalgamation: advancing crop productivity with PGPR consortia solutions. *Curr Agri Res* 13(1). https://doi.org/10.12944/CARJ.13.1.02.

Kellogg, J., and Kang, S. (2020) Metabolomics, an essential tool in exploring and harnessing microbial chemical ecology. *Phytobiomes J* 4, 195–210.

Khoso, M. A., et al. (2024) Impact of plant growth-promoting rhizobacteria (PGPR) on plant nutrition and root characteristics: current perspective. *Plant Stress* 11, 100341.

Kong, Z., et al. (2025) Priority effects of inoculation timing of plant growth-promoting microbial inoculants: role, mechanisms and perspectives. *Plant Soil* 513, 1675–1687. https://doi.org/10.1007/s11104-025-07291-z.

Korenblum, E., et al. (2022) Plant–microbe interactions in the rhizosphere via a circular metabolic economy. *The Plant Cell* 34(9), 3168–3182.

Kumar, A., et al. (2020) Plant growth-promoting bacteria: biological tools for the mitigation of salinity stress in plants. *Front Microbiol* 11, 1216.

Kuzyakov, Y., and Razavi, B. S. (2024) Rhizosphere microbiome and root exudates: Mechanisms and implications for soil health. *New Phytol* 241, 12–29.

Li, C., Chen, X., Jia, Z., et al. (2024) Meta-analysis reveals the effects of microbial inoculants on the biomass and diversity of soil microbial communities. *Nat Ecol Evol* 8, 1270–1284. https://doi.org/10.1038/s41559-024-02437-1.

Liu, C., et al. (2026) Predation by soil protists shifts bacterial metabolism from competitive to cooperative interactions. *Cell Host Microbe* 34, 201–211.e6.

Liu, H., et al. (2024) Microbiome-mediated plant stress tolerance in agriculture. *Trends Microbiol* 32, 1–15.

Maciel-Rodríguez, M., et al. (2025) The role of plant growth-promoting bacteria in soil restoration. A strategy to promote agricultural sustainability. *Microorganisms* 13, 1799.

Mellidou, I. M., et al. (2021) Comparative transcriptomics and metabolomics reveal an intricate priming mechanism involved in PGPR-mediated salt tolerance in Tomato. *Front Plant Sci* 12: 713984.

Mendes, R., et al. (2022) Harnessing the soil microbiome for sustainable crop production. *Science*, 376, 139–144.

Miller, S. E., et al. (2023) Metaproteomics reveals functional partitioning and vegetational variation among permafrost-affected Arctic soil bacterial communities. *mSystems8*, e01238-22. https://doi.org/10.1128/msystems.01238-22.

Mitchell, C., and Bellows, B. (2020) *Payments for Ecosystem Services. ATTRA/Sustainable Agriculture Research and Education (SARE).* Texas Institute for Applied Environmental Research.

Mitter, E. K., et al. (2021) A meta-analysis of microbial inoculants in agriculture. *Soil Biol Biochem* 162, 108402.

Mmotla, K., et al. (2025) Exploring the intricacies of plant growth promoting rhizobacteria interactions: an omics review. *Ann Microbiol* 75, 5.

Molefe, R. R., et al. (2023) Communication between plant roots and the soil microbiome; involvement in plant growth and development. *Symbiosis* 90, 231–239.

Moore, J. A. M., et al. (2022) Ecosystem consequences of introducing plant growth promoting rhizobacteria to managed systems and potential legacy effects. *New Phytol* 234, 1070–1087.

Moukanni, N., et al. (2022) Optimizing carbon sequestration through cover cropping in mediterranean agroecosystems: synthesis of mechanisms and implications for management. *Front Agron* 4, 844166.

Renoud, S., et al. (2022) Effect of inoculation level on the Impact of the PGPR *Azospirillum lipoferum* CRT1 on selected microbial functional groups in the rhizosphere of field maize. *Microorganisms* 10, 325.

Roesti, D., et al. (2006) Plant growth stage, fertiliser management and bio-inoculation of arbuscular mycorrhizal fungi and plant growth promoting rhizobacteria affect the rhizobacterial community structure in rain-fed wheat fields. *Soil Biol Biochem* 38, 1111–1120.

Rubin, R. L., et al. (2017) Plant growth promoting rhizobacteria are more effective under drought: a meta-analysis. *Plant Soil* 416, 309–323.

Santoyo, G., et al. (2021) Plant growth-promoting bacterial endophytes. *Microbiol Res* 242, 126608.

Savastano, N., and Bais, H. (2024) Synergism or antagonism: do arbuscular mycorrhizal fungi and plant growth-promoting rhizobacteria work together to benefit plants? *Int J Plant Biol*, *15*, 944-958.

Song, Q., et al. (2022) Plant growth-promoting rhizobacteria promote growth of seedlings, regulate soil microbial community, and alleviate damping-off disease caused by *Rhizoctonia solani* on *Pinus sylvestris* var. *mongolica*. *Plant Dis* 106, 2730–2740. https://doi.org/10.1094/PDIS-11-21-2562-RE.

Tang, L., et al. (2023) Effects of plant-growth-promoting rhizobacteria on soil bacterial community, soil physicochemical properties, and soil enzyme activities in the rhizosphere of alfalfa under field conditions. *Diversity* 15, 537. https://doi.org/10.3390/d15040537.

Topa, D.-C., et al. (2025) Sustainable practices for enhancing soil health and crop quality in modern agriculture: a review. *Agriculture* 15, 998.

Tripathi, A., et al. (2024) An updated review on significance of PGPR-induced plant signalling and stress management in advancing sustainable agriculture. *J Agric Food Res* 16, 101169.

Trivedi, P., et al. (2020) Plant–microbiome interactions: from community assembly to plant health. *Nat Rev Microbiol* 20, 607–621. https://doi.org/10.1038/s41579-020-0412-1.

Trivedi, P., et al. (2021) Enabling sustainable agriculture through understanding and enhancement of microbiomes. *New Phytol* 230, 2129–2147. https://doi.org/10.1111/nph.17319.

Vurukonda, S. S. K. P., et al. (2023) Plant growth-promoting rhizobacteria: current perspectives and future challenges. *Front Microbiol* 14, 1198452.

Wahab, A., et al. (2024) Plant growth-promoting rhizobacteria biochemical pathways and their environmental impact: a review of sustainable farming practices. *Plant Growth Regula* 104, 637–662. https://doi.org/10.1007/s10725-024-01218-x.

Wang, C., and Kuzyakov, Y. (2024) Mechanisms and implications of bacterial–fungal competition for soil resources. *The ISME J* 18(1), wrae073. https://doi.org/10.1093/ismejo/wrae073.

Wang, H., et al. (2021) Fungi dominated the incorporation of ^{13}C-CO_2 into microbial biomass in tomato rhizosphere soil under different CO_2 concentrations. *Microorganisms* 9, 2121. https://doi.org/10.3390/microorganisms9102121.

Wen, Z., et al. (2022) Linking root exudation to belowground economic traits for resource acquisition. *New Phytol* 233, 1620–1635.

Yan, W., et al. (2025) Mechanisms and regulation of interactions between AM fungi and hyphosphere bacteria in organic phosphorus mineralization. *Bull Bot Res* 45, 345–351.

Yang, P., et al. (2024) Utilising plant growth-promoting rhizobacteria (PGPR) to advance sustainable agriculture. *Bacteria* 3(34), 434–451.

Yu S, et al. (2025) Growth-promoting effects and mechanisms of synthetic plant growth-promoting rhizobacteria on maize seedlings. *Microorganisms* 13, 2460. https://doi.org/10.3390/microorganisms13112460.

Zhang, X., et al. (2022) Robust acquisition of spatial transcriptional programs in tissues with immunofluorescence-guided laser capture microdissection. *Front Cell Dev Biol* 10, 853188. https://doi.org/10.3389/fcell.2022.853188.

Zhang, S., et al. (2025a) Broad-spectrum applications of plant growth-promoting rhizobacteria across diverse crops and intricate planting systems. *Microbiol Spectr* 13, e01879-24.

Zhang, X., et al. (2025) A new strategy for sustainable agricultural development: Meta-analysis of the efficient interaction of plant growth-promoting rhizobacteria with nanoparticles. *Plant Physiol Biochem* 223, 109845.
Zhao, S-Y., et al. (2025) Rhizosphere microbiome metagenomics in PGPR-mediated alleviation of combined stress from polypropylene microplastics and Cd in hybrid Pennisetum. *Front Microbiol* 16, 1549043.
Zhong, J., et al. (2023) Long-term effects of ecological restoration projects on ecosystem services and their spatial interactions. *Environ Manage* 73, 493–508.

Chapter 11

Optimising functionality of phosphate-solubilising microorganisms in promoting crop nutrient use efficiency

Nelly S. Raymond and Frederik J. T. van der Bom, University of Copenhagen, Denmark

1 Introduction

Phosphorus (P) is a crucial element for life on Earth. Although most soils contain ample P, low P availability often limits ecosystem productivity (Cakmak, 2002). In agriculture, P is therefore applied mostly as mineral fertilisers to increase plant availability. This P comes from phosphate rock, a non-renewable resource distributed in a non-uniform manner across the planet (MacDonald et al., 2011; Geissler et al., 2018). Despite the recent discovery of large phosphate rock deposits such as in Norway, the resource remains finite, and new uses for P, such as in batteries, could lead to additional competition and demand, threatening long-term food security.

A significant amount of agricultural soils have built up large P reserves due to the historical over-application of mineral P fertilisers, exceeding the amounts needed for optimal crop growth. In many parts of Europe, high soil P levels are also attributed to the use of animal manure or other organic fertilisers, which

http://dx.doi.org/10.19103/AS.2025.0166.14

were applied to meet the nitrogen (N) requirements of crops. The imbalance between N and P in these fertilisers often resulted in excessive P application (Sharpley et al., 2001). However, these accumulated soil P reserves are generally not readily available to plants, particularly in soils with high sorption capacities. Consequently, there is a growing interest in finding solutions to better utilise this 'legacy P' for crop nutrition (Rowe et al., 2016).

Readily available P in soil ranges from 1.5% to 11% with the majority of P being adsorbed on minerals (adsorbed on iron (Fe) (e.g. hematite, goethite), aluminium (Al) (hydr)oxides (e.g. gibbsite), calcium carbonates ($CaCO_3$), and clay minerals), precipitated (i.e. with Al, Fe, calcium, etc.), or organically bound (Menezes-Blackburn et al., 2018) (Fig. 1). On average, plants utilise only about 20% of the added P. The remaining P rapidly interacts with soil surfaces and becomes unavailable, building legacy P.

Overall, it is imperative that we alter our future P management strategies, focusing on alternative pathways and means to mobilise legacy P to deliver P to plants.

Soil microorganisms are the key driver of soil P cycling, governing short- (days-seasons) and long-term (years-centuries) changes in available and non-available P pools. Microbial P processing includes (a) storing P within their biomass, (b) mineralising non-plant available organic P, and (c) solubilising inorganic P forms. Phosphate-solubilising microorganisms (PSM) are microorganisms that can 'solubilise' otherwise non-plant available forms

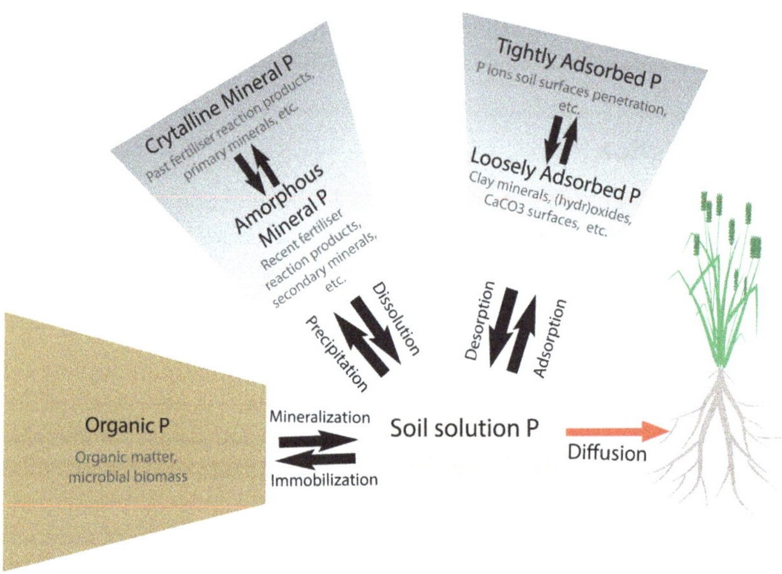

Figure 1 Phosphorus conceptual pools and equilibriums in soil (adapted from Moody et al. (2013)).

of soil P. In this chapter, we define PSM as microorganisms mobilising P from inorganic P forms. This means that PSM may desorb P from soil mineral surfaces or dissolve phosphate minerals. Mineralisation of organic P is not considered, as enzymatic processes are distinct. Moreover, the mechanisms of solubilisation of inorganic P and the mineralisation of organic P belong to two distinct P pools, which are consistently differentiated within the P cycle. Arbuscular mycorrhizal fungi from a PSM perspective are also excluded as they can, to some degree, be considered to function as an 'extension' of the root system, such that they can improve plant P access through greater soil exploration (Jakobsen et al., 2005).

Due to the ability of PSM to solubilise P from inorganic origin (i.e. often legacy P), PSM have been proposed as a mean to enhance P uptake by plants, particularly from legacy P or other poorly soluble P sources. The enrichment of PSM population in soil, via inoculation, has attracted significant interest over the years from both applied and basic science perspectives. This interest is driven by the push towards sustainable agriculture, aiming to reduce reliance on chemical-based synthetic P fertilisers and make better use of rock-phosphate products or recycled P resources like biochar. While many reviews (Khan et al., 2009; Sharma et al., 2013; Mącik et al., 2020) are optimistic, more recent research presents a mixed view (Raymond et al., 2021; O'Callaghan et al., 2022; Barrow, 2024). Despite their promise, PSM are not widely adopted by farmers (Khan et al., 2009; Owen et al., 2015).

Since Gerretsen's pioneering study in 1948 on the potential of inoculating plants with microorganisms to enhance P nutrition, numerous PSM, including both bacteria and fungi, have been tested across various environments and conditions. Despite this extensive research, there remains a lack of scientific consensus on the effectiveness of PSM for crop production. There is a notable gap between the positive effects of PSM on crop growth reported in controlled studies and their limited use in practical farming (Raymond et al., 2021). Field experiments often fail to replicate the positive results seen in controlled conditions, yet few studies have investigated the conditions that lead to inconsistent outcomes (Meyer et al., 2017; Gómez-Muñoz et al., 2018; Meyer et al., 2019). This chapter includes: (a) a short review on the ecological role of PSM; (b) an evaluation of the performance of PSM and the potential barriers to their success in soil-plant environments; (c) guidelines for robust testing of PSM inoculants; and (d) a view of the ecological function of PSM for P cycling as a component of the wider microbial community.

2 The ecological role of phosphate-solubilising microorganisms

Phosphate-solubilising microorganisms, including bacteria, fungi, and yeast, are typically found in soil and form a crucial part of the microbial community.

These microorganisms are distinguished by their ability to solubilise sparingly soluble phosphate compounds to satisfy their needs. This capacity actually constitutes a significant portion of the total microbial community (Kucey, 1983), but some microorganisms exhibit superior performance in *in vitro* laboratory assays, and these are conventionally termed PSM. These microorganisms are crucial within the microbial community as they can supply P that can be assimilated and stored in the microbial biomass, where it fulfils an essential role for various cellular functions, including energy production, gene expression, DNA and RNA synthesis, protein function, and maintaining cell wall integrity.

At the ecosystem level, PSM are essential for introducing 'new P' into the soil over time. In natural ecosystems, P is cycled through soil microorganisms before being released into the plant-available P pool (Tamburini et al., 2012). This process is particularly important during soil development, where PSM play a critical role in making P available from minerals (Turner et al., 2013) (Fig. 2).

The mechanisms by which PSM solubilise P are well-documented *in vitro* (Oburger et al., 2011; O'Callaghan et al., 2022). These processes involve the dissolution of inorganic soil P minerals, such as calcium (Ca), iron (Fe), and aluminium (Al) phosphates, as well as the desorption of orthophosphate anions bound to Al and Fe oxides through (a) proton secretion and (b) the production of organic anions (e.g. citrate, oxalate, gluconate). Secreted protons acidify the environment around the PSM, leading to the dissolution of inorganic P minerals, primarily Ca-P minerals. Released organic anions can either complex metal cations and dissolve mineral and precipitated forms of P (including Fe, Al, and Ca complexes) through ligand-promoted mechanisms, or they can directly desorb orthophosphate P from soil minerals (e.g. clays and silicates) and Al- and Fe-oxides via ligand exchange (Jones & Edwards, 1998; Sharma et al., 2013). However, the acidification of the soil environment through proton secretion is considered the primary mechanism behind phosphate solubilisation (Barrow & Lambers, 2022).

In the context of inoculating PSM to enhance crop P acquisition, an ideal scenario involves PSM solubilising non-plant available P forms to meet plant demands. In this case, PSM release more P than they need for their own metabolism, directly supplying the plant with the necessary P for growth. Much research has focused on reinforcing this function of PSM in cultivated ecosystems where P is a limiting nutrient, by increasing the number of microorganisms with superior P solubilisation capabilities. By inoculating PSM into the soil or onto plant seeds, it is hoped that PSM establish themselves in soil niches or the rhizosphere, from where they will provide a continuous supply of P to the plant in synchronisation with its needs.

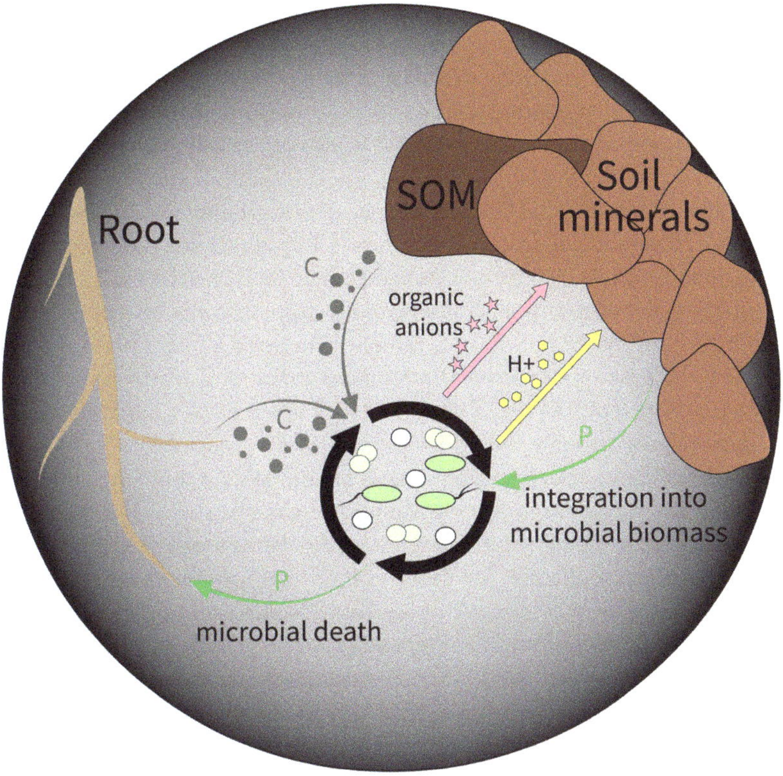

Figure 2 Overview of the conceptual pathway for phosphate-solubilising microorganism(s) (PSM). PSM solubilise P to sustain their own need and serve the system over the long term by releasing P from their biomass turnover. A carbon-rich environment (e.g. soil organic matter (SOM), root exudates) will facilitate microbial biomass turnover that is likely to be even more evident in P-limited environments (i.e. stronger driver for the PSM effect). In a P-rich environment, soil microorganisms will use available P, whereas in a P-limited environment it will drive P-solubilisation activity within the microbial biomass. Modified from Raymond et al. (2021).

3 The need of robust testing when evaluating phosphate-solubilising microorganism inoculants in soils

The development of PSM inoculants often begins with the screening of candidate microorganisms, typically isolated from soils. During laboratory *in vitro* assays, candidate PSMs are evaluated for their ability to solubilise unavailable forms of P in a nutrient medium containing an excess of nutrients except P (e.g. NBRIP – National Botanical Research Institute's phosphate growth medium). Often, the pH of the medium and organic anions production are assessed to elucidate the mechanisms of P solubilisation (Efthymiou et al.,

2018; Zutter et al., 2022). Once microorganisms with superior P solubilisation abilities are identified, they are tested under controlled conditions. These experiments involve growing crops in a controlled environment (temperature, light, humidity) with managed soil conditions (moisture, nutrients, sterilisation). Typically, soil with low plant-available P is used, with or without added fertiliser, which can be a soluble source of P (e.g. mineral fertiliser such as triple super phosphate) or a poorly available source (e.g. phosphate rock, biochar, ashes, compost). PSM inoculation can be performed by direct soil inoculation, by P source application, or by the plant via seed inoculation or seedling root dipping into a microbial suspension. Subsequently, plant growth (biomass) and/or shoot P content are monitored. If a satisfying increase in plant performance is observed, the assay is performed under 'field conditions', which refers to trials conducted in the field where environmental factors are not controlled, and soil and water management is limited to common farmer practices.

The basis for conducting plant-based experiments is that inoculated PSM will establish themselves in the rhizosphere or specific soil niches. Here, they are expected to exhibit P-solubilisation traits similar to those observed *in vitro*, releasing orthophosphate P for roots to then take up. However, confirming that the inoculated microorganisms express P-solubilising traits necessitates a precise experimental design in plant-soil systems. Firstly, for plants to show a productivity response to added P, P must be the primary limiting growth factor, according to the Sprengel-Liebig Law of the Minimum (van der Ploeg et al., 1999). All other potentially limiting factors must be controlled, including adequate levels of other essential nutrients, water availability, light, and optimal temperature for plant growth, as well as the removal of physical (e.g. pH, soil compaction) and biological constraints (e.g. pests and diseases). While such conditions can be achieved in controlled environments like glasshouse pot trials, these factors can vary significantly in field conditions, where P is often not the most limiting factor for plant growth.

When soil P availability is the primary limiting factor for plant growth, a response should be observed following inoculation with PSM. This response can occur either without additional P (indicating improved access to native soil P) or at intermediate levels of applied P (indicating enhanced efficiency in acquiring both applied and native P), which influences the 'P response curve' (Fig. 3). As P limitation is alleviated, the plant's response should diminish until its P demand is fully met. Therefore, PSM-induced increases in P availability through solubilisation are expected to show effects only within the 'responsive' range of the response curve (Fig. 3). If PSMs continue to enhance P uptake when P is no longer limiting, other mechanisms, such as phytohormone production or alleviation of other growth constraints, must be considered. The mechanisms behind improved plant productivity can be identified by using

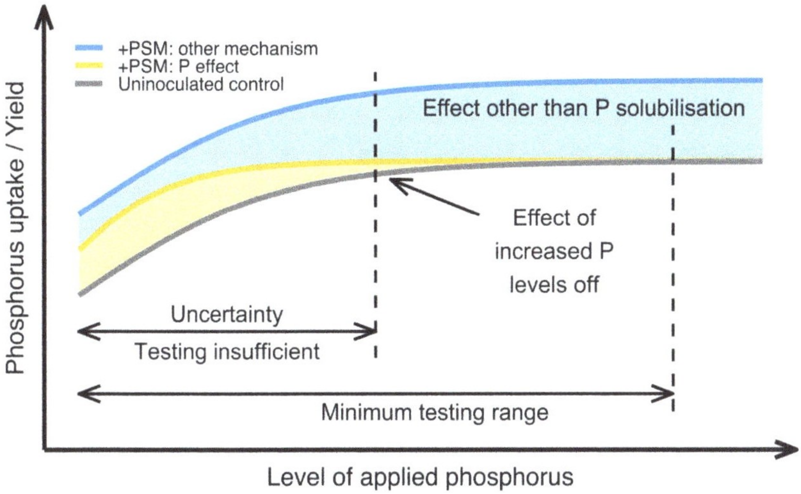

Figure 3 Crop productivity (P uptake, yield or biomass) response curve to estimate the success or the failure of P solubilisation by PSM inoculant from Raymond et al. (2021).

various P fertiliser levels or conducting mechanistic studies with radioisotopes (Meyer et al., 2017; Gómez-Muñoz et al., 2018; Sarabia et al., 2018).

It is essential that experimental designs are rigorous enough to conclusively determine that P solubilisation is the mechanism responsible for increased crop productivity. Confounding factors and the failure to eliminate observed effects when P is non-limiting can lead to incorrect conclusions about P solubilisation. Therefore, experiments must be carefully designed to distinguish whether enhanced crop productivity is due to P solubilisation or other mechanisms. Barrow (2024) and Raymond et al. (2021) emphasised that increased production at a single phosphate level does not necessarily indicate P solubilisation. To confirm that P solubilisation has occurred, it is necessary to measure phosphate response curves both in the presence and absence of such treatments. Some important points to consider in experimental design when evaluating the effect of PSM on crop productivity are:

- P must be the most limiting factor for plant growth;
- A full phosphate response curve must be included;
- Plant root biomass and architecture may be evaluated to test potential root hormonal effects; and
- PSM persistence in soil – Are they there, are they alive, are they active?

Including these experimental factors allows the researcher to isolate the P nutritional effect (if present), as well as rule out and deduce PSM activities.

4 Phosphate-solubilising microorganism inoculants: from controlled conditions to field performances

One of the primary challenges with PSM inoculation is translating the effects observed *in vitro* to *in vivo* conditions. While numerous studies have reported positive effects of PSM inoculation on crop productivity, and several reviews have extensively covered this topic (Kucey et al., 1989; Gyaneshwar et al., 1998; Richardson, 2001; Khan et al., 2009; Kishore et al., 2015; Rafi et al., 2019), the results remain inconsistent. Numerous studies have shown that PSM inoculation often does not significantly impact plant productivity and P uptake, both in controlled environments and in field conditions (Androsoff et al., 1991; Fernández et al., 2007; Karamanos et al., 2010; Lekfeldt et al., 2016; Meyer et al., 2017; Raymond et al., 2018; Raymond et al., 2019a; Raymond et al., 2019b; Raymond et al., 2021). The variability in crop response to PSM inoculation is well illustrated by a 7-year study conducted by Karamanos et al. (2010), which evaluated wheat inoculated with *Penicillium bilaiae* across 26 locations in North America. In this study, significant yield increases were observed in only 5 out of 47 trials. Other studies have similarly reported inconsistent effects of PSM under field conditions, with no clear correlation between increased plant biomass and P mobilisation by PSM (Bullock et al., 1990; Whitelaw et al., 1997; Krey et al., 2013; Leggett et al., 2015). Overall, it is evident that most positive responses to PSM inoculation are observed in controlled condition studies, while 'no effect' or even 'negative effects' are more common in field conditions. This highlights the need to better understand how the effects observed in controlled environments can be consistently replicated in the field, emphasising the importance of identifying factors that contribute to these inconsistencies.

For PSM to express their traits and mediate P availability, they must successfully colonise and persist in soil and on plant roots. PSMs are often assumed to be capable of colonising and thriving in a wide range of environmental conditions and soil types. However, their ability to occupy and function within specific ecological niches, such as the rhizosphere, is crucial (Wakelin et al., 2007). Poor performance of PSMs in enhancing crop productivity and P uptake is often attributed to inadequate soil colonisation, primarily due to competition with other microorganisms (Richardson, 2001; Meyer et al., 2017; Raymond et al., 2019a). Competition with native soil microorganisms can inhibit inoculated microorganisms from expressing P solubilisation traits. The inoculated microorganisms may compete for resources necessary for basic metabolic activities to colonise soil, as well as for more complex processes, such as P solubilisation. This topic is often speculative due to the lack of definitive studies on the colonisation of PSM. Even when PSMs can effectively colonise various soil environments, certain soil conditions may be more conducive to their persistence and P-solubilisation activity (Wakelin et al., 2007). The ecology of inoculated PSM is a crucial factor in optimising the effectiveness of PSM inoculation.

Understanding the colonisation and persistence of PSM in soil, particularly in the rhizosphere, requires deeper insights into the environmental conditions that favour or suppress P-solubilisation activity. Additionally, it is crucial to determine whether the P-solubilisation potential of these microorganisms is fully realised in agro-ecological environments. Within this, some potential key factors that may prevent or enhance PSM effectiveness in soil conditions are listed. These will often be unique to the local soil environment, may occur simultaneously, and can interact with each other and/or the broader soil environment:

- **Soil pH.** PSMs have been extensively studied in alkaline and calcareous soils (Wakelin et al., 2007; Meyer et al., 2017) and are often found to be more effective in alkaline soils compared to neutral or acidic soils, where acidification is a key mechanism for P solubilisation (Barrow, 2024). However, identifying the true effect of soil pH is challenging because the soils used in such studies also differ in other properties, such as texture and nutrient availability. Additionally, the optimal pH for the growth and metabolic activity of different PSM strains can vary, affecting their ability to solubilise P in a strain-dependent manner.
- **Soil P status.** The availability of soil P, including its accessibility and the forms in which it is present (such as complexation with cations and occlusion within soil minerals), may significantly influence the performance of PSM. It is generally believed that low soil P availability favours the expression of P solubilisation traits in PSM, enabling them to meet their P needs. However, some studies suggest that a minimum P threshold is necessary to elicit a response in plants (Sánchez-Esteva et al., 2016; Gómez-Muñoz et al., 2018). These findings are further supported by a meta-analysis from Schütz et al. (2018), which indicated that PSM induced greater yield increases under conditions of moderate P availability in soil, but were less effective in more P-deficient environments.
- **Rhizosphere conditions.** The rhizosphere is a critical environment for PSM activity and plant P uptake (Jakobsen et al., 2005). The rhizosphere has a higher microbial abundance and metabolic activity compared to bulk soil (Sørensen et al., 2001; Jakobsen et al., 2005), due to root exudates that contain easily metabolisable hexose sugars, organic acids, and amino acids (Richardson et al., 2021). However, the rhizosphere is also significantly depleted of available orthophosphate in solution, leading to intense competition for P between plants and microorganisms (Jakobsen et al., 2005). Additionally, these beneficial conditions may not always be present in the rhizosphere due to anoxic conditions or competition

with other microorganisms, where exudates, signalling compounds, or defensive secondary metabolites can influence community dynamics (Werra et al., 2009; Compant et al., 2019).

- **Plant hosts.** Plants recruit specific microbial communities from the soil through root exudates which differ in composition between plant species (Richardson et al., 2021). Consequently, it becomes important to consider whether an inoculant strain is suited to colonise a given plant species (Compant et al., 2019). Currently, our understanding of the performance of PSM across different host species is relatively limited (Khan et al., 2009).

Another important consideration when comparing *in vitro* laboratory conditions to field performance is the potential discrepancy introduced by *in vitro* laboratory assays. In laboratory settings, PSM can solubilise over 90% of the total P added to the culture system within a relatively short period (Efthymiou et al., 2018). Such amounts of orthophosphate released *in vitro* by PSM are substantial and exceed their own metabolic requirements. However, soil conditions are often carbon-limited, have poor nutrient availability, and are highly buffered against pH change, which means that the ability of microorganisms to solubilise P does not translate directly between lab assays and soil conditions. Microorganisms that display greater P solubilisation potential *in vitro* do not always show an advantage in plant-soil system (Zutter et al., 2022).

Contrary to laboratory cultures, C is typically a limiting factor for microbial activity in soils (Soong et al., 2020). This limitation of C in soil can restrict the production of organic anions by PSM due to the high-energy cost associated with their production. Further, PSM may prioritise energy for other activities, such as defence against other microorganisms (Jones et al., 2003; Werra et al., 2009; Meyer et al., 2019). In addition, organic anions produced by PSM in soil can also serve as an energy source for other soil microorganisms and have high turnover rates (typically with a half-life of 1–3 hours). This prevents their accumulation and reduces their efficacy in mobilising P (Jones et al., 2003). The capacity of different soils to adsorb organic anions is also significant, as interaction with the soil solid phase can lower their effective concentration in soil solution (Jones & Edwards, 1998). Lastly, while it is widely assumed that organic anions are specifically released by PSM to mobilise P, they may serve a multitude of other purposes in microbial habitats, such as recruitment or defence interactions, with P solubilisation as an indirect outcome.

Apart from C limitation, the chemistry of soil also differs significantly from that of growth media, particularly the pH buffering capacity, which is typically much greater in soil. The comparatively large buffering capacity in soil can significantly reduce the effect of acidification, a mechanism for P solubilisation. Specifically, inoculated soil microorganisms are unlikely to produce enough acid to significantly alter the pH beyond their immediate niches, such as soil micropores. Furthermore, if P solubilisation does occur, the orthophosphate made available must be immediately utilised by plant roots or microorganisms in the vicinity, otherwise it will quickly undergo re-adsorption or precipitation in the soil.

Overall, we suggest that the mechanisms of organic anion secretion and acidification by PSM, which are considered important in laboratory cultures, are likely to be significantly less effective in soil environments. While possible excess of solubilised P under certain specific conditions cannot be ruled out, current data do not demonstrate consistent performance, particularly at the field scale. Therefore, the relevance of laboratory screening methods and the selection of PSM candidates, as well as the significance of the proposed solubilisation mechanisms in soil, can be questioned. These methods do not appear to adequately reflect the functioning of PSM in soil environments.

5 Promoting the functional role of phosphate-solubilising microorganisms in agricultural systems

Optimising the functional role of PSM to enhance agricultural production and reduce the need for P fertiliser inputs remains a significant challenge. We here give some research targets that may promote the functional role of PSM in future agricultural systems:

- **Farming practices:** Practices that promote the proportional representation of PSM and their function within complex microbial communities may be one of the next research targets. Increasing C and possibly N availability in soil systems can enhance microbial biomass, turnover, and associated microbial P mobilisation (Simpson et al., 2011; Spohn & Kuzyakov, 2013). It was, for example, shown that adding C in the form of glucose can boost the abundance of specific taxonomic groups capable of solubilising and making P available (Jing et al., 2017; Meyer et al., 2019). Organic inputs with high labile C have been shown to promote P cycling through faster microbial turnover (Oehl et al., 2004; Oberson & Joner, 2005; Bünemann et al., 2007). These findings suggest that readily metabolisable C supports efficient P acquisition and turnover through microbial biomass. A recent study by Widdig et al. (2019) indicated that adding N to soils decreased the abundance of phosphate-solubilising organisms while increasing phosphatase activities, suggesting that microbial P mobilisation processes may shift towards the mineralisation of organic P forms. Managing the content and availability of soil C and N through the use of C-rich or N-rich residue inputs may provide strategies to influence microbial P processing functions.

- **Microbial consortia:** Most commercial inoculants consist of a single species or strain of microorganisms. However, O'Callaghan et al. (2022) reviewed that there is now a general consensus that using a single microorganism formulation across diverse environmental conditions may not be optimal. Soil matrices, particularly the rhizosphere, are complex and competitive environments. A consortium of various microorganisms may enhance their establishment and persistence in the soil. The development of multi-strain consortia with redundant capabilities is becoming increasingly common, with many microbial inoculant products now containing two or more genera or species (O'Callaghan et al., 2022), even for PSM. However, the success of inoculation with microbial consortia remains variable and is mostly evaluated under controlled conditions. For example, in artificial plant growth media, specific phosphate-solubilising consortia have been shown to increase maize P uptake (Zutter et al., 2022), and the combination of arbuscular mycorrhizal fungi and phosphate-solubilising bacteria in intercropping systems has shown improved crop performances (Liu et al., 2024). The association of *Penicillium bilaiae* and *Bacillus simplex* increased P uptake in low-P soil under controlled conditions (Hansen et al., 2020). Despite the recent increase in studies involving microbial consortia, most are conducted under controlled conditions and/or in artificial/sterile media. The inoculation of multiple microbial strains and species may represent a promising strategy that may enhance the success of PSM inoculation, but convincing results under field conditions are still necessary.
- **Enhancing phosphate-solubilising microbial communities through artificial selection:** Artificial selection methods, which involve host-mediated artificial selection to engineer plant microbiomes for beneficial traits (Faller et al., 2024), remain at an early stage. Currently, these methods have only been tested *in vitro*. For instance, Faller et al. (2024) explored two artificial selection methods – environmental perturbation and propagation – in a hydroponic system to construct phosphate-solubilising microbial communities.
- **4R strategy:** The application of strategies such as the 4R strategy – using the Right product for a specific purpose, applied at the Right rate to be effective, at the Right time when it is most beneficial, and at the Right location rather than the bulk soil – as suggested by McGuire (2022), may improve inoculation success at the field scale. However, many aspects of the 4R strategy remain poorly defined for PSM inoculants.

6 Conclusion

In natural ecosystems, microorganisms are the primary drivers of soil P cycling, both in the short term (e.g. over a growing season) and long term, influencing changes in soil P pools. Promoting this role in agricultural systems to support crop production could reduce reliance on non-renewable mineral P inputs and mitigate associated negative environmental impacts. However, inoculating single strains of microorganisms has not proven to be a reliable strategy under field conditions, possibly due to a lack of rigorous testing in strain selection and an incomplete understanding of their interactions. Recent advances suggest that inoculating multiple strains may increase the success of these efforts. However, the inoculation of microorganisms to soil environment remains a challenging process, and simple solutions have not seemed to be effective and reliable. There are several research avenues that need to be explored such as soil management practices that could help select for beneficial microbial functions. The functional redundancy of soil microorganisms means that the necessary functions are likely already present in the soil. The environment, shaped by management practices, determines which microorganisms will thrive.

7 Where to look for further information

For readers interested in exploring this topic in greater depth, a comprehensive overview of PSM can be found in Raymond et al. (2018), where we review the current understanding, mechanisms, and practical considerations surrounding PSM in agricultural systems. Evaluating the true efficacy of PSM inoculants can be challenging, particularly because many conventional methods fail to capture the complexity of P dynamics in soil. For more precise and mechanistic assessments, we recommend consulting studies that employ isotopic techniques, which allow a clearer quantification of phosphorus fluxes and transformations. In particular, gross P flux estimation: Meyer et al. (2017). These approaches provide valuable tools for disentangling the contributions of microbial activity, fertiliser behaviour, and soil processes, and they offer a more robust framework for testing PSM performance under realistic conditions.

8 References

Androsoff, G., et al. 1991. Yield of PB50 inoculated and phosphorus fertilized wheat. In: *Soils and Crops Workshop Proceedings*.

Barrow, N.J. 2024. Phosphate solubilizing microorganisms; the modern philosopher's stone. *Plant and Soil*, 508, 21–28. https://doi.org/10.1007/s11104-024-06785-6.

Barrow, N.J. & Lambers, H. 2022. Phosphate-solubilising microorganisms mainly increase plant phosphate uptake by effects of pH on root physiology. *Plant and Soil*, 476, 397–402.

Bullock, P., et al. 1990. Penicillium bilaji (PB50) and phosphorus fertilizer responses of yield of wheat and barley grown on stubble and summerfallow. In: *Soils and Crops Workshop Proceedings*.

Bünemann, E.K., et al. 2007. Measuring rates of gross and net mineralisation of organic phosphorus in soils. *Soil Biology and Biochemistry*, 39, 900–913.

Cakmak, I. 2002. Plant nutrition research: Priorities to meet human needs for food in sustainable ways. *Plant and Soil*, 247, 3–24.

Compant, S., et al. 2019. A review on the plant microbiome: ecology, functions, and emerging trends in microbial application. *Journal of Advanced Research*, 19, 29–37.

Efthymiou, A., et al. 2018. Augmentation of the phosphorus fertilizer value of biochar by inoculation of wheat with selected Penicillium strains. *Soil Biology and Biochemistry*, 116, 139–147.

Faller, L., et al. 2024. Enhancing phosphate-solubilising microbial communities through artificial selection. *Nature Communications*, 15, 1649.

Fernández, L.A., et al. 2007. Phosphate-solubilization activity of bacterial strains in soil and their effect on soybean growth under greenhouse conditions. *Biology and Fertility of Soils*, 43, 805–809.

Geissler, B., et al. 2018. Clearing the fog on phosphate rock data – Uncertainties, fuzziness, and misunderstandings. *Science of The Total Environment*, 642, 250–263.

Gerretsen, F.C. 1948. The influence of microorganisms on the phosphate intake by the plant. *Plant and Soil*, 1, 51–81.

Gómez-Muñoz, B., et al. 2018. Effects of Penicillium bilaii on maize growth are mediated by available phosphorus. *Plant Soil*, 431, 159–173.

Gyaneshwar, P., et al. 1998. Effect of buffering on the phosphate-solubilizing ability of microorganisms. *World Journal of Microbiology and Biotechnology*, 14, 669–673.

Hansen, V., et al 2020. Seed inoculation with Penicillium bilaiae and Bacillus simplex affects the nutrient status of winter wheat. *Biology and Fertility of Soils*, 56, 97–109.

Jakobsen, I., et al. 2005. Rhizosphere Microorganisms and Plant Phosphorus Uptake. *Phosphorus: Agriculture and the Environment*, 437–494.

Jing, Z., et al. 2017. Response and feedback of C mineralization to P availability driven by soil microorganisms. *Soil Biology and Biochemistry*, 105, 111–120.

Jones, D., et al. 2003. Organic acid behavior in soils–misconceptions and knowledge gaps. *Plant and Soil*, 248, 31–41.

Jones, D.L. & Edwards, A.C. 1998. Influence of sorption on the biological utilization of two simple carbon substrates. *Soil Biology and Biochemistry*, 30, 1895–1902.

Karamanos, R.E., et al. 2010. Re-visiting use of Penicillium bilaii with phosphorus fertilization of hard red spring wheat. *Canadian Journal of Plant Science*, 90, 265–277.

Khan, M.S., et al. 2009. Role of Phosphate Solubilizing Microorganisms in Sustainable Agriculture-A Review. In: *Sustainable Agriculture*, pp. 551–570. Springer, New Delhi.

Kishore, N., et al. 2015. Phosphate-Solubilizing Microorganisms: A Critical Review. In: *Plant Biology and Biotechnology: Volume I: Plant Diversity, Organization, Function and Improvement* (eds. Bahadur, B., Venkat Rajam, M., Sahijram, L. & Krishnamurthy, K.V.), pp. 307–333. Springer India, New Delhi.

Krey, T., et al 2013. Effects of long-term phosphorus application and plant-growth promoting rhizobacteria on maize phosphorus nutrition under field conditions. *European Journal of Soil Biology*, 55, 124–130.

Kucey, R.M.N. 1983. Phosphate-solubilizing bacteria and fungi in various cultivated and virgin Alberta soils. *Canadian Journal of Soil Science*, 63, 671–678.

Kucey, R.M.N., et al. 1989. Microbially Mediated Increases in Plant-Available Phosphorus. In: *Advances in Agronomy* (ed. Brady, N.C.), pp. 199–228. Academic Press, London.

Leggett, M., et al. 2015. Maize yield response to a phosphorus-solubilizing microbial inoculant in field trials. *Plant and Soil*, 153, 1464–1478.

Lekfeldt, J.D.S., et al. 2016. Effect of bioeffectors and recycled P-fertiliser products on the growth of spring wheat. *Chemical and Biological Technologies in Agriculture*, 3, 22.

Liu, Y., et al. 2024. Phosphorus solubilizing bacteria rather than arbuscular mycorrhizal fungi drive maize/faba bean intercropping advantages. *Plant and Soil,* 511, 753–766. https://doi.org/10.1007/s11104-024-07018-6.

MacDonald, G.K., et al. 2011. Agronomic phosphorus imbalances across the world's croplands. *Proceedings of the National Academy of Sciences*, 108, 3086–3091.

Mącik, M., et al. 2020. Chapter Two - Biofertilizers in agriculture: An overview on concepts, strategies and effects on soil microorganisms. In: *Advances in Agronomy* (ed. Sparks, D.L.), pp. 31–87. Academic Press, London.

McGuire, A. 2022. A Problem with Soil Inoculants In: *Sustainable Practices and Technology* (ed. Sustainability, P.o.). Washington State University, Center for Sustaining Agriculture and Natural Resources, Washington.

Menezes-Blackburn, D., et al. 2018. Opportunities for mobilizing recalcitrant phosphorus from agricultural soils: a review. *Plant and Soil*, 427, 5–16.

Meyer, G., et al. 2017. Gross phosphorus fluxes in a calcareous soil inoculated with Pseudomonas protegens CHA0 revealed by 33P isotopic dilution. *Soil Biology Biochemistry*, 104, 81–94.

Meyer, G., et al. 2019. Pseudomonas protegens CHA0 does not increase phosphorus uptake from 33P labeled synthetic hydroxyapatite by wheat grown on calcareous soil. *Soil Biology & Biochemistry*, 131, 217–228.

Moody, P.W., et al. 2013. Soil phosphorus tests I: What soil phosphorus pools and processes do they measure? *Crop Pasture Science*, 64, 461–468.

O'Callaghan, M., et al. 2022. Soil microbial inoculants for sustainable agriculture: Limitations and opportunities. *Biology and Fertility of Soils*, 38, 1340–1369.

Oberson, A. & Joner, E. 2005. Microbial turnover of phosphorus in soil. *Phosphorus in Plant Nutrition: Organic and Inorganic Sources*, 133–164.

Oburger, E., et al. 2011. Phosphorus saturation and pH differentially regulate the efficiency of organic acid anion-mediated P solubilization mechanisms in soil. *Plant and Soil*, 341, 363–382.

Oehl, F., et al. 2004. Basal organic phosphorus mineralization in soils under different farming systems. *Soil Biology and Biochemistry*, 36, 667–675.

Owen, D., et al. 2015. Use of commercial bio-inoculants to increase agricultural production through improved phosphrous acquisition. *Applied Soil Ecology*, 86, 41–54.

Rafi, M.M., et al. 2019. Chapter 17 - Phosphate-Solubilizing Microorganisms and Their Emerging Role in Sustainable Agriculture. In: *Recent Developments in Applied Microbiology and Biochemistry* (ed. Buddolla, V.), pp. 223–233. Academic Press, London.

Raymond, N.S., et al. 2021. Phosphate-solubilising microorganisms for improved crop productivity: a critical assessment. *Plant and Soil*, 229, 1268–1277.

Raymond, N.S., et al. 2018. Enhancing the phosphorus bioavailability of thermally converted sewage sludge by phosphate-solubilising fungi. *Ecological Engineering*, 120, 44–53.

Raymond, N.S., et al. 2019a. Fertilising effect of sewage sludge ash inoculated with the phosphate-solubilising fungus Penicillium bilaiae under semi-field conditions. *Biology and Fertility of Soils*, 55, 43–51.

Raymond, N.S., et al. 2019b. Biotic strategies to increase plant availability of sewage sludge ash phosphorus. *Waste Management*, 182, 175–186.
Richardson, A.E. 2001. Prospects for using soil microorganisms to improve the acquisition of phosphorus by plants. *Functional Plant Biology*, 28, 897–906.
Richardson, A.E., et al. 2021. Root Microbiome Structure and Microbial Succession in the Rhizosphere. In: *Rhizosphere Biology: Interactions Between Microbes and Plants* (eds. Gupta, V.V.S.R. & Sharma, A.K.), pp. 109–128. Springer Singapore, Singapore.
Rowe, H., et al. 2016. Integrating legacy soil phosphorus into sustainable nutrient management strategies for future food, bioenergy and water security. *Nutrient Cycling in Agroecosystems*, 104, 393–412.
Sánchez-Esteva, S., et al. 2016. The effect of Penicillium bilaii on wheat growth and phosphorus uptake as affected by soil pH, soil P and application of sewage sludge. *Chemical and Biological Technologies in Agriculture*, 3, 21.
Sarabia, M., et al. 2018. Rhizosphere yeasts improve P uptake of a maize arbuscular mycorrhizal association. *Applied Soil Ecology*, 125, 18–25.
Schütz, L., et al. 2018. Improving crop yield and nutrient use efficiency via biofertilization—a global meta-analysis. *Frontiers in Plant Science*, 8, Article 2204.
Sharma, S.B., et al. 2013. Phosphate solubilizing microbes: sustainable approach for managing phosphorus deficiency in agricultural soils. *Springer Plus*, 2, 587.
Sharpley, A.N., et al. 2001. Innovative management of agricultural phosphorus to protect soil and water resources. *Communications in Soil Science and Plant Analysis*, 32, 1071–1100.
Simpson, R.J., et al. 2011. Strategies and agronomic interventions to improve the phosphorus-use efficiency of farming systems. *Plant and Soil*, 349, 89–120.
Soong, J.L., et al. 2020. Microbial carbon limitation: the need for integrating microorganisms into our understanding of ecosystem carbon cycling. *Global Change Biology*, 26, 1953–1961.
Sørensen, J., et al. 2001. Soil and rhizosphere as habitats for Pseudomonas inoculants: new knowledge on distribution, activity and physiological state derived from micro-scale and single-cell studies. *Plant and Soil*, 232, 97–108.
Spohn, M. & Kuzyakov, Y. 2013. Phosphorus mineralization can be driven by microbial need for carbon. *Soil Biology and Biochemistry*, 61, 69–75.
Tamburini, F., et al. 2012. Oxygen isotopes unravel the role of microorganisms in phosphate cycling in soils. *Environmental Science & Technology*, 46, 5956–5962.
Turner, B.L., et al. 2013. Soil microbial biomass and the fate of phosphorus during long-term ecosystem development. *Plant and Soil*, 367, 225–234.
van der Ploeg, R.R., et al. 1999. On the origin of the theory of mineral nutrition of plants and the law of the minimum. *Soil Science Society of America Journal*, 63, 1055–1062.
Wakelin, S.A., et al. 2007. The effect of Penicillium fungi on plant growth and phosphorus mobilization in neutral to alkaline soils from Southern Australia. *Biology and Fertility of Soils*, 53, 106–115.
Werra, P.D., et al. 2009. Role of gluconic acid production in the regulation of biocontrol traits of pseudomonas fluorescens CHA0. *Applied and Environmental Microbiology*, 75, 4162–4174.
Whitelaw, M.A., et al. 1997. Plant growth promotion of wheat inoculated with Penicillium radicum. *Soil Research*, 35, 291–300.

Widdig, M., et al. 2019. Nitrogen and phosphorus additions alter the abundance of phosphorus-solubilizing bacteria and phosphatase activity in grassland soils. *Frontiers in Environmental Science*, 7, 552.

Zutter, N.D., et al. 2022. Innovative rhizosphere-based enrichment under P-limitation selects for bacterial isolates with high-performance P-solubilizing traits. *Microbiology Spectrum*, 10, e02052–02022.

Chapter 12

Understanding and optimizing the role of cyanobacteria in promoting crop health and nutrient use efficiency

Noureddine Bouaïcha, University of Paris-Saclay, France; and Mohammed Loudiki, Cadi Ayyad University, Morocco

1 Introduction

The large-scale use of chemical agricultural fertilizers, while they can have significant short-term impacts on crop yields, contributes to long-term greenhouse gas emissions (Chen et al., 2019; Rouwenhorst et al., 2021; Rabbai et al., 2024) and also leads to contamination of water sources with nitrates and nitrites, which is considered a growing problem threatening both human health and ecosystems (; Mateo-Sagasta et al., 2018;). On the one hand, a growing number of observational studies have established associations between nitrate levels in drinking water and human pathologies, including certain forms of cancer (Warld et al., 2018; Picetti et al., 2022). On the other hand, nitrogen fertilizers are the main source of water-soluble nitrate and nitrite compounds in soil that can be washed by surface runoff into lakes, rivers, and reservoirs, leading to eutrophication and harmful cyanobacterial blooms. Consequently, the frequency and global distribution of these toxic cyanobacterial blooms in different aquatic ecosystems appear to have increased in recent years, leading to animal and human poisoning (Benayache et al., 2019; Chorus et al., 2021; Manganelli et al., 2023). Therefore, environmentally friendly solutions are needed to address this recurring phenomenon. The potential benefits of

http://dx.doi.org/10.19103/AS.2025.0166.15

recycling this nutrient-rich, free-living cyanobacterial biomass in agricultural land as biofertilizers for soil fertilization and crop nutrition in a 'circular economy' could be considered a sustainable solution.

Due to their morphological and physiological adaptations (extracellular matrix, gelatinous sheath, heterocysts, akinetes, and resting cells), cyanobacteria are an important component of soils and the most abundant in terms of biomass and species richness (Johansen and Rushforth, 1985; Flechtner et al., 2008; Hu et al., 2012). They are a major component of the formation of biological soil crusts (BSCs) communities in drylands, where they play a fundamental role in primary productivity and biochemical cycling (Garcia-Pichel et al., 2016; Büdel et al., 2016; Belnap et al., 2016). These ubiquitous photosynthetic microorganisms are pioneers on many soils and substrates under water stress and nutrient deficiencies. They enhance the productivity and fertility of terrestrial ecosystems through carbon and nitrogen fixation, oxygen production, and biomass production (Belnap et al., 2016; Büdel et al., 2016; Baumann and Jung, 2023). They stabilize soil structure through the production of exopolysaccharides (EPS), improve soil texture, and increase its porosity and water-holding capacity (Singh et al., 2016; Liu et al., 2017). They can also promote the creation of more stable organo-mineral layers, thereby reducing soil erosion and promoting the establishment of vascular plants (Lin et al., 2013; Rossi and De Philippis, 2015). Cyanobacteria are well known for their agronomic importance, improving soil fertility by adding organic matter, fixing nitrogen, solubilizing phosphorus, and improving crop yields by providing micronutrients, vitamins, amino acids, and phytohormones (Whitton and Potts, 2007; Singh et al., 2014b; Rossi et al., 2017). According to these multiple beneficial effects and potential applications, terrestrial cyanobacteria have attracted increasing interest as a potential source of biofertilizers, plant growth promoters, and biological control agents (Rai et al., 2019; Joshi et al., 2020; Pandey et al., 2020; Alvarez et al., 2024). In recent decades, the number of publications on terrestrial cyanobacteria has increased considerably, focusing not only on their diversity and ecology but also on their potential biotechnological applications, particularly in agriculture (Fig. 1) (Pentecost and Whitton, 2012; Hu et al., 2012; Gaysina et al., 2018; Davydov and Patova, 2018; Kumar et al., 2019).

Despite the fact that much research has been conducted to assess the value of several species of cultivated microalgae and harvested or cultivated macroalgae, particularly in agriculture, as well as in the food and pharmaceutical industries, few studies have focused on cyanobacteria with the exception of the non-toxin-producing species, *Spirulina*. In fact, several species of cyanobacteria, producing dense blooms in aquatic ecosystems, can also release secondary metabolites, such as cyanotoxins, which can compromise the quality of agricultural products and pose a potential risk to human and animal health (Huisman et al 2018; Bouaïcha et al., 2019; Chorus & Welker,

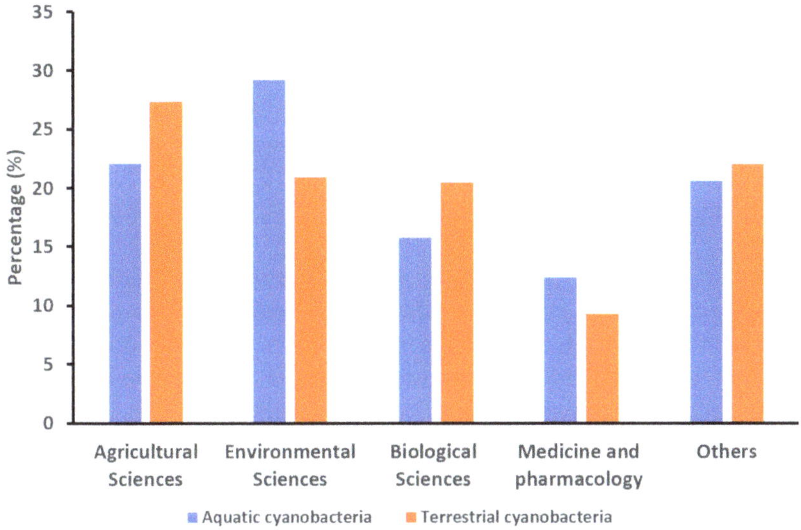

Figure 1 Percentage of publications assigned to Scopus on aquatic and terrestrial cyanobacteria by subject area. The search queries used for retrieving data were constructed as follows: aquatic and cyanobacteria, terrestrial and soil and cyanobacteria.

2021). Therefore, further research is still needed to explore the potential use of wild-harvested cyanobacterial biomasses as soil amendment, biostimulant, or biofertilizer for sustainable agriculture. As soil cyanobacteria have unique characteristics and promising applications in sustainable agriculture and land restoration, this chapter summarizes and updates current knowledge on their diversity, functional characteristics, and important role in restoring degraded soils, improving fertility, promoting crop health, and nutrient use efficiency. Furthermore, this chapter highlights the limitations and knowledge gaps that hinder the recycling in agriculture as cyanofertilizers of cyanobacterial biomasses present in large quantities in nature and which can be harvested without production costs.

2 Soil cyanobacterial community: diversity, taxonomic composition and functional characteristics

2.1 Diversity and taxonomic composition

The diversity of cyanobacteria has been characterized by their morphological, physiological, biochemical, ecological, and molecular attributes, which enable them to occupy a wide range of terrestrial habitats, including extreme environments. Cyanobacteria vary significantly in size and shape, from unicellular to colonial or filamentous, including simple or complex branched forms with the formation of specialized cells (heterocysts, hormogonia, akinetes). Due to this wide variation, several monographs and treatises have proposed identification keys using morphological and ecological features for identifying and describing the cyanobacterial diversity (Fremy, 1929; Geitler, 1932; Elenkin, 1938; Desikachary, 1959; Starmach, 1966; Kondrateva, 1968). This classical taxonomy approach, widely accepted until the late twentieth century, described 1500 species belonging to 150 genera, 20 families, 5 orders, and 1 single class (cyanophyceae) (Geitler, 1932; Anagnostidis and Komárek, 1988, 1990; Komárek and Anagnostidis, 1986, 1989, 1998). In spite of this traditional taxonomic system, which has produced a description of a significant cyanobacterial taxa and publication of several regional monographs, the morphology-based classification has not been agreed upon and was generally considered less discriminating and does not reflect the real genetic diversity and phylogenetic relationships (Alvarenga et al., 2015; Anand et al., 2019; Mishra 2020). Several studies highlighted the high morphological and phenotypic plasticity of cyanobacteria and the capacity of many taxa of natural samples to modify their morphological and physiological traits (ecotypes and ecophenes species), without changes in their genotype, in response to environmental factors and culture conditions (Cameron and Blank, 1966; Pearson and Kingsbury, 1966; Drouet 1973; Castenholz and Waterbury,1989; Otsuka et al., 1999; Komárek and Komarkova, 2004). Some researchers have attempted to assess the biochemical-level diversity of a number of aquatic cyanobacterial genera by characterizing their fatty acid profiles (Kenyon et al., 1972; Liu et al., 2003; Romano et al., 2000; Li and Watanabe, 2004; Shukla et al., 2012). These two morpho- and chemo-taxonomy methods have shown several limitations due to phenotypic and ecotypic variations. The morphological taxonomy can induce a high level of error and an overestimation of the diversity of cyanobacteria. Komárek and Anagnostidis (1989) reported that up to 50% of the strains in culture collections are misidentified or assigned to the wrong taxonomic group. In the last decades, to overcome these challenges for a more accurate cyanobacterial identification, several researchers have proposed modern approaches to the classification system for cyanobacteria (Wilmotte, 1994; Komárek and Anagnostidis, 1998; 2005; Boone et al., 2001; Hoffmann et al., 2005; Büdel and Kauff, 2012; Mishra et al., 2013; Singh et al., 2013). These studies suggested stable and reliable taxonomic characters as ultramorphological features and molecular methods such as DNA barcoding, metagenomic analysis, and quantitative polymerase chain reaction (PCR).

Nowadays, modern classification of cyanobacteria integrates both conventional (morphological, ecological, and cytological) features and molecular methods in order to combine classical taxonomy with the results of phylogenetic studies (Komárek, 2010, 2014, 2016, 2018). Based on this modern polyphasic approach, Komárek et al. (2014) subdivided the phylum of cyanobacteria into eight different orders (Chroococcales,

Chroococcidiopsidales, Gloeobacterales, Nostocales, Oscillatoriales, Pleurocapsales, Spirulinales, and Synechococcales). In recent years, the molecular-based classification, using polyphasic taxonomic assessment with 16S rRNA as the standard phylogenetic marker, has provided new insights into cyanobacterial composition and diversity with the description of several new taxa in different types of habitats (Shih et al., 2013; Singh et al., 2016; Singh et al., 2017; Kumari and Rai, 2020). These recent methodological developments have improved the molecular diversity assessment of cyanobacteria, especially in aquatic habitats and extreme environments such as geothermal systems (Ward et al., 2012; Komárek, 2018; Osorio-Santos et al., 2014; Rott et al., 2018; Palinska et al., 2018; Anand et al., 2019). Every year, an increasing number of new taxa are added to the cyanobacteria database, highlighting the huge potential of cyanobacterial biodiversity.

Historically, the taxonomy of terrestrial cyanobacteria was studied along with the eukaryotic microalgae in the botanical tradition based on morphological characteristics. According to Peterson (1935), the first reference known to a soil alga (probably a species of *Nostoc*) was reported by Lyngbye (1819). In his first compilation of a species list of soil algae in the nineteenth century, Graebner (1895) included descriptions of soil cyanobacterial taxa such as *Oscillatoria tenerrima, Phormidium vulgare, Gleocapsa livida, and Lyngbya lateritia* considered as pioneer species in the poor soils. The earliest reviews on soil algae were published in the twentieth century as monographs including several genera and species list of cyanobacteria occurring in terrestrial habitats (Fritsch, 1922; Petersen, 1935; Shields and Durell, 1964; Hoffman, 1989; Gollebarch and Shtina, 1969). Because of the growing interest in soil algae, several species lists including soil cyanobacteria were published, and several taxa were described from different regions of the world (Friedmann et al., 1967; Johansen, 1984; Friedmann, 1980; Garcia-Pichel et al., 2001; Flechtner et al., 2005). Most of the literature on the floristic, ecological parameters, and distribution of soil cyanobacteria has been summarized by Metting (1981). This publication included a list of 37 genera of cyanobacteria with species inventoried in the soil. Whitton and Potts (2000), in their book 'The Ecology of Cyanobacteria,' compiled the whole literature to the 2000s and reported several examples of studies with detailed taxonomic and/or floristic information on cyanobacteria in soil or rice fields from different regions. Whitton (2000) reported that many rice field soils contain a high density of cyanobacteria and typically about half of the genera present are heterocytous forms (*Anabaena, Aulosira, Calothrix, Cylindrospermum, Fischerella, Gloeotrichia, Nostoc, Scytonema, Tolypothrix, Wollea*). In fact, particular attention has been paid to the cyanobacteria of rice paddies in several countries, especially in India (Singh RN, 1961; Venkataraman, 1972, 1975; Anand and Hopper, 1987; Singh NI, 2002; Tiwari et al., 2007; Singh et al., 2014a), Japan (Watanabe, 1965; Kobayashi et al., 1967; Watanabe and Yamamoto, 1967) and other localities in southeast Asia (Watanabe and Yamamoto, 1971). These authors reported that nitrogen-fixing species of cyanobacteria were widely distributed in both tropical and temperate areas but were more abundant in the tropics (Metting, 1981). The most common tropical species belong to the genera *Anabaena, Aulosira, Nostoc, Scytonema, and Tolypothrix*. This high occurrence of cyanobacteria in tropical soils has been explained as being mainly due to the relatively higher optimal growth temperatures than for most eukaryotic algae (Castenholz and Waterbury, 1989). The number of species of soil cyanobacteria capable of nitrogen fixation has been estimated at 25 (Singh, 1961) to 38 (Shtina, 1972). Tiwari and Singh (2005) observed that the heterocytous forms grow better in rice fields with less nitrogen and plenty of water. Nevertheless, fields that are not submerged and have no application of fertilizer promote the growth of many nitrogen-fixing terrestrial cyanobacteria. Singh (1961) also reported a list of dominant cyanobacterial species and their successional trends in rice soils of India. Although several species of *Nostoc* (e.g. *N. muscorum, N. punctiforme*) and *Anabaena* (*A. variabilis, A. cylindrica*) are ubiquitous and have been reported to be universal (Venkataraman, 1975), *Aulosira fertilissima* has been found to be the most widespread and important cyanobacteria associated with rice soils (Singh, 1961).

Other taxa of cyanobacteria such as *Nostoc commune, Microcoleus autumnalis, Leptolyngbya fragilis, L. tenuis, Nostoc linckia,* and *Scytonema* hyalinum are able to colonize arid salted soils in diverse geographical areas (Gollerbakh and Shtina, 1969; Vinogradova and Darienko, 2008; Khaibullina and Gaisina, 2008; Mehda, 2022). In addition, while many cyanobacteria species (e.g. *Nostoc commune, N. sphaericum, Fischerella muscicola, Tolypothrix tenuis*) are typical inhabitants of undisturbed soils (Metting, 1981), several other resistant genera (e.g. *Lyngbya, Oscillatoria, Phormidium, Microcoleus, Nostoc*) can colonize lifeless soils, urban and disturbed soils, including mine soils (Gaysina et al., 2019; Hakkoum et al., 2021). Several researchers have also detected cyanobacterial species such as *Microcoleus vaginatus, M. autumnalis, Nostoc muscorum, N. calcicola, N. commune* and representatives of the genera *Leptolyngbya, Lyngbya, Phormidium, Oscillatoria, Schizothrix* in reclaimed soils and in areas disturbed by coal mining (Lukešová, 2001; Starks and Shubert, 1982).

Due to recent progress in identifying cyanobacteria using modern molecular and genetic analysis, a significant proportion of the huge hidden biodiversity of terrestrial cyanobacteria has been estimated, especially in arid and desert regions (Mishra, 2020; Kumari and Rai, 2020). Most of the works published to date on terrestrial cyanobacteria refer much more to the biological soil crusts that colonize the first few centimeters of the soil surface. These biocrusts are present in most environments of the world, but they are particularly important in arid and desert areas, where they make up a large part of the living soil cover (Belnap et al., 2003). Although several microorganisms (green algae, heterotrophic bacteria, fungi, mosses, lichens, etc.) colonize biocrusts, cyanobacteria form the main microbial group especially in arid regions and dry deserts (Wynn-Williams, 2000; Hu et al., 2012; Büdel et al., 2016). Early taxonomic surveys of desert cyanobacteria have a floristic focus with emphasis on diverse habitats (lithosols, chasmoliths, and endoliths) of hot desert from central Asia, the Arctic, the Negev and south-western United States, and Mexico (Cameron and Blank, 1966; Friedmann, 1972; Friedmann and Galun, 1974). Wynn-Williams (2000) reviewed to that date the diversity of cyanobacteria of diverse soil and lithic habitats from different hot and cold deserts in the world. He mentioned 33 and 49 cyanobacterial species commonly found, respectively, in selected hot deserts and Antarctic cold deserts, both in soil crusts and endolithic communities. Even though the taxonomic composition of desert cyanobacteria varied between different microhabitats, the filamentous genera such as *Microcoleus, Nostoc, Phormidium, Plectonema, Lyngbya, Leptolyngbya, Scytonema*, *Schizothrix,* and *Trichocoleus* were the most frequently dominant in the soils (Friedmann and Galun, 1974; Wynn-Williams, 2000). Among these filamentous taxa, *Microcoleus vaginatus* is the most widespread species commonly inventoried in biological soil crusts around the world (Johansen and Shubert, 2001; Castenholz & Garcia-Pichel, 2012; Hu et al., 2012; Hagemann et al., 2015; Zhang et al., 2015, 2021a; Fernandes et al., 2018; Dulic et al., 2017; Munoz-Martın et al., 2019; Mehda et al., 2021; Sosa-Quintero et al., 2022). Indeed, *M. vaginatus* can widely thrive in arid desert regions due to its high resistance to multiple abiotic stresses and its adaptation to extreme conditions such as extreme temperature and pH (high or low), low humidity, and nutrient-poor substrates (Chen et al., 2003; Rossi et al., 2017;). Other characteristic species such as *Nostoc punctiforme, N. paludosum, N. commune, Tolypothrix distorta, Tolypothrix tenuis, Scytonema javanicum, Schizothrix calcicola, Schizothrix fragile, Phormidium minnesotense,* and *Phormidium tenue* were also frequently found in the desert crusts (Johansen, 1993; Hu et al., 2012). Whereas, filamentous forms predominate in crusts, the unicellular coccoid cyanobacterial genera such as *Synechococcus* and *Synechocystis* were less abundant (Zhao et al., 2019). Among the representative taxa of this group, *Chroococcidiopsis* was reported as the most dominant form genus and the most closely adapted to the desert endoliths (Wessels and Büdel, 1995; Caiola et al., 1996; Hu et al., 2012).

More recently, a number of papers have been published on the diversity and ecology of cyanobacterial soil crusts in semi-arid and arid regions, and their potential applications in desert reclamation, soil rehabilitation, and fertilization have been highlighted (Hu et al., 2012; Büdel et al., 2016; Belnap et al., 2016; Dulic et al., 2017; Cano-Díaz, 2018; Hakkoum et al., 2020; Mehda et al., 2022; Rossi et al., 2022). Using morphological, molecular taxonomy, or chemotaxonomic markers for identification, over 320 species belonging to more than 70 genera have been inventoried in biological soil crusts worldwide (Büdel et al., 2016). The typical and most widespread cyanobacterial genera within these soil BSCs are *Chroococcidiopsis* in unicellular form as well as *Microcoleus, Nostoc, and Scytonema* in filamentous form (Büdel et al., 2016). Other cyanobacteria are present only sporadically and may originate from other habitats, such as aquatic environments or symbiosis with lichens (e.g. *Chroococcus, Gleocapsa, Gloeocapsopsis, Cylindrospermum, Phormidium*).

In spite of their important ecological functions and possibilities for valorization, the cyanobacterial diversity occurring in soils and many other terrestrial habitats has not been thoroughly explored yet (Gaysina et al., 2019; Mishra, 2020; Kumari and Rai, 2020). For instance, the cyanobacterial communities of rhizospheric soils, which are inhabited and influenced by plant roots, remain insufficiently explored. Several researchers have estimated that a large part of terrestrial cyanobacterial biodiversity is still hidden, particularly in geographical regions that have been less studied (Nabout et al., 2013; Gaysina et al., 2018). On a continental scale, the cyanobacteria of BSCs have been relatively well studied in certain continents (Büdel et al., 2016). Several works have focused on arid and desert steppes of Southwestern United States (Johansen, 1993; Garcia-Pichel, 2002; Belnap, 2003; Flechtner et al., 2005; Williams et al., 2012), Mexico (Flechtner et al., 1998), Australia (Williams et al., 2014), Russia (Temraleeva, 2018), western Europe (Williams et al., 2016, 2018), China (Zhang et al., 2016, 2021b; Chen et al., 2023), India (Kumar and Adhikary, 2015), Southern Africa (Dojani et al., 2014; Büdel et al., 2009), and polar regions (Pointing et al., 2015). Elster (2002) reported that the BSCs are also common in high altitude regions of the Arctic territories of Russia, Canada, European Arctic, and Greenland. The majority of studies have been carried out in Europe to reveal the cyanobacterial composition of telluric communities in temperate regions (Hoffman, 1989; Schulz et al., 2016; Langhans et al., 2010; Dvořák et al., 2017). The most recent study of Belnap et al. (2016) compiled floristic records of terrestrial cyanobacteria living in BSCs on different continents. In total, 98 taxa were recorded in Europe, 53 in North and Central America, 40 in South America, 83 in Africa, 82 in Asia, 58 in Australia–New Zealand, and 65 in the Antarctic and Arctic regions. The largest number of soil cyanobacterial species has been described in Europe, the most studied continent over the last three centuries (Dvořák et al., 2017), even though Asia and Africa are the two largest continents with a greater diversity of arid and semi-arid landscapes (Büdel et al., 2016). However, reliable comparisons of soil cyanobacterial diversity between different geographical regions cannot be made due to a lack of studies and data in some parts of the world, such as Africa (Dvořák et al., 2017).

Unlike the southern territories of Africa (Büdel et al., 2009; Dojani et al., 2014), in North Africa and the Sahara Desert, little is known about soil cyanobacterial communities, which are still underexplored and whose ecology remains poorly understood (Büdel et al., 2016). More recently, the research on biocrusts of these arid and desert areas has gained more interest due to their potential use as a solution against erosion and desertification (Li et al., 2010; Wu et al., 2013; Lan et al., 2014). Malam Issa et al. (1999) reported that microbiotic soil crusts of the tiger bush of Niger are formed mainly by filamentous cyanobacteria dominated by *Schizothrix,* along with other common genera such as *Scytonema, Lyngbya, Microcoleus, Nostoc,* and *Phormidium*. Furthermore, Hakkoum et al. (2020) have studied the cyanobacterial diversity of biological soil crusts and soil physicochemical properties along an altitudinal gradient in the arid area of Marrakech (Morocco). A first cyanobacterial inventory of 26 morphotypes was recorded in which the filamentous genera such as *Phormidium, Leptolyngbya, and Nostoc* were the most common. Recently, sampling of biological soil crusts in other sites in the same region has completed the inventory to 267 morphotypes of cyanobacteria (personal data not yet published). The most frequent taxa were *Phormidium autumnale, Leptolyngbya foveolarum, Pseudanabaena Starmachii, Nostoc commune, Phormidium paulsenianum, Microcoleus vaginatus, Nostoc punctiforme, N. minutum, Scytonema javanicum,* and *Aphanocapsa muscicola.* Based on both morphological identification and phylogenetic analysis (16S rRNA gene), Mehda et al. (2021) reported that cyanobacterial community in biocrusts from the Sahara Desert (Algeria) is composed mainly of characteristic filamentous species such as *Microcoleus steenstrupii, Microcoleus vaginatus, Scytonema hyalinum, Tolypothrix distorta, and Calothrix sp.* whereas the coccoid forms are represented by *Chroococcidiopsis* and *Acaryochloris*. In addition, these authors have also studied lithic cyanobacterial communities in the polyextreme Sahara Desert where the conditions are more severe. Seven taxa were identified in the hypolithic communities including six filamentous species: *Nostoc commune, Trichocoleus sociatus, Microcoleus steenstrupii, Pseudophormidium* sp., *Trichocoleus desertorum, Nodosilinea* sp., and one coccoid *Pseudoacaryochloris sahariense*, a newly described taxon. Eight species of endolithic cyanobacteria were inventoried, and coccoid forms such as *Chroococcidiopsis, Gloeocapsopsis, and Pseudoacaryochloris* were dominant besides other filamentous taxa such as *Scytonema hyalinum, Leptolynbya* sp., *Albertania* sp., and *Nodosilinea* sp.

2.2 Functional characteristics

2.2.1 Cyanobacteria in biological soil crusts: significance in drylands

Drylands are the world's largest terrestrial biome that represent 40% of the Earth's land surface and include arid, semi-arid, desert, alpine, and Polar Regions

(Belnap 2006; Pointing and Belnap, 2012). The soil microorganisms living in these arid environments are exposed to various abiotic stresses, including large fluctuations in temperature, ultraviolet (UV) radiation, low moisture, desiccation, and salinity, which determine their composition, abundance, and diversity (Belnap et al., 2016). Only life forms that have developed adaptation mechanisms and protection strategies such as cyanobacteria are able to inhabit these hyperarid soils. These microorganisms should be tolerant to a wide range of extreme temperatures, low humidity, solar radiation, low nutritional status, and physical erosion (Pointing, 2016; Wierzchos et al., 2018). For example, the predominance of cyanobacteria in arid areas and deserts results from their diverse adaptive strategies to resist environmental stressors such as extreme daily fluctuations in temperature and humidity, evaporation, and high UV radiation, as well as their ability to survive repeated desiccation/rehydration cycles (Raanan et al., 2015; Xu et al., 2021). Despite their harsh environmental conditions, the soils of arid regions can be frequently colonized by crust-forming microorganisms, known as biological soil crusts, which are a complex association between soil particles, microalgae, lichens, bryophytes, microfungi, and heterotrophic bacteria (Belnap et al., 2011; Weber et al., 2022). These communities are widespread in arid and semi-arid regions where they cover approximately 70% of the soil surface and play a key role in its stabilization, biotic composition, and functioning (Belnap et al., 2016; Bowker et al., 2018; Buis et al., 2009). According to the dominant biological components and their functions, BSCs can be divided into cyanobacterial crusts, algal crusts, lichen crusts, and moss crusts (Weber et al., 2016). The structure, appearance, species composition, and biomass of biocrusts are affected by several environmental factors (e.g. soil texture and structure, chemical composition, pH, moisture, potential evapotranspiration, salinity, rainfall, disturbances) and vary considerably across climatic regimes. Different types of biocrusts have been described based on their floristic composition and external morphology (Belnap, 2006; Colesie et al., 2016). In dry arid and hyperarid areas with higher evapotranspiration potential, biocrusts are flat and dominated by cyanobacteria, especially filamentous taxa, which form the biocrust matrix and grow in the upper millimeters of the soil (Belnap, 2006). Pocs (2009) have described two main cyanobacterial crusts. The epiterranean type consists of cyanobacteria growing above the soil surface (thin crusts only a few millimeters thick), and the subterranean type has at least one layer of cyanobacteria below the soil surface. Many studies have focused on cyanobacteria in biological crust communities, particularly in arid regions and deserts (Hu et al., 2012; Büdel et al., 2016; Cano-Díaz et al., 2018; Gaysina et al., 2019). These studies have highlighted the fundamental role of these phototrophic microorganisms as primary producers in arid habitats and, in some cases, the only ones in the extreme arid environments such as deserts (Wynn-Williams,

2000; Mehda et al., 2022). Cyanobacteria are pioneers in the colonization of biocrusts in arid areas due to their ability to grow photoautotrophically and fix nitrogen via heterocysts. Due to their morphological and physiological characteristics (gelatinous sheath, extracellular matrix, heterocyst, akinete, resting cells), filamentous cyanobacteria are the main and most abundant microorganisms forming biocrusts (Johansen and Rushforth, 1985; Flechtner et al., 2008). Cyanobacteria living in biocrusts are very diverse, but only a limited number of taxa are actually responsible for their formation. The main crust-forming cyanobacteria are filamentous genera with distinct sheaths, such as *Microcoleus*, *Nostoc*, *Scytonema*, *Lyngbya*, *Trichocoleus*, *Plectonema*, *Schizothrix*, and *Stigonema* (Büdel et al., 2016; Gaysina et al., 2019). Among these typical genera, non-heterocysted filamentous cyanobacteria such as *Microcoleus vaginatus*, *Trichocoleus*, and *Schizothrix* are a major component of many BSCs worldwide (Büdel et al., 2016; Muñoz-Martin et al., 2019; Xu et al., 2021). While species such as *Nostoc* sp., *Scytonema* sp., *Tolypothrix* sp., and *Spirirestis* sp. are the dominant nitrogen-fixing cyanobacteria in BSCs of drylands. Motile filamentous cyanobacteria such as *Microcoleus vaginatus* are the first colonizers of bare soil areas in arid zones that dominate early successional biocrusts. Indeed, the filamentous form is a key characteristic that allows cyanobacteria to colonize physically unstable substrates and play a pioneering role in the biostabilization process (Belnap 2006; Garcia-Pichel and Wojciechowski, 2009; Büdel et al., 2016). Therefore, cyanobacteria living in BSCs have developed adaptive morphological, physiological, and metabolic mechanisms as described in Fig. 2 (Gao and Garcia-Pichel, 2011; Pentecost and Whitton, 2012; Rastogi et al., 2016; Mackelprang et al., 2022; Xu et al., 2021).

Understanding the survival strategies of cyanobacteria in xeric environments has progressed due to recent advances in metagenomics and omics analysis. These experimental studies have provided new insights into the physiological, biochemical, and metabolomic aspects, as well as the dynamic responses that allow cyanobacteria to survive desiccation and tolerate abiotic stress in these extreme environments (Xu et al., 2021). Unlike their aquatic relatives, BSCs cyanobacteria are highly adapted to arid terrestrial environments. These adaptations allow them to colonize unstable substrates under changing conditions and enhance the ecological functions of the biological crust, including carbon and nitrogen fixation, bio-weathering activities, and diverse microbial interactions (Garcia-Pichel et al., 2016; Büdel et al., 2016). As photosynthetic microorganisms, biocrust cyanobacteria are an important source of carbon in arid terrestrial areas and desert oligotrophic ecosystems (Belnap, 2006; Sharma et al., 2012). These photoautotrophic communities in the Earth's crust can absorb CO_2, especially during the cool periods of autumn, winter, and spring (Elbert et al., 2012). They increase the soil organic carbon pool by excreting carbon in the form of exopolysaccharides and promoting

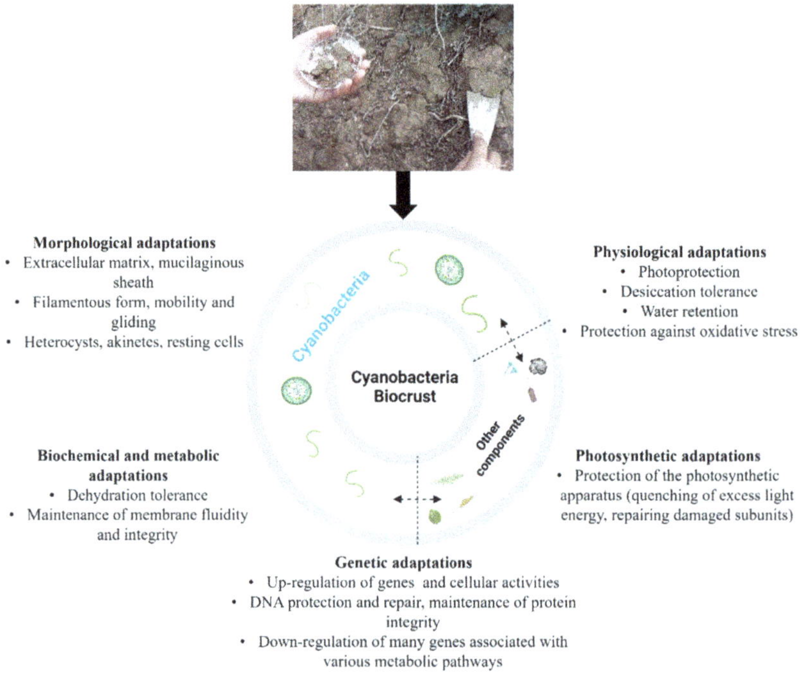

Figure 2 Adaptive strategies of cyanobacteria biocrusts to harsh dryland conditions.

the growth of heterotrophic bacteria, microflora, and microfauna. They also increase soil organic content through the degradation and decomposition of their biomass (Durall and Lindblad, 2015; Renuka et al., 2018). Diazotrophic filamentous cyanobacteria can fix and transform gaseous nitrogen into organic compounds that can be readily assimilated by heterotrophic soil communities and plants (Win et al., 2018). Therefore, their presence in soil increases the availability of nitrogen, one of the essential nutrients for the growth of living organisms (Sharma et al., 2012; Renuka et al., 2018). Nitrogen fixed by cyanobacteria is released either through secretion or through degradation of cyanobacterial cells into ammonia, polypeptides, amino acids, vitamins, and auxin-like substances (Jhala et al., 2017).

2.2.2 Role in improving soil structure and stability

Soils in arid areas form particularly slowly, as it can take thousands of years to form a few centimeters of soil (Pillans, 1997). Many arid soils are highly erodible due to their fine texture (sand and fine particles) and low moisture content, organic matter, and vascular vegetation. Wind and water are the primary drivers of erosion in arid and desert areas. Sparse vegetation leaves large areas of bare soil without plant protection. Bare soils and open spaces between plants are usually stabilized by cyanobacteria living in the BSCs, protecting them from wind and water erosion (Belnap, 2003; Li et al., 2021; Sivakumar, 2007). Therefore, BSCs play a major role in arid ecosystems: they prevent soil erosion, retain moisture, and contribute to the restoration of disturbed lands. Their role in maintaining nutrient balance, texture, fertility, and soil health is widely recognized (Hooda et al., 2023). Filamentous cyanobacteria secrete an exopolysaccharide (EPS) matrix that promotes soil stabilization and improves the nutritional and water conditions of the microhabitat for the colonization of other cyanobacteria, green algae, and remaining components of the biocrust (Mager and Thomas, 2010). Mucilaginous sheaths and polysaccharides exuded by cyanobacteria bind soil particles together and create aggregates, increasing particle size. The formation of larger and heavier aggregates reduces sediment loss due to erosive pressures (Belnap, 2003; Belnap, 2006; Bowker et al., 2018; Li et al., 2021). EPS production further enhances surface colonization and the formation of biofilms and microbial mats, leading to the formation of more complex multilayered structures (Zhang et al., 2015). Besides its role in the physical formation of soil crusts, EPS modifies soil hydrology through water retention, potentially promoting local runoff (Benard et al., 2019). Moreover, the decomposition of cyanobacterial biomass and exopolysaccharides increases the organic matter content of the soil, thus improving its structure and stability (Chen et al., 2014).

Numerous studies have shown that cyanobacterial-dominated biocrust influences soil cohesion and significantly increases its resistance to wind and water erosion in arid regions. However, this influence depends on biocrust characteristics, such as the composition and morphology of cyanobacterial species, the developmental stage and amount of biomass of the community, and its previous moisture content (Belnap and Büdel, 2016). Therefore, soil structure, morphology, and topography are strongly determined by the size and shape of the cyanobacterial species present. For example, Hu et al. (2002) tested the ability of some filamentous cyanobacteria such as *Microcoleus vaginatus*, *Phormidium tenue*, *Scytonema javanicum*, and *Nostoc* sp. to withstand wind speed, showing that *Microcoleus vaginatus* was the most effective species for soil stabilization. From a functional point of view, Büdel et al. (2016) classified BSCs cyanobacteria into three different groups. Group 1 contains filamentous cyanobacteria, such as *Microcoleus*, which stabilize soils by gluing soil particles together and thus form soil aggregates due to the presence of extracellular matrix. These cyanobacteria are responsible for the formation of biocrusts and are also the most abundant cyanobacterial species in biocrusts. Group 2 includes the cyanobacteria *Scytonema, Stigonema*, and *Chroococcidiopsis* that live in the biological crust environment, reinforcing their ecological role through their contribution to the carbon and nitrogen cycle. Group 3 includes cyanobacteria that are only stochastically present in biological crusts and may originate from other habitats, such as the aquatic environment or symbiosis with lichens (e.g. *Chroococcus, Gleocapsa, Gloeocapsopsis, Cylindrospermum, Phormidium, Tolypothrix*).

2.2.3 Role in improving soil fertility

In addition to their role as soil stabilizers, cyanobacteria are a source of carbon and nitrogen in oligotrophic arid environments. Due to their distinctive characteristics and biological activities such as carbon and nitrogen fixation, organic matter synthesis, and phosphate solubilization, crust-dwelling cyanobacteria contribute significantly to soil fertility and, consequently, to soil maintenance in these nutrient-poor environments (Belnap 2006; Elbert et al., 2012; Strauss et al., 2012).

2.2.3.1 Carbon sequestration and organic matter production

Cyanobacteria are the only prokaryotic microorganisms capable of absorbing atmospheric carbon dioxide and converting it into organic biomass and calcium carbonate through oxygenic photosynthesis using the Calvin–Benson–Bassham cycle (C3 cycle) (Pandey at al., 2025; Sharma et al., 2012). Unlike other photosynthetic plants, they possess the carboxysome, a cell microcompartment which concentrates CO_2 around the enzyme ribulose-1,5-bisphosphate carboxylase/oxygenase (RuBisCO) – a key enzyme responsible for catalyzing the first major step of carbon fixation, creating a favorable microenvironment for enhanced carbon assimilation (Rae et al., 2013; Garcia et al., 2025). Via their efficient bioconversion of light into chemical energy, these Gram-negative prokaryotes can efficiently sequester billions of tons of CO_2 due to an efficient carbon concentrating mechanism (CCM) and transform it to energy-rich compounds using their flexible metabolic system (Pandey et al., 2025). For every kilogram of biomass produced, cyanobacteria can sequester approximately 1.3 kg of CO_2 (Sarwer et al., 2022). Their efficient carbon absorption is 10 to 50 times higher than that of terrestrial plants (Ashour et al., 2024). In fact, cyanobacteria have a greater light-to-biomass conversion efficiency and photosynthetic yield (typically between 1 and 3%) which can approach the theoretical limit of 12% under optimal conditions (Al-Najjar et al., 2012; Ooms et al., 2016). Certain species such as *Spirulina platensis* can achieve CO_2 uptake rates of 160–175 mg of biomass per liter per day and up to 96% CO_2 removal efficiency (Kim et al., 2025; Ighalo et al., 2022).

Due to their high carbon sequestration efficiency, cyanobacteria play a crucial role in reducing greenhouse gas emissions, including hydrocarbons (HC), sulfur dioxide (SO_2), and nitrogen oxides (NOx), making them a valuable tool for mitigating climate change and promoting environmental sustainability (Hoque et al., 2025). In addition to their role as a potent CO_2 sink, cyanobacteria can also mitigate methane production by modulating microbial methane metabolism and boosting aerobic methane oxidation by methanotrophs (Pandey et al., 2025).

These microorganisms have gained immense interest as alternative biotechnological solutions to CO_2 capture for global warming and climate change mitigation. Their metabolism contributes not only to climate change mitigation but also to reducing the rate of carbon turnover, increasing carbon retention and storage in the soil (Piatka et al., 2022). In fact, as primary producers, cyanobacteria increase the soil's organic carbon pool by excreting carbon in the form of EPS, which benefits saprophytic soil microorganisms and promotes soil microbial biomass, microflora, and microfauna growth. They also increase the organic content of the soil through the degradation and decomposition of their biomass (Durall and Lindblad, 2015; Renuka et al., 2018). Cyanobacteria are an important source of organic matter in the soil, playing a central role in microbial and enzymatic activities, nutrient dynamics, carbon and N-cycling, and improving the quality and fertility of soil. This organic matter comprises a diverse array of compounds, such as amino acids, vitamins, auxins, and the biomass of the cyanobacteria themselves (Bao et al., 2021).

Thanks to their release of EPS and photosynthates rich in fresh carbon into the soil, various filamentous cyanobacteria such as *Microcoleus*, *Nostoc*, and *Anabaena* are able to trigger a priming effect in nutrient-poor soils, stimulating heterotrophic bacteria – such as nitrogen-fixing diazotrophs – leading to increased decomposition and consumption of both recent and older carbon (Garcia et al., 2025). These authors highlighted a mutualistic interaction (C and N trading relationship) between filamentous cyanobacteria (e.g. *Microcoleus vaginatus*) and microorganisms inhabiting the cyanosphere that enables the survival of low nutrient and degraded soils, supporting biocrust formation and promoting plant cultivation.

2.2.3.2 Nitrogen fixation

Despite its abundance in the atmosphere (78%), nitrogen gas (N_2) is inaccessible to plants, which can only use nitrogen in its reduced or oxidized forms. Various free-living or symbiotic bacteria and cyanobacteria species are capable of converting atmospheric nitrogen gas into bioavailable form for plants and soil heterotrophic communities via a process known as biological nitrogen fixation (Kant, 2018; Mącik et al., 2020). It is well established that cyanobacteria are the major regulators of earth's nitrogen cycle. Many filamentous genera (*Anabaena*, *Nostoc*, *Scytonema*, etc.) have the ability to differentiate heterocysts, specialized cells in the fixation of N_2 and its conversion into assimilable nitrogen using oxygen-sensitive nitrogenase (Kumar et al., 2010; Win et al., 2018). This enzyme complex reduces N_2 into ammonia, which is converted into glutamine and glutamate amino acids via the GS–GOGAT pathway (Sarma and Prakash, 2020). The heterocysts provide fixed nitrogen to the vegetative cells, which in turn serve as a source of carbon and reductant for the proper functioning of the heterocysts. Thus, the processes of nitrogen fixation and production of oxygen are spatially separated. Heterocysts perform the fixation of atmospheric nitrogen, whereas oxygenic photosynthesis is located in vegetative cells (Kumar et al., 2010). The aerobic N_2 fixation during darktime has also been reported in several nonheterocystous unicellular (e.g. *Synechococcus, Aphanothece, Gloeocapsa,* etc.) and filamentous cyanobacteria (*Oscillatoria, Plectonema,* etc.) (Mishra et al., 2019; Sarma & Prakash, 2020). The main role of rhizospheric cyanobacteria is to convert atmospheric nitrogen to make it available for plant growth and soil nitrogen storage. Several free-living or endosymbiotic cyanobacteria such as *Nostoc and Anabaena* can supply large amounts of nitrogen in the form of ammonia or amino acids to plant roots, via cyanobacterial cell lysis after they die, or to tissues of host plants via direct secretion in their living state (Meeks and Elhai 2002; Issa et al., 2014; Jhala et al., 2017). It was reported that cyanobacteria can fix nitrogen at 20–30 kg/ha/season (Renuka et al., 2017; Sarma and Prakash, 2020).

Nitrogen plays a crucial role in the survival and growth of bacterial and plant communities through the synthesis of essential compounds such as proteins, nucleic acids, and urea. Several cyanobacteria are considered potent biofertilizers because of their ability to fix atmospheric nitrogen, secrete secondary metabolites, and improve soil fertility and structure. Numerous studies have demonstrated that cyanobacterial extract and inoculation enhance soil microbial activity, improve soil carbon storage, and the rhizosphere nutrient status by influencing the mobility and uptake of macro- and micronutrients such as Zn and Fe (Renuka et al., 2017; Alvarez et al., 2024; Minaoui et al., 2024; Hakkoum et al., 2025; Li et al., 2025). These cyanobacteria-based biofertilizers promote not only crop productivity and quality but also contribute to soil fertility by improving its organic matter and micronutrient content. They are being increasingly considered the best alternative eco-friendly source of nitrogen to reduce agriculture dependence on chemical fertilizers.

2.2.3.3 Phosphate solubilization

Although either phosphorus is present naturally or from chemical fertilizer in sufficient quantity in soil, it is predominantly in an insoluble form, making it unavailable for absorption and uptake by plants (Ray et al., 2013). Only a negligible amount, about 0.2% (i.e. $<10\ \mu M$) is used by plants and the rest is precipitated by metal cations in soil such as Fe, Al, Mg, Ca, etc. (Islam et al., 2019). Phosphorus-solubilizing microorganisms especially various strains of bacteria (e.g. actinobacteria) or arbuscular mycorrhizal fungi are known for their ability to dissolve mineral and/or organic phosphorus to make it available to plants (Soumare et al., 2020). Among the key physiological traits and mechanisms that promote plant growth, cyanobacteria are capable of solubilizing phosphate, thus increasing the mobility and availability of phosphorus for plant uptake. Certain species of cyanobacteria play a vital role in enhancing plant nutrition by improving the bioavailability of phosphate, and mobilizing insoluble form in the soil using phosphatase enzymes (Singh et al., 2016). Several genera including *Anabaena*, *Calothrix*, *Nostoc,* and *Scytonema* have shown the ability to improve the solubility of less soluble forms of phosphorus as tricalcium phosphate ($Ca_3(PO_4)_2$), ferric phosphate ($FePO_4$), aluminum phosphate ($AlPO_4$), and hydroxyapatite ($Ca_5(PO_4)_3(OH)$) (Singh et al., 2016). In addition, Ca^{2+}-bound P can be dissociated either by the formation of a Ca^{2+}- specific chelator or by the formation of organic acids (Cameron and Julian, 1988). Subsequently, cyanobacteria recover soluble P and incorporate it into their cellular biomass for their own metabolism, which is then available to plants through its slow release by secretion, autolysis, or microbial decomposition of dead cyanobacterial cells (Jhala et al., 2017).

Previous studies have demonstrated that phosphate solubilization capacity may vary among different cyanobacterial species. Among 28 strains tested by Toribio et al. (2020), only two strains of *Nostoc* spp. showed significant phosphate solubilization. Furthermore, Afkairin et al. (2021) demonstrated that a strain of *Anabaena* sp. exhibited higher phosphorus solubilization than a commercially available bacterial consortium product.

2.2.3.4 Micronutrient mobilization and translocation in plants

Some terrestrial cyanobacteria such as *Anabaena* and *Nostoc* are rich in nitrogen, phosphorus, potassium, amino acids, and micronutrients. These compounds improve soil structure, promote microbial communities, and boost nutrient availability (Gonçalves et al., 2023). Many studies have demonstrated that soil cyanobacterial inoculation enhances microbial activity and increases soil nutrient content by secreting sticky, negatively charged polysaccharides that bind and prevent the loss of positively charged nutrients (Singh et al., 2019). These polysaccharide sheaths also retain dust, thus improving the physicochemical properties of the soil as well as its water retention capacity (Belnap, 2003; Zhao et al., 2010; Adessi et al., 2018). In addition, cyanobacteria also increase soil fertility by increasing soil pore size and the content of micronutrients (Zn, Cu, Fe, etc.) and macronutrients (C, N, P, K) in the soil (Alvarez et al., 2021). Renuka et al. (2017) have shown that cyanobacterial-based consortia formulation enhances the soil micronutrient availability and uptake in wheat crop, improving micronutrient enrichment in grains, and facilitating enhanced micronutrient (Zn, Fe, Cu, and Mn) dynamics in soil and plant. Furthermore, a recent study by Hakkoum et al. (2025) revealed a significant enhancement in the mineral content (P, N, K, Ca, and Na) and mineral status of tomato plants treated with soil cyanobacterial bio-inoculants.

2.2.4 Role in degraded soil rehabilitation

Several studies have shown that natural and anthropogenic disturbances, including erosion, salinization, urbanization, and pollution, reduce the biomass and development stage of biocrusts and, consequently, the soil's resistance to loss (Belnap and Büdel, 2016). Chamizo et al. (2017) analyzed the influence of biocrust type and its removal on soil organic carbon losses due to water erosion. They emphasized that biocrust loss can lead to increased sediment losses and, consequently, a loss of soil organic carbon and other soil resources. Biocrust removal was accompanied by a significant increase in dissolved organic carbon and sediment organic carbon mobilization. Due to their potential use in agricultural biofertilization, restoration, and rehabilitation of soil environments, cyanobacteria are associated with a wide range of useful environmental processes and biotechnological applications such as degraded land restoration and bioremediation (Singh et al., 2014b; Rossi et al., 2017; Wang et al., 2025). In fact, the restoration of cyanobacterial biocrusts is emerging as an important bioremediation strategy for soil restoration and rehabilitation in arid environments (Acea et al., 2003; Malam Issa et al., 2007; Rossi et al., 2022). One of the main approaches adopted is the identification and isolation of new cyanobacterial strains that can be cultured and then inoculated to produce artificial biocrusts (Giraldo-Silva et al., 2019). Identification and isolation of soil cyanobacteria strains thriving under stressful conditions and multiomics approaches would help in understanding the complex interactions among plants, soil, and cyanobacteria and enable the design of robust and well-integrated biotechnological tools for soil remediation (Hooda et al., 2023).

Soil inoculation with cyanobacteria has been studied as a valuable approach to promote soil stabilization and fertilization and counteract erosion in marginal soils (Mugnai et al., 2024). Due to their contribution to ecosystem functions and global relevance to terrestrial ecosystems (Rodriguez-Caballero et al., 2018), they are receiving scientific attention as valuable biotechnological tools for land rehabilitation and restoration (Rossi et al., 2017; Antoninka et al., 2020; Chamizo et al., 2020). Hu et al. (2002) reported that the application of a cyanobacterial consortium composed of 80% *Microcoleus vaginatus* and 5% *Phormidium tenue*, *Scytonema javanicum*, *Nostoc* sp., and *Desmococcus olivaceus* increased soil cohesion and reduced wind erosion. Numerous studies conducted over the past decade have demonstrated the potential value of selected cyanobacterial consortia as a useful tool for the restoration of degraded lands and the remediation of saline soils (Li et al., 2019; Kheirfam et al., 2020; Hooda et al., 2023; Duan et al., 2023). Moreover, inoculation of cyanobacterial consortia resulted in significant improvement in soil fertility, including exopolysaccharides, organic carbon, organic matter, phosphorus content, and total soil porosity (Maurya et al., 2025).

In recent years, soil inoculation with cyanobacteria has become one of the most promising biotechnological strategies to restore the functionality of degraded soils, prevent erosion, and combat desertification processes in arid regions (Maestre et al., 2017; Rossi et al., 2017). Cyanobacteria promote soil aggregation and structural stability by trapping coarse particles in their filament network and cementing fine particles with their polysaccharide secretions (Wubs et al., 2016; Chamizo et al., 2019; Eldridge et al., 2020). This leads to an increase in soil resistance to water and wind erosion, which are major contributors to desertification (Rossi et al., 2017). To date, cyanobacteria are the only microbial group that has been used for large-scale inoculation experiments to promote biocrust development (Zhou et al., 2020) and the only type of microbial inoculant for which the long-term effects of inoculation have been evaluated by monitoring studies (Wang et al., 2009; Lan et al., 2014). Artificially induced cyanobacterial biocrusts are considered a promising eco-friendly biotechnological tool to combat land desertification (Bowker et al., 2018; Chamizo et al., 2018; Zhao et al., 2019). Nevertheless, several technical challenges need to be addressed to improve the applicability of these tools in various environments in a sustainable manner and make cyanobacteria a realistic and viable bioactive (Hooda et al., 2023).

2.2.5 Role in restoring soil physicochemical properties

Dryland soils are generally light and low in organic matter, which explains their low water retention capacity. Cyanobacteria, one of the main components of soil biocrusts, have been studied for their role as remediation agents for highly degraded and polluted soils. Several laboratory experiments were conducted to evaluate the beneficial effects of cyanobacteria on the physicochemical properties of dryland soils. The results highlighted the considerable impact of cyanobacteria on soil chemical properties (Fig. 3). They significantly increase soil organic matter, total nitrogen, available phosphorus, available sulfur, exchangeable calcium, and CEC and decrease soil acidity, exchangeable potassium, and sodium concentrations (Hossain et al., 2023). Cyanobacteria can significantly contribute to the mobility of essential minerals, redox status, soil water retention, soil texture, and improvement of soils deeply affected

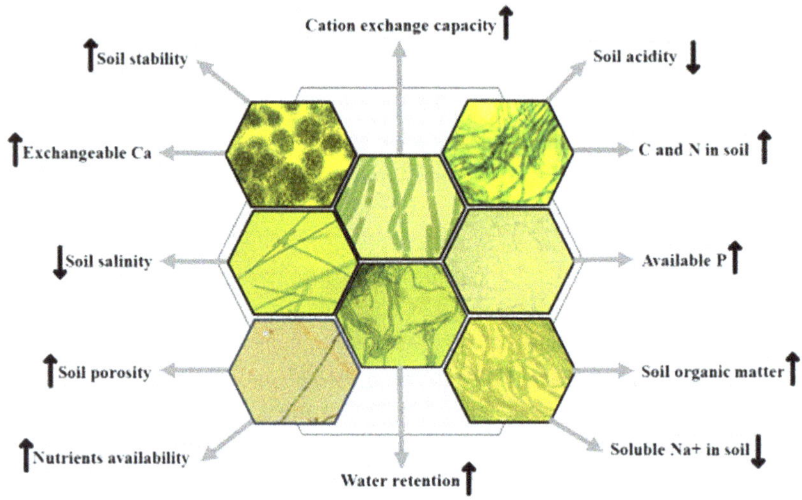

Figure 3 Roles and impacts of soil cyanobacteria in restoring soil physicochemical properties.

by high salt content. Asghari et al. (2022) demonstrated that inoculation of cyanobacteria resulted in a significant increase in the soil moisture characteristic curve of loamy and sandy loam soils.

Considering the growing awareness of the vital ecological role of cyanobacteria in arid areas as major promoters of BSCs formation, their inoculation into soils has been proposed as a novel sustainable biotechnological technique for the restoration of degraded soils in arid regions (Rossi et al., 2017; Chamizo et al., 2018, Román et al., 2018). Indeed, previous studies have suggested that co-inoculation of *Microcoleus vaginatus* with *Scytonema javanicum* has shown positive and promising results in increasing soil fertility and successfully restoring soil physicochemical properties (Wang et al., 2009). Cyanobacteria contribute to nutrient cycling, production of plant growth-promoting substances, desalination, and degradation of various organic and inorganic contaminants. They utilize various bioremediation mechanisms such as biosorption, bioaccumulation, and biotransformation of contaminants in addition to the specific mechanism employed (Hooda et al., 2023). Several studies have explored the application of cyanobacteria in the remediation of heavy metal-contaminated soils. Liu et al. (2025) examined the effects of living consortia of cyanobacteria and green algae (*Anabaena azotica*, *Anabaena variabilis*, *Chlorella vulgaris*, and *Scenedesmus obliquus*) on cadmium levels in soil and wheat, and revealed that the application of cyanobacteria significantly reduced exchangeable Cd in soil and limited Cd accumulation in wheat roots and stems, even at low doses.

2.2.6 Cyanobacterial-plant symbioses: signaling and development

Due to their high physiologically adaptive strategies, unicellular or filamentous cyanobacteria are unique prokaryotes capable of forming effective symbiotic associations with a wide and diverse range of eukaryotic hosts, including fungi, plants, and animals (Rai, 1990; Adams, 2000; Mishra, 2020). This symbiotic capacity is due to their photoautotrophy and photosynthetic efficiency, which allows them to carry out this activity even in very low light, which is the case for most symbiotic relationships (Usher, 2008). In these symbioses, the cyanobacterial partner (cyanobiont) provides the host with fixed nitrogen in exchange for fixed carbon, nutrients, and relatively protected habitats, free from predation and harsh environmental conditions (Adams et al., 2012). The host's cellular environment likely provides a micro-oxic niche for the cyanobacterial partner to efficiently fix nitrogen (Devaprakash et al., 2024). In the majority of these associations, the contribution of cyanobacterial symbionts is the product of nitrogen fixation, allowing hosts, such as plants and lichens, to occupy nitrogen-deficient areas as well as human-disturbed environments. However, cyanobionts are also capable of providing fixed carbon to non-photosynthetic hosts such as lichen fungi (Adams et al., 2006). Furthermore, through their symbiotic association with plants, cyanobacteria also assist plants in their metabolic processes (Gonçalves, 2021; Gonçalves et al., 2023). Although the majority of cyanobacterial symbionts are photoautotrophic, many also have facultative photo- or chemoheterotrophic activities, allowing them to inhabit symbiotic structures, such as plant roots, which receive little or no light and where photosynthetic hosts can provide them with fixed carbon (Adams et al., 2012).

An extensive literature, including numerous reviews and chapters, has addressed symbiotic cyanobacteria, host-cyanobiont interactions, and their environmental impacts (Rai, 2002; Adams 2002; Adams et al., 2012; Bustos-Díaz et al., 2019). For example, Adams (2000, 2012) published two illustrated chapters focusing on many known cyanobacterial symbioses and discussing host-cyanobiont interactions before and after infection. Although early research focused on plant associations such as those with *Gunnera* (angiosperm), *Azolla* (pteridophyte), and bryophytes (liverworts and hornworts), more recently, there has been a growing interest in other groups of host organisms. These studies have demonstrated the important ecological role of cyanobacteria, which interact in symbiosis with a wide variety of plants, covering a broad phylogenetic spectrum within the plant kingdom (Adams et al., 2012).

Cyanobacteria can form a wide variety of nitrogen-fixing symbiotic relationships with several plant representatives such as algae, fungi, bryophytes, pteridophytes, gymnosperms, and angiosperms, allowing host organisms to thrive in regions where they might otherwise not survive at all (Gaysina et al., 2019). Considerable attention has been paid to symbiotic associations between cyanobacteria and plants, particularly in rice paddies, due to their potential to increase plant growth, improve crop productivity, and promote sustainable agricultural practices. Symbiosis with the pteridophyte *Azolla* is well studied, mainly due to the prospective use of *Azolla* as an excellent biofertilizer in rice paddies (van Hove and Lejeune, 2002). *Azolla* cyanobionts, described as *Trichormus azollae* (formerly known as *Nostoc* or *Anabaena azollae*) (Baker et al., 2003), are hosted in specialized cavities in the fronds of the water fern (Lechno-Yossef and Nierzwicki-Bauer, 2002). Inside these cavities, the percentage of heterocysts rises to 25–30% and the whole cyanobiont becomes photosynthetically inactive (Peters, 1991). Unlike most cyanobionts, *A. azollae* is unable to grow independently of its *Azolla* host and remains intimately associated with the plant throughout its life cycle (Adams et al., 2012). The obligate cyanobiont is inherited from the parent plant, as it is attached to the fern's spores (Rasmussen and Johansson, 2002). In the context of rice cultivation, this permanent *Azolla*-cyanobiont symbiotic association has been used as a biofertilizer and has also been applied to the improvement of other crops, including banana, wheat, tomato, and taro (Vaishampayan et al., 2001; van Hove and Lejeune, 2002). This cyanobacterial inoculation has shown promising results in terms of improved yields and reduced dependence on synthetic fertilizers (Massey and Davis, 2023; Prasanna et al., 2021).

Several studies have shown that cyanobionts that successfully form plant symbioses are capable of forming motile filaments called hormogonia, which are essential for successful plant infection, and heterocysts, which are necessary for nitrogen fixation and the establishment of a functional symbiosis (Adams et al., 2012). They have also reported that many host plants secrete chemical signals that stimulate hormogonia production and serve as chemoattractants to guide hormogonia toward symbiotic structures. Once infected, host plants secrete hormogonia-repressing factors to ensure that the cyanobiont returns to vegetative growth and produces heterocysts to fix nitrogen for the host (Adams et al., 2012). Several studies have explored the complex relationships between cyanobacteria and plants, focusing on new insights and new research frontiers in this field (Gupta et al., 2021; Ramakrishnan et al., 2023). These investigations revealed two types of symbiotic associations with plants that play a crucial role in their growth, particularly through nitrogen fixation. Endophytic cyanobacteria live inside plant tissues, often forming well-defined symbiotic relationships (endosymbiosis), and epiphytic cyanobacteria, which grow on the surface of plants (exosymbiosis). The dominance of cyanobacteria in epiphytic microbial communities has been well established using light microscopy, molecular (16S rRNA) analysis, and genetic analysis (Hamisi et al., 2013). Although some epiphytic associations can establish mutually beneficial interactions, not all necessarily form true symbioses, as the degree of interdependence and mutual benefit can vary considerably (Nawaz et al., 2024). Many endophytic cyanobacterial symbionts are heterocystic forms and in many symbioses, particularly with plant hosts, the frequency of heterocysts is considerably high, as is the rate of N_2 fixation (Adams et al., 2012).

Except for *Azolla–Anabaena* symbiosis, the unique example of a hereditary plant–cyanobacterial association in which the cyanobiont is transferred from one generation to the next, most of the diversity studies of the cyanobionts from the individual plant hosts have shown high strain diversity both within and among different host species (Gaysina et al., 2019). The most common and prolific symbiotic partners with many plants belong to the heterocystic genus *Nostoc* that fixes N_2 and produces motile hormogonia, which serve for dispersal and as infectious units in symbiosis (Rai et al., 2002; Bergman et al., 2008). Other heterocystic genera such as *Anabaena*, *Aphanizomenon*, and *Trichormus*, also members of the family Nostocaceae, have been reported as cyanobionts more closely related to *Nostoc* (Singh et al., 2015). Many species of endophytic cyanobacteria, such as *Nostoc punctiforme*, *Nostoc azollae*, and A*nabaena cycadae*, inhabit specialized extracellular compartments or reside intracellularly in various plant species, from non-vascular plants such as hornworts and liverworts to vascular plants such as ferns and cycads (Nawaz et al., 2024). For example, *Nostoc azollae* lives in perpetual symbiosis in the leaf cavities of the *Azolla* fern, while *A. cycadae* inhabits the specialized coralloid roots of cycads. Epiphytic cyanobacteria, such as *Gloeothece* sp., *Leptolyngbya* sp., and *Phormidium* sp., form colonies on the surfaces of roots, stems, and leaves, contributing to plant health through nutrient exchange and protection against pathogens (Alam et al., 2023; Dey et al., 2022; Nawaz et al., 2024). While other species such as *Calothrix* sp. and *Scytonema* sp. can exist as both endophytes and epiphytes, highlighting their adaptability. These relationships demonstrate the remarkable versatility of cyanobacteria and their importance in plant ecology, spanning a wide range of plant species and providing essential benefits to their hosts (Nawaz et al., 2024). One of the most remarkable advantages of cyanobacterial strains is their ability to fix nitrogen (N_2), facilitated by the enzyme nitrogenase, and provide bioavailable nitrogen compounds to the host plant. Recent research has demonstrated that nitrogenase expression and activity can be influenced by the host plant, suggesting a bidirectional regulation of nitrogen fixation in a symbiotic context (Carrell et al., 2022; Pandharikar, 2020). Moreover, cyanobacteria–plant symbiosis can confer various other benefits to their hosts, such as enhanced nutrient uptake, production of plant growth-promoting substances, and improved tolerance to abiotic stresses such as drought, salinity, and heavy metal toxicity. These beneficial effects are mediated by a complex interplay of metabolic pathways and signaling cascades, which are currently the subject of extensive research

(Nawaz et al., 2024). Other recent studies have shown the role of molecular mechanisms underlying the establishment and maintenance of these symbiotic relationships (Ho-Plagaro and García-Garrido, 2022; Alvarez et al., 2023). These works have highlighted specific signaling molecules and receptors involved in the recognition and communication between cyanobacteria and host plants. For example, the production of phytohormones, such as auxins and cytokinins, by cyanobacteria has been shown to play a crucial role in modulating plant growth and development. Moreover, the exchange of small RNA molecules between symbiotic partners has emerged as a novel communication mechanism, regulating gene expression and facilitating the adaptation of both organisms to the symbiotic lifestyle (Yang et al., 2022).

2.2.7 Role in improving plant growth and stress tolerance

The potential of cyanobacteria as a novel source of biofertilizers has been well documented for improving plant growth, crop yield, modulation of soil microbial activity, and nutrient characteristics (Renuka et al., 2018). In recent years, cyanobacteria have been widely explored for their role as biofertilizers in improving plant growth and stress tolerance. Moreover, several studies have revealed the potential of cyanobacterial-based biostimulants as plant growth promoters and stress mitigating agents, highlighting their promising potential applications in agriculture (Roque et al., 2023; Chabili et al., 2024; Jimenez-Ríos et al., 2024; Bibi et al., 2024; Nawaz et al., 2024; Sathyananth and Raj 2024; Li et al., 2025). Cyanobacteria promote plant growth through carbon sequestration, nitrogen fixation, phosphorus mobilization, exopolysaccharides, and phytohormones production, indicating their great potential as valuable inputs in agricultural systems (Hakkoum et al., 2024). Through their efficient bioconversion of solar energy, cyanobacteria fix billions of tons of CO_2 in the form of biomass and energy-rich molecules (proteins, lipids, polyunsaturated fatty acids, alkanes, carbohydrates, polysaccharides, pigments, etc.), secondary metabolites (vitamins, amino acids, etc.), and phytohormones (cytokinins, auxins, gibberellins, jasmonic acid, etc.). These bioactive metabolites, such as phytohormones and polysaccharides, play a key role in plant development and can stimulate growth, improve crop yield and quality, and mitigate the effects of salt stress (Fig. 4) (Tarakhovskaya et al., 2007; El Arroussi et al., 2016; Chanda et al., 2021).

Soil inoculation with live cyanobacteria can enhance nitrogen supply through biological fixation (Win et al., 2018) and soil organic matter through biofilm formation (Renuka et al., 2018). Nitrogen-fixing cyanobacteria have recently been highlighted as promising plant growth-promoting bacteria (PGPB) (Iniesta-Pallares et al., 2021; Zhang et al., 2021b; Singh et al., 2016). Nitrogen fixation occurs in specialized cells called heterocysts, facilitated by

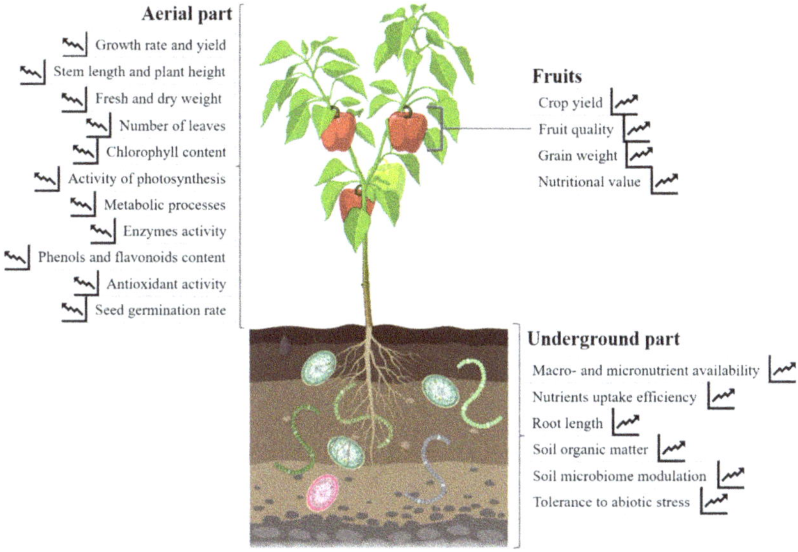

Figure 4 Benefits and effects of cyanobacteria on crop growth, yield, and stress tolerance.

the action of the nitrogenase enzyme complex (Kumar et al., 2010). These diazotrophic cyanobacteria have been widely used as biofertilizers, mainly for rice crops. Several species of cyanobacteria, such as *Nostoc muscorum*, *Anabaena variabilis,* and *Tolypothrix tenuis,* have been found to be effective biostimulants. Many Asian countries, such as China, Vietnam, India, etc., have used these cyanobacterial species in rice cultivation as an alternative to nitrogen fertilizers (Singh, 1961; D'Acqui, 2016). It has been reported that nitrogen availability to plants is increased by the application of cyanobacteria in agricultural ecosystems, especially rice paddies (Azari et al., 2011; Singh et al., 2014b). The positive effect of cyanobacteria is not only due to their nitrogen-fixing activity but also due to the production of bioactive substances and their stimulatory influence on soil heterotrophic bacteria (Singh et al., 2016; Prasanna et al., 2017; Singh et al., 2022). Other studies have reported the positive effect of cyanobacteria on other crops, such as wheat, pea, maize, bean, sugar beet, pumpkin, cucumber, potato, and tomato, highlighting the increasing expansion in the use of cyanobacterial-based biofertilizers (Osman et al., 2010, Rana et al., 2012, Prasanna et al., 2015, 2017; Shariatmadari et al., 2013; Swarnalakshmi et al., 2013).

Water scarcity and soil salinity are major constraints in arid and desert regions and negatively impact plant growth and development. Due to their multiple adaptive capacities, cyanobacteria have proven tolerant to desiccation, water stress, and salt stress. They produce secondary metabolites that improve soil condition and nutrient bioavailability, and protect plants against biotic and abiotic stresses. In addition to their role as growth promoters for various crops, cyanobacteria can also be used as anti-stress agents. Several studies have demonstrated that cyanobacteria can produce various compounds (antioxidants, carotenoids, tocopherols, stress enzymes) that improve plant tolerance to drought and salt stress, leading to better crop performance under adverse environmental conditions (Chanda et al., 2021; Bello et al., 2022; Alharbi et al., 2023). Furthermore, other compounds (phenolics and thiamine) that regulate the antioxidant properties of cyanobacteria are enhanced under abiotic stress (Assunçao et al., 2023). The use of cyanobacterial cell suspensions or extracts revealed their potential for stress adaptation as they helped stimulate defense mechanisms and prepare against potential sources of stress (Chabili et al., 2024; Hakkoum et al., 2024). It has been shown that cyanobacterial polysaccharides can mitigate ROS toxicity in plants by enhancing the production of antioxidant enzyme activities (Gonçalves, 2021; Chanda et al., 2019).

2.2.8 Role in biocontrol of pathogens and pests

Considering the harmful effects of using chemical pesticides to prevent crop losses due to pests and pathogens, biological control of plant diseases offers a safe alternative to reduce the adverse effects of chemicals on the environment, human, and animal health (Alvarez et al., 2021; Costa et al., 2019). Cyanobacteria are important sources of antifungal and partially antibacterial compounds (Borowitzka, 1995; Burja et al., 2001). They can function as biological control agents against soil-borne pathogens and facilitate the establishment of agriculturally important rhizospheric communities as well as plant protection against phytopathogens (Solomon et al., 2023). They have an antagonistic effect against many plant pathogens such as bacteria, fungi, and nematodes (Renuka et al., 2018; Alvarez et al., 2021).

There are more publications on the antimicrobial activity of marine and freshwater cyanobacteria than on terrestrial cyanobacteria due to the higher shot number (Kadam et al., 2013). Cyanobacteria are well known to produce allelopathic chemicals – secondary metabolites – that can be used as algicides, fungicides, herbicides, insecticides, and nematicides. Some cyanobacterial compounds also have antimicrobial, antiviral, and antiprotozoal activities (Berry et al., 2008; Hernández-Carlos and Gamboa-Angulo, 2011; Holajjer et al., 2013). In addition, cyanobacteria, especially *Nostoc*, *Anabaena*, and *Tolypothrix* strains, are highly rich in bioactive compounds, including polyphenols, biocidal compounds such as benzoic acid and majusculonic acid, oils, and pigments with antimicrobial properties (Chaudhary et al., 2012; Gupta et al., 2013). These compounds can eradicate pathogenic microbes by disrupting the cytoplasmic membrane and inhibiting protein synthesis (Swain et al., 2017).

Extensive literature data confirm the antifungal effect of many cyanobacterial species against plant pathogens. Stewart et al. (1988) examined 700 strains of cyanobacteria isolated from aquatic and terrestrial ecosystems, and 9% of the strains were antifungal against at least one pathogen. The extracellular products of *Nostoc muscorum* inhibit *Rhizoctonia solani* (Caire et al., 1990). According to Burja et al. (2001), biologically active cyanobacteria, including fungicides, mainly belong to the Oscillatoriales and Nostocales. The filtrate of *Spirulina platensis* and *Nostoc muscorum* inhibits the growth of *Fusarium oxysporum* and *Rhizoctonia solani* (Tantawy, 2011).

Other cyanobacterial strains of *Oscillatoria*, *Lyngbya*, *Anabaena*, *Phormidium*, and *Synechococcus* have also shown effective activity against at least one pathogen, including *Fusarium oxysporum* (Pawar and Puranik, 2008). Other studies have shown that cyanobacterial extracts can enhance plant protection against pathogens by releasing antifungal substances that exert phytopathogenic biocontrol (Swain et al., 2017). Some cyanobacteria can secrete HCN, which functions as an antibacterial and antifungal substance, enabling the biocontrol of plant pathogens (Arafa and Abd El-All, 2013).

Cyanobacteria-based biopesticides provide effective pest control through various mechanisms, including inhibiting the growth, nutrition, development, or reproduction of a pest or pathogen. To achieve this, cyanobacterial strains release enzymes and endotoxins capable of degrading the cell wall of invasive microorganisms and even causing viral infections in pests (Costa et al., 2019). Research on the biocidal effects of cyanobacteria has revealed unprecedented potential for the development of new pest control methods. Further research is needed to validate their spectrum and applications for commercial purposes (Win et al., 2018).

3 Cyanobacterial biomass as a potential resource for sustainable agriculture

3.1 Biochemical biomass characterization

Reports of massive freshwater bloom-forming cyanobacteria have increased across the world, and some of the factors leading to the formation of such blooms are still unsolved. The dense scums observed during the proliferation of some cyanobacterial species such as *Microcystis* spp., *Planktothrix* sp., and *Dolichospermum* spp. on the surface of the water of lakes and reservoirs could reach tens of tons (Fig. 5). This dense biomass, a source of many nutrients, particularly nitrogen and phosphorus, produced ecologically and free of charge, is therefore easy to harvest for possible use as a sustainable biofertilizer. However, since this dense biomass is often considered harmful, few studies have focused on its biochemical characterization. Indeed, approximately

Figure 5 A cyanobacterial *Microcystis* sp. bloom forming scums in freshwater.

25–70% of cyanobacterial blooms are toxigenic (Rogers et al., 2005) producing different classes of cyanotoxins.

Many studies revealed the effects of cyanotoxins at cellular, histological, and whole-organism levels on agriculturally important crops (peas, lentils, alfalfa, lettuce, white mustard, maize, wheat, rice, etc.) via contaminated irrigation water (Weiss et al., 2000; Pflugmacher, 2002; Mitrovic et al., 2005; Máthé et al., 2007; Peuthert et al., 2007; Saqrane et al., 2008; Corbel et al., 2014; Bouaïcha and Corbel, 2016). These findings highlight the imperative for further comprehensive investigations to elucidate the implications of utilizing cyanotoxin-containing cyanobacterial biomass as biofertilizers in agricultural practices, particularly when elevated toxin concentrations are detected. Once cyanotoxins are in the soil, they can accumulate in the plants, which poses a potential risk to humans and animals (Bouaïcha and Corbel, 2016). Consequently, the majority of global studies on cyanobacterial biomass have focused on the detection and chemical characterization of the cyanotoxins and other bioactive compounds as secondary metabolites. However, to our knowledge, no study has addressed the biochemical characterization of primary metabolites such as proteins, lipids, carbohydrates, and essential minerals of environmental cyanobacterial bloom samples. The amount of produced cyanotoxins, which are considered secondary metabolites, relative to the total biomass remains low compared to other primary metabolites (e.g. proteins, carbohydrates, lipids) and essential minerals that can be used as a sustainable source of organic and inorganic fertilizers. For example, the total contents of microcystins (MCs) in dried dominant *Microcystis* bloom samples, which are considered the most reported cyanotoxin class worldwide (63%) (Svirčev et al., 2019), reached in the UK a value of 3.57 mg MC-LR equivalents/g (Ward et al., 1997), in Germany a value of 2.763 mg MC-LR equivalents/g (Fastner et al., 1999), and in Algeria a value of 1.12 mg MC-LR equivalents/g (Nasri et al., 2008); corresponding to 0.35, 0.28, and 0.11% dry weight biomass, respectively. However, chemical analysis of the dry biomass of *Spirulina platensis*, a non-toxic cyanobacterial species produced in polyethylene transparent containers (1000 liters capacity) under continuous aeration, at 30°C and pH 9.00, revealed that it contains 6.7; 2.47; 1.14% on a dry basis of nitrogen (N), phosphorus (P), and potassium (K), respectively (Aly and Esaewy, 2008). Recently, Baracho and lambardi (2023) evaluated 20 cyanobacterial strains belonging to different genera (*Gloeothece, Aphanizomenon, Geitlerinema, Nostoc, Inacoccus, Aphanocapsa, Microcystis, Trichormus, Synechococcus, Sphaerocavum, Phormidium, Rhabdoderma, Leptolyngbya*) regarding their biochemical compositions. The organisms were cultivated in cylindrical photobioreactors for 144 h and the levels of proteins, carbohydrates, and lipids were determined in the dried biomass. The percentage of proteins, carbohydrates, and lipids in dry biomass after 144 h of cultivation ranged between 10.3–77.9, 1.7–27.7, and 2.2–15.2% among the cultivated species, respectively. The two species *Geitlerinema unigranulatum* and *Geitlerinema* sp. produced the highest amounts of total proteins, ranging from 75.8 to 77.9%, respectively. The highest total carbohydrate contents were found in *M. navacekii* (27.7%) and *Aphanizomenon* sp. (22.2%). While for lipid contents, the species *Phormidium* sp. and *Sphaerocavum brasiliense* showed the highest values with 14.8 and 15.2% of dry biomass, respectively.

A similar study was carried out for the evaluation of the levels of carbohydrates, lipids, and proteins of 12 strains of filamentous nitrogen-fixing cyanobacteria belonging to the genera *Anabaena* (actually *Dolichospermum*), *Anabaenopsis*, *Nodularia*, and *Nostoc* grown under diazotrophic conditions (Vargas et al., 1998). The percent composition of proteins, carbohydrates, and lipids in the dry biomass of these filamentous species ranged between 37 and 52%; 16 and 37%; and 8 and 13%, respectively. In addition, nucleic acids, which are a very important source of carbon, nitrogen, and phosphorus, represent 8–11% of the dry weight, thus making cyanobacteria suitable for several applications, including agriculture. Although the biochemical composition and content of these macromolecules are not often reported for field-collected bloom samples, it is important to note that several studies carried out on many species of cyanobacteria grown under controlled laboratory conditions have shown that cyanobacteria remain a very important source of nutrients that could be exploited as biofertilizers. Among these species that have been cultivated under controlled conditions and whose richness in proteins, carbohydrates and lipids has been highlighted, some belonging to the genera *Microcystis*, *Anabaena*, *Nostoc*, Phormidium, *Nodularia, Spirulina,* and *Aphanizomenon*, which are often responsible for the formation of dense blooms in aquatic ecosystems (Bernard et al., 2016; Benayache et al., 2019).

For example, *Spirulina* (previously known as *Arthrospira* and actually as *Limnospira*) harvested in Lake Texcoco (Mexico) and in Lake Chad has been used as food or medicine by humans in Asian, African, and South American countries for centuries (Ciferri, 1983; Johnson et al., 2008; Qiu et al., 2005; Lafarga et al., 2020). Its nutritional richness, notably for protein content (57–70% of dry weight), beneficial fatty acids (α-linoleic and palmitic acids), vitamins and vitamin precursors (B12, astaxanthin, zeaxanthin, and β-carotene), and minerals (iron, potassium, calcium, zinc, and selenium), has led the Food and Agriculture Organization of the United Nations (FAO) to consider it as a food of major interest in combating malnutrition and responding to food emergencies (Capelli and Cysewski, 2010). Commercially available food supplements based on cyanobacteria often do not come from only harvesting from natural lakes but are also produced in open ponds around the world for a total of 56,000 tons fresh weight in 2019 (FAO, 2019; Furmaniak et al., 2017; Soni et al., 2017). Two other species of the genus *Nostoc*, *Nostoc commune,* and *Nostoc flagelliforme* have also been consumed by humans for centuries in China (Gao, 1998; Takenaka et al., 1998; Potts, 2000), in the island of Java, Indonesia (Zaneveld, 1959) and by the indigenous people of the Peruvian highlands (Jonhson et al., 2008). Since 1983, another species, *Aphanizomenon flos-aquae*, collected from Upper Klamath Lake in southern Oregon (United States), has been marketed as a dietary supplement and is commonly consumed in the United States, Canada, and Europe for its purported beneficial effects (Gilroy et al., 2000). Therefore, the presence of several molecules such as complex carbohydrates, lipids, vitamins, amino acids, peptides, and minerals in the scums formed by cyanobacterial blooms highlights the interest in using this dense environmental biomass as a sustainable source of biofertilizers, returning both nutrients and carbon to the soil, potentially improving soil quality, crop growth, and nutrition.

3.2 Role as biofertilizers over traditional organic and chemical fertilizers

Global population growth and climate change are putting increasing pressure on the agricultural system, posing a lasting threat to nutritional security for population worldwide. To meet their nutritional needs, most countries around the world have relied on conventional mineral fertilizers for decades. For example, total global agricultural consumption of elemental N from synthetic fertilizers reached 107.7 Mt in 2018, with China, India, the United States, the EU28, and Brazil accounting for 68% of the total N use (Menegat et al.,

2022). However, the excessive use of these chemical fertilizers, often under uncontrolled conditions, contributes to carbon dioxide, nitrous oxide, and methane emissions, which pose a serious threat to human and environmental health (Jain, 2023; Menegat et al., 2022; Chen et al., 2019; Rouwenhorst et al., 2021).

Furthermore, to reduce dependence on mineral fertilizers, organic fertilizers such as animal manure, biosolids from human waste, anaerobic digestate, biochar, and crop residues are used as alternatives (Farrell et al., 2014; Rady, 2011; Walsh et al., 2012). Among these fertilizers, manure, biosolids, and digestate have been considered for several decades and continue to be the potentially most important sources of nutrients in agriculture. Nevertheless, after a few years of excessive use, these complex materials have become less and less recommended in many countries for some of them, such as activated sludge from wastewater treatment plants. These are linked to the presence of drug residues (Habimana and Sauvé, 2025) and certain growth-promoting metals, such as copper and zinc, in pig feed (Ciesinksi et al., 2018) and other toxic metals such as arsenic, cadmium, lead, and mercury (Heimann et al., 2015; Wuang et al., 2016; Zhang et al., 2015), which present ecological risks and threaten food safety. Therefore, the use of algae-based biofertilizers in agricultural management programs reduces the over-reliance on all these conventional fertilizers and offers an alternative strategy for sustainable and resilient agriculture (Sharma et al., 2014; Shukla et al., 2025).

For example, seaweeds (brown, green, and red) have been used successfully for thousands of years, mainly in the form of extracts, as potent organic agricultural inputs for better plant growth, yield, and nutrient use efficiency (Arioli et al., 2015; Milledge et al., 2016; Katakula et al., 2020; Mendez et al., 2023). Offering, therefore, beneficial biological effects for all agricultural plants, as well as for animals and humans. As recently reviewed by Shukla et al. (2025), 97% of harvested seaweed biomass comes from cultivation, representing a volume of 36 million tons of wet weight, while the remaining 3% (1.1 million tons of wet weight) of global production comes from the exploitation of wild populations washed up on beaches. These wild algae proliferate relating to anthropogenic eutrophication, such as green tides of Ulva or red tides of Solieria, and then end up depositing on various beaches, mainly due to the influence of currents, winds, and tides (; Van Alstyne et al., 2015; Harb and Chow, 2022). Therefore, they can represent an abundant and accessible biomass without production costs, potentially exploitable as biofertilizers in agriculture. However, these wild marine harvested seaweeds are often co-mixed and contain an excess of salts, which can contribute to soil salinization resulting in loss of their fertility and therefore harmfully impact the environmental functions of cultivated agricultural land (Fu et al., 2020; Wei et al., 2019;). In addition, many aspects such as climate change as well as seasonal and environmental factors often influence the chemical composition and biochemical characteristics of this wild-harvested macroalgae biomass, making their standardization and reliability as biofertilizer unsustainable and difficult to manage (Jansen et al., 2022). Therefore, research on microalgae-based biostimulants is becoming increasingly popular. Among the most commonly cultivated microalgae species are *Chlorella* spp. and *Dunaliella* spp. (Spolaore et al., 2006). Interest in using microalgae in agriculture has been justified for several reasons such as their rapid growth under controlled conditions, their positive effects on crop development, growth and yields, and the provision of reliable sources of biomass with predictable composition, necessary to achieve economic viability (Carvalho et al., 2006; Colla and Rouphael, 2020; Dagnaisser et al., 2022; Gonzalez-Perez et al., 2022; Prisa and Spagnolo, 2023; Ronga et al., 2019;). Although known to produce a raw material homogeneous in its composition in bioactive compounds and suitable for many applications in agriculture, microalgae have a much more limited commercial implementation, perhaps due to their higher production costs. In contrast, the large amount of cyanobacterial biomass that could be harvested during massive blooms, which has a protein, lipid, carbohydrate, and mineral composition comparable to microalgae, constitutes an alternative source of biofertilizers without production costs (Anandakumar, 2023). The exploitation of this cyanobacterial biomass as 'cyanofertilisant' also constitutes a viable circular option as part of a broader management strategy aimed at managing harmful cyanobacterial blooms in freshwater ecosystems while producing low-cost sustainable fertilizers.

Various organic and inorganic contaminants present in organic fertilizers such as animal manure and activated sludge from wastewater treatment plants can have significant short- and long-term effects on soil microbial communities. Therefore, the application of cyanofertilizers should be further explored, due to their availability in large quantities at low cost, as well as their richness in nitrogen, phosphorus, and essential minerals. For example, Ritika and Utpal (2014) reported that inoculation with cyanobacteria could save 25–40% of chemical nitrogen fertilizers in soil. However, the potential benefits of recycling this cyanobacterial biomass on agricultural land, due to its potential impact on soil microbial biomass and fungi for improved soil health, in addition to its benefits for agricultural production and the environment in a 'circular economy,' have received little attention from researchers and farmers. The fact that cyanobacterial blooms are often toxic has led to limiting their possible exploitation in agriculture as biofertilizers. However, several studies have reported that the filamentous non-toxic cyanobacterial species *Arthrospira platensis* when inoculated in *Phaseolus aureus* seeds (Bhowmik et al., 2010) or applied as a biofertilizer in *Begonia semperflorens* (Jowkar et al., 2017), rice (Dineshkumar et al., 2018), tomato (Supraja et al., 2020), or groundnut (Sivalingam, 2020), increases seed germination and plant growth. In another study, Khurshid et al. (2017) reported that the non-nitrogen-fixing unicellular cyanobacterium *Synechocystis* sp., which grows in soil and also proliferates massively in aquatic environments but without producing toxins, induces high production of phytohormones in wheat when inoculated into its rhizosphere.

Recently, Andrade et al. (2024) conducted field experiments to compare nitrogen fertilization with 100 kg/ha N versus single inoculation with *Bradyrhizobium arachidis*, co-inoculation with *Arthrospira platensis* or *Synechocystis* sp., on groundnut growth, nodulation, tissue nitrogen accumulation, grain protein concentration, and grain yield. Seeds were treated with a cell suspension of each bacterial or cyanobacterial species, containing approximately 10^6 cells/mL at 12 mL/kg of seeds, and then sown manually at 60 kg of seeds/ha. The results obtained during this study showed that inoculation treatment and nitrogen fertilization had no effect on shoot or root dry weight. However, in a clay soil, co-inoculation of the seed plant with *B. arachidis* and the two cyanobacterial species *Arthrospira platensis* and *Synechocystis* sp. increased grain productivity by an average of 19% compared to the uninoculated control. Furthermore, in this high-P clay soil, whether co-inoculated with *B. arachidis* or the two cyanobacterial species or inoculated alone, grain productivity was on average 16% higher than that resulting from nitrogen fertilizer application. These two non-nitrogen-fixing cyanobacteria can colonize many sections of groundnut tissues, such as roots and shoots, prompting *B. arachidis*, a microsymbiotic nitrogen-fixing bacterium that nodulates legume roots, to fix nitrogen and solubilize phosphorus. Thus, the plant's development, nutritional status, and defense mechanisms, as well as soil fertility, are improved. In addition to being rich in nitrogenous products such as proteins, many filamentous diazotrophic cyanobacterial species are capable of fixing dinitrogen present in the atmosphere in specialized cells known as heterocysts and using it as a source of nitrogen in soil for plant growth. For example, Guo et al. (2020) reported that inoculating rice plants with nitrogen-fixing filamentous cyanobacteria reduced chemical fertilizer use by 50% while maintaining grain yield and quality. Another study also showed that a biofertilizer based on the nitrogen-fixing species *Nostoc entophytum* could improve the nutritional value of pea plants while reducing chemical fertilizer use by 50% (Osman et al., 2010).

3.3 Role as potential bioasset for restoring degraded land

The United Nations, in its Sustainable Development Goals (SDGs) adopted in 2015, included land degradation among the international targets to be achieved by 2030. The SDG 15.3 states that 'combat desertification, restore degraded land and soil, including land affected by desertification, drought and floods, and strive to achieve a land degradation-neutral world' (United Nations,

2015). The intensification of agricultural production in response to the global growing population and increasing food demand has led to the increasing use of chemical fertilizers, which are considered among the main factors of land degradation and declining fertility of arable soils. Climate change exacerbates these processes by increasing heat stress and drought, and intensifying precipitation and wind action, leading to erosion of the most fertile surface layer. Currently, to minimize the negative effects of these chemical fertilizers, modern agriculture aims to improve crop quantity and quality by favoring the use of biofertilizers, which offer several advantages over conventional fertilizers. Several studies have shown that these biofertilizers are an important source of nitrogen, phosphorus, essential minerals, and bioactive compounds capable of improving soil fertility and, therefore, stimulating plant growth and yield by providing protection against biotic and abiotic stresses (Alvarez et al., 2021; Kumar et al., 2022; Guo et al., 2020; Çakirsoy et al., 2022; Fernandes et al., 2020; Gonçalves et al., 2023; Osorio-Reyes et al., 2023). In recent decades, the use of biofertilizers based on microalgae produced under controlled conditions or harvested marine macroalgae resulting from eutrophication has grown to reduce dependence on synthetic fertilizers (Sharma et al., 2014; Shukla et al., 2025). However, despite these advantages, the success of microalgae-based biofertilizers depends on the cost-effectiveness of biomass production and processing, as well as the implementation of energy-efficient technologies in the production process.

However, cyanobacteria, due to their capacity for excessive proliferation in aquatic ecosystems, can form abundant, accessible, and inexpensive biomass, which is rich in nitrogen, phosphorus, and mineral elements, offering numerous benefits for improving soil fertility, optimizing nutrient management, and reducing dependence on conventional fertilisants (Passos et al., 2023). Furthermore, as pioneer colonizers of poor and highly degraded soils in arid zones, they play a key role in the development of the biological crust. Asghari et al. (2022) investigated the impact of cyanobacteria inoculation on soil hydraulic properties to improve soil water behavior, in order to ensure better irrigation management on farms. To this end, *Nostoc* sp. and *Oscillatoria* sp. were inoculated on loam and sand textures at a small cylinder lab-scale, and after 35 days, the moisture characteristic curve and saturated hydraulic conductivity were measured. The results showed that the inoculation of cyanobacteria led to a significant increase in the moisture characteristic curve of the loam and sand soils, reaching 36–65% and 16–31%, respectively. However, in the control treatment, the volumetric water content of the loam and sand textures was 28–58% and 8–24%, respectively. However, cyanobacteria inoculation to promote biocrust formation and improve soil properties against erosion has shown positive results in laboratory studies but limited success when applied under field conditions. For example, the application of this technique has shown successful results in laboratory studies for different cyanobacterial species and on a variety of soil types, including desert soils (Roman et al., 2018), burned soils (Chamizo et al., 2020), and agricultural soils (Maqubela et al., 2009). However, field establishment of these laboratory-grown organisms can be challenging. Indeed, large-scale biomass production of cyanobacteria and their successful field application have only been documented for a few species, such as *Microcoleus vaginatus*, *Scytonema javanicum*, and *Phormidium tenue*, found on sand dunes in China (Chen et al., 2006; Park et al., 2017; Wang et al., 2009; Xie et al., 2007).

Soil salinization is also one of the arable soil degradations that occur mainly in arid and semi-arid areas and result from a shallow groundwater table, poor drainage systems, and the use of high salinity irrigation water (D'Odorico et al., 2013; Sahab et al., 2021). Therefore, drought and salinization of arable soils, caused by human activity and climate change, could potentially affect soil fertility, reduce agricultural yields, and thus exacerbate risks to global food security (Singh, 2021). In addition to their potential use as biofertilizers to fertilize soils and thus increase plant growth and crop yield, cyanobacteria may also have a potential role in the restoration of degraded soils. Cyanobacteria, due to their ability to cope with drought, high temperatures, and high salt concentrations, could thus be used to restore soils degraded by drought and salinization. Many of these benefits are related to their ability to synthesize and release large amounts of exopolysaccharides, which act by binding soil particles, increasing soil stability, and reducing soil erosion during periods of drought (Roman et al., 2020). In addition to promoting biocrust formation, cyanobacteria have the ability to actively export ions via K^+/Na^+ channels and Na^+/H^+ antiporters, and to produce defense enzymes, which are possible mechanisms for their remediation of salt-affected soils (Singh et al., 2016; Li et al., 2019). Some cyanobacterial strains also have the ability to tolerate salt by secreting extracellular polysaccharides into the soil to protect their cells from salt stress (Chen et al., 2006; Lan et al., 2010). In recent years, numerous studies have demonstrated the effectiveness of cyanobacteria for the remediation of saline soils, in laboratory and field studies (Apte and Thomas, 1997; Jaiswal et al., 2010; Rady et al., 2018; Roman et al., 2018; Singh et al., 2011; Singh and Singh, 2015). For example, a pot experiment using different strains of cyanobacteria collected from the cotton fields of Navoi and Samarkand regions in Uzbekistan was conducted to evaluate their restorative capabilities in degraded soils affected by salt (Taira et al., 2021). Results of this study showed that inoculating cotton crops with these strains reduced salt stress as well as leaching by lowering soil Na+ content and electrical conductivity. In addition, inoculating these salt-tolerant strains also provided mineral and organic nitrogen to the soil and promoted cotton growth. Abdel El-Baky et al. (2010) reported that aqueous extract of *Spirulina* spp. improves wheat tolerance to salinity and induces antioxidant capacity and protein content of whole grains. In another study, Jesus et al. (2015) showed that the combination of salt-tolerant plants and cyanobacteria could remove salts from the soil and improve its fertility in a sustainable manner. The main advantage of this approach to remediating saline soils using cyanobacteria is its environmental friendliness compared to chemical methods (Li et al., 2019).

Despite the potential of cyanobacteria in restoring degraded soils, the need to produce large amounts of biomass under controlled conditions, the difficulty of extrapolating laboratory data to field conditions, and the lack of understanding of the effects and mechanisms on soil and plants under various conditions continue to hamper their widespread use in agricultural settings. In addition, the majority of studies are based on the inoculation of live species of cyanobacteria produced under controlled conditions. However, few studies have focused on the exploitation of large biomass from natural blooms. These biomasses can be available in large quantities and without production costs, but their contamination by cyanotoxins currently presents a barrier to their potential exploitation in agriculture. Some non-toxin-producing species, such as *Spirulina*, which is often harvested from the wild, have found widespread application as a valuable biological resource in many sectors, including the food and pharmaceutical industries. In contrast, proliferations of other species often considered toxic have not yet attracted interest in agriculture, although they are considered a promising and sustainable alternative to agrochemicals, offering numerous advantages for improving soil fertility, optimizing nutrient management, and reducing dependence on synthetic fertilizers. Recently, Massa et al. (2024) investigated the potential of using freeze-dried biomass of a non-cylindrospermopsin-producing strain of *Raphidiopsis raciborskii* and a cylindrospermopsin-producing strain as a sustainable resource for lettuce cultivation. The results obtained showed that no significant difference in photosystem II efficiency was observed with the addition of the two strains. However, the fresh weight of shoots significantly increased in lettuce plants grown with the addition of cyanobacterial biomass, especially with the toxic strain. Moreover, when incorporating the toxic strain (up to 40 μg cylindrospermopsin/g), the concentration of cyanotoxins in edible tissues did not exceed the WHO-proposed tolerable daily intake of 0.03 μg/kg/day. This recent study highlights the potential of the available low-cost toxic and non-toxic cyanobacterial biomasses as a sustainable and environmentally friendly alternative to conventional fertilizers. Nevertheless, caution is warranted regarding the potential use of cyanobacterial biomass containing cyanotoxins as biofertilizers. Therefore, further research is needed on high cyanotoxin concentrations or repeated exposures to more accurately assess the benefit/risk ratio of these biomasses of cyanobacteria available at low cost as cyanofertilizers in agriculture.

Excessive nitrogen input from chemical fertilizers into the soil could alter the rate of various nitrogen transformation processes, leading to acidification and decreased soil fertility by affecting microbial transformations and consequently reducing crop growth and yield (Bolan et al., 2003). However, *Spirulina platensis* biomass, added to soil as a biofertilizing agent, has been shown to improve soil macronutrients (nitrogen, phosphorus, and potassium) and increase crop growth (Aly and Esawy, 2008). For example, Wuang et al. (2016) demonstrated that adding 5 g of *Spirulina* biomass in 500 g of soil increased the height of Bayam rouge (red spinach) by 58.3% and the fresh and dry weights by 110.1% and 155.8%, respectively, compared to the control group. Recently, Alobwede et al. (2019) tested the effects of *Arthrospira platensis* (Spirulina) additions at 0.2, 2, and 4 g/kg on arable soil in greenhouse-grown pea plots and at 24 g/m^2 in the field on spring wheat on plant growth and nutrition and on soil physical and chemical properties. In the greenhouse study and at the highest application rate (4 g/kg), Spirulina increased total soil nitrogen and nitrate levels, as well as available phosphorus. Similarly, in the field study, the addition of spirulina increased soil inorganic nitrogen levels, confirming its potential to recycle mineralizable nitrogen in agricultural soils; however, no significant effect was observed on wheat yield at this application rate.

3.4 Optimizing the use of cyanobacteria-based biofertilizers

To meet the ever-growing needs of the sustainable biofertilizer market, fresh seaweed harvested from the foreshore or coast and microalgae cultivated under controlled conditions currently represent the main source of raw materials for the production of these biofertilizers. Various commercial entities utilize different extraction procedures for the production of algae-based biostimulants in liquid or soluble powder form prepared from both dry and wet biomass (Kadam et al., 2013; Michalak and Chojnacka, 2015; Shukla et al., 2025). In the biostimulant market, total algal extracts represent the largest category of biostimulants produced worldwide (i.e. 40%), followed by humic and fulvic acids (36%), microbial biostimulants (10%), amino acids (10%), and minerals and others (4%) (Shukla et al., 2025). However, the main challenges in developing seaweed-based biostimulants depend on the freshness of the harvest, its accessibility, pretreatment (preservation such as sun-drying, freezing, and storage conditions), extraction processes, and the origin of the seaweed (age, biochemical composition, seasonality). These different factors must be taken into account, as each can have a significant impact on the bioactive molecule composition of the extract and its fertility potential for soil/plant systems. Therefore, to sustainably meet market needs, predictable and reliable access to low-cost, wild-harvested or cultivated algal biomass with stable biochemical composition is both a prerequisite and a limiting factor in meeting the growing demand for biofertilizers in agriculture. Nevertheless, the increasingly observed global cyanobacteria blooms resulting from eutrophication and global warming may represent abundant, freely available, and accessible biomass, potentially exploitable in sustainable agriculture. The benefits of these cyanobacterial blooms as biofertilizers include:

- Their contribution to soil fertility by releasing nitrogen, phosphorus, and other micronutrients, thereby enriching the overall nutrient content and restoring degraded arable soils;
- Their positive impact on the formation of the biocrust, facilitating water retention, aeration, and drainage, thus promoting optimal plant growth conditions and increasing crop yields;
- Their richness in bioactive compounds, which act as natural agents to suppress soil pathogens and strengthen plant resistance to disease; and
- Their abundance, free production, and cost-effective application provide a cost-effective alternative to conventional fertilizers and minimize their environmental impact.

Despite these various advantages, the presence of cyanotoxins in some natural blooms limits their use as a source of biostimulants in agriculture. If these cyanotoxins present in a cyanofertilizer are introduced into a soil ecosystem, they can affect plants at the cellular, histological, and whole-organism levels, inhibiting their growth and thus reducing agricultural yields (Saqrane et al., 2008; Corbel et al., 2014; Nowruzi et al., 2021). In addition, several studies have shown that microcystins could be absorbed through roots and accumulated in grains and/or fruits, which pose a potential risk to humans and animals (Corbel et al., 2014; Backovi and Tokodi, 2024; Haida et al., 2024). Once these cyanotoxins are present in the soil, they can also be transported to other bodies of water through leaching, runoff, or drainage, contaminating therefore drinking water supplies (Bouaicha and Corbel 2016). However, microcystins, the most commonly occurring cyanotoxins in natural blooms, are subject to biodegradation by a range of species of bacteria and fungi (Li et al., 2017; Dexter et al., 2021). The genera *Sphingomonas*, *Sphingopyxis*, *Pseudomonas*, *Bacillus*, and *Burkholderia* are well-documented among microcystin-degrading bacteria (Massey and Yang 2020). The terrestrial *Trichaptum abietinum* – white rot fungus – has been characterized as capable of efficiently resisting and degrading the most toxic and commonly occurring variant of microcystins, microcystin-LR (Jia et al., 2012). All these terrestrial microcystins-tolerant microorganisms produce enzymes, which reduce the stable cyclic peptide to a linear non-toxic metabolite and, therefore, a more degradable molecule (Bourne et al., 1996, 2001). This linear non-toxic peptide and several other molecules such as complex carbohydrates, lipids, proteins, vitamins, and amino acids released during cyanobacterial lysis in soil can be used as a sustainable biofertilizer. However, post-harvest treatments are essential before use to preserve the cyanobacterial biomass, prevent deterioration of the compounds of interest, and, most importantly, mitigate cyanotoxin levels. For example, these treatments can include the addition of a bacterial suspension containing genes coding for enzymes capable of degrading cyanotoxins, or an extract containing these enzymes, extracted from these bacteria. This treatment could be performed directly on the freshly harvested biomass or after drying/dehydration. Drying processes, such as solar drying, which prioritize energy savings and aim to produce a high-quality product that can be used year-round, could be considered.

4 Conclusion and future trends in research

Research into large-scale algal biomass production has intensified in recent years for a variety of applications, including biofuels, animal feed, dietary supplements, pharmaceuticals, and agriculture. This has paved the way for the development of algae-based fertilizers that can replace or complement the costly and energy-intensive chemical fertilizers currently used, and thus potentially contribute to a more sustainable circular economy for nutrients in environmentally friendly and renewable agricultural systems. This chapter provides a literature review summarizing recent research on the roles of cyanobacteria in soil aggregation, biofertilization, abiotic/biotic stress tolerance, yield enhancement, carbon sequestration, disease control, and bioremediation. Although several groups of microalgae, such as green algae and diatoms, can colonize topsoil, cyanobacteria are the most common, playing a major role in the formation of biological soil crust communities in arid areas. Due to their morphological and physiological adaptations (extracellular matrix, gelatinous sheath, heterocysts, and akinetes), cyanobacteria can improve soil structure, provide fixed nitrogen, mitigate salt stress, increase crop yields by supplying micronutrients, vitamins, amino acids, and phytohormones, and sequester carbon. In addition, cyanobacteria are distinguished by their ability to form symbiotic associations with plants and microbes. In most cases, the benefit to the host is obvious and invariably consists of a metabolic contribution, primarily in the form of combined nitrogen, or occasionally carbon.

However, translating laboratory research findings on these many benefits of cyanobacteria into effective inoculants in the field remains a challenge. Only the non-toxic genus *Spirulina* (*Arthrospira*), widely recognized as a dietary supplement, has also seen its potential as a large-scale biofertilizer for agricultural soils well established. Other species of cyanobacteria, whose massive proliferation in different aquatic ecosystems is currently increasing worldwide, mainly due to eutrophication following the use of chemical fertilizers, could also be a promising source of biofertilizers. However, the fact that these blooms often contain cyanotoxins, which can compromise the quality of agricultural products and pose a potential risk to human health, has led to their possible use in agriculture as biofertilizers being limited. Therefore, the use of this abundant and cost-effective biomass, whose cyanotoxins represent no more than 0.1% of the total biomass, as a cyanofertilizer should be further explored, given its richness in essential nutrients beneficial for agricultural production and the restoration of soil fertility, even after cell death and decomposition.

5 References

Abdel El-Baky, H., et al. (2010). Enhancing antioxidant availability in wheat grains from plants grown under seawater stress in response to microalgae extract treatments. *J. Sci. Food Agric.* 90(2), 299–303.

Acea, M. (2003). Cyanobacterial inoculation of heated soils: effect on microorganisms of C and N cycles and on chemical composition in soil surface. *Soil Biol. Biochem.* 35(4), 513–524.

Adams, D.G. (2000). Symbiotic interactions. In: Whitton, B.A. and Potts, M. (eds) *Ecology of cyanobacteria: their diversity in time and space*. Kluwer Academic Publishers, Dordrecht, pp. 523–561, 669 pp.

Adams, D.G. (2002). Cyanobacteria in symbiosis with hornworts and liverworts. In: Rai, A.N., et al. (eds) *Cyanobacteria in symbiosis*. Kluwer Academic Publishers, Dordrecht, pp. 117–135.

Adams, D.G., et al. (2006). Cyanobacterial-plant symbioses. In: Dworkin, M., et al. (eds) *The prokaryotes. A handbook on the biology of bacteria*, vol 1, 3rd edn, Symbiotic associations, biotechnology, applied microbiology. Springer, New York, pp. 331–363.

Adams, D.G., et al. (2012). Cyanobacterial symbioses. In: Whitton, B.A. (ed) *Ecology of cyanobacteria II: their diversity in space and time, Chapter 23*. Springer Science+Business Media B.V., Dordrecht, pp. 593–636.

Adessi, A., et al. (2018). Microbial extracellular polymeric substances improve water retention in dryland biological soil crusts. *Soil Biol. Biochem.* 116, 67–69.

Afkairin, A., et al. (2021). Solubilization of organic phosphorus sources by cyanobacteria and a commercially available bacterial consortium. *Appl. Soil Ecol.* 162, 103900.

Alam, M.S., et al. (2023). Environment-friendly nitrogen management practices in wetland paddy cultivation. *Front. Sustain. Food Syst.* 7, 1020570.

Alharbi, K., et al. (2023). Mitigating osmotic stress and enhancing developmental productivity processes in cotton through integrative use of vermicompost and cyanobacteria. *Plants* 12, 1872.

Al-Najjar, M.A.A., et al. (2012). Light utilization efficiency in photosynthetic microbial mats. *Environ. Microbiol.* 14, 982–992.

Alobwede, E., et al. (2019). Circular economy fertilization: testing micro and macro algal species as soil improvers and nutrient sources for crop production in greenhouse and field conditions. *Geoderma* 334, 113–123.

Alvarenga, D.O., et al. (2015). Cyanobacteria in mangrove ecosystems. *Biodivers. Conserv.* 24, 799–817.

Alvarez, A.L., et al. (2021). Microalgae, soil and plants: a critical review of microalgae as renewable resources for agriculture. *Algal Res.* 54, 102200.

Alvarez, A.L., et al. (2024). Cyanobacteria-based soil amendments in the soil-plant system: Effects of inoculations on soil nutrient and microbial dynamics under spring wheat growth. *Algal Res.* 77, 103326.

Alvarez, C., et al. (2023). Symbiosis between cyanobacteria and plants: from molecular studies to agronomic applications. *J. Exp. Bot.* 74(19), 6145–6157.

Aly, M.S. and Esawy, M.A. (2008). Evaluation of Spirulina platensis as biostimulatory for organic farming systems. *Genet. Eng. Biotechnol. J.* 6, 1–7.

Anagnostidis, K. and Komárek, J. (1988). Modern approach to the classification system of cyanophytes 3 - oscillatoriales. *Algol. Stud.* 50, 327–472.

Anagnostidis, K. and Komárek, J. (1990). Modern approach to the classification system of the cyanophytes 5, Stigonematales. *Algol. Stud.* 86, 1–74.

Anand, N. and Hopper, R.S.S. (1987). Blue-green algae from rice fields in Kerala state, India. *Hydrobiologia* 144, 223–232.

Anand, N., et al. (2019). Cyanobacterial taxonomy: morphometry to molecular studies. In: Mishra A.K., et al. (eds) *Cyanobacteria from basic science to applications, Chapter 3*. Academic Press, London, pp. 43–64.

Anandakumar, A. (2023). The role of cyanobacteria on sustainable agriculture. *Agric. Food E-Newslett.* 5(12), 522–524.

Andrade, D.S., et al. (2024). Does coinoculation with bradyrhizobia and cyanobacteria improve ground nut growth and yield? *Exp. Agric.* 60(9), 1–11.

Antoninka, A., et al. (2020). Biological soil crusts in ecological restoration: emerging research and perspectives. *Restor. Ecol.* 28(S1), S8.

Apte, S.K. and Thomas, J. (1997). Possible amelioration of coastal soil salinity using halotolerantnitrogen-fixing cyanobacteria.*Plant Soil*, 189, 205–211.

Arafa, A.S. and Abd El-All, A.A. (2013). Evaluation of the technological properties of the bio-organic colored cotton. *J. Agron.* 12(2), 78–85.

Arioli, T., et al. (2015). Applications of seaweed extracts in Australian agriculture: past, present and future. *J. Appl. Phycol.* 27, 2007–2015.

Asghari S., et al. (2022). The impact of cyanobacteria inoculation on soil hydraulic properties at the lab-scale experiment. *Agric. Water Manag.* 272, 107865.

Ashour, M., et al. (2024). Usage of chlorella and diverse microalgae for CO2 capture-towards a bioenergy revolution. *Front. Bioeng. Biotechnol.* 12, 1387519.

Assunçao, J., et al. (2023). Effects of temperature, pH, and NaCl concentration on biomass and bioactive compound production by Synechocystis salina. *Life (Basel)* 13(1), .

Azari, A.M., et al. (2011). Seasonal changes in phytoplankton community structure in relation to physico-chemical factors in Bukan dam reservoir (northwest Iran). *Turk. J. Bot.* 35(1), 77–84.

Backovi, D.D. and Tokodi, N. (2024). Blue revolution turning green? A global concern of cyanobacteria and cyanotoxins in freshwater aquaculture: a literature review. *J. Environ. Manage.* 360, 121115.

Baker, J.A., et al. (2003). The cyanobiont in an Azolla fern is neither Anabaena nor Nostoc. *FEMS Microbiol. Lett.* 229, 43–47.

Bao, J., et al. (2021). Potential applicability of a cyanobacterium as a biofertilizer and biopesticide in rice fields. *Plant Soil* 463, 97–112.

Baracho, D.H. and Lombardi, A.T. (2023). Study of the growth and biochemical composition of 20 species of cyanobacteria cultured in cylindrical photobioreactors. *Microb. Cell Fact.* 22, 36.

Baumann, K. and Jung, P. (2023). Soil algae encyclopedia of soils in the environment (Second Edition) 1, 61–69.

Bello, A.S., et al. (2022). Application of Cyanobacteria (Roholtiella sp.) Liquid extract for the alleviation of salt stress in bell pepper (Capsicum annuum L.) plants grown in a soilless system. *Plants* 11, 104.

Belnap J. (2003). Biological soil crusts and wind erosion. In: Belnap, J. and Lange, O.L. (eds) *Biological soil crusts: structure, function, and management*. Ecological Studies 150. Springer, Berlin.

Belnap, J. (2006). The potential roles of biological soil crusts in dryland hydrologic cycles. *Hydrol. Process. Int. J.* 20(15), 3159–3178.

Belnap, J. and Büdel, B. (2016). Biological soil crusts as soil stabilizers. In: Weber, B., et al. (eds) *Biological soil crusts: an organizing principle in Drylands, Chapter 16*, Ecological Studies 226, pp. 305–320.

Belnap, J., et al. (2003). Biological soil crusts: characteristics and distribution. In: Belnap, J. and Lange, O.L. (eds) *Biological soil crusts: structure, function, and management*. Springer-Verlag, Berlin, pp. 3–30.

Belnap, J., et al. (2011). Aeolian and fluvial processes in dryland regions: the need for integrated studies. *Ecohydrology* 4(5), 615–622.

Belnap, J., et al. (2016). Biological soil crusts as an organizing principle in drylands. In: Weber, B., et al. (eds), *Biological soil crusts: an organizing principle in drylands*. Ecological studies 226, Springer International Publishing, Switzerland, pp. 3–13.

Benard, P., et al. (2019). Microhydrological niches in soils: how mucilage and EPS alter the biophysical properties of the rhizosphere and other biological hotspots. *Vadose Zone J.* 18(1), 180211.

Benayache, N.Y., et al. (2019). An overview of cyanobacteria harmful algal bloom (CyanoHAB) issues in freshwater ecosystems. In: Gokce, D. (ed) *Limnology-some new aspects of inland water ecology*. IntechOpen, London, pp. 1–25.

Bergman, B., et al. (2008). Cyanobacterial-plant symbioses: signaling and development. In: Herrero, A. and Flores, E. (eds) *The cyanobacteria: molecular biology, genomics and evolution*. Caister Academic Press, Norfolk, pp. 447–473.

Bernard, C., et al. (2016). Appendix 2: cyanobacteria associated with the production of cyanotoxins. In: *Handbook of cyanobacterial monitoring and cyanotoxin analysis*. John Wiley & Sons, Inc., Chichester, UK, pp. 501–525.

Berry, J.P., et al. (2008). Cyanobacterial toxins as allelochemicals with potential applications as algaecides, herbicides and insecticides. *Marine Drugs* 6(2), 117–146.

Bhowmik, D., et al. (2010). Evaluating potential of spirulina as innoculant for pulses. *Acad. J. Plant Sci.* 3, 161–164.

Bibi, S., et al. (2024). Applications, advancements, and challenges of cyanobacteria-based biofertilizers for sustainable agro and ecosystems in arid climates. *Bioresour Technol. Rep.* 25, 101789.

Bolan, N.S., et al. (2003). Soil acidification and liming interactions with nutrient and heavy metal transformation and bioavailability. *Adv. Agron.* 78, 215–272.

Boone, D.R., et al. (2001). *Bergey's manuals of systematic bacteriology*. Springer, New York.

Borowitzka, M.A. (1995). Microalgae as sources of pharmaceuticals and other biologically active compounds. *J. Appl. Phycol.* 7:3–15.

Bouaïcha, N. and Corbel, S. (2016). Cyanobacterial toxins emerging contaminants in soils: a review of sources, fate and impacts on ecosystems, plants and animal and human health. In: Larramendy, M.L. and Soloneski, S. (eds) *Soil contamination-current consequences and further solutions*. TechOpen, London. https://doi.org/10.5772/64940.

Bouaïcha, N., et al. (2019). Structural diversity, characterization and toxicology of microcystins. *Toxins* 11, 714.

Bourne, D.G., et al. (1996). Enzymatic pathway for the bacterial degradation of the cyanobacterial cyclicpeptide toxin microcystin-LR. *Appl. Environ. Microbiol.* 62, 4086–4094.

Bourne, D.G., et al. (2001). Characterisation of a gene cluster involved in bacterial degradation of the cyanobacterial toxin microcystin-LR. *Environ. Toxicol.* 16, 523–534.

Bowker, M.A., et al. (2018). Biocrusts: the living skin of the earth. *Plant Soil* 429, 1–7.

Büdel, B., et al. (2016). Cyanobacteria and algae of biological soil crusts. In Weber, B., et al. (eds), *Biological soil crusts: an organizing principle in drylands. Ecological studies 226*, Springer International Publishing, Switzerland, pp. 55–80.

Büdel, B. and Kauff, F. (2012). Blue-green algae. In: Frey, W. (ed) *Syllabus of plant families, Engler's syllabus der Pflanzenfamilien, Part VI*. Borntraeger, Stuttgart, pp. 5–39.

Büdel, B., et al. (2009). Southern African biological soil crusts are ubiquitous and highly diverse in drylands, being restricted by rainfall frequency. *Microb. Ecol.* 57, 229–247.

Buis, E., et al. (2009). Controls on plant functional surface cover types along a precipitation gradient in the Negev Desert of Israel. *J. Arid Environ.* 73(1), 82e90.

Burja, A.M., et al. (2001). Marine cyanobacteria a prolific source of natural products. *Tetrahedron* 57, 9347–9377.

Bustos-Díaz, E.D., et al. (2019). Cyanobacteria in nitrogen-fixing symbioses. In: Mishra, A.K., et al. (eds) *Cyanobacteria from basic sciences to applications, Chapter 2*. Academic Press, London, pp. 29–42.

Caiola, M.G., et al. (1996). Effect of dessication on envelopes of the cyanobacterium Chroococcidiopsis sp. (Chroococcales). *Eur. J. Phycol.* 31, 97–105.

Caire G., et al. (1990). Antimycotic products from the cyanobacterium Nostoc muscorum against Rhizoctonia Solani. cabdirect.org. https://www.cabdirect.org/cabdirect/abstract/19916775307.

Çakirsoy, I., et al. (2022). Physiology of microalgae and their application to sustainable agriculture: a mini-review. *Front. Plant Sci.* 13, 1005991.

Cameron, H.J. and Julian, G.R. (1988). Utilization of hydroxyapatite by cyanobacteria as their sole source of phosphate and calcium. *Plant Soil* 109(1), 123–124.

Cameron, R.E. and Blank, G.B. (1966). *Desert algae: soil crusts and diaphanous substrata as algal habitats*. California Institute of Technology, Pasadena, 1–41pp.

Capelli, B. and Cysewski, G.R. (2010). Potential health benefits of Spirulina microalgae. *Nutrafoods* 9, 19–26.

Carrell, A.A., et al. (2022). Novel metabolic interactions and environmental conditions mediate the boreal peatmoss-cyanobacteria mutualism. *ISME J.* 16 (4), 1074–1085.

Carvalho, A.P., et al. (2006). Microalgal reactors: a review of enclosed system designs and performances. *Biotechnol. Prog.* 22, 1490–1506.

Castenholz, R.W. and Waterbury, J.B. (1989). Group I. Cyanobacteria. In: Krieg, N.R. and Holt, J.G. (eds) *Bergey's manual of systematic bacteriology*, vol. 3. Williams & Wilkins, Baltimore, pp. 1710–1728.

Castenholz, R.W. and Garcia-Pichel, F. (2012). Cyanobacterial responses to UV radiation. In Whitton, B. A. (ed) *Ecology of cyanobacteria II: their diversity in space and time, Chapter 19*. Springer-Verlag, Netherlands, pp. 481–499.

Chabili, A., et al. (2024a). A comprehensive review of microalgae and cyanobacteria-based biostimulants for agriculture uses. *Plants* 13, 159.

Chamizo, S., et al. (2020). Cyanobacteria inoculation as a potential tool for stabilization of burned soils. *Restor. Ecol.* 28, S106–S114.

Chamizo, S., et al. (2019). Soil type and cyanobacteria species influence the macromolecular and chemical characteristics of the polysaccharidic matrix in induced biocrusts. *Microb. Ecol.* 78:482–493.

Chamizo, S., et al. (2018). Cyanobacteria inoculation improves soil stability and fertility on different textured soils: gaining insights for applicability in soil restoration. *Front. Environ. Sci.* 6, 49.

Chamizo, S., et al. (2017). Effects of biocrust on soil erosion and organic carbon losses under natural rainfall. *Catena* 148, 117–125.

Chanda, M., et al. (2019). Microalgae polysaccharides: the new sustainable bioactive products for the development of plant bio-stimulants? *World J. Microbiol. Biotechnol.* 35, 177.

Chanda, M., et al. (2021). Microalgae-cyanobacteria–based biostimulant effect on salinity tolerance mechanisms, nutrient uptake, and tomato plant growth under salt stress. *J. Appl. Phycol.* 33, 3779–3795.

Chaudhary, V., et al. (2012). Bioefficacy of novel cyanobacteria-amended formulations in suppressing damping off disease in tomato seedlings. *World J. Microbiol. Biotechnol.* 28(12), 3301–3310.

Chen, L., et al. (2003). Salt tolerance of *Microcoleus vaginatus* Gom., a cyanobacterium isolated from desert algal crust, was enhanced by exogenous carbohydrates. *J. Arid Environ.* 55(4), 645–656.

Chen, L., et al. (2014). Macromolecular and chemical features of the excreted extracellular polysaccharides in induced biological soil crusts of different ages. *Soil Biol. Biochem.* 78, 1–9.

Chen, L.Z., et al. (2006). Effect of salt stress on carbohydrate metabolism in desert soil alga *Microcoles vaginatus* Gom. *J. Integr. Plant Biol.* 48, 914–919.

Chen, Q., et al. (2023). Cyanobacterial diversity of biological soil crusts and soil properties in karst desertification area. *Front. Microbiol.* 14, 1113707.

Chen, S., et al. (2019). Electrochemical dinitrogen activation: to find a sustainable way to produce ammonia. *Stud. Surf. Sci. Catal.* 178, 31–46.

Chorus, I., et al. (2021). Cyanobacteria and cyanotoxins in a changing environment: concepts, controversies, challenges. *Water* 13, 2463.

Chorus, I. and Welker, M. (eds) (2021). *Toxic cyanobacteria in water: a guide to their public health consequences, monitoring and management* (2nd ed.). CRC Press on Behalf of World Health Organisation, Geneva, Switzerland, p. 858.

Ciesinksi, L., et al. (2018). High dietary zinc feeding promotes persistence of multi-resistant E. coli in the swine gut. *PLoS One* 13, 1–18.

Ciferri, O. (1983). Spirulina, the edible microorganism. *Microbiol. Rev.* 47, 551.

Colesie, C., et al. (2016). Composition and macrostructure of biological soil crusts. In *Biological soil crusts: an organizing principle in drylands*. Springer, Cham, pp. 159–172.

Colla, G. and Rouphael, Y. (2020). Microalgae: new source of plant biostimulants. *Agronomy* 10, 1240.

Corbel, S., et al. (2014). Cyanobacterial toxins: modes of actions, fate in aquatic and soil ecosystems, phytotoxicity and bioaccumulation in agricultural crops. *Chemosphere* 96, 1–15.

Costa, J.A.V., et al. (2019). Potential of microalgae as biopesticides to contribute to sustainable agriculture and environmental development. *J. Environ. Sci. Health* 54(5), 366–375.

D'Acqui, L.P. (2016). Use of indigenous cyanobacteria for sustainable improvement of biogeochemical and physical fertility of marginal soils in semiarid tropics. *Bioformul. Sust. Agric.* 2016, 213–232.

D'Acqui, L.P. (2016). Use of indigenous cyanobacteria for sustainable improvement of biogeochemical and physical fertility of marginal soils in semiarid tropics. In: Arora, N., et al. (eds) *Bioformulations: for sustainable agriculture*. Springer, New Delhi, India, pp. 213–232.

Dagnaisser, L.S., et al. (2022). Microalgae as bio-fertilizer: a new strategy for advancing modern agriculture, wastewater bioremediation, and atmospheric carbon mitigation. *Water Air Soil Pollut.* 233, 477.

Davydov, D. and Patova, E. (2018). The diversity of Cyanoprokaryota from freshwater and terrestrial habitats in the Eurasian Arctic and Hypoarctic. *Hydrobiologia* 811, 119–137.

Desikachary, T.V. (1959). *Cyanophyta*. Indian Council of Agricultural Research, New Delhi, p. 686.

Devaprakash, M., et al. (2024). *Cyanobacterial interactions and symbiosis*. Cyanobacteria, Elsevier, pp. 425–489.

Dey, M., et al. (2022). Promoting crop growth with symbiotic microbes in agro-ecosystems–I. In: *Microbes and microbial biotechnology for green remediation*. Elsevier, Heidelberg, pp. 117–133.

Dexter, J., et al. (2021). Microcystinase a-review of the natural occurrence, heterologous expression, and biotechnological application of MlrA. *Water Res.* 189, 116646.

Dineshkumar, R., et al., (2018). Microalgae as bio-fertilizers for rice growthand seed yield productivity. *Waste and Biomass Valorization*, 9, 793–800.

Dojani, S., et al. (2014). Genotypic and phenotypic diversity of cyanobacteria in biological soil crusts of the Succulent Karoo and Nama Karoo of Southern Africa. *Environ. Microbiol.* 67(2), 286e301.

Drouet, F. (1973). *Revision of the Nostocaceae with cylindrical trichomes*. Hafner Press, New York.

Duan, H., et al. (2023). Improving saline-alkali soil and promoting wheat growth by co-applying potassium-solubilizing bacteria and cyanobacteria produced from brewery wastewater. https://doi.org/10.3389/fenvs.2023.1170734.

Dulic, T., et al. (2017). Cyanobacterial diversity and toxicity of biocrusts from the Caspian Lowland loess deposits, North Iran. *Quat. Int.* 429, 74e85.

Durall, C. and Lindblad, P. (2015). Mechanisms of carbon fixation and engineering for increased carbon fixation in cyanobacteria. *Algal Res.* 11, 263–270.

Dvořák, P., et al. (2017). Diversity of the cyanobacteria. In: *Modern topics in the phototrophic prokaryotes*. Springer, Cham, pp. 3–46.

El Arroussi, H., et al. (2016). Microalgae polysaccharides a promising plant growth biostimulant. *J. Algal Biomass Utln.* 7, 55–63.

Elbert, W., et al. (2012). Contribution of cryptogamic covers to the global cycles of carbon and nitrogen. *Nat. Geosci.* 5, 459–462.

Eldridge, D.J., et al. (2020). The pervasive and multifaceted influence of biocrusts on water in the world's drylands. *Global Change Biol.* 26(10), 6003–6014.

Elenkin, A.A. (1938). *Monographia algarum cyanophycearum aquidulcium et terrestrium infinibus URSS inventarum, pars generalis, pars specialis I. et II.* vol. 1. Russian Akad Nauk SSSR, Moscow and Leningrad, USSR, pp. 1–1908.

Elster, J. (2002). Ecological classification of terrestrial communities of polar environment. In: Beyer, L. and Boelter, M. (eds) *Geoecology of antarctic ice-free coastal landscapes*, vol. 154. Springer, Berlin, pp. 303–326.

FAO Fisheries division–yearbook of fishery and aquaculture statistics. Available online: https://www.fao.org/fishery/static/Yearbook/YB2019_USBcard/navigation/index_intro_e.htm. (Accessed 20 April 2023).

Farrell, M., et al. (2014). Biochar and fertiliser applications influence phosphorus fractionation and wheat yield. *Biol. Fertil. Soils.* 50, 169–178.

Fastner, J., et al. (1999). Microcystins hepatotoxic heptapeptides in German fresh water bodies. *Environ. Toxicol.* 14, 13–22.

Fernandes, I., et al. (2020). Perspective application of the circular economy in the blue biotechnology: microalgae as sources of health promoting compounds. *Glob. J. Nutr. Food Sci.* 3(1).

Fernandes, V.M.C., et al. (2018). Exposure to predicted precipitation patterns decreases population size and alters community structure of cyanobacteria in biological soil crusts from the Chihuahuan Desert. *Environ. Microbiol.* 20, 259–269.

Flechtner, V.R., et al. (2008). The biological soil crusts of the San Nicolas Island: enigmatic algae from a geographically isolated ecosystem. *Western North Am. Nat.* 68(4), 405–436.

Flechtner, V.R., et al. (2005). Algal diversity in North American desert soils. In: Richard, B., et al. (eds) *Astrobiology and planetary missions. Proc. of SPIE 5906*, 590602–1.

Flechtner, V.R., et al. (1998). Algal composition of microbiotic crusts from the central desert of Baja California, Mexico. *Great Basin Nat.* 58, 295–311.

Fremy, P. (1929). Les Myxophycees de l'Afrique Equatoriale Francaise. *Arch. Bot. Mem. Caen.* 3(2), 1–508.

Friedmann, E.I. (1972). Ecology of lithophytic algal habitats in middle, eastern, and north American deserts. In: Bodin, L.F. (ed) *Eco-physiological foundation of ecosystems productivity in arid zone*. U.S.S.R. Academic Science, Nauka, Leningrad.

Friedmann, E.I. (1980). Endolithic microbial life in hot and cold deserts. In: *Limits of life*. Springer, Dordrecht, pp. 33–45.

Friedmann, E.I. and Galun, M. (1974). Desert algae, lichens, and fungi. In: Brown, G.W. (ed) *Desert biology II*. Academic Press, New York.

Friedmann, E.I., et al. (1967). Desert algae of the Negev (Israel). *Phycologia* 6, 185–196.

Fritsch, F.E. (1922). The terrestrial alga. *J. Ecol.* 10, 220–236.

Fu, Z., et al. (2020). Composition, seasonal variation, and salinization characteristics of soil salinity in the chenier island of the Yellow River delta. *Glob. Ecol. Conserv.* 24, e01318.

Furmaniak, M.A., et al. (2017). Edible cyanobacterial genus arthrospira: actual state of the art in cultivation methods, genetics, and application in medicine. *Front. Microbiol.* 8, 2541.

Gao, Q. and Garcia-Pichel, F. (2011). Microbial ultraviolet sunscreens. *Nat. Rev. Microbiol.* 9, 791–802.

Gao, K. (1998). Chinese studies of the edible blue-green alga, *Nostoc flagelliforme*: a review. *J. Appl. Phycol.* 10, 37–49.

Garcia-Pichel, F., et al. (2016). Microstructure and weathering processes within biological soil crusts. In: Weber, B., et al. (eds) *Biological soil crusts: an organizing principle in Drylands*. Ecological Studies 226. Springer, Cham, pp. 237–255.

Garcia-Pichel, F. (2002). Desert environments: biological soil crusts. In: Bitton, G. (ed) *Encyclopedia of environmental microbiology*. Wiley, New York, pp. 1019–1023.

Garcia-Pichel, F. and Wojciechowski, M.F. (2009). *The evolution of a capacity to build supra-cellular ropes enabled filamentous cyanobacteria to colonize highly erodible substrates*. *PLoS One* 4(11), e7801.

Garcia-Pichel, F., et al. (2001). Phylogenetic and morphological diversity of cyanobacteria in soil desert crusts from the Colorado Plateau. *Appl. Environ. Microbiol.* 67(4), 1902–1910.

Garcia, M., et al. (2025). Cyanobacteria and soil restoration: bridging molecular insights with practical solutions. *Microorganisms* 13, 1–20.

Gaysina, L.A., et al. (2018). Biodiversity of terrestrial cyanobacteria of the South Ural region. *Cryptogamie Algol.* 39(2), 1–32.

Gaysina, L.A., et al. (2019). Cyanobacteria in diverse habitats. In: Mishra, A.K., et al. (eds) *Cyanobacteria from basic sciences to applications, Chapter 1*. Academic Press, London, pp. 1–28.

Geitler, L. (1932). Cyanophyceae. In: Kolkwitz, R. (ed) *Rabenhorst's Kryptogamenflora von Deutschland, Osterreich und der Schweiz*, vol. 14. Akademische Verlagsgesellschaft, Leipzig, Germany.

Gilroy, D.J., et al. (2000). Assessing potential health risks from microcystin toxins in blue-green algae dietary supplements. *Environ. Health Perspect.* 108, 435–439.

Giraldo-Silva, A., et al. (2019). Nursing biocrusts: isolation, cultivation, and fitness test of indigenous cyanobacteria. *Restor. Ecol.* 27(4), 793–803.

Gollerbakh, M.M. and Shtina, E.A. (1969). *Soil algae*. Academic Science U.S.S.R., Moscow.

Gonçalves, A.L. (2021). The use of microalgae and cyanobacteria in the improvement of agricultural practices: a review on their biofertilising, biostimulating and biopesticide roles. *Appl. Sci.* 11(2), 871.

Gonçalves, J., et al. (2023). Microalgae as biofertilizers: a sustainable way to improve soil fertility and plant growth. *Sustainability* 15(16), 12413.

Gonzalez-Perez, B.K., et al. (2022). Microalgae as biostimulants: a new approach in agriculture. *World. J. Microbiol. Biotechnol.* 38, 4.

Graebner, P. (1895). Studien uber die norddeutsche heide. *Bot. Jahrb. Syst.* 20, 500.

Guo, S., et al. (2020). Microalgae as biofertilizer in modern agriculture. In: Alam, M., et al. (eds) *Microalgae biotechnology for food, health and high value products*. Springer, Singapore. https://doi.org/10.1007/978-981-15-0169-2_12.

Gupta, C., et al. (2021). Cyanobacteria as sustainable microbiome for agricultural industries. In: Dar, G.H., et al. (eds) *Microbiota and biofertilizers*, vol. 2. Springer Nature, Switzerland, AG, pp. 267–291.

Gupta, C., et al. (2021). Cyanobacteria as sustainable microbiome for agricultural industries. *Micro Biofertil.* 2, 267–291.
Gupta, V., et al. (2013). New insights into the biodiversity and applications of cyanobacteria (blue-green algae)–prospects and challenges. *Algal Res.* 2(2), 79–97.
Habimana, E. and Sauvé, S. (2025). A review of properties, occurrence, fate, and transportation mechanisms of contaminants of emerging concern in sewage sludge, biosolids, and soils: recent advances and future trends. *Front. Environ. Chem.* 6, 1547596.
Hagemann, M., et al. (2015). Cyanobacterial diversity in biological soil crusts along a precipitation gradient, Northwest Negev Desert, Israel. *Microbial Ecol.* 70, 219e230.
Haida, M., et al. (2024). Microcystin contamination in irrigation water and health risk. *Toxins* 16, 196.
Hakkoum, Z., et al. (2025). Impact of a soil cyanobacteria consortium-based bioinoculant on tomato growth, yield, and fruit quality. *Plants* 14, 2034.
Hakkoum, Z., et al. (2020). Diversity and spatial distribution of soil cyanobacteria along an altitudinal gradient in Marrakesh area (Morocco). *Appl. Ecol. Env. Res.* 18(4), 5527–5545.
Hakkoum, Z., et al. (2021). Impact of human disturbances on soil cyanobacteria diversity and distribution in suburban arid area of Marrakesh, Morocco. *Ecological Process.* 10, 16.
Hakkoum, Z., et al. (2024). Biofertilizing effect of soil cyanobacterium anabaena cylindrica–based formulations on wheat growth, physiology, and soil fertility. *Agriculture* 15, 20.
Hamisi, M., et al. (2013). Epiphytic cyanobacteria of the seagrass Cymodocea rotundata: diversity, diel nifH expression and nitrogenase activity. *Environ. Microbiol. Rep.* 5(3), 367.
Harb, T.B. and Chow, F. (2022). An overview of beach-cast seaweeds: potential and opportunities for the valorization of underused waste biomass. *Algal Res.* 62, 102643.
Heimann, L., et al., (2015). Nutrients and pollutants in agricultural soils in the peri-urban region of Beijing: status and recommendations. *Agric. Ecosyst. Environ.* 209, 74–88.
Hernández-Carlos, B. and Gamboa-Angulo, M.M. (2011). Metabolites from freshwater aquatic microalgae and fungi as potential natural pesticides. *Phytochem. Rev.* 10(2), 261–286.
Hoffmann, L., et al. (2005). System of cyanoprokaryotes (cyanobacteria) state in 2004. *Algological Stud.* 117(1), 95–115.
Hoffmann, L. (1989). Algae of terrestrial habitats. *Bot. Rev.* 55(2), 77–105.
Holajjer, P., et al. (2013). Potential of cyanobacteria for biorational management of plant parasitic nematodes: a review. *Crop Prot.* 53, 147–151.
Hooda, S., et al. (2023). Cyanobacteria as a potential bioasset for restoring degraded land. *Land Degrad Dev.* 1–16. DOI:10.1002/ldr.4711
Ho-Plagaro, T. and García-Garrido, J.M. (2022). Molecular regulation of arbuscular mycorrhizal symbiosis. *Int. J. Mol. Sci.* 23(11), 5960.
Hoque, M.M., et al. (2025). Microalgae: green engines for achieving carbon sequestration, circular economy, and environmental sustainability–a review based on last ten years of research. *Bioengineering* 12, 909.
Hossain, M.S., et al. (2023). Effects of cyanobacteria on soil properties in forest and cultivated lands of madhupur tract. *Bangladesh J. Agri.* 48(1), 112–122.

Hu, C., et al. (2012). Semi-arid regions and deserts. In: Whitton, B.A. (ed) *Ecology of cyanobacteria II: their diversity in space and time, Chapter 12*. Springer-Verlag, Netherlands, pp. 345–369.

Hu, C.X., et al. (2002). Effect of desert soil algae on the stabilization of fi ne sands. *J. Appl. Phycol.* 14, 281–292.

Huisman, J., et al. (2018). Cyanobacterial blooms. *Nat. Rev. Microbiol.* 16, 471.

Ighalo, J.O., et al. (2022). Progress in microalgae application for CO2 sequestration. *Clean. Chem. Eng.* 3, 100044.

Iniesta-Pallares, M., et al. (2021). Sustaining rice production through biofertilization with N2-fixing cyanobacteria. *Appl. Sci.* 11, 4628.

Islam, M.K., et al. (2019). Isolation and molecular characterization of phosphate solubilizing filamentous fungi from subtropical soils in Okinawa. *Appl. Ecol. Env. Res.* 17, 9145–9157.

Issa, A.A., et al. (2014). Nitrogen fixing cyanobacteria: future prospect. *Adv. Biol. Ecol. Nitr. Fixat.* 2, 24–48.

Jain, A.K. (2023). Greenhouse gas emissions from nitrogen fertilizers. *Nat. Food*. Published online. https://doi.org/10.1038/s43016-023-00706-z. (Accessed 9 February 2023).

Jaiswal, P., et al. (2010). Evaluating the potential of N. calcicola and its bicarbonate resistant mutant as bioameleorating agents for 'usar' soil. *Indian J. Microbiol.* 50, 12–18.

Jansen, H M., et al. (2022). Seasonal variation in productivity, chemical composition and nutrient uptake of *Ulva* spp. (Chlorophyta) strains. *J. Appl. Phycol.* 34, 1649–1660.

Jesus, J.M., et al. (2015). Phytoremediation of salt-affected soils: a review of processes, applicability, and the impact of climate change. *Environ. Sci. Pollut. Res.* 22, 6511–6525.

Jhala, Y.K., et al. (2017). Cyanobacteria: source of organic fertilizers for plant growth. In: *Microorganisms for green revolution*. Springer, Singapore, pp. 253–264.

Jia, Y., et al., (2012). A fungus capable of degrading microcystin-LR in the algal culture of *Microcystis aeruginosa* PCC7806. *Appl. Biochem. Biotechnol.* 166, 987–996.

Jimenez-Ríos, L., et al. (2024). Emerging nitrogen-fixing cyanobacteria for sustainable cotton cultivation. *Sci. Total Environ.* 924, 171533.

Johansen, J.R. (1984). Response of soil algae to a hundred – year storm in the Great Basin Desert, USA. *Phycos* 23(1&2), 51–54.

Johansen, J.R. (1993). Cryptogamic crusts of semiarid and arid lands of North America. *J. Phycol.* 29(2), 140–147.

Johansen, J.R. and Rushforth, S.R. (1985). Cryptogamic soil crusts: seasonal variation in algal populations in the Tintic Mountains, Juab County, Utah. *Great Basin Nat.* 45, 14–21.

Johansen, J.R. and Shubert, L.E. (2001). Algae in soil. *Nova Hedwig. Beih.* 123, 297–306.

Johnson, H.E., et al. (2008). Cyanobacteria (Nostoc commune) used as a dietary item in the Peruvian highlands produce the neurotoxic amino acid BMAA. *J. Ethnopharmacol.* 118, 159–165.

Joshi, H., et al. (2020). Cyanobacteria as a source of biofertilizers for sustainable agriculture. In: Singh, P.K., et al. (eds) *Advances in cyanobacterial biology, Chapter 25*. Academic Press, London, Pp. 385–396.

Jowkar, A., et al. (2017). The effect of soil fertilization and foliar spray of semperflorens begonia (Begonia semperflorens) by Spirulina cyanobacterium biomass. *J. Sci. Technol. Greenh. Cult.* 8, 65–74.

Kadam, S.U., et al. (2013). Application of novel extraction technologies for bioactives from marine algae. *J. Agric. Food Chem.* 61, 4667–4675.

Kant, S. (2018). Understanding nitrate uptake, signaling and remobilisation for improving plant nitrogen use efficiency. *Semin Cell Dev. Biol.* 74, 89–96.

Katakula, A.A.N., et al. (2020). The potential fertilizer value of namibian beach-cast seaweed (*Laminaria pallida* and *Gracilariopsis funicularis*) biochar as a nutrient source in organic agriculture. *Sci. Afr.* 10, e00592.

Kenyon, C.N., et al. (1972). Fatty acid composition and physiological properties of some filamentous blue-green algae. *Arch. Microbiol.* 83, 216–230.

Khaibullina, L.S. and Gaisina, L.A. (2008). Effect of salinization on the species composition and morphological features of soil algae. *Eurasian Soil Sci.* 41(2), 215–221.

Kheirfam, H., et al. (2020). Soil conservation in an abandoned agricultural rain-fed land through inoculation of cyanobacteria. *Catena* 187, 104341.

Khurshid, S., et al. (2017). Indoleacetic acid production and chromium reduction by cyanobacteria Synechocystis sp. P2A (Chroococcales) immobilized in alginate beads. *Biosci. J.* 33, 1592–1600.

Kim, K.H., et al. (2025). Microalgae-integrated building enclosures: a nature-based solution for carbon sequestration. *Front. Built Environ.* 11, 1574582.

Kobayashi, M., et al. (1967). Distribution of nitrogen-fixing microorganisms in paddy soils of southeast *Asia. Soil Sci.* 104, 113–118.

Komárek, J. (2010). Recent changes (2008) in cyanobacterial taxonomy based on a combination of molecular background with phenotype and ecological consequences (genus and species concept). *Hydrobiologia* 639, 245–259.

Komárek, J. (2014). Modern classification of cyanobacteria. In: Sharma, N.K., et al. (eds) *Cyanobacteria: an economic perspective*. Wiley Blackwell, Oxford, pp. 21–39.

Komárek, J. (2016). A polyphasic approach for the taxonomy of cyanobacteria, principles and applications. *Eur. J. Phycol.* 51, 346–353.

Komárek, J. (2018). Several problems of the polyphasic approach in the modern cyanobacterial system. *Hydrobiologia* 811(1), 7–17.

Komárek, J., et al. (2014). Taxonomic classification of cyanoprokaryotes (cyanobacterial genera) using a polyphasic approach. *Preslia* 86, 295–335.

Komárek, J. and Anagnostidis, K. (1986). Modern approach to the classification system of the cyanophytes 2, Chroococcales. *Algol. Stud.* 43, 157–226.

Komárek, J. and Anagnostidis, K. (1989). Modern approach to the classification system of the cyanophytes 4, Nostocales. *Algol. Stud.* 56, 247–345.

Komárek, J. and Anagnostidis, K. (1998). Cyanoprokaryota 1. Teil, chroococcales. In: Ettl, H., et al. (eds) *Susswasserflora von Mitteleuropa 19/1*. Gustav Fischer, Jena/Stuttgart/ Lübeck/Ulm, p. 548.

Komárek, J. and Anagnostidis, K. (2005). Cyanoprokaryota 2. Oscillatoriales. In: Büdel, B., et al. (eds) *Süsswasserflora von Mitteleuropa 19/2*. Elsevier/Spektrum, Heidelberg, 759 pp.

Komárek, J. and Komarkova, J. (2004). Taxonomic review of the cyanoprokaryotic genera Planktothrix and Planktothricoides. *Czech Phycol.* 4, 1–18.

Kondrateva, N.V. (1968). Sino-zeleni vodorosti-Cyanophyta (Blue-green algae – Cyanophyta). *Viznanik prisnovodnych Vodorostej Ukrajinskoj RSR*. 1, 1–524.

Kumar, J., et al. (2019). Cyanobacteria: applications in biotechnology. In: Mishra, A.K., et al. (eds) *Cyanobacteria from basic sciences to applications, Chapter 16*. Academic Press, London, pp. 327–346.

Kumar, D. and Adhikary, S.P. (2015). Diversity, molecular phylogeny, and metabolic activity of cyanobacteria in biological soil crusts from Santiniketan (India). *J. Appl. Phycol.* 27, 339–e349.

Kumar, K., et al. (2010). Cyanobacterial heterocysts. Cold Spring Harb. *Perspect. Biol.* 2, a000315.

Kumar, S., et al. (2022). Biofertilizers: an ecofriendly technology for nutrient recycling and environmental sustainability. *Curr. Res. Microbiol. Sci.* 3, 100094.

Kumari, N. and Rai, L.C. (2020). Cyanobacterial diversity: molecular insights under multifarious environmental conditions. In: Singh, P.K., et al. (eds) *Advances in cyanobacterial biology, Chapter 2*. Academic Press, London, pp. 17–33.

Lafarga, T., et al. (2020). Spirulina for the food and functional food industries. *Food Res. Int.* 137, 109356.

Lan, S., et al. (2010). Effect of drought and salt stresses on man-made cyanobacterial crusts. *Eur. J. Soil Biol.* 46, 381–386.

Lan, S., et al. (2014). Artificially accelerating the reversal of desertification: cyanobacterial inoculation facilitates the succession of vegetation communities. *Environ. Sci. Technol.* 48, 307e315.

Langhans, T.M., et al. (2010). Regeneration processes of biological soil crusts, macro-cryptogams and vascular plant species after fine-scale disturbance in a temperate region: recolonization or successional replacement? *Flora, Morphol. Distrib. Funct. Ecol Plants* 205(1), 46–60.

Lechno-Yossef, S. and Nierzwicki-Bauer, S.A. (2002). Azolla - Anabaena symbiosis. In: Rai, A.N., et al. (eds) *Cyanobacteria in symbiosis*. Kluwer Academic Publishers, Dordrecht, pp. 153–178.

Li, S., et al. (2025). Enhancement of rice production and soil carbon sequestration utilizing nitrogen-fixing cyanobacteria. *Appl. Soil Ecol.* 207, 105940.

Li, R. and Watanabe, M.M. (2004). Fatty acid profiles and their chemotaxonomy in planktonic species of Anabaena (Cyanobacteria) with coiled trichomes exhibited a significant taxonomic value. *Curr. Microbiol.* 49, 376–380.

Li, Z., et al. (2021). Synergistic effect of cyanobacteria and nano-sand-stabilizer on biocrust formation and sand fixation. *J. Environ. Chem. Eng.* 9, 104887.

Li, X.R., et al. (2010). Do biological crusts determine vegetation changes in sandy deserts? Implications for managing artificial vegetation. *Hydrol. Process.* 24, 3621e3630.

Li, H., et al. (2019). Current states and challenges of saltaffected soil remediation by cyanobacteria. *Sci. Total Environ.* 669, 258–272.

Li, J., et al. (2017). Current research scenario for microcystins biodegradation–a review on fundamental knowledge, application prospects and challenges. *Sci. Total Environ.* 595, 615–632.

Lin, C.S., et al. (2013). Biodiversity of soil algae in the farmlands of mid-Taiwan. *Bot. Stud.* 54(1), 41.

Liu, Y., et al. (2017). Effect of biological soil crusts on microbial activity in soils of the Tengger Desert (China). *J. Arid. Environ.* 144, 201–211.

Liu, J., et al. (2025). Application of microalgae in remediation of heavy metal-contaminated soils and its stimulatory effect on wheat growth. *Algal Res.* 88, 103995.

Liu, X., et al. (2003). Differentiation of Nostoc flagelliforme and its neighbouring species using fatty acid profile as a chemotaxonomic tool. *Curr. Microbiol.* 47, 467–474.

Lukešová, A. (2001). Soil algae in brown coal and lignite post-mining areas in central Europe (Czech Republic and Germany). *Restor. Ecol.* 9, 341–350.

Lyngbye, H.C. (1819). *Tentamen hydrophytologiae danicae*. Hafniae, Hafniae.

Mącik, M., et al. (2020). Biofertilizers in agriculture: an overview on concepts, strategies and effects on soil microorganisms. *Adv. Agron.* 162, 31–87.

Mackelprang, R., et al. (2022). Adaptation to environmental extremes structures functional traits in biological soil crust and hypolithic microbial communities. *Am. Soc. Microbiol.* 7(4), 16.

Maestre, F.T., et al. (2017). Microbial biotechnology as a tool to restore degraded drylands. *Microbial Biotechnol.* 10(5), 1250–1253.

Mager, D.M. and Thomas, A.D. (2010). Carbohydrates in cyanobacterial soil crusts as a source of carbon in the southwest Kalahari, Botswana. *Soil Biol. Biochem.* 42, 313–318.

Manganelli, M., et al. (2023). Co-occurrence of taste and odor compounds and cyanotoxins in cyanobacterial blooms: emerging risks to human health? *Microorganisms* 11, 872.

Malam Issa, O.M., et al. (1999). Morphology and microstructure of microbiotic soil crusts on a tiger bush sequence (Niger, Sahel). *Catena* 37, 175–196.

Malam Issa, O., et al. (2007). Effects of the inoculation of cyanobacteria on the microstructure and the structural stability of a tropical soil. *Plant Soil* 290(1–2), 209–219.

Maqubela, M.P., et al. (2009). Nostoc cyanobacterial inoculation in South African agricultural soils enhances soil structure, fertility, and maize growth. *Plant Soil* 315, 79–92.

Massa, A., et al. (2024). Toxic and non-toxic cyanobacterial biomass as a resource for sustainable agriculture: a lettuce cultivation experiment. *Environ. Res.* 262, 119942.

Massey, I.Y. and Yang, F. (2020). A mini review on microcystins and bacterial degradation. *Toxins* 12, 268.

Massey, M.S. and Davis, J.G. (2023). Beyond soil inoculation: Cyanobacteria as a fertilizer replacement. *Nitrogen* 4(3), 253–262.

Máthé, C., et al., (2007). Microcystin-LR, a cyanobacterial toxin, induces growth inhibition andhistological alterations in common reed (*Phragmitesaustralis*) plants regenerated from embryogenic calli. *New Phytol.* 176, 824–835.

Maurya, N., et al. (2025). Enhancement of soil fertility utilizing cyanobacteria-bacteria consortia. *J. Appl. Phycol.* 37(2), 1059–1074.

Meeks, J. and Elhai, J. (2002). Regulation of cellular differentiation in filamentous cyanobacteria in freeliving and plant-associated symbiotic growth states. *Microbiol. Mol. Biol. Rev.* 66, 94–121.

Mehda, S., et al. (2022). Lithic cyanobacterial communities in the polyextreme Sahara Desert: implications for the search for the limits of life. *Environ. Microbiol.* 24(1), 451–474.

Mehda, S., et al. (2021). Microenvironmental conditions drive the differential cyanobacterial community composition of biocrusts from the Sahara Desert. *Microorganisms* 9, 27.

Mendez, A., et al. (2023). Synergism of microorganisms and seaweed extract on vegetative growth, yield and quality of cucumber fruit. *Not. Bot. Horti. Agrobot. Cluj-Napoca* 51, 12888–12888.

Menegat, S., et al. (2022). Greenhouse gas emissions from global production and use of nitrogen synthetic fertilisers in agriculture. *Sci. Rep.* 12, 14490.

Metting, B. (1981). The systematics and ecology of soil algae. *Bot. Rev.* 47, 195–312.

Michalak, I. and Chojnacka, K. (2015). Algae as production systems of bioactive compounds. *Eng. Life Sci.* 15, 160–176.

Milledge, J.J. and Harvey, P.J. (2016). Golden tides: problem or golden opportunity? The valorisation of Sargassum from beach inundations. *J. Mar. Sci. Eng.* 4, 60.

Minaoui, F., et al. (2024). Biostimulant effect of green soil microalgae chlorella vulgaris suspensions on germination and growth of wheat (Triticum aestivum Var. Achtar) and soil fertility. *Algal Res.* 82, 103655.

Mishra, S. (2020). Cyanobacterial imprints in diversity and phylogeny. In: Singh, P.K., et al. (eds) *Advances in cyanobacterial biology, Chapter 1*. Academic Press, London, pp. 1–15.

Mishra, A.K., et al. (2013). Phylogenetic comparison among the heterocystous cyanobacteria based on a polyphasic approach. *Protoplasma* 250(1), 77–94.

Mishra, A., et al. (2019). Nitrogenase and hydrogenase: enzymes for nitrogen fixation and hydrogen production in cyanobacteria. In: Mishra, A.K., et al. (eds) *Cyanobacteria from basic science to applications, Chapter 8*. Academic Press, London, pp. 173–191.

Mitrovic, S.M., et al. (2005). Bioaccumulation and harmful effects of microcystin-LR in the aquatic plants *Lemna minor* and *Wolffia arrhiza* and the filamentous alga *Chladophora fracta*. *Ecotoxicol. Environ. Saf.* 61, 345–352.

Mugnai, G., et al. (2024). The contribution of the phototrophic fraction in the fertility of different successional stages of induced biological soil crusts. *Biol. Fert. Soils.* https://doi.org/10.1007/s00374-024-01840-x.

Muñoz-Martin, M.A., et al. (2019). Cyanobacterial biocrust diversity in Mediterranean ecosystems along a latitudinal and climatic gradient. *New Phytol.* 221, 123–141.

Nabout, J.C., et al. (2013). How many species of Cyanobacteria are there? Using a discovery curve to predict the species number. *Biodivers. Conserv.* 22(12), 2907–2918.

Nasri, H., et al. (2008). First reported case of turtle deaths during a toxic *Microcystis* spp. bloom in Lake Oubeira, Algeria. *Ecotoxicol. Environ. Saf.* 71, 535–544.

Nawaz, T., et al. (2024). Harnessing the potential of nitrogen-fixing cyanobacteria: A rich bio-resource for sustainable soil fertility and enhanced crop productivity. *Environ. Technol. Innovat.* 36, 103886.

Nowruzi, B., et al. (2021). Plant-cyanobacteria interactions: beneficial and harmful effects of cyanobacterial bioactive compounds on soil-plant systems and subsequent risk to animal and human health. *Phytochem.* 192, 112959.

Ooms, M.D., et al. (2016). Photon management for augmented photosynthesis. *Nat. Commun.* 7, 12699.

Osman, M.E.H., et al. (2010). Effect of two species of cyanobacteria as biofertilizers on some metabolic activities, growth, and yield of pea plant. *Biol. Fertil. Soils* 46, 861–875.

Osorio-Santos, K., et al. (2014). Seven new species of Oculatella (Pseudanabaenales, Cyanobacteria): taxonomically recognizing cryptic diversification. *Eur. J. Phycol.* 49(4), 450–470.

Otsuka, S., et al. (1999). Characterization of morphospecies and strains of the genus Microcystis (Cyanobacteria) for a reconsideration of species classification. *Phycol. Res.* 47, 189–197.

Palinska, K.A., et al. (2018). Biodiversity analysis of the unique geothermal microbial ecosystem of the Blue Lagoon (Iceland) using next-generation sequencing (NGS). *Hydrobiologia* 811(1), 93–102.

Pandey, P., et al. (2025). Cyanobacterial green chemistry: a blue-green approach for a sustainable environment, energy, and chemical production. *RSC Sust.* 3, 661–675.

Pandey, S.N., et al. (2020). Cyanobacteria: potential source of biofertilizer and synthesizer of metallic nanoparticles. In: Singh, P.K., et al. (eds) *Advances in cyanobacterial biology, Chapter 23*. Academic Press, London, pp. 351–367.

Pandharikar, G. (2020). *Nitrogen-fixing symbiosis versus nitrogenous mineral nutrition: consequence on the interaction between Medicago truncatula and the Acyrthosiphon pisum* (Université Côte d'Azur).

Park, C.H., et al. (2017). Rapid development of cyanobacterial crust in the field for combating desertification. *PLoS One* 12(6), e0179903.

Passos, L.S., et al. (2023). Content of lipids, fatty acids, carbohydrates, and proteins in continental cyanobacteria: a systematic analysis and database application. *Appl. Sci.* 13, 3162.

Pawar, S.T. and Puranik, P.R. (2008). Screening of terrestrial and freshwater halotolerant cyanobacteria for antifungal activities. *World J. Microbiol. Biotechnol.* 24, 1019–1025.

Pearson, E. and Kingsbury, J.M. (1966). Culturally induced variation in four morphologically diverse blue green algae. *Am. J. Bot.* 53, 192–200.

Pentecost, A. and Whitton, B.A. (2012). Subaerial cyanobacteria. In: Whitton, B.A. (ed) *Ecology of cyanobacteria II: their diversity in space and time, Chapter 10*. Springer-Verlag, Netherlands, pp. 291–316.

Peters, G.A. (1991). Azolla and other plant-cyanobacteria symbioses: aspects of form and function. *Plant Soil* 137, 25–36.

Peterson, J.B. (1935). Studies on the biology and taxonomy of soil algae. *Dansk. Bot. Arkiv.* 8, 1–180.

Peuthert, A., et al. (2007). Uptake of microcystins-LR and -LF (cyanobacterial toxins) in seedlings of several important agricultural plant species and the correlation with cellular damage (lipid peroxidation). *Environ. Toxicol.* 22, 436–442.

Pflugmacher, S. (2002). Possible allelopathic effects of cyanotoxins, with reference to microcystin-LR, in aquatic ecosystems. *Environ. Toxicol.* 17, 407–413.

Piatka, D.R., et al. (2022). Balance of carbon species combined with stable isotope ratios show critical switch towards bicarbonate uptake during cyanobacteria blooms. *Sci. Total Environ.* 807, 151067.

Picetti, R., et al. (2022). Nitrate and nitrite contamination in drinking water and cancer risk: a systematic review with meta-analysis. *Environ. Res.* 210, 112988.

Pillans, B. (1997) Soil development at snail's pace: evidence from a 6 Ma soil chronosequence on basalt in north Queensland, Australia. *Geoderma* 80, 117–128.

Pocs, T. (2009). Cyanobacterial crust types, as strategies for survival in extreme habitats. *Acta Bot. Hung.* 51, 147–178.

Pointing, S., et al. (2015). Biogeography of photoautotrophs in the high polar biome. *Front. Plant Sci.* 6, 692.

Pointing, S.B. and Belnap, J. (2012). Microbial colonization and controls in dryland systems. *Nat. Rev. Microbiol.* 10(8), 551–562.

Pointing, S.B. (2016). Hypolithic communities. In: *Biological soil crusts: an organizing principle in drylands*. Springer, Cham, pp. 199–213.

Prasanna, R., et al. (2021). Natural and constructed cyanobacteria-based consortia for enhancing crop growth and soil fertility. *Role Microb. Commun. Sustain.* 333–362.

Prasanna, R., et al. (2017). Cyanobacterial and rhizobial inoculation modulates the plant physiological attributes and nodule microbial communities of chickpea. *Arch. Microbiol.* 199, 1311–1323.

Prasanna, R., et al. (2015). Prospecting cyanobacterial formulations as plant-growth-promoting agents for maize hybrids. *South African J. Plant Soil* 32(4), 199–207.

Prisa, D. and Spagnuolo, D. (2023). Plant production with microalgal biostimulants. *Horticulturae* 9, 829.

Qiu, L., et al. (2005). Differentially expressed proteins in the pancreas of diet-induced diabetic mice. *Mol. Cell. Proteomics* 4, 1311–1318.

Raanan, H., et al. (2015). Three-dimensional structure and cyanobacterial activity within a desert biological soil crust. *Environ. Microbiol.* 18(2), 372–83.

Rabbai, A., et al. (2024). Fertilization induced greenhouse gas emissions partially offset carbon sequestration during afforestation. *Soil Biol. Biochem.* 199, 109577.

Rady, M.M. (2011). Effects on growth, yield, and fruit quality in tomato (Lycopersicon esculentum Mill.) using a mixture of potassium humate and farmyard manure as an alternative to mineral-N fertilizer. *J. Hortic. Sci. Biotechnol.* 86, 249–254.

Rady, M.M., et al. (2018). Integrative application of cyanobacteria and antioxidants improves common bean performance under saline conditions. *Sci. Hortic.* 233, 61–69.

Rae, B.D., et al. (2013). Cyanobacterial carboxysomes: microcompartments that facilitate CO_2 fixation. *J. Mol. Microbiol. Biotechnol.* 23, 300–307.

Rai, A.N., et al. (2019). Plant growth-promoting abilities in cyanobacteria. In: Mishra, A.K., et al. (eds) *Cyanobacteria from basic science to applications, Chapter 23.* Academic Press, London, pp. 459–476.

Rai, A.N. (1990). *CRC handbook of symbiotic cyanobacteria*. CRC Press, Boca Raton, 253 pp.

Rai, A.N. (2002). Cyanolichens: nitrogen metabolism. In: Rai, A.N., et al. (eds) *Cyanobacteria in symbiosis*. Kluwer Academic Publishers, Dordrecht, pp. 97–115.

Ramakrishnan, B., et al. (2023). Potential of microalgae and cyanobacteria to improve soil health and agricultural productivity: a critical view. *Environ. Sci. Adv.* 2(4), 586–611.

Rana, A., et al. (2012). Biofortification of wheat through inoculation of plant growth promoting rhizobacteria and cyanobacteria. *Eur. J. Soil Biol.* 50, 118–126.

Rasmussen, U. and Johansson, C. (2002). Diversity and specificity in cyanobacterial symbioses. *Biol. Environ. Proc. R. Irish Acad.* 102, 53–56.

Rastogi R.P., et al. (2016) Characterization and antioxidant functions of mycosporine-like amino acids in the cyanobacterium Nostoc sp. R76DM. *Algal Res.* 16, 1108.

Ray, K., et al. (2013). A way to curb phosphorus toxicity in the environment: use of polyphosphate reservoir of cyanobacteria and microalga as a safe alternative phosphorus biofertilizer for Indian agriculture. *Environ. Sci. Technol.* 47(20), 11378–11379.

Renuka, N., et al. (2018). Microalgae as multi-functional options in modern agriculture: current trends, prospects and challenges. *Biotechnol. Adv.* 36(4), 1255–1273.

Renuka, N., et al. (2017). Wastewater grown microalgal biomass as inoculants for improving micronutrient availability in wheat. *Rhizosphere* 3, 150–159.

Ritika, B. and Utpal, D. (2014). Biofertilizer, a way towards organic agriculture: a review. *Afr. J. Microbiol. Res.* 8, 2332–2343.

Rodriguez-Caballero, E., et al. (2018). Dryland photoautotrophic soil surface communities endangered by global change. *Nat. Geosci.* 11, 185–189.
Rogers, E.H., et al. (2005). Potential developmental toxicity of anatoxin-a, a cyanobacterial toxin. *J. Appl. Toxicol.* 25(6), 527–534.
Román, J.R., et al. (2018). Restoring soil functions by means of cyanobacteria inoculation: Importance of soil conditions and species selection. *Land Degrad. Dev.* 29(9), 3184–3193.
Román, J.R., et al. (2020). Assessing the viability of cyanobacteria pellets for application in arid land restoration. *J. Environ. Manage.* 270, 110795.
Romano, I., et al. (2000). Lipid profile a useful chemotaxonomic marker for classification of new cyanobacterium in Spirulina genus. *Phytochemistry* 54, 289–295.
Ronga, D., et al. (2019). Microalgal biostimulants and biofertilisers in crop productions. *Agronomy* 9, 192.
Roque, J., et al. (2023). Isolation and characterization of soil cyanobacteria and microalgae and evaluation of their potential as plant biostimulants. *Plant Soil* 493, 115–136.
Rossi, F. and De Philippis, R. (2015). Role of cyanobacterial exopolysaccharides in phototrophic biofilms and in complex microbial mats. *Life* 5, 1218–1238.
Rossi, F., et al. (2017). Cyanobacterial inoculation (cyanobacterisation): perspectives for the development of a standardized multifunctional technology for soil fertilization and desertification reversal. *Earth-Sci. Rev.* 171: 28–43.
Rossi, F., et al. (2022). Cyanobacterial biocrust induction: a comprehensive review on a soil rehabilitation-effective biotechnology. *Geoderma* 415, 115766.
Rott, E., et al. (2018). Introduction: recent developments in cyanobacterial research with special reference to aquatic habitats, molecular ecology, and phylogenetic taxonomy. *Hydrobiologia* 811, 1–6.
Rouwenhorst, K.H.R., et al. (2021). *Ammonia production technologies, techno-economic challenges of green ammonia as an energy vector*. Elsevier, Heidelberg, pp. 41–83.
Sahab, S., et al. (2021). Potential risk assessment of soil salinity to agroecosystem sustainability: current status and management strategies. *Sci. Total Environ.* 764, 1–15.
Saqrane, S., et al. (2008). Effects of cyanobacteria producing microcystins on seed germination and seedling growth of several agricultural plants. *J. Environ. Sci. Health, Part B* 43, 443–451.
Sarma, R. and Prakash, P. (2020). Physiological aspects of cyanobacterial nitrogen fixation and its applications in modern sciences. In: Singh, P.K., et al. (eds) *Advances in cyanobacterial biology, Chapter 14*. Academic Press, London, pp. 205–217.
Sarwer, A., et al. (2022). Algal biomass valorization for biofuel production and carbon sequestration: a review. *Environ. Chem. Lett.* 20, 2797–2851.
Sathyananth, M. and Raj, T.L.S. (2024). An overview of cyanobacterial contributions to agriculture. *Asian Res. J. Agric.* 17(2): 363–380.
Schulz, K., et al. (2016). Biological soil crusts from coastal dunes at the Baltic Sea: cyanobacterial and algal biodiversity and related soil properties. *Microbial Ecol.* 71, 178–193.
Shariatmadari, Z., et al. (2013). Plant growth promoting cyanobacteria and their distribution in terrestrial habitats of Iran. *Soil Sci. Plant Nut.* 59(4), 535–547.
Sharma, H.S.S., et al. (2014). Plant biostimulants: a review on the processing of macroalgae and use of extracts for crop management to reduce abiotic and biotic stresses. *J. Appl. Phycol.* 26, 465–490.

Sharma, R., et al. (2012). Role of algae and cyanobacteria in sustainable agriculture system. *Wudpecker J. Agric. Res.* 1(9), 381–388.

Shields, L.M. and Durrell, L.W. (1964). Algae in relation to soil fertility. *Bot. Rev.* 30, 92–128.

Shih, P.M., et al. (2013). Improving the coverage of the cyanobacterial phylum using diversity-driven genome sequencing. *Proc. Natl. Acad. Sci. U.S.A.* 110, 1053–1058.

Shtina, E.A. (1972). Some peculiarities of the distribution of nitrogen fixing blue-green algae in soils. In: Desikachary, T.V. (ed) *Taxonomy and biology of Blue-green Algae*. Madras University Press, Madras, pp. 294–295.

Shukla, E., et al. (2012). Chemotaxonomy of heterocystous cyanobacteria using FAME profiling as species markers. *Protoplasma* 249(3), 651–661.

Shukla, P.S., et al. (2025). Phycobiostimulants: next-generation sustainable agricultural inputs. *ACS Agric. Sci. Technol.* 5(4), 424–453.

Singh, S., et al. (2019). Cyanobacterial exopolysaccharides: composition, biosynthesis, and biotechnological applications. In: Mishra, A.K., et al. (eds) *Cyanobacteria from basic science to applications, Chapter 17*. Academic Press, London, pp. 347–358.

Singh, A. (2021). Soil salinity: a global threat to sustainable development. *Soil Use Manage.* 38(1), 39–67.

Singh, R.N. (1961). *Role of blue green algae in nitrogen economy of Indian agriculture*. Indian Council of Agricultural Research, New Delhi.

Singh, N.I. (2002). *Studies on cyanobacterial microflora isolated from rice-fields of N.E. India*. Regional Biofertilizer Development Centre, Imphal, pp. 72–83.

Singh, V. and Singh, D.V. (2015). Cyanobacteria modulated changes and its impact on bioremediation of saline-alkaline soils. *Bangladesh J. Bot.* 44, 653–658.

Singh, S.S., et al. (2014a). Diversity and distribution pattern analysis of cyanobacteria isolated from paddy fields of Chhattisgarh. India. *J. Asia Pac. Biodivers.* 7, 462–470.

Singh, H., et al. (2014b). Cyanobacteria and agricultural crops. *Vegetos* 27(1), 37–44.

Singh, P., et al. (2013). Molecular phylogeny, population genetics, and evolution of heterocystous cyanobacteria using nifH gene sequences. *Protoplasma* 250(3), 751–764.

Singh, P., et al. (2016). New species of Nostoc (cyanobacteria) isolated from Pune, India, using morphological, ecological and molecular attributes. *Plant Syst. Evol.* 302(10), 1381–1394.

Singh, P., et al. (2017). A new species of Scytonema isolated from Bilaspur, Chhattisgarh, India using the polyphasic approach. *Plant Syst. Evol.* 303(2), 249–258.

Singh, P., et al. (2015). Molecular phylogeny and evogenomics of heterocystous cyanobacteria using rbcL gene sequence data. *Ann. Microbiol.* 65(2), 799–807.

Singh, J.S., et al. (2011). Efficient soil microorganisms: a new dimension for sustainable agriculture and environmental development. *Agric. Ecosyst. Environ.* 140, 339–353.

Singh, P.K., et al. (2022). Cyanobacteria as biostimulants in the paddy fields. In: *Cyanobacterial lifestyle and its applications in biotechnology*. Academic Press, New York, pp. 281–306.

Sivakumar, M.V.K. (2007). Interactions between climate and desertification. *Agric. For Meteorol.* 142, 143–155.

Sivalingam, K.M. (2020). Isolation, identification and evaluation of Spirulina platensis for its effect on seed germination of ground nut (Arachis hypogaea L.), Wolaita Sodo, Southern Ethiopia. *J. Algal Biomass Util.* 11, 34–42.

Solomon, W., et al. (2023). Harnessing the synergy of the cyanobacteria-plant growth promoting bacteria for improved maize (Zea mays) growth and soil health. *Sustainability* 15(24), 16660.

Soni, R.A., et al. (2017). Spirulina–from growth to nutritional product: a review. *Trends Food Sci. Technol.* 69, 157–171.

Soumare, A., et al. (2020). From Isolation of Phosphate Solubilizing Microbes to Their Formulation and Use as Biofertilizers: Status and Needs, Front. Bioeng. Biotechnol. 7.

Sosa Quintero, J., et al. (2022). Biocrusts in Mexican deserts and semideserts: a review of their species composition, ecology, and ecosystem function. *J. Arid Environ.* 199, 104712.

Spolaore, P., et al. (2006). Commercial applications of microalgae. *J. Biosci. Bioeng.* 101, 87–96.

Starks, T.L. and Shubert, L.E. (1982). Colonization and succession of algae and soil-algal interactions associated with disturbed areas. J. Phycol. 18, 99–107.

Starmach, K. (1966). Cyanophyta-sinice, Glaucophyta. Glaukofity (Polish). Flora slodkowedna polski. *Pol. Acad. Sci.* 2, 1–807.

Stewart J.B., et al. (1988). Cytotoxic, fungicidal nucleosides from blue green algae belonging to the Scytonemataceae. *J. Antibiot. (Tokyo).* 61, 1048–1056.

Strauss, S.L., et al. (2012). Nitrogen cycling in desert biological soil crusts across biogeographic regions in the Southwestern United States. *Biogeochemistry* 108, 171–182.

Supraja, K.V., et al. (2020). Performance evaluation of hydroponic system for co-cultivation of microalgae and tomato plant. *J. Clean. Prod.* 272, 122823.

Svirčev, Z., et al. (2019). Geographical and historical overview of cyanotoxin distribution and cyanobacterial poisonings. *Arch. Toxicol.* 93, 2429–2481.

Swain, S.S., et al. (2017). Antibacterial, antifungal and antimycobacterial compounds from cyanobacteria. *Biomed Pharmacoth.* 90, 760–776.

Swarnalakshmi, K., et al. (2013). Evaluating the influence of novel cyanobacterial biofilmed biofertilizers on soil fertility and plant nutrition in wheat. *Eur. J. Soil Biol.* 55, 107–116.

Taira, H., et al. (2021). Chemical characteristics of degraded soils in Uzbekistan and remediation by cyanobacteria. *Nutr. Cycl. Agroecosyst.* 120, 193–203.

Takenaka, H., et al. (1998). Safety evaluation of *Nostoc flagelliforme* (Nostocales, Cyanophyceae) as a potential food. *Food Chem. Toxicol.* 36, 1073–1077.

Tantawy, S.T. (2011). Biological potential of cyanobacterial metabolites against some soil pathogenic fungi. *J. Food, Agric. Environ.* 9, 663–666.

Tarakhovskaya, E.R., et al. (2007). Phytohormones in Algae. *Russian J. Plant Physiol.* 54(2), 163–170

Temraleeva, A.D. (2018). Cyanobacterial diversity in the soils of russian dry steppes and semideserts. Microbiology 87, 249–260.

Tiwari, G., et al. (2007). Distribution, diversity and characterization of cyanobacteria of rice-fields. *Proc. Natl. Acad. Sci.* 77 B(IV), 387–402.

Tiwari, O.N. and Singh, H.T. (2005). Biodiversity of cyanobacteria in Loktak lake and rice fields of Manipur, India having acidic properties. *Proc Natl Acad Sci India B Bio Sci* 75, 209–213.

Toribio, A.J., et al. (2020). Prospection of cyanobacteria producing bioactives substances and their application as potential phytostimulating agents. *Biotechnol. Rep.* 26, e0049.

United Nations. (2015). A/RES/70/1–transforming our world: the 2030 agenda for sustainable development. https://www.un.org/en/development/desa/population/migration/generalassembly/docs/globalcompact/A_RES_70_1_E.pdf. (Accessed 27 March 2025).
Usher, K.M. (2008). The ecology and phylogeny of cyanobacterial symbionts in sponges. *Mar. Ecol.* 29(2), 178–192.
Vaishampayan, A., et al. (2001). Cyanobacterial biofertilizers in rice agriculture. *Bot. Rev.* 67, 453–516.
Van Alstyne, K.L., et al. (2015). Environmental chemistry and chemical ecology of "green tide" seaweed blooms. *Integr. Comp. Biol.* 55, 518–532.
Vankatarmann, G.S. (1972). *Algal biofertilizers and rice cultivation*. Today and Tomorrow Printers and Publishers, New Delhi, India, p. 75.
Vankatarmann, G.S. (1975). The role of blue-green algae in tropical rice cultivation. In: Stewart, W.D.P. (ed) *Nitrogen fixation by free-living microorganisms*. IBP Programme 6, Cambridge University Press, Cambridge.
van Hove, C. and Lejeune, A. (2002) The Azolla - Anabaena symbiosis. *Biol. Environ. Proc. R. Ir. Acad.* 102B, 23–26.
Vargas, M., et al. (1998). Biochemical composition and fatty acid content of filamentous nitrogen-fixing cyanobacteria. *J. Phycol.* 34, 812–817.
Vinogradova, O.N. and Darienko, T.M. (2008). Algae of Azovo-Syvashsky National Nature Park (Ukraine). *Int. J. Algae* 8, 163–178.
Walsh, J.J., et al. (2012). Replacing inorganic fertiliser with anaerobic digestate may maintain agricultural productivity at less environmental cost. *J. Plant Nutr. Soil Sci.* 175, 840–845.
Wang, Z., et al. (2025). *Microcoleus vaginatus*: a novel amendement for constructing artificial soil from tailings. *Environ. Technol. Innovat.* 37, 103939.
Wang, W., et al. (2009). Feasibility of cyanobacterial inoculation for biological soil crusts formation in desert area. *Soil Biol. Biochem. Sci. Goes Underground China* 41(5), 926–929.
Ward, C.J., et al. (1997). Colorimetric protein phosphatase inhibition assay of laboratory strains and natural blooms of cyanobacteria: comparison with high performance liquid chromatography analysis for microcystins. *FEMS Microbiol. Lett.* 153, 465–473.
Ward, D.M., et al. (2012). Cyanobacteria in Geothermal Habitats. In Whitton, B.A. (ed) *Ecology of cyanobacteria II: their diversity in space and time, Chapter 3*. Springer-Verlag, Netherlands, pp. 39–63.
Warld, M.H., et al. (2018). Drinking water nitrate and human health: an updated review. *Int. J. Environ. Res. Public Health* 15, 1557.
Watanabe, A. (1965). Studies on the blue-green algae as green manure in Japan. *Proc. Nat. Acad. Sci. India* 35: 361–369.
Watanabe, A. and Yamamoto, Y. (1967). Heterotrophic nitrogen-fixation by the blue-green alga Anabaenopsis circularis. *Nature* 214, 5089.
Watanabe, A. and Yamamoto, Y. (1971). Algal nitrogen fixation in the tropics. In: Lie, T.A. and Mulder, E.G. (eds) *Biological nitrogen fixation in natural and agricultural habitats*. Plant and Soil Special Volume.
Weber, B., et al. (2016). Synthesis on biological soil crust research. In *Mediterranean-type ecosystems*. Springer, Berlin, Germany, pp. 527–534.

Weber, B., et al. (2022). What is a biocrust? A refined, contemporary definition for a broadening research community. *Biol. Rev.* 43, 1768–1785.

Wei, Y., et al. (2019). Updated information on soil salinity in atypical oasis agroecosystem and desert-oasis ecotone: case study conducted along the Tarim River, China. *Sci. Total Environ.* 716, 135387.

Weiss, J., et al. (2000). Influence of microcystin-RR on growth and photosynthetic capacity of the duckweed *Lemna minor L. J. Appl. Bot.* 74, 100–105.

Wessels, D.C.J. and Büdel, B. (1995). Epilithic and cryptoendolithic cyanobacteria of Clarens sandstone cliffs in the Golden Gate Highlands National Park, South Africa. *Botanica Acta* 108, 220–226.

Whitton, B.A. (2000). Soils and rice-fields. In: Whitton, B.A. and Potts, M. (eds) *Ecology of cyanobacteria: their diversity in time and space.* Kluwer Academic Publishers, Dordrecht, pp. 235–255, 669 pp.

Whitton, B.A. and Potts, M. (2007). *The ecology of cyanobacteria: their diversity in time and space.* Springer Science & Business Media, Dordrecht, pp. 669.

Whitton, B.A. and Potts, M. (2000). Introduction to the cyanobacteria. In: Whitton, B.A. and Potts, M. (eds) *Ecology of cyanobacteria: their diversity in time and space.* Kluwer Academic Publishers, Dordrecht, pp. 1–11, 689 pp.

Wierzchos, J., et al. (2018). Endolithic microbial habitats as refuges for life in polyextreme environment of the Atacama Desert. *Curr. Opinion Microbiol.* 43, 124–131.

Williams A.J.J. and Beyene, M.A. (2012). Biological soil crusts in the Mojave Desert, USA: micromorphology and Pedogenesis. *Soil Sci. Soc. Am. J.* 76, 1685–1698.

Williams, L., et al. (2016). Cyanobacterial diversity of western European biological soil crusts along a latitudinal gradient. *FEMS Microbiol. Ecol.* 92(10), 1–9.

Williams, W.J., et al. (2014). Cyanobacteria in the Australian northern savannah detect the difference between intermittent dry season and wet season rain. *Biodiv. Conser.* 23(7), 1827–1844.

Williams, L., et al. (2018). Assessing recovery of biological soil crusts across a latitudinal gradient in Western Europe. *Restorat. Ecol.* 26, 543–554. https://doi.org/10.1111/rec.12579

Wilmotte, A. (1994). Molecular evolution and taxonomy of the cyanobacteria. In: Bryant, A. (ed) *The molecular biology of cyanobacteria.* Kluwer Academic Publishers, The Netherlands, pp. 1–25.

Win, T.T., et al. (2018). Algal biofertilizers and plant growth stimulants for sustainable agriculture. *Indust. Biotechnol.* 14(4), 203–211.

Wu, Y., et al. (2013). Development of artificially induced biological soil crusts in fields and their effects on top soil. *Plant Soil* 370 (1), 115e124.

Wuang, S.C., et al. (2016). Use of spirulina biomass produced from treatment of aquaculture wastewater as agricultural fertilisers. *Algal Res.* 15, 59–64.

Wubs, E.R., et al. (2016). Soil inoculation steers restoration of terrestrial ecosystems. *Nat. Plants* 2(8), 1–5.

Wynn-Williams, D.D. (2000). Cyanobacteria in deserts – life at the limit? In: Whitton, B.A. and Potts, M. (eds) *The ecology of cyanobacteria, Chapter 13.* Kluwer Academic Publishers, Dordrecht, The Netherlands, pp. 341–366.

Xie, Z.M., et al. (2007). Relationships between the biomass of the algal crusts in field and their compressive strength. *Soil Biol. Biochem.* 39, 567–572.

Xu, H.F., et al. (2021). Reading and surviving the harsh conditions in desert biological soil crust: the cyanobacterial viewpoint. *FEMS Microbiol Rev.* 45, fuab036.

Yang, J., et al. (2022). Mechanisms underlying legume–rhizobium symbioses. *J. Integr. Plant Biol.* 64(2), 244–267.

Zaneveld, J.S. (1959). The utilization of marine algae in tropical South and East Asia. *Econ. Bot.* 13, 89–131.

Zhang, B.C., et al. (2015). Responses of microbial activities and soil physico-chemical properties to the successional process of biological soil crusts in the Gurbantunggut Desert, Xinjiang. *J. Arid Land* 7, 101–e109.

Zhang, B., et al. (2016). Cyanobacterial composition and spatial distribution based on pyrosequencing data in the Gurbantunggut Desert, Northwestern China. *J-. Basic Microbiol.* 56(3), 308–320.

Zhang, J., et al. (2021a). Effect of substituting nitrogen fertilizer with nitrogen-fixing cyanobacteria on yield in a double-rice cropping system in southern China. *J. Appl. Phycol.* 33, 2221–2232.

Zhang, X.C., et al. (2021b). Temporal shifts in cyanobacterial diversity and their relationships to different types of biological soil crust in the southeastern Tengger Desert. *Rhizosphere* 17, 100322.

Zhang, Y. and Liu, B. (2024). Biological soil crusts and their potential applications in the sand land over Qinghai-Tibet Plateau. *Res. Cold Arid Reg.* 16, 20–29.

Zhao, Y., et al. (2019). Towards stopping land degradation in drylands: Water saving techniques for cultivating biocrusts in situ. *Land Degrad. Dev.* 30, 2336–2346.

Zhao, Y., et al. (2010). Potential nitrogen fixation activity of different aged biological soil crusts from rehabilitated grasslands of the hilly Loess Plateau, China. *J. Arid Environ.* 74(10), 1186–1191.

Zhou, X., et al. (2020). Practices of biological soil crust rehabilitation in China: experiences and challenges. *Restor. Ecol.* 28, S45–S55.

Chapter 13

Understanding and optimizing the use of amino acid-based biostimulants in promoting crop nutrient use efficiency and health

Wenli Sun and Mohamad Hesam Shahrajabian, Chinese Academy of Agricultural Sciences, China

1 Introduction

Biostimulants are sustainable and eco-friendly products that can promote enzymatic and metabolic processes in plants to improve their productivity, quality, and final yields (Shahrajabian et al., 2021a, 2021b). Key advantages of using biostimulants include (Sun and Shahrajabian, 2023; Shahrajabian et al., 2023; Sun et al., 2023):

- accelerated seedling establishment;
- increased cation exchange;
- deeper roots;
- improved stress tolerance;
- a stimulated immune system and increased natural plant toxins to resist pathogens and pests;
- better nutrient content, color, and improved taste; and
- increased harvest index and yield.

Amino acids for the production of biostimulants are obtained from chemical synthesis of plant proteins as well as from animal proteins by chemical or

http://dx.doi.org/10.19103/AS.2025.0166.16

enzymatic hydrolysis. Amino acids used to formulate biostimulant products are added either as free amino acids or as parts of hydrolyzed proteins (Soleymani and Shahrajabian, 2012; Khoshkharam et al., 2021).

Amino acid-based biostimulants have been found to increase crop yields by from 5% to up to 20%, depending on variables such as crop type and environmental conditions (Popko et al., 2018; Sun et al., 2024b). Positive impacts of application of amino acids on both quantitative and qualitative characteristics of crops have been reported, including for a range of herbs and spices such as the following (Mehrabi et al., 2013; Rafiee et al., 2013; Aghaye Noroozlo et al., 2019; Ayyat et al., 2021; Elsayed et al., 2022):

- *Satureja hortensis* L. (Summer savory);
- *Calendula officinalis* L. (Common marigold);
- *Mentha piperita* (Peppermint);
- *Urtica pilulifera* L. (Roman nettle);
- *Ocimum basilicum* (Sweet basil);
- *Nigella sativa* L. (Black cumin);
- *Achillea millefolium* L. (Yarrow);
- *Coriandrum sativum* L. (Coriander); and
- *Foeniculum vulgare* Mill (Fennel).

The goal of this chapter is to investigate the impact of application of biostimulants containing amino acids on growth, yield, nutrient content (including flavonoid and phenolic content), and quality of different crops.

2 Amino acids: mechanisms of action

Amino acids are organic components, which include carboxyl C(=O)OH) and amine (-NH_2) functional groups together with a side chain (R group). Amino acids can promote synthesis of proteins and stimulate enzymes and gene expression (Rai, 2002; Calvo et al., 2014; Poorghadir et al., 2020). Amino acids have been found to increase the activity of important enzymes such as capsaicin synthase, phenylalanine ammonia lyase, and peroxidase (Zamljen et al., 2023).

Amino acids can act as building blocks for proteins, which can then promote cell growth and thus boost plant vigor and help protect them from the effects of abiotic stresses such as drought or salinity (Lonnerdal, 2000; Abdelkader et al., 2021). Their important function in metabolite synthesis due to their structure as protein units is reported in a number of studies (Rodriguez-Morgado et al., 2015; Orts et al., 2019).

Amino acids can also act as signaling molecules, for example in inducing stomatal opening and closure (affecting plant water status). Through mechanisms such as scavenging reactive oxygen species (ROS) and metal chelation, they can also improve stress tolerance, e.g. to water deficit and high salinity (Oosten et al., 2017; Alfosea-Simon et al., 2020; Seyedi et al., 2023).

Amino acids also have an important function in C_4 photosynthetic metabolism and ammonium fixation (Amin et al., 2011; Popko et al., 2018; Sun et al., 2025). In part by acting as precursors for different secondary and primary metabolites, amino acids can increase vitamin and mineral content in plant tissues and boost nitrogen status (Meister, 1988; Das et al., 2002; Khan et al., 2019; Sun and Shahrajabian, 2025). Amino acid-based biostimulants have been found to boost ripening and quality characteristics of grape (Mian et al., 2022).

Amino acids can also promote the biosynthesis of different components, such as alkaloid derivatives, acylated polyamine, phenylpropenes, lignans, lignins, catechins, proanthocyanidins, suberin, sporopollenins, cutin, aurones, stilbenes, isoflavonoids, and flavonoids (Yang et al., 2020; Ayyanath et al., 2023; Forlani et al., 2023; Shahrajabian and Petropoulos, 2024; Sun et al., 2024a, 2024b). Amino acids for example have been found to boost phenolic and flavonoid content of soybean, as well as plant height, pod, and seed numbers (Kocira, 2019).

3 Amino acids as biostimulants

Both amino acids and protein hydrolysates are commonly accessible because of the abundance of raw materials for processing at relatively low cost (Martinez-Alvarez et al., 2015; Ebinezer et al., 2020; Makhaye et al., 2021). Protein-based biostimulants are often derived from the hydrolysis of protein-rich agro-wastes via enzymatic, thermal, and chemical processes, or their combination (Calvo et al., 2014; Brown and Saa, 2015; Baqer et al., 2019; Caruso et al., 2019). Amino acid biostimulants can be derived from hydrolysis of plant proteins (such as corn or soybean), animal proteins, and algae (Cosovanu et al., 2022; Islam et al., 2022; Peslerbes et al., 2022). The type of hydrolysis affects the amount and purity of free amino acid content (Ertani et al., 2014; Luo et al., 2020).

Common amino acids used in biostimulants include proline, alanine, arginine, glycine, and glutamic acid. Others include lysine, methionine, and tyrosine (Abdelkader et al., 2023). Products combining the amino acids valine + proline, and alanine + proline, e.g. have been found to boost growth and reduce drought stress in chickpea (Khalesi et al., 2023). Phenylalanine, proline, methionine, and tyrosine have been found to enhance leaf area, root length, shoot length, fresh and dry weight of spinach plants (Kausar et al., 2023). The usage of phenylalanine at a 100 mg L^{-1} concentration increased the essential oil content of mint (Velicka et al., 2022). Phenylalanine and tyrosine solutions can increase the total amount of phenols in a range of crops (Reham et al., 2016; Noviyanti et al., 2017; Baharlou et al., 2019). Phenylalanine can increase total chlorophyll contents, shoot length, and yield (Roman et al., 2020; Ramzan et al., 2023).

The amino acid-based Phytostim® biostimulant, derived from extracts of *Moringa oleifera* Lam., contains over 20 amino acids, notably glutamic acid, leucine, and alanine (as well as serine, phenylalanine, isoleucine, threonine, glycine, and proline), and has been found to increase growth and yield in lettuce (Mpai et al., 2022). The biostimulant Amino Expert® Impuls has been found to increase plant height, head diameter, seed oil content, mass, and yield in sunflower (Neshev et al., 2022). Actium® includes amino acids that increase phenylalanine, total monosaccharides, and carotenoids of pepper plants (Barrajon-Catalan et al., 2020),

AminoPrim and AminoHort, which contain 15% and 20% amino acids at 1.0 L/ha and 1.25 L/ha, have been found to promote copper, sodium, calcium, and molybdenum content in winter wheat grains (Hammad and Ali, 2014; Popko et al., 2014). Amino acid-based biostimulants such as Delfan Plus, Natural Crops SL, Tecamin Max, Tecamin Brix, Agrocean B, Metalosate Calcium, and Metalosate Fe improve photosynthesis, increase co-enzyme formation, and enhance the resistance of plants to environmental stresses (Seadh et al., 2008; Amin et al., 2011; Popko et al., 2018; Toscano et al., 2023). Methyl jasmonate can increase the content of volatile compounds (Portu et al., 2015; Roman et al., 2020). Perfectose™ can enhance the nutritive quality and yield of a range of plants (Al-Karaki and Othman, 2023). Phytostim® can also enhance growth and yield (Mpai et al., 2022). Codasil® can enhance resistance to drought stress (Hidalgo-Santiago et al., 2021). Amino Expert® can enhance yield (Neshev et al., 2022). Actium® can boost carotenoids and other quality parameters (Barrajon-Catalan et al., 2020). Bonamid® and Algabon® can increase biomass, leaf water content, and chlorophyll content (Popko et al., 2014; Hosseini et al., 2023; Kausar et al., 2023).

4 Applying amino acid-based biostimulants

A common means of using amino acid-based biostimulants is foliar application of amino acid liquid fertilizer, which e.g. increased the yield of cowpea (Wang et al., 2019). The foliar application of amino acid biostimulant (Perfectose™, liquid) increased the yield and nutritive value of lettuce (Al-Karaki and Othman, 2023). Foliar application of 100 mg/L aspartic acid enhanced oil content and yield of peanut (Sadak et al., 2023).

It has also been reported that the combined use of amino acids and humic acids enhances strawberry nutritional traits such as phenolic components and commercial traits such as firmness and color (Soppelsa et al., 2019). The combined application of humic acids and amino acids also improved tomato yields (Klokic et al., 2020), as well as enhancing flavonoid, phenol, and overall antioxidant content in tomato plants (Niu et al., 2022). Amino acids have also been found to increase the accumulation of total soluble sugars in tomato plants (Almadi et al., 2020).

The combined usage of humic acid, fulvic acid, and amino acids significantly enhances the oil content and the quality of olives (Nargesi et al., 2022). Application of zinc-enriched amino acids (Zn-AAC) enhanced quality, chlorophyll content, and final yield of rice plants (Raza et al., 2023). The combined application of amino acids and potassic fertilizer enhanced both the quality and the yield of rice (Mirtaleb et al., 2021).

The biostimulant Codasil®, which blends silicon, peptides, and amino acids, has been found to enhance lettuce growth and increase resistance to water stress (Hidalgo-Santiago et al., 2021). Terra Sorb® radicular and Terramin® Pro, which have a high percentage of amino acids, have also been reported to enhance lettuce development in nitrogen-limiting areas (Navarro-Leon et al., 2022). Sorb®, Radicular, and Terramin® can also play a role in increasing the yield of plants in nitrogen-limiting regions (Navarro-Leon et al., 2022). The combined application of two biostimulants, Algabon® at 0.5 g/L and Bonamid® at 2 g/L, was found to significantly increase biomass and leaf water content in weeping alkaligrass (Hosseini et al., 2023).

5 Declaration of interests

This research was also supported by the National Key R&D Program of China (Grant No. 2019YFA0904700). This research was funded by the Natural Science Foundation of Beijing, China (Grant No. M21026).

The authors declare no conflicts of interest.

Author contributions: W.S.: writing-original draft preparation; M.H.S.: writing-original draft preparation and editing. All authors have read and agreed to the published version of the manuscript.

6 References

Abdelkader, M. M., et al. (2021), The influence of biostimulants on tomato plants cultivated under hydroponic systems, *J. Hortic. Res.* 29: 107–116.

Abdelkader, M., et al. (2023), Biostimulants-based amino acids augment physio-biochemical responses and promote salinity tolerance of lettuce plants (*Lactuca sativa* L.), *Horticulturae*. 9: 807.

Aghaye Noroozlo, Y., et al. (2019), Stimulation effects of foliar application of glycine and glutamine amino acids on growth and quality of sweet basil, *Adv. Hortic. Sci.* 33: 495–501.

Alfosea-Simon, M., et al. (2020), Effect of foliar application of amino acids on the salinity tolerance of tomato plants cultivated under hydroponic system, *Sci. Hortic.* 272: 109509.

Al-Karaki, G. N. and Othman, Y. (2023), Effect of foliar application of amino acid biostimulants on growth, macronutrient, total phenol contents and antioxidant activity of soilless grown lettuce cultivars, *S. Afr. J. Bot.* 154: 225–231.
Almadi, L., et al. (2020), A biostimulant based on protein hydrolysates promotes the growth of young olive trees, *Agriculture*. 10: 618.
Amin, A. A., et al. (2011), Physiological response of onion plants to foliar application of putrescine and glutamine, *Sci. Hortic.* 129: 353–360.
Ayyanath, M.-M., et al. (2023), Straw mulch induced indoleamines alleviate reproductive depression in cold sensitive hazelnut cultivars, *Plants*. 12(13): 2577. https://doi.org/10.3390/plants12132577
Ayyat, A. M., et al. (2021), Improving growth, productivity, and oil yield of *Nigella sativa* L. plants by foliar spraying with some stimulants, *J. Plant Prod.* 12: 339–344.
Baharlou, M. J., et al. (2019), Effect of different concentrations of L-phenylalanine on chemical compositions and yield of essential oil of lemon balm (*Melissa officinalis*), *J. Herb. Drugs*. 10: 175–183.
Baqer, H. A. A.-R., et al. (2019), The role and importance of amino acids within plants: a review, *Plant Arch.* 19: 1402–1410.
Barrajon-Catalan, E., et al. (2020), Metabolomic analysis of the effects of a commercial complex biostimulant on pepper crops, *Food Chem.* 310: 125818.
Brown, P. and Saa, S. (2015), Biostimulants in agriculture, *Front. Plant Sci.* 6: 671.
Calvo, P., et al. (2014), Agricultural uses of plant biostimulants, *Plant Soil*. 383(1–2): 3–41.
Caruso, G., et al. (2019), Protein hydrolysate or plant extract-based biostimulants enhanced yield and quality performances of greenhouse perennial wall rocket grown in different seasons, *Plants*. 8: 208.
Cosovanu, D., et al. (2022), Rendered-protein hydrolysates as a low-cost nitrogen source for the fungal biotransformation of 5-hydroxymethylfurfural, *Catalysts*. 12: 839.
Das, C., et al. (2002), Involvement of kinetin and spermidine in controlling salinity stress in mulberry (*Morus alba* L. cv. S1), *Acta Physiol. Plant.* 24: 53–57.
Ebinezer, L. B., et al. (2020), Quantitative proteomics of maize roots treated with a protein hydrolysate: a comparative study with transcriptomics highlights the molecular mechanisms responsive to biostimulants, *J. Agric. Food Chem.* 68: 7541–7553.
Elsayed, A. A. A., et al. (2022), Changes in bitter fennel essential oils exposed to foliar spray with L-phenylalanine, *Egypt. J. Bot.* 62: 241–253.
Ertani, A., et al. (2014), *Capsicum chinensis* L. growth and nutraceutical properties are enhanced by biostimulants in a long-term period: Chemical and metabolomic approaches, *Front. Plant Sci.* 5: 375. https://doi.org/10.3389/fpls.2014.00375
Forlani, G., et al. (2023), The levels of DAHP synthase, the first enzyme of the shikimate pathway, are related to free aromatic amino acids and glutamine content in *Nicotiana plumbaginifolia* cell cultures, *Plants*. 12(13): 2524. https://doi.org/10.3390/plants12132524
Hammad, S. A. R. and Ali, O. A. M. (2014), Physiological and biochemical studies drought tolerance of wheat plants by application of amino acids and yeast extract, *Annals Agric. Sci.* 59: 133–145.
Hidalgo-Santiago, L., et al. (2021), The application of the silicon-based biostimulant codasil®offset water deficit of lettuce plants, *Sci. Hortic.* 285: 110177.

Hosseini, S., et al. (2023), Foliar spray of commercial seaweed and amino-acid derived biostimulants promoted phytoremediation potential and salinity stress tolerance in halophytic grass, *Puccinellia distans*, *Int. J. Phytoremed.* 25: 415–429.

Islam, M., et al. (2022), Influence of the degree of hydrolysis on functional properties and antioxidant activity of enzymatic soybean protein hydrolysates, *Molecules*. 27: 6110.

Kausar, A., et al. (2023), Modulation of growth and biochemical responses in spinach (*Spinacia oleraceae* L.) through foliar application of some amino acids under drought conditions, *S. Afr. J. Bot.* 158: 243–253.

Khalesi, A., et al. (2023), Effect of foliar application of amino acids on grain yield and physiological traits of chickpea under drought stress. *Gesunde Pflanz*, 75: 1705–1718.

Khan, S., et al. (2019), Exogenous application of amino acids improves the growth and yield of lettuce by enhancing photosynthetic assimilation and nutrient availability, *Agronomy*, 9: 266.

Khoshkharam, M., et al. (2021), The effects of methanol and amino acid glycine betaine on qualitative characteristics and yield of sugar beet (*Beta vulgaris* L.) cultivars, *Not Sci Biol*. 13(2): 1–13. https://doi.org/10.15835/nsb13210949

Klokic, I., et al. (2020), Biostimulants' influence on tomato fruit characteristics at conventional and low-input NPK regime, *Acta Agric. Scand. Sec. B Soil Plant Sci*. 70: 233–240.

Kocira, S. (2019), Effect of amino acid biostimulant on the yield and nutraceutical potential of soybean, *Chil. J. Agric. Res.* 79: 17–25.

Lonnerdal, B. (2000), Dietary factors influencing zinc absorption, *J. Nutr.* 130: 1378S-1383S.

Luo, Y., et al. (2020), Synergistic effects of plant protein hydrolysates and xanthan gum on the short- and long-term retrogradation of rice starch, *Int. J. Biol. Macromol.* 144: 967–977. https://doi.org/10.1016/j.ijbiomac.2019.09.174

Makhaye, G., et al. (2021), Bioprinting with seaweed extract and microbial-based commercial biostimulants influences seed germination of five *Abelmoschus esculentus* genotypes, *Plants*. 10: 1327.

Martinez-Alvarez, O., et al. (2015), Protein hydrolysates from animal processing by-products as a source of bioactive molecules with interest in animal feeding: a review, *Food Res. Int.* 73: 204–212.

Mehrabi, S., et al. (2013), Clarifying the role of methanol and amino acids application on savory (*Satureja hortensis* L.), *Ann. Biol. Res.* 4: 190–195.

Meister, A. (1988), Glutathione metabolism and its selective modification, *J. Biol. Chem.* 263: 17205–17208.

Mian, G., et al. (2022), Effects of a triacontanol-rich biostimulant on the ripening dynamic and wine must technological parameters in *Vitis vinifera* cv. Ribolla Gialla, *Plant Physiol. Biochem.* 188: 60–69.

Mirtaleb, S. H., et al. (2021), Foliar spray of amino acids and potassic fertilizer improves the nutritional quality of rice, *J. Plant Nutr.* 44: 2029–2041.

Mpai, S., et al. (2022), Untargeted metabolites and chemometric approach to elucidate the response of growth and yield attributes on different concentrations of an amino acid based biostimulant in two lettuce cultivars, *Sci. Hortic.* 306: 111478.

Nargesi, M. M., et al. (2022), Effect of foliar application of amino acid, humic acid, and fulvic acid on the soil content and quality of olive, *Saudi J. Biol. Sci.* 29: 3473–3481.

Navarro-Leon, E., et al. (2022), Effect of L-amino acid-based biostimulants on nitrogen use efficiency (NUE) in lettuce plants, *J. Sci. Food Agric.* 102: 7098–7106.
Neshev, N., et al. (2022), Is the plant biostimulant application ameliorative for herbicide-damaged sunflower hybrids? *Ind. Crops Prod.* 182: 114926.
Niu, C., et al. (2022), Biostimulants alleviate temperature stress in tomato seedlings, *Sci. Hortic.* 293: 110712.
Noviyanti, R., et al. (2017), Biomass and flavonoids production of gynura procumbens adventitious roots induced by sucrose, phenylalanine, and tyrosine, *Biosci. Res.* 14: 934–941.
Oosten, M. J. V., et al. (2017), The role of biostimulants and bioeffectors as alleviators of abiotic stress in crop plants, *Chem. Biol. Technol. Agric.* 4(5): 1–12. https://doi.org/10.1186/s40538-017-0089-5.
Orts, A., et al. (2019), Production of biostimulants from Okara through enzymatic hydrolysis and fermentation with Bacillus licheniformis: Comparative effect on soil biological properties, *Environ. Technol.* 40: 2073–2084.
Peslerbes, M., et al. (2022), Manufacture of whey protein hydrolysates using plant enzymes: effect of processing conditions and simulated gastrointestinal digestion on angiotensin-I-converting enzyme (ACE) inhibitory activity, *Foods.* 11: 2429.
Poorghadir, M., et al. (2020), Interactions of amino acids (proline and phenyl-alanine) and biostimulants (salicylic acid and chitosan) and biostimulants (Salicylic acid and chitosan) on the growth and essential oil components of savory (*Satureja hortensis* L.), *Biocatal. Agric. Biotechnol.* 30: 101815.
Popko, M., et al. (2014), New amino acid biostimulators based on protein hydrolysate of keratin, *Przem. Chem.* 93: 1012–1015.
Popko, M., et al. (2018), Effect of the new plant growth biostimulants based on amino acids on yield and grain quality of winter wheat, *Molecules.* 23: 470.
Portu, J., et al. (2015), Changes on grape phenolic composition induced by grapevine foliar applications of phenylalanine and urea, *Food Chem.* 180: 171–180.
Rafiee, H., et al. (2013), Phytochemical, agronomical, and morphological response of pot marigold (*Calendula officinalis* L.) to foliar application of biostimulators (Bioactive amino acid compounds), *J. Med. Plants.* 12: 48–61.
Rai, V. K. (2002), Role of amino acids in plant responses to stresses, *Biol. Plant.* 45(4): 481–487. https://doi.org/10.1023/A:1022308229759
Ramzan, T., et al. (2023), Phenylalanine supply alleviates the drought stress in mustard (*Brassica campestris*) by modulating plant growth, photosynthesis, and antioxidant defense system, *Plant Physiol. Biochem.* 201: 107828.
Raza, S., et al. (2023), Effect of zinc-enriched amino acids on rice plants (*Oryza sativa* L.) for adaptation in saline-sodic soil conditions: Growth, nutrient uptake and biofortification of zinc, *S. Afr. J. Bot.* 162: 370–380.
Reham, M. S., et al. (2016), Influence of foliar spray with phenylalanine and nickel on growth, yield quality, and chemical composition of genoveser basil plant, *Afr. J. Agric. Res.* 14: 934–941.
Rodriguez-Morgado, B., et al. (2015), Obtaining edaphic biostimulants/biofertilizers from different sewage sludges, effects on soil biological properties, *Environ. Technol.* 36: 2217–2226.
Roman, S. M.-S., et al. (2020), Foliar application of phenylalanine plus methyl jasmonate as a tool to improve Grenache grape aromatic composition, *Sci. Hortic.* 272: 109515.

Sadak, M. S., et al. (2023), Amino acids foliar application for maximizing growth, productivity, and quality of peanut grown under sandy soil, *Braz. J. Biol.* 6: e256338.

Seadh, S. R., et al. (2008), Effect of foliar nutrition with humic and amino acids under N-levels on wheat productivity and quality of grains and seeds, *Egypt. J. Appl. Sci.* 23: 543–558.

Seyedi, A., et al. (2023), The effect of biostimulants based on free amino acids on some growth and physiological parameters of *Dracocephalum moldavica* L. under salinity stress, *J. Med. Plants By-Prod.*

Shahrajabian, M. H. and Petropoulos, S. A. (2024), Editorial for the special issue on plant biostimulants in sustainable horticulture and agriculture: development, function, and applications, *Plants*. 13(17): 1–3. https://doi.org/10.3390/plants13172342.

Shahrajabian, M. H., et al. (2021a), Biostimulants application: a low input cropping management tool for sustainable farming of vegetables, *Biomolecules*. 11(5): 698. https://doi.org/10.3390/biom11050698.

Shahrajabian, M. H., et al. (2021b), Sustainable agriculture systems in vegetable production using chitin and chitosan as plant biostimulants, *Biomolecules*. 11(6): 819. https://doi.org/10.3390/biom11060819.

Shahrajabian, M. H., et al. (2023). Survey of the influences of microbial biostimulants on horticultural crops: case studies and successful paradigms, *Horticulturae*. 9(193): 1–24. https://doi.org/10.3390/horticulturae9020193.

Soleymani, A. and Shahrajabian, M. H. (2012), The effects of Fe, Mn, and Zn foliar application on yield, ash, and protein percentage of *Forage sorghum* in climatic condition of Esfahan, *Int J Biol*. 4(3): 92–96. https://doi.org/10.5539/ijb.v4n3p92.

Soppelsa, S., et al. (2019), Foliar applications of biostimulants promote growth, yield, and fruit quality of strawberry plants grown under nutrient limitation, *Agronomy*. 9: 483.

Sun, W. and Shahrajabian, M. H. (2023), The application of arbuscular mycorrhizal fungi as microbial biostimulant, sustainable approaches in modern agriculture, *Plants*. 12(17): 3101. https://doi.org/10.3390/plants12173101.

Sun, W. and Shahrajabian, M. H. (2025), Biostimulant and beyoung: Bacillus spp. the important plant growth-promoting rhizobacteria (PGPR)-based biostimulant for sustainable agriculture, *Earth Systems and Environment*. Advance online publication. https://doi.org/10.1007/s41748-024-00552-4.

Sun, W., et al. (2023), Developing sustainable agriculture systems in medicinal and aromatic plant production by using chitosan and chitin-based biostimulants, *Plants*. 12(13): 2469. https://doi.org/10.3390/plants12132469.

Sun, W., et al. (2024a), The roles of plant-growth-promoting rhizobacteria (PGPR)-based biostimulants for agricultural production systems, *Plants*. 13(613): 1–38. https://doi.org/10.3390/plants13050613.

Sun, W., et al. (2024b), Amino acids biostimulants and protein hydrolysates in agricultural sciences, *Plants*. 13: 210. https://doi.org/10.3390/plants13020210.

Sun, W., et al. (2025), A study of the different strains of the genus Azospirillum spp. on increasing productivity and stress resilience in plants, *Plants*. 14(2): 267. https://doi.org/10.3390/plants14020267.

Toscano, S., et al. (2023), Physiological and biochemical changes in response to *Moringa oleifera* biostimulant in petunia plants under water deficit, *Sci. Hortic*. 319: 112187.

Velicka, A., et al. (2022), Impact of foliar application of amino acids on essential oil content, odor profile, and flavonoids content of different mint varieties in field conditions, *Plants*. 11: 2938.

Wang, D., et al. (2019), Effects of foliar application of amino acid liquid fertilizers, with or without *Bacillus amyloliquefaciens* SQR9, on cowpea yield and leaf microbiota, *PLoS One*. 14: e0222048.

Yang, Q., et al. (2020), Connections between amino acid metabolisms in plants: Lysine as an example, *Front. Plant Sci*. 11: 928. https://doi.org/10.3389/fpls.2020.00928.

Zamljen, T., et al. (2023), Biostimulative effect of amino acids on the enzymatic and metabolic response of two *Capsicum annuum* L. cultivars grown under salt stress, *Sci. Hortic*. 309: 111713.

Chapter 14

Promoting soil microbial activity and soil organic matter accumulation: inorganic and organic fertilizer effects

John Havlin, North Carolina State University, USA

1 Introduction

Numerous diverse bacterial and fungal microorganisms are intimately involved in bio-geochemical processes and transformations essential to nutrient availability and soil health (Fierer et al., 2021; Ferris and Tuomisto, 2021). Sustaining healthy soil-plant ecosystem functions is the foundation of global food security, while providing vital natural resource protection (Zereyesus et al., 2023; Fischer et al., 2014; CAST, 2013). A robust microbially mediated cycling of essential plant nutrients is intimately connected to sustaining soil carbon (C) reserves (Dondini et al., 2023; Baveye et al., 2020; Lal, 2014). Unfortunately, human activity (agriculture, deforestation, etc.) has substantially degraded soil productivity predominately related to topsoil and soil C loss (Bridges and Oldeman, 2010). As a result, global efforts to understand soil C cycling and develop soil/crop management technologies and practices to enhance soil C reserves have occurred over the last six decades (Beillouin et al., 2023; Ma et al., 2023; Smith et al., 2019; Post et al., 2004). Using a global meta-analysis, Oldfield et al. (2019) reported that nearly 70% of world corn and wheat

http://dx.doi.org/10.19103/AS.2025.0166.17

production occurs on soils with <2% soil organic carbon (SOC), while little or no yield responses were predicted above 2% SOC. If global SOC on these soils were increased to 2%, corn and wheat yield would increase by 18% and 30% with a 8% and 4% reduction in fertilizer N rates, respectively. Although many soil properties distinguish soil health, increasing soil organic matter (SOM) increases soil microbial biomass and microbial function to impart improved aggregation, soil structure, and infiltration, which increases plant available water (PAW) and minimizes potential water stress-related yield losses. For example, when averaged over all soil textures, Hudson (1994) demonstrated PAW nearly doubled as SOM increased from 0.5% to 3%.

Although difficult to quantify the contribution of synthetic fertilizers, Stewart et al. (2005) estimated 30–60% of crop yield is attributable to commercial fertilizer nutrients, where nitrogen (N), phosphorus (P), and potassium (K) inputs dominate global fertilizer use (Ludeman et al., 2022). Erisman et al. (2008) reported enhanced crop yields from N fertilization sustain nearly half of the global population. While long-term use of commercial fertilizers has substantially enhanced global food security, increasing nutrient use (mostly N and P) has had effects on selected soil chemical and biological properties. One of the most important properties studied is long-term soil and crop management effects on SOM (Singh, 2018). The positive effects of increasing SOM on many soil physical and chemical properties (>water-holding capacity (WHC), < bulk density (Bd), < compaction, > infiltration, > cation exchange capacity (CEC), etc.) are well documented (Oldfield et al., 2019; Havlin et al., 2014).

The effects of increasing N rate on biomass production, and increasing or stabilizing SOM, have been documented (Jagadamma et al., 2007; Alvarez, 2006; Glendining and Powlson, 1995; Paustian et al., 1992). For example, with 240 kg N ha^{-1} applied annually over 15 years, total crop biomass yield increased 44% compared to no N, while SOC increased by 6% or 0.5 Mg ha^{-1} $year^{-1}$ (Mazzoncini et al., 2011). A thorough review of long-term field experiments in China reported surface SOM increased in most regions due to increased N rates, biomass yield, crop residue returned, and increased use of conservation tillage practices (Miao et al., 2011). Hao et al. (2017) reported similar increases in SOC and N after 27 years with fertilizer N + P compared to N only on a P-deficient soil. In long-term continuous dryland winter wheat experiments, Aula et al. (2016) reported similar increases in SOC (0–15 cm depth) with increasing N rate, especially with annual applications at or above 90 kg N ha^{-1}.

Although many long-term studies document significant positive correlations between SOM and synthetic fertilizer N applied and increasing crop residue returned, several studies document SOM loss, especially if the cumulative N applied exceeds grain N removal (Mulvaney et al., 2009; Khan et al., 2007). Several reasons have been forwarded including: (a) N rates > crop N removal lowers residue C:N which enhances heterotrophic bacterial activity and increases SOM decomposition and (b) enhanced SOM decomposition increases N loss by leaching and denitrification (Powlson et al., 2010; Johnston et al., 2009). The positive effect of long-term N fertilizer on increased SOM or decreased decomposition was also supported by Ladha et al. (2011) and Li et al. (2017).

With an emphasis on N, the objectives are to review (a) how long-term synthetic fertilizers may influence the diversity and activity of soil microbes important to nutrient transformations that affect nutrient supply and SOM and (b) how N fertilizers combined with imported C as residue or wastes may enhance microbial activity and SOM accumulation.

2 Tillage and crop rotation effects

Over the last six decades producer conversion from conventional tillage (CT, <30% surface residue cover) to no-tillage (NT) has improved soil health through substantial reduction in soil erosion of SOM-rich topsoil, reduced SOC oxidation to CO_2, and increased CO_2 capture (Baveye et al., 2020; Lal, 2006). For example, using several long-term (60 years) field studies comparing CT and NT to nearby undisturbed forest land (FL), Mishra et al. (2010) reported SOC (0–25 cm) in NT was 50% greater than in CT, whereas FL was 23% and 85% greater than NT and CT, respectively. Comparing NT to CT on three long-term (>21 years) field experiments over a range in annual precipitation (440 - 890 mm year^{-1}), SOC was greater and Bd was lower with NT compared to CT (0–15 cm), where differences increased with increasing annual precipitation (Blanco-Canqui et al., 2011). In a detailed review of over 350 long-term (>10 years) tillage experiments, Haddaway et al. (2017) documented nearly twofold increase in SOC (0–15 cm) with NT compared to more intensive tillage systems. Significant increases in SOC were also documented at 30 cm soil depth, especially in longer duration studies.

If both tillage and crop rotation are included in field experiments, significant treatment interactions are commonly observed. For example, in semiarid regions, increasing cropping frequenting (reducing dependency on fallowing) generally increases residue produced and left on the field, which increases SOC, especially under NT (Wood et al., 1992). Including legumes in the rotation often enhances SOM due to increased residue N associated with legumes (Drinkwater et al., 1998, Havlin et al., 1990). In a long-term study (~60 years), Deiss et al. (2021) reported greater SOM under continuous corn and corn-perennial forage legume rotations compared to soybean in the rotation, especially with NT compared to CT. As expected, the SOM increase was greater in surface soil depths, although ~10% increase in SOM (0–30 cm) was observed. They documented enhanced microbial production of organic functional compounds (aliphatic, phenol) in NT important to SOM accumulation, concluding that rotations with perennial grasses and forage legumes under limited tillage increased SOM. Similarly, in a long-term (25 years) corn-soybean study, SOC (0–20 cm) was 27% greater under NT than CT (St. Luce et al., 2021). Using five long-term (6–30 years) tillage studies, Chatterjee and Lal (2009) documented 30–67% greater SOC (0–5 cm) with NT compared to CT, where no differences were observed below 30 cm sampling depth. They also sampled similar soil series in nearby wooded areas, which exhibited greater SOC compared to NT in three of the five sites. As documented earlier, in most studies increased SOC is generally significant in surface soil (0–10 cm), where increasing SOC with depth occurs frequently with increasing treatment duration. In addition, depending on rotation, increasing residue inputs reduces soil Bd and compaction due to increased SOM and aggregate stability (Hussain et al., 2021; Mehra et al., 2018; Liu et al., 2006; Havlin and Schlegel, 1997).

3 Inorganic nitrogen fertilizer effects

Comprised of salts, synthetic fertilizers added to soils will have an influence on soil chemical, biological, and physical properties depending on rate and method of application and soil properties (Haj-Amor et al., 2022). Depending on specific fertilizer salt, mismanagement can increase soil salinity with potential negative effects on plant growth and yield (Mortvedt, 2001). Generally, problems occur more frequently with products containing NH_4^+ and K^+ salts, where most problems occur with subsurface band (near or with the seed) compared to broadcast applications (Havlin et al., 2014). Guidelines for crop sensitivities to soil salinity, source and rate limits for in-row application at planting, and other agronomic principles and practices are well documented (Makaza and Khiari, 2023; Evangelou, 1983; US Laboratory Staff, 1954). Unless excessive rates of high N and/or K sources are surface or subsurface band applied, soil salinity generally will not exceed 4 dS m^{-1}, which is not sufficient to reduce crop growth. Using NH_4^+/K^+ and SO_4^{-2}/Cl^- salts over a wide range in salinity levels (0.5–60 dS m^{-1}), Rath et al. (2016) concluded bacteria were less tolerant than fungi to elevated salt concentrations, while C/N mineralization and nitrification processes were not affected. In addition, microbial respiration was suppressed more by salts containing Cl^- than SO_4^{-2}. An extensive review by Rath and Rousk (2015) documented that increasing soil salinity reduced microbial biomass, microbial respiration, and decomposer microorganism function. In contrast, a recent review by Dincă et al. (2022) concluded microbial growth, diversity, and function were stimulated by added nutrients (inorganic salts). Negative effects on selected microbes and enzyme activities are generally short-lived (< 4–5 weeks), restricted to the application zone, in low SOM soils, and are more commonly observed at elevated salinity levels (> 4 dS m^{-1}) (Singh and Ryan, 2015). With crop requirement-based rates, inorganic fertilizer salts generally would not increase salinity above 4 dS m^{-1}. Obviously, adding inorganic fertilizers may increase soil salinity, depending on source and rate applied; however, at typical N and K rates, the increase in salinity is minor when broadcast applied, especially in humid compared to semiarid regions where greater nutrient uptake and dilution/percolation potential occurs. In addition, enhancing SOM through NT and use of organic fertilizers often enhances soil structure and increases WHC, which reduces the risk of soil salinization.

One of the most potentially toxic N fertilizers is liquified anhydrous ammonia (NH_3), which must be subsurface band applied. As NH_3 rapidly reacts with soil H_2O in the application zone, the resulting reaction product ($NH_4^+ + OH^-$) temporarily increases soil pH ≥ 9 depending on soil buffering capacity (Sawyer, 2019). Other than increased soil acidity from nitrification, no deleterious effects of NH_3 use on soil physical properties (Intrawech et al., 1982; Eno and Blue, 1954), SOM (Tomasiewicz and Henny, 1985), or soil microbes (Biederbeck et al., 1996; Eno and Blue, 1954) have been established. Any negative effects on soil microbe populations and functions generally occurred at high N rates and dissipated with time. Although Deng et al. (2006) reported soils receiving NH_3 exhibited reduced microbial activities, especially with organisms and enzymes important to soil N cycling, soil sampling time after application was not reported. The review by Geisseler and Scow (2014) suggests long-term N fertilization generally exhibits more positive than detrimental effects on nitrifying microorganisms.

Long-term use of NH_4^+ or NH_4^+-forming synthetic fertilizers is widely recognized to reduce soil pH, the degree to which is highly variable depending on N rate and source, soil buffering capacity, and other soil chemical and physical properties (Dikir, 2023; Schroder et al., 2011). In a recent review of long-term N management experiments in China, Miao et al. (2011) reported 0.5–2.0 decrease in soil pH with application of N fertilizers, especially when annual N rates exceeded crop N demand (Daba et al., 2021; Guo et al., 2010). In this review, 1.5–2.0 kg $CaCO_3$ ha^{-1} were needed to neutralize the acidity produced from high N rates in intensive cropping systems. Similar results were reported in India (Hemalatha et al., 2013). After nearly 140 years, soil pH decreased 1.5 units with 96 kg N ha^{-1} as $(NH_4)_2SO_4$ compared to a slight increase in soil pH (+0.5 pH) with the same N rate as $NaNO_3$ (Johnston and Poulton, 2018). Most of the decrease in soil pH occurred within the first 40–50 years. After 109 years of conventionally tilled, continuous dryland winter wheat production, soil pH declined from pH 5.5 (unfertilized) to pH 4.8 with 37 kg N ha^{-1} annually applied until 1968 and 67 kg N ha^{-1} applied thereafter (Davis et al.,

2003). Soil pH decreased with increasing N fertilization, especially at the highest N rates in long-term continuous dryland winter wheat experiments (Aula et al., 2016). Decreasing soil pH with increasing NH_4^+-based or -forming N sources will generally decrease exchangeable basic cations, increase exchangeable Al^{+3}, and increase solubility/availability of selected micronutrients (Dikir, 2023; Havlin et al., 2014; Barak et al., 1997).

Fierer and Jackson (2006) reported soil pH was positively correlated with soil bacterial diversity across a wide variation in soils and crops/vegetation. Rousk et al. (2010) reported bacterial diversity doubled over soil pH 4.0 to 8.3, whereas only marginal increases in fungal diversity were observed. Similar results were found by Geisseler and Scow (2014), who also summarized long-term N fertilization effects on soil microbial biomass composition and activity were minimal in agroecosystems unless soil pH dropped below 5. A recent summary by Sabir et al. (2021) also showed decreased soil pH with N fertilization reduced the abundance of bacteria, while increasing fungal growth. Using soil samples from a long-term (35 years) study of N fertilization in continuous corn and corn/soybean rotations, Huang et al. (2019) reported increased populations of nitrifiers and fungal communities, no effect on denitrifier populations, but reduced non-symbiotic N_2 fixing bacteria, suggesting enhanced dependency on synthetic N sources in intensive corn production systems. In addition, long-term use of synthetic N enhanced bacterial and fungal diversity and function and increased mineralization of native SOM, commonly referred to as the 'priming effect' (Dijkstra et al., 2013; Blagodatskaya and Kuzyakov, 2011; Hamer and Marschner, 2005).

In a highly pH-buffered soil, increasing N rate (0-360 kg ha^{-1}) decreased rhizosphere pH 0.1 and 0.3 units after 1 and 10 years of annual N in continuous wheat, respectively (Ren et al., 2020). The highest wheat yields and abundance of acidobacteria were found with the intermediate N rate; however, the authors did not account for allelopathic species that likely affected continuous wheat yield decline over time (Lam et al., 2012). Similarly, Sun et al. (2019) reported increased abundance and diversity of ureolytic bacteria with long-term (25 years) urea application, suggesting an increased NH_3 volatilization potential.

4 Organic fertilizer effects

Despite tillage, rotation, and inorganic N effects on SOM, microbial diversity and function in increasing SOM is primarily determined by the rate and quality of organic nutrient inputs (Hartmann et al., 2015). Organic wastes added to soil commonly increase microbial activity and SOM, while potentially degrading native SOM (priming effect) (Abdalla et al., 2022; O'Brien and Hatfield, 2019). Enhanced respiration associated with added digestible C, under limited N, results in mineralization of native SOC through increased soil microbial N demand. With higher waste rates and/or residue N (low C:N or added N), native SOC may be conserved. Regardless of organic waste C and N content, expanding global population will increase the organic waste supply, which must be utilized to enhance global soil health (Termorshuizen et al., 2004).

The net effect of organic waste amendments on SOM and related soil properties is highly variable depending on source and rate, soil physical properties, climate, and numerous soil and crop management factors. Detailed reviews are found in Brichi et al. (2023), Aytenew and Bore (2020), Hossain et al. (2017), Diacono and Montemurro (2010), and Liu et al. (2006). These reviews document organic waste amendments increased SOM, which increases soil WHC and PAW and reduces soil Bd, while increasing aggregate stability, soil porosity, and infiltration. These effects enhance soil surface resistance to soil erosion forces and increase subsoil PAW, which may enhance plant growth during periods of drought. In addition, significantly reduced soil Bd and increased soil WHC generally were positively correlated to soil texture (% sand) and increased SOC level (Eden et al., 2017; Rasool et al., 2008; Khaleel et al., 1981). In a literature review of over 500 studies, Shi et al. (2024) concluded inorganic fertilizers decreased microbial species richness (diversity and function) compared to organic sources, which decreased soil Bd and increased CEC and WHC.

Using soils from three long-term (17 years) field studies, Hu et al. (2018) documented that animal wastes and green manure crops increased C input by an average of 0.6–1.0 Mg C ha^{-1} year^{-1}, which translated into a net average SOC increase (+0.4 Mg C ha^{-1} year^{-1}). Bulluck et al. (2002) quantified selected soil microbe species in fields with organic sources compared to synthetic nutrient sources and reported higher populations of beneficial fungi (Trichoderma) and thermophilic and enteric bacteria, whereas plant pathogens *Phytophthora* and *Pythium* were lower with organic nutrient amendments compared to synthetic sources.

Sustainable nutrient management depends on SOM mineralizable N contributing to total crop N demand. For example, increasing SOM from 1% to 2% (0–15 cm) increases estimated mineralizable N from 10 kg N ha^{-1} to 20 kg N ha^{-1}, assuming 1% mineralization rate and 5% N in SOM (Havlin et al, 2014). Spargo et al. (2011) quantified N mineralization from both long-term (15 years) CT, NT, and organic crop rotations and documented a ~30% increase in potentially mineralizable N between CT and organic systems. They concluded inorganic N rates should be adjusted (decreased) in fields with a history of legume-based, organic waste-amended, and/or NT cropping systems. Over 15 years, Liang et al. (2012) added 15 Mg ha^{-1} year^{-1} of manure to a wheat-corn cropping system and documented that SOC (0–20 cm) increased nearly twofold over urea (equivalent N rate). Unfortunately, the study did not include a manure + urea-N treatment, which likely would have further enhanced SOC. In addition to plant and animal wastes, industrial and municipal wastes are increasingly available and produce similar effects on soil biological, physical, and chemical properties (Bastida et al., 2008), which are especially important in reclaiming severally degraded urban and other soils (Larney and Angers, 2012).

Although an organic nutrient source, biofertilizers generally represent carrier-based preparations containing specific dormant, viable microorganisms to enhance nutrient supply through symbiotic and/or associative N_2 fixing bacteria, decomposing organic wastes, and diverse plant growth-promoting substances (Kumar et al., 2021). Although considered relatively new technology, and increasingly available in the marketplace, rhizobium inoculants used to enhance symbiotic N_2 fixation represent a 'biofertilizer' and have been commonly used over the last six decades, although the first use was in 1887 (Deaker et al., 2004). Excellent reviews of biofertilizer technologies are available (Daniel et al., 2022; MItter et al., 2021). Many biofertilizer products have been developed to influence the microbial diversity and function in the rhizosphere to enhance N mineralization, solubilization of P-bearing minerals, and associative and non-symbiotic N_2 fixation (Fasusi et al., 2021; Vega, 2007). When products contain dormant (live) organisms, careful storage and handling are essential to maintain microbial viability; thus, the shelf-life of microbial-based biofertilizers is highly variable. While microorganism activity may be enhanced in the soil zone of application (subsurface band) or in the rhizosphere (applied with seed at planting), the contribution to N availability is highly variable and generally much smaller than crop requirements, whereas plant-, animal-, or industrial-waste-based biofertilizers (low C:N) may be applied at rates sufficient to meet crop N demand. These biofertilizer sources behave similarly in soils to enhance SOC as composts and animal waste sources. For example, Malik et al. (2014) reported decreased soil Bd and increased WHC in surface soil (0–15 cm) with biofertilizer (containing Azospirillum and P-solubilizing bacteria) compared to urea and superphosphate. Since increasing SOM is the primary objective in enhancing soil health and agricultural sustainability, microbial biofertilizers may contribute to SOC content without substantial crop biomass response (Debska et al., 2016). Just et al. (2024) reviewed nearly 60 studies and reported an average increase in SOC of 0.44 g C kg^{-1} (0.2–0.7 g C kg^{-1}) with biofertilizers containing mixtures of organisms (mycorrhizal fungi, N_2 fixing bacteria, and cyanobacteria), where the SOC increases were positively correlated to crop yield increases. Despite the above constraints with biofertilizers containing live/dormant organisms, biofertilizer development and markets are rapidly expanding (Santos et al., 2024), due to increasing producer interest in enhancing soil health and increased native plant nutrient supply.

5 Combined organic with inorganic nitrogen fertilizer effects

With approximately 8:1 and 10:1 C:N in soil microbial biomass and SOM, respectively, decomposition of residue with > 25:1 C:N will be limited by insufficient residue N available to the diverse decomposer microflora (Liu et al., 2021; Havlin et al., 2014). Although urea is an organic compound, adding urea or other synthetic N sources enhances microbial degradation of C-rich, N-limiting materials added to soil, reduces the priming effect, and enhances microbial-mediated processes producing SOM. For example, Yu et al. (2024) evaluated decomposition of 6000 kg ha^{-1} corn residue combined with urea (150 kg N ha^{-1}) and reported enhanced diversity and function of decomposing bacteria, which increased residue decomposition rate compared to no urea applied. They also demonstrated enhanced residue decomposition throughout the growing season if slow-release N sources composed ~50% if the added N.

On a low pH, P, and SOM soil, combining animal manure with urea N fertilizer resulted in a positive interaction on Cucumber yield, where yield was greater when N sources were combined compared to being applied separately (Nweke and Nsoanya, 2015). Mahmood et al. (2017) reported similar results where corn grain yield with manure + urea was 19–22% greater than fertilizer N and manure applied separately. Interestingly, SOC was about 20% higher with manure applied alone compared to manure + urea N, likely because the latter manure rate (manure + N) was about half that applied with manure alone. In a short-term (3 year) study, Muhammad et al. (2023) reported higher SOM when total applied N comprised of 75% manure and 25% urea-N.

Using a long-term (59 years) wheat-sugar beet field trial, Šimon and Czakó (2014) documented that manure and urea applied together increased SOC and microbial activity (0–20 cm) compared to urea alone. Although organic nutrient sources varied in their effect on SOC, when significant increases were observed with added urea, they included composts and/or manure. In a review of long-term field trials comparing organic and mineral fertilizers in India, Hemalatha et al. (2013) reported higher SOC (0–20 cm) at sites where organic sources were used compared to inorganic sources; however, when both sources were combined, SOC levels were the highest and maintained original or native SOC levels. Compared to separate urea or plant compost treatments, Habteweld et al. (2020) documented enhanced nematode diversity and activity effects on soil food web function with 1:1 urea:compost ratio.

In continuous production since 1888, the Sanborn Field includes tillage, rotation, and nutrient source treatments (Johnson et al., 1989). Treatment effects on five selected soil enzymes were correlated with microbial biomass and SOC, where each enzyme activity was significantly higher under NT compared to CT corn plots (Eivazi et al., 2003). In wheat plots, four of the five enzyme activities were significantly higher with manure + urea compared to manure only, while in the corn-wheat-clover rotation with manure treatment exhibited greater enzyme activity compared to N only. These data suggest including a high biomass forage legume may substitute for inorganic N in increasing microbial biomass and SOC. Despite two animal waste sources with £ 12:1 C:N, rice yields and N recovery were highest with treatments combining urea with waste, where urea-N contributed 40–50% of total N applied (Ismael et al., 2021). Using a granulated poultry waste + urea fertilizer product, Yeasmin et al. (2024) reported 12%, 28%, and 17% greater corn yield, N use efficiency, and SOC (0–10 cm), respectively. After 10 years of cotton rotations, SOC (0–20 cm) increased over threefold (140® 460 kg C ha^{-1} $year^{-1}$) with poultry litter annually added to the crop residue compared to NH_4NO_3 (Sainju et al., 2008). In addition, remediation of soils contaminated with high C:N industrial waste sources can also be enhanced with supplemental inorganic N. For example, land application of urea N added with compost enhanced microbial activity and nearly doubled the degradation of petroleum contaminants in soil (Bao et al., 2024).

6 Estimating soil organic matter accumulation

Clearly, soil and crop management effects on increasing SOM are highly variable and site-specific. At some locations, conversion from CT to NT may not increase SOM compared to importing additional C over that returned with the cropping system (Govaerts et al., 2009). In addition, while including legumes generally will increase SOC accumulation, hairy vetch and other forage legumes, for example, will have a greater effect on SOM compared to soybean and other low biomass grain legumes (Sisti et al., 2004). Govaerts et al. (2009) also concluded that enhancing SOC stocks requires both crop residue C and an external N supply, the quantity dependent on residue mass and C:N ratio.

de Rosa et al. (2017) used response data from a short-term (3 year) study comparing composted poultry manure and urea applied separately on selected vegetable crops to simulate SOC accumulation and optimum N rates using the SALUS crop model (Basso and Ritchie, 2015). Simulations over 12 years showed nearly 50% reduction in urea N use when combined with manure, while maintaining the initial SOC level. Using process-based modeling (Century model) Oelbermann et al. (2023) predicted ~84% increase in SOC after 115 years of biennial applications of manure + urea to corn.

The quantity of SOM produced through adding organic wastes depends on many soil, environment, and waste source factors (Manlay et al., 2007). While several sophisticated mechanistic models have been developed to estimate SOM accumulation (Gollany et al., 2009; Metherell et al., 1993), a simple procedure to estimate SOM accumulation from crop residue or organic waste application provides a producer with an estimated quantity of residue or waste and time required to increase SOM levels (Table 1). The SOM estimates are based on several assumptions:

- Residue C (C_{RES}) = 40% residue mass;
- Residue N (N_{RES}) from C:N and $C_{RES;}$
- 80% N_{RES} converted to organic matter N (N_{OM});
- Microbial C (C_{MIC}) represents 35% of C_{RES};
- C_{MIC}:N_{MIC} = 8:1;
- C_{OM}:N_{OM} = 10:1;
- SOM contains ~50% C_{OM} (Pribyl, 2010);
- Estimated SOM (kg ha^{-1}) = (N_{RES} x 0.80) = N_{OM} --> (N_{OM}) (10 C_{OM}/N_{OM}) (1 SOM/0.5 C_{OM}).

Regardless of the residue mass, with 20:1 C_{RES}:N_{RES}, the quantity of N required by the decomposing microbes (N_{MIC}) is approximately equal to the N content in the added residue (N_{RES}). With C_{RES}:N_{RES} >20:1, the microbial N demand (N_{MIC}) > N_{RES}, which substantially reduces estimated SOM accumulation rate. Estimated SOM (Table 1) is similar to measured values (Sainju et al., 2008), especially if adjusted for annual N mineralization. These SOM estimates illustrate increasing residue mass increases SOM; however, increasing C_{res}:N_{res} decreases potential

Table 1 Estimated SOM* (0–15 cm) produced by adding crop residue with variable mass and C:N content.

Residue composition	**Residue mass**		
	kg ha⁻¹		
	2000	4000	6000
C_{RES} (40% mass)	800	1600	2400
C_{MIC} (35% C_{RES})	*280*	*560*	*840*
N_{MIC} (8:1 C_{MIC}:N_{MIC})	*35*	*70*	*105*
C_{RES}: N_{RES}		**N_{RES}**	
20:1	40	80	120
60:1	13.3	26.7	40
120:1	6.7	13.3	20
		SOM*	
20:1	640	1280	1920
60:1	213	427	640
120:1	107	213	320

*SOM estimates should be reduced for potential annual mineralization, where 1% SOM (0–15 cm) at 1% mineralization rate represents ~200 kg SOM ha^{-1} $year^{-1}$.

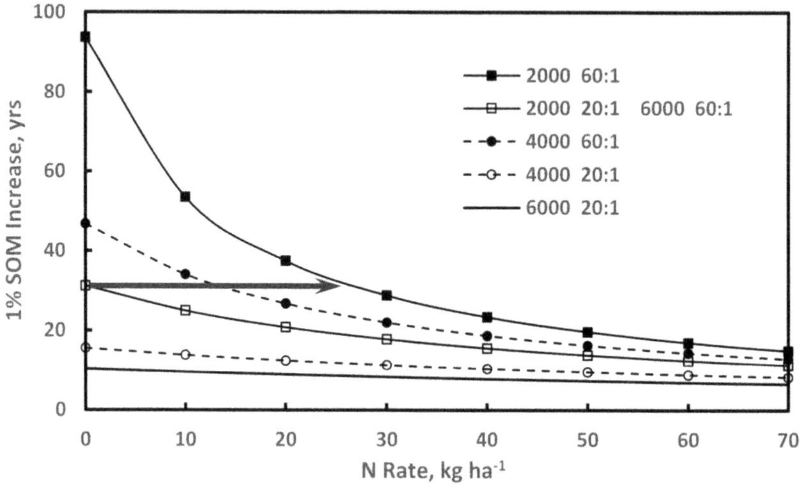

Figure 1 Estimated number of years to increase SOM 1% (0–15 cm depth) depends on residue mass, C_{res}:N_{res}, and supplemental N rate added to residue (1% increase in SOM represents ~20,000 kg SOM ha^{-1} / annual SOM added (kg SOM ha^{-1} $year^{-1}$ (Table 1) = # years).

SOM accumulation because N is more limiting to diverse microbes with ~8:1 C_{mic}:N_{mic} (Havlin et al., 2014), which is the primary rationale for adding N with organic wastes to enhance SOM accumulation (Fig. 1).

For example, with no N added, ~95 years of annual application of 2000 kg ha^{-1} (60:1 C:N) increases SOM (0–15 cm) 1%, whereas adding 2000 kg ha^{-1} (20:1 C:N) would take only ~30 years (Fig. 1). Adding ~25 kg N ha^{-1} with 2000 kg ha^{-1} (60:1 C:N) residue reduces the time required to increase SOM 1% from ~95 to 30 years (see *arrow* Fig. 1). Using the same parameters, Table 1 shows N needed by decomposing microbes is ~22 kg N_{MIC} ha^{-1} greater than the residue N content (35 kg N_{MIC} ha^{-1} - 13.3 kg N_{RES} ha^{-1}). Using this simple estimate, SOM accumulation rate was similar between 2000 kg ha^{-1} (20:1) and 6000 kg ha^{-1} (60:1), illustrating that supplemental N added to low N residues is essential to enhancing SOM accumulation.

While there are many variables that affect the accuracy of the SOM estimates (Table 1), the simplified approach facilitates practical appreciation for essential principles of SOM management to increase SOM and improve soil health (Fig. 2). Depending on soil properties, climate, and native vegetation (soil forming factors), virgin SOM content (C gain = C loss) declines approximately 50% in the first 50 years of cultivation (Matson et al, 1997; Lal, 2006). The resultant steady-state degraded SOM content (C balance) depends on soil properties, climate, and crop and soil management. Recovering SOM loss requires significant changes in soil and crop management practices resulting in C gain > C loss. Specifically, with degraded, low SOM soils, sustainable soil

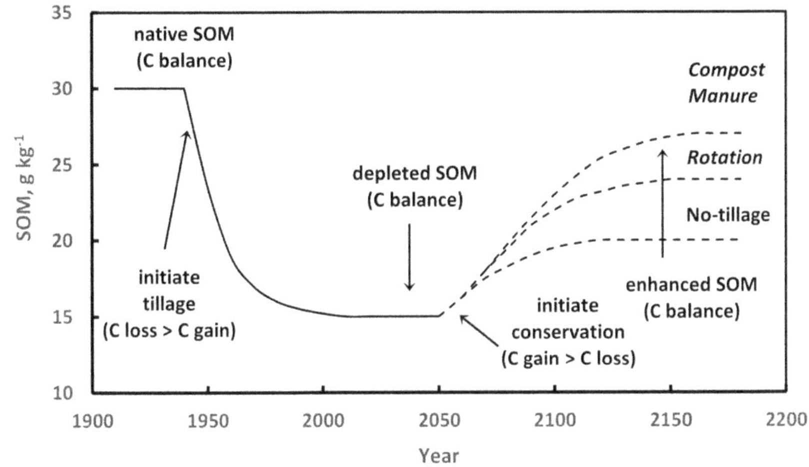

Figure 2 Dynamic changes in SOM with soil/crop management practices that degrade or enhance SOM.

and crop management practices needed to regenerate soil health include the following: (a) reduce or eliminate tillage; (b) include rotations of high biomass crops (CO_2 fixation) and retain all residues in soil; (c) include legumes, especially forage, in rotation to provide N (N_2 fixation); (d) import additional C (composts, manures, organic wastes); (e) utilize supplemental N to maximize marketable yield, and enhance high C:N residue/waste conversion to SOM; and (f) where economically viable, utilize organic, biofertilizers.

7 Conclusion

Sustaining global food security requires enhancing soil health by increasing SOM levels through integrated management of soils, crops, and nutrient sources. Research results from numerous long-term field studies demonstrate that judicious use of synthetic fertilizers increases microbial diversity and biomass, which sustains or enhances SOM and soil health. Synthetic N applied above the economic optimum rate often increases soil acidity and reduces bacterial diversity and function, which may reduce SOM through increased nitrification and mineralization of native SOM. Depending on inorganic fertilizer rate and percolation potential (rainfall regime), increasing soil salinity also impacts microbial functions degrading soil health. Long-term integrated management practices that build and maintain SOM include residue management (NT), diverse crop rotations (high biomass crops with legumes), C imports (manures, cover crops, etc.), and supplemental N to enhance residue C decomposition and conversion to SOC. In addition, the resultant increase in SOM enhances native nutrient supply through increased mineralizable organic nutrients and soluble/exchangeable nutrients, reducing dependency on nutrient inputs. The rate of SOM accumulation depends on source and quantity of C input. With high C:N residues, additional N is required to accelerate residue decomposition and SOM formation. Regardless of C source and quantity, management practices focused on increasing SOM will reduce effects of soil degrading practices and enhance soil health.

8 Where to look for further information

For general background information on nutrient cycling in soil and nutrient management see:

- Havlin, J.L., et al. (2014). *Soil Fertility and Nutrient Management: An Introduction to Nutrient Management.* 8th Edition. 516 p. Pearson. Upper Saddle River, NJ.

For specific background on inorganic fertilizer salts on soil biology, organic matter, and soil health see:

- Geisseler, D. and Scow, K. (2014), Long-term effects of mineral fertilizers on soil microorganisms - A review, Soil Biol. Biochem. 75: 54-63, ISSN 0038-0717, doi.org/10.1016/j.soilbio.2014.03.023.
- Glendining, M.J. and Powlson, D.S. (1995), The effects of long-continued applications of inorganic nitrogen fertilizer on soil organic nitrogen–A review. In *Soil Management: Experimental Basis for Sustainability and Environmental Quality*; Advances in Soil Science Series; Lal, R., Stewart, B.A., Eds.; Lewis: Boca Raton, FL, USA, pp. 385–446.
- Rath, K. and Rousk, J. (2015), Salt Effects on the Soil Microbial Decomposer Community and Their Role in Organic Carbon Cycling: A Review. *Soil Biol. Biochem*. 81:108–123. doi.org/10.1016/j.soilbio.2014.11.001.
- Sabir, M., et al. (2021), Comparative effect of fertilization practices on soil microbial diversity and activity: An overview. Curr Microbiol. 78(10):3644–3655. DOI: 10.1007/s00284-021-02634-2.

For organic and biofertilizer effects on soil organic matter and soil health:

- Just, B., et al. (2024), 'Biofertilization increases soil organic carbon concentrations: results of a meta-analysis', *Internat. J. Agric. Sustain*. 22:1, 2361578, DOI: 10.1080/14735903.2024.2361578.
- Santos, F., et al. (2024), Biofertilizer use in the United States: definition, regulation, and prospects. *Appl Microbiol Biotechnol*. Nov 12; 108(1):511. DOI: 10.1007/s00253-024-13347-4.

- Shi, T., et al. (2024), A global meta-analysis on the effects of organic and inorganic fertilization on grasslands and croplands. *Nat. Commun.* 15, 3411. doi.org/10.1038/s41467-024-47829-w.

For importance of long-term studies on soil organic matter and soil health:

- Johnston, A. and Poulton, P. (2018), 'The importance of long-term experiments in agriculture: their management to ensure continued crop production and soil fertility; the Rothamsted experience'. *European J. Soil Sci.* 69, 113–125. doi.org/10.1111/ejss.12521.
- Miao, Y., et al. (2011), 'Long-term experiments for sustainable nutrient management in China: A review', *Agron. Sustain. Dev.* 31, 397–414.

9 References

Abdalla, K., et al. (2022), 'Long-term continuous farmyard manure application increases soil carbon when combined with mineral fertilizers due to lower priming effects', *Geoderma*. 428.https://doi.org/10.1016/j.geoderma.2022.116216.

Alvarez, R. (2006), 'A review of nitrogen fertilizer and conservative tillage effects on soil organic carbon storage', *Soil Use Manage*. 21, 38–52.

Aula, L., et al. (2016), 'Effect of fertilizer nitrogen (N) on soil organic carbon, total N, and soil pH in long-term continuous winter wheat (*Triticum Aestivum* L.). *Comm. Soil Sci. Plant Anal. 47*, 863–874. https://doi.org/10.1080/00103624.2016.1147047.

Aytenew, M. and Bore, G. (2020), 'Effects of organic amendments on soil fertility and environmental quality: a review', *J. Plant Sci.* 8(5): 112–119. https://doi.org/10.11648/j.jps.20200805.12.

Bao, J., et al. (2024), 'Urea addition as an enhanced strategy for degradation of petroleum contaminants during co-composting of straw and pig manure: Evidences from microbial community and enzyme activity evaluation', *Biores. Tech.* 393, 130135. https://doi.org/10.1016/j.biortech.2023.130135.

Barak, P., et al. (1997), 'Effects of long-term soil acidification due to nitrogen fertilizer inputs in Wisconsin', *Plant Soil* 197, 61–69. https://doi.org/10.1023/A:1004297607070.

Basso, B. and Ritchie, J. (2015), *Simulating crop growth and bio-geochemical fluxes in response to land management using the SALUS model. The ecology of agricultural landscapes: long-term research on the path to sustainability*. Oxford University Press, New York. pp. 252–274.

Bastida, F., et al. (2008), 'Long-term effect of municipal solid waste amendment on microbial abundance and humus-associated enzyme activities under semiarid conditions', *Microbial Ecol.* 55, 651–661. http://www.jstor.org/stable/40605729.

Baveye, P., et al. (2020), 'Soil organic matter research and climate change: Merely re-storing carbon versus restoring soil functions', *Front. Environ. Sci.* 8, 579904. https://doi.org/10.3389/fenvs.2020.579904.

Beillouin, D., et al. (2023), 'A global meta-analysis of soil organic carbon in the Anthropocene', *Nat. Commun.* 14, 3700. https://doi.org/10.1038/s41467-023-39338-z.

Biederbeck, V., et al. (1996), 'Soil microbial and biochemical properties after ten years of fertilization with urea and anhydrous ammonia', *Can. J. Soil Sci.* 76, 7–14. https://doi.org/10.4141/cjss96-002.

Blagodatskaya, E. and Kuzyakov, Y. (2011), 'Priming effects in relation to soil conditions-Mechanisms. In: Gliński, J., et al. (eds) *Encyclopedia of Agrophysics. Encyclopedia of Earth Sciences Series*. Springer, Dordrecht. https://doi.org/10.1007/978-90-481-3585-1_128.

Blanco-Canqui, H., et al. (2011), 'Soil-profile distribution of carbon and associated properties in no-till along a precipitation gradient in the central Great Plains', *Agric. Ecosyst. Environ.* 144, 107–116.

Brichi, L., et al. (2023), 'Organic residues and their impact on soil health, crop production and sustainable agriculture: a review including bibliographic analysis', *Soil Use Manag.* 39, 686–706. https://doi.org/10.1111/sum.12892.

Bridges, M. and Oldeman, L. (2010), 'Global assessment of human-induced soil degradation (GLASOD). *Arid Soil Res. Rehab.* 13(4), 319–325. https://doi.org/10.1080/089030699263212.

Bulluck, R., et al. (2002), 'Organic and synthetic fertility amendment influence soil microbial physical and chemical properties on organic and conventional farms', *Applied Soil Ecol.* 19: 147–160. https://doi.org/10.1016/S0929-1393(01)00187-1.

CAST. (2013), *Food, fuel, and plant nutrient use in the future*', Issue Paper 51. Council for Agric. Sci. and Technol., Ames, IA.

Chatterjee, A. and Lal, R. (2009), 'On farm assessment of tillage impact on soil carbon and associated soil quality parameters', *Soil Till. Res.* 104, 270–277. https://doi.org/10.1016/j.still.2009.03.006.

Daba, N., et al. (2021), 'Long-term fertilization and lime-induced soil pH changes affect nitrogen use efficiency and grain yields in acidic soil under wheat-maize rotation. *Agronomy* 11(10), 2069. https://doi.org/10.3390/agronomy11102069.

Daniel, A., et al. (2022), 'Biofertilizer: the future of food security and food safety', *Microorganisms* 10(6), 1220. https://doi.org/10.3390/microorganisms10061220.

Davis, M., et al. (2003), 'Carbon storage along a stand development sequence in a New Zealand Nothofagus forest', *For. Ecol. Manage.* 177: 313–321.

Deaker, R., et al. (2004), 'Legume seed inoculation technology-a review', *Soil Biol. Biochem.* 36, 1275–1288. https://doi.org/10.1016/j.soilbio.2004.04.009.

Debska, B., et al. (2016), 'The impact of a bio-fertilizer on the soil organic matter status and carbon sequestration-results from a field-scale study', *J Soils Sediments* 16, 2335–2343. https://doi.org/10.1007/s11368-016-1430-5.

Deiss, D., et al. (2021), 'Does crop rotation affect soil organic matter stratification in tillage systems? *Soil Till. Res.* 209, 104932. https://doi.org/10.1016/j.still.2021.104932.

Deng, S., et al. (2006), 'Animal manure and anhydrous ammonia amendment alter microbial carbon use efficiency, microbial biomass, and activities of dehydrogenase and amidohydrolases in semiarid agroecosystems', *Appl. Soil Ecol.* 33, 258–268.

de Rosa, D., et al. (2017), 'Can organic amendments support sustainable vegetable production?. *Agron. J.* 109, 1856–1869. https://doi.org/10.2134/agronj2016.12.0739.

Diacono, M. and Montemurro, F. (2010), 'Long-term effects of organic amendments on soil fertility', *Review. Agron. Sustain. Dev.* 30, 401–422. https://doi.org/10.1051/agro/2009040.

Dijkstra, F., et al. (2013), 'Rhizosphere priming: a nutrient perspective', *Front Microbiol.* 4, 216. https://doi.org/10.3389/fmicb.2013.00216.

Dikir, W. (2023), 'Effect of long-term chemical fertilizer application on soil chemical properties: a review, *J. Biol. Agric. Healthcare.* 13(5). https://doi.org/10.7176/JBAH/13-5-02.

Dincă, L., et al. (2022). 'Fertilization and soil microbial community: a review', *Appl. Sci.* 12(3), 1198. https://doi.org/10.3390/app12031198.

Dondini, M., et al. (2023), *Global assessment of soil carbon in grasslands – From current stock estimates to sequestration potential*', FAO Animal Production and Health Paper No. 187. FAO, Rome.

Drinkwater, L., et al. (1998), 'Legume-based cropping systems have reduced carbon and nitrogen losses', *Nature* 396, 262–265. https://doi.org/10.1038/24376.

Eden, M., et al. (2017), 'Organic waste recycling in agriculture and related effects on soil water retention and plant available water: a review. *Agron. Sustain. Dev.* 37, 11. https://doi.org/10.1007/s13593-017-0419-9.

Eivazi, F., et al. (2003), 'Select soil enzyme activities in the historic Sanborn Field as affected by long-term cropping systems', *Comm. Soil Sci. Plant Anal.* 34, 2259–2275. https://doi.org/10.1081/CSS-120024062.

Eno, C. and Blue, W. (1954), 'The effect of anhydrous ammonia on nitrification and microbiological population in sandy soils', *Soil Sci. Soc. Am. Proc.* 18, 178–181.

Erisman, J., et al. (2008), 'How a century of ammonia synthesis changed the world? *Nat. Geosci.* 1, 636–639.

Evangelou, V. (1983), 'Effect of fertilizer salts on crop production', Soil Sci News Views, 164. uknowledge.uky.edu/pss_views/164.

Fasusi, O., et al. (2021), 'Agricultural sustainability: Microbial biofertilizers in rhizosphere management', *Agriculture* 11(2), 163. https://doi.org/10.3390/agriculture11020163.

Ferris, H. and Tuomisto, H. (2021), 'Unearthing the role of biological diversity in soil health', *Soil Bio. Biochem.* 85, 101–109. https://doi.org/10.1016/j.soilbio.2015.02.037.

Fierer N. and Jackson, R. (2006), 'The diversity and biogeography of soil bacterial communities', *Proc. Nat. Acad. Sci.* 103, 626–631.

Fierer, N., et al. (2021), 'How microbes can, and cannot, be used to assess soil health', *Soil Bio. Biochem.* 153. https://doi.org/10.1016/j.soilbio.2020.108111.

Fischer R., et al. (2014), *Crop yields and global food security: will yield increase continue to feed the world?* ACIAR Monograph No. 158. Australian Centre for International Agricultural Research, Canberra. xxii + 634 pp.

Geisseler, D. and Scow, K. (2014), 'Long-term effects of mineral fertilizers on soil microorganisms - A review', *Soil Biol. Biochem.* 75, 54–63. https://doi.org/10.1016/j.soilbio.2014.03.023.

Glendining, M. and Powlson, D. (1995), 'The effects of long-continued applications of inorganic nitrogen fertilizer on soil organic nitrogen–A review'. In Lal, R. and Stewart, B.A. (eds) *Soil management: experimental basis for sustainability and environmental quality*, Advances in Soil Science Series; Lewis: Boca Raton, FL, USA, pp. 385–446.

Gollany, H., et al. (2009), 'CQESTR simulation of soil organic matter dynamics in long-term agricultural experiments across USA', Geophysical Research Abstracts EGU2009-1096-2, 11.

Govaerts, B., et al. (2009), 'Conservation agriculture and soil carbon sequestration: between myth and farmer reality', *Critical Rev. Plant Sci.* 28, 97–122. https://doi.org/10.1080/07352680902776358.

Guo J., et al. (2010), 'Significant acidification in major Chinese croplands', *Science* 327, 1008–1010.

Habteweld, A., et al. (2020), 'Effects of integrated application of plant-based compost and urea on soil food web, soil properties, and yield and quality of a processing carrot cultivar', *J. Nematol.* 52, e2020–111. https://doi.org/10.21307/jofnem-2020-111.

Haddaway H., et al. (2017), 'How does tillage intensity affect soil organic carbon? A systematic review', *Environ. Evid.* 6, 30. https://doi.org/10.1186/s13750-017-0108-9.

Haj-Amor, Z., et al. (2022), 'Soil salinity and its associated effects on soil microorganisms, greenhouse gas emissions, crop yield, biodiversity and desertification: a review', *Sci. Total Environ.* 843, 156946. https://doi.org/10.1016/j.scitotenv.2022.156946.

Hamer, U. and Marschner, B. (2005), 'Priming effects in different soil types induced by fructose, alanine, oxalic acid and catechol additions', *Soil Biol. Biochem.* 37, 445–454. https://doi.org/10.1016/j.soilbio.2004.07.037.

Hao, Y., et al. (2017), 'Effects of long-term fertilization on soil organic carbon and nitrogen in a highland agroecosystem', *Pedosphere* 27, I725–736. https://doi.org/10.1016/S1002-0160(17)60386-2.

Hartmann, M., et al. (2015), 'Distinct soil microbial diversity under long-term organic and conventional farming', *The ISME J.* 9, 1177–1194. https://doi.org/10.1038/ismej.2014.210.

Havlin, J., et al (2014), *Soil fertility and nutrient management: an introduction to nutrient management.* 8th Edition. 516 p. Pearson, Upper Saddle River, NJ.

Havlin, J. and Schlegel, A. (1997), 'Dryland conservation technologies: enhancing agricultural productivity and profitability. *Ann. Arid Zone* 36, 291–303.

Havlin, J., et al. (1990), 'Crop rotation and tillage effects on soil organic carbon and nitrogen', *Soil Sci. Soc. Am. J.* 54, 448–452.

Hemalatha, S., et al. (2013), 'Influence of long term fertilization on soil fertility: a review, *Res. Rev. J. Agric. Allied Sci.* 2, 30–36.

Hossain, Z., et al. (2017), 'Effect of different organic wastes on soil properties and plant growth and yield: a review', *Sci. Agric. Bohemica* 48, 224–237. https://doi.org/10.1515/sab-2017-0030.

Hu, T., et al. (2018), 'Soil carbon varies between different organic and conventional management schemes in arable agriculture', *European J. Agron.* 94, 79–88. https://doi.org/10.1016/j.eja.2018.01.010.

Huang, L., et al. (2019), 'Long-term N fertilization imbalances potential N acquisition and transformations by soil microbes', *Sci. Total Environ.* 691, 562–571. https://doi.org/10.1016/j.scitotenv.2019.07.154.

Hudson, B. (1994), 'Soil organic matter available water capacity', *J. Soil Water Cons.* 49, 189–194.

Hussain, S., et al. (2021), 'Carbon sequestration to avoid soil degradation: a review on the role of conservation tillage', *Plants* 10(10), 2001. https://doi.org/10.3390/plants10102001.

Jagadamma, S., et al. (2007), 'Nitrogen fertilization and cropping systems effects on soil organic carbon and total nitrogen pools under chisel-plow tillage in Illinois', *Soil Till. Res.* 95, 348–356. https://doi.org/10.1016/j.still.2007.02.006.

Johnson, A., et al. (1989), *Proceedings of the sanborn field centennial: a celebration of 100 years of agricultural research'*, Miscellaneous Reports 257842. University of Missouri, Columbia. https://doi.org/10.22004/ag.econ.257842.

Johnston, A., et al. (2009), 'Soil organic matter: its importance in sustainable agriculture and carbon dioxide fluxes', *Advances Agron.* 101, 1–57. https://doi.org/10.1016/S0065-2113(08)00801-8.

Johnston, A. and Poulton, P. (2018), 'The importance of long-term experiments in agriculture: their management to ensure continued crop production and soil fertility; the Rothamsted experience', *Eur. J. Soil Sci.* 69, 113–125. https://doi.org/10.1111/ejss.12521.

Intrawech, A., et al. (1982), 'Influence of fertilizer nitrogen source on soil physical and chemical properties', *Soil Sci. Soc. Am. J.* 46, 832–836.

Ismael, F., et al. (2021), 'New fertilizer strategies combining manure and urea for improved rice growth in Mozambique. *Agronomy* 11, 783. https://doi.org/10.3390/agronomy 11040783.

Just, B., et al. (2024), 'Biofertilization increases soil organic carbon concentrations: results of a meta-analysis', *Internat. J. Agric. Sustain.* 22:1, 2361578. https://doi.org/10.1080/14735903.2024.2361578.

Khaleel, R., et al. (1981), 'Changes in soil physical properties due to organic waste applications: a review', *J. Environ. Qual.* 10, 133–141.

Khan, S., et al. (2007), 'The myth of nitrogen fertilization for soil carbon sequestration', *J. Environ. Qual. 36,* 1821–1832.

Kumar, S., et al. (2021), 'Biofertilizers: an ecofriendly technology for nutrient recycling and environmental sustainability', *Curr. Res. Microb. Sci.* https://doi.org/10.1016/j.crmicr.2021.100094.

Ladha, J., et al. (2011), 'Role of nitrogen fertilization in sustaining organic matter in cultivated soils', *J. Environ. Qual. 40*, 1756–1766.

Lal, R. (2014), 'Societal value of soil carbon', *J. Soil Water Cons.* 69, 86A–192A. https://doi.org/10.2489/jswc.69.6.186A.

Lal, R. (2006), 'Enhancing crop yields in the developing countries through restoration of the soil organic carbon pool in agricultural lands', *Land Degrad. Develop.* 172, 197–209.

Lam, Y., et al. (2012), 'Research on the allelopathic potential of wheat', *Agric. Sci.* 3, 979–985. https://doi.org/10.4236/as.2012.38119.

Larney, F. and Angers, D. (2012), 'The role of organic amendments in soil reclamation: a review', *Can. J. Soil Sci.* 92, 19–38. https://doi.org/10.4141/CJSS2010-064.

Li, X., et al. (2017), 'Nitrogen fertilization decreases the decomposition of soil organic matter and plant residues in planted soils', *Soil Biol. Biochem.* 112, 47–55. https://doi.org/10.1016/j.soilbio.2017.04.018.

Liang, Q., et al. (2012), 'Effects of 15 years of manure and inorganic fertilizers on soil organic carbon fractions in a wheat-maize system in the North China Plain', *Nutr Cycl. Agroecosyst.* 92, 21–33. https://doi.org/10.1007/s10705-011-9469-6.

Liu, J., et al. (2021), 'Wheat straw decomposition patterns and control factors under nitrogen fertilization. *Nutrition* 21, 3110–3121. https://doi.org/10.1007/s42729-021-00592-z.

Liu, X., et al. (2006), 'Effects of agricultural management on soil organic matter and carbon transformation: a review', *Plant Soil Environ.* 52, 531–543. https://doi.org/10.17221/3544-PSE.

Ludemann, C., et al. (2022), 'Global data on fertilizer use by crop and by country', *Sci Data* 9, 501. https://doi.org/10.1038/s41597-022-01592-z.

Ma, Y., et al. (2023), 'Global crop production increase by soil organic carbon', *Nat. Geosci.* 16, 1159–1165. https://doi.org/10.1038/s41561-023-01302-3.

Mahmood, F., et al. (2017), 'Effects of organic and inorganic manures on maize and their residual impact on soil physico-chemical properties', *J. Soil Sci. Plant Nutr.* 17, 22–32. https://doi.org/10.4067/S0718-95162017005000002.

Makaza, W. and Khiari, L. (2023), 'Too salty or toxic for use: a tale of starter fertilizers in agronomic cropping systems. *Agronomy* 13(11), 2690. https://doi.org/10.3390/agronomy13112690.

Malik, S., et al. (2014), 'Influence of organic and synthetic fertilizers on soil physical properties', *Int. J. Curr. Microbiol. App. Sci.* 3(8), 802–810.

Manlay, R., et al. (2007), 'Historical evolution of soil organic matter concepts and their relationships with the fertility and sustainability of cropping systems', *Agric. Ecosys. Environ.* 119: 217–233. https://doi.org/10.1016/j.agee.2006.07.011.

Matson, P., et al. (1997), 'Agricultural intensification and ecosystem properties', *Sci.* 277, 504–509. https://doi.org/10.1126/science.277.5325.504.

Mazzoncini, M., et al. (2011), 'Long-term effect of tillage, nitrogen fertilization and cover crops on soil organic carbon and total nitrogen content', *Soil Till. Res.* 114, 165–174. https://doi.org/10.1016/j.still.2011.05.001.

Mehra, P., et al. (2018), 'A review of tillage practices and their potential to impact the soil carbon dynamics'. In Sparks, D. L. (ed.) *Adv. Agron.* 150, 185–230. Academic Press, New York. https://doi.org/10.1016/bs.agron.2018.03.002.

Metherell, A., et al. (1993), *CENTURY soil organic matter model environment.* USDA-ARS, Fort Collins, CO.

Miao, Y., et al. (2011), 'Long-term experiments for sustainable nutrient management in China: a review', *Agron. Sustain. Dev.* 31, 397–414.

Mishra, D., et al. (2010), 'Tillage effects on soil organic carbon storage and dynamics in Corn Belt of Ohio USA', *Soil Till. Res.* 107, 88–96. api.semanticscholar.org/CorpusID: 128686522.

Mitter, E., et al. (2021), 'Rethinking crop nutrition in times of modern microbiology: innovative biofertilizer technologies. *Front. Sustain. Food Syst.* 5. https://doi.org/10.3389/fsufs.2021.606815.

Mortvedt, J. (2001), 'Calculating salt index', *Fluid J.* 9(2), 8–11.

Muhammad, I., et al. (2023), 'Integrated use of manure and urea fertilizer economically improve wheat production and nitrogen Indices in a range of tillage practices', *J. Plant Growth Reg.* 43, 589–600.

Mulvaney, R., et al. (2009), 'Synthetic nitrogen fertilizers deplete soil nitrogen: a global dilemma for sustainable cereal production', *J. Environ. Qual.* 38, 2295–2314.

Nweke, I. and Nsoanya, L. (2015), 'Effect of cow dung and urea fertilization on soil properties, growth, and yield of cucumber (Cucumis sativus L)', *J. Agric. Ecol. Res. Intern.* 3(2), 81–88.

O'Brien, P. and Hatfield J. (2019), 'Dairy manure and synthetic fertilizer: a meta- analysis of crop production and environmental quality. agrosystems. *Geosci Environ.* 2(1), 1–12.

Oelbermann, M., et al. (2023), 'Predicting changes in soil organic carbon after a low dosage and one-time addition of biochar blended with manure and nitrogen fertilizer', *Front. Soil Sci.* 3, 1209530. https://doi.org/10.3389/fsoil.2023.1209530.

Oldfield, E., et al. (2019), 'Global meta-analysis of the relationship between soil organic matter and crop yields', *SOIL*, 5, 15–32. https://doi.org/10.5194/soil-5-15-2019.

Paustian, K., et al. (1992), 'Modeling soil organic matter in organic amended and N-fertilized long-term plots', *Soil Sci. Soc. Am. J.* 56, 476–488.
Post, W., et al. (2004), 'Enhancement of carbon sequestration in US soils. *BioScience* 54, 895–908.
Powlson, D., et al. (2010), 'Comments on "synthetic nitrogen fertilizers deplete soil nitrogen: A global dilemma for sustainable cereal production," by R.L. Mulvaney, S.A. Khan, and T.R. Ellsworth in the journal of environmental quality 2009 38, 2295–2314', *J. Environ. Qual.* 39(2), 749–756.
Pribyl, D. (2010), 'A critical review of conventional SOC to SOM conversion factor', *Geoderma* 156, 75–83.
Rasool, R., et al. (2008), 'Soil organic carbon and physical properties as affected by long-term application of FYM and inorganic fertilizers in maize-wheat system', *Soil Tillage Res.* 101(1–2), 31–36. https://doi.org/10.1016/j.still.2008.05.015.
Rath, K. and Rousk, J. (2015), 'Salt effects on the soil microbial decomposer community and their role in organic carbon cycling: a review', *Soil Biol. Biochem.* 81, 108–123. https://doi.org/10.1016/j.soilbio.2014.11.001.
Rath, K., et al. (2016), 'Comparative toxicities of salts on microbial processes in soil', *Appl Environ Microbiol.* 82(7), 2012–2020. https://doi.org/10.1128/AEM.04052-15.
Ren, N., et al. (2020), 'Effects of continuous nitrogen fertilizer application on the diversity and composition of rhizosphere soil bacteria', *Front. Microbiol.* 21, 1948. https://doi.org/10.3389/fmicb.2020.01948.
Rousk, J., et al. (2010), 'Soil bacterial and fungal communities across a pH gradient in an arable soil', *ISME J.* 4(10), 1340–1351. https://doi.org/10.1038/ismej.2010.58.
Sabir, M., et al. (2021), 'Comparative effect of fertilization practices on soil microbial diversity and activity: an overview', *Curr. Microbio.* 78, 3644–3655. https://doi.org/10.1007/s00284-021-02634-2.
Sainju, U., et al. (2008), 'Soil carbon and nitrogen sequestration as affected by long-term tillage, cropping systems, and nitrogen fertilizer sources', *Agric. Ecosyst. Environ.* 127, 234–240. https://doi.org/10.1016/j.agee.2008.04.006.
Santos, F., et al. (2024), 'Biofertilizer use in the United States: definition, regulation, and prospects', *Appl. Microbiol. Biotechnol.* 108(1), 511. https://doi.org/10.1007/s00253-024-13347-4.
Sawyer, J. (2019), 'Understanding anhydrous ammonia application in soil', Iowa State Univ. Ext. Outreach-Integrated Crop Management. crops.extension.iastate.edu/cropnews/2019/03/.
Schroder, J., et al. (2011), 'Soil acidification from long-term use of nitrogen fertilizers on winter wheat', *Soil Sci. Soc. Am. J.* 75, 957–964.
Shi, T., et al. (2024), 'A global meta-analysis on the effects of organic and inorganic fertilization on grasslands and croplands', *Nat. Commun.* 15, 3411. https://doi.org/10.1038/s41467-024-47829.
Šimon, T. and Czakó, A. (2014), 'Influence of long-term application of organic and inorganic fertilizers on soil properties', *Plant Soil Environ.* 60, 314–319. https://doi.org/10.17221/264/2014-pse.
Singh, B. (2018), 'Are nitrogen fertilizers deleterious to soil health?' *Agronomy* 8(4), 8. https://doi.org/10.3390/agronomy8040048.
Singh, B. and Ryan, J. (2015), *Managing fertilizers to enhance soil health.* International Fertilizer Industry Association (IFA), Paris, France.

Sisti, C., et al. (2004), 'Change in carbon and nitrogen stocks in soil under 13 years of conventional or zero tillage in southern Brazil', *Soil Tillage Res.* 74, 39–58. https://doi.org/10.1016/S0167-1987(03)00196-X.

Smith, P., et al. (2019), 'Which practices co-deliver food security, climate change mitigation and adaptation, and combat land degradation and desertification?', *Glob. Change Biol.* 26(3), 1532–1575. https://doi.org/10.1111/gcb.14878.

Spargo, J., et al. (2011), 'Mineralizable soil nitrogen and labile soil organic matter in diverse long-term cropping systems', *Nutr. Cycl. Agroecosys.* 90, 253–266. https://doi.org/10.1007/s10705-011-9426-4.

St. Luce, M., et al. (2021). 'Long-term effects of tillage and nitrogen fertilization on soil C and N fractions in a corn–soybean rotation', *Can. J. Soil Sci.* 102, 277–292. https://doi.org/10.1139/CJSS-2021-0129.

Stewart, W., et al. (2005), 'The contribution of commercial fertilizer nutrients to food production', *Agron. J.* 97, 1–6.

Sun, R., et al. (2019), 'Long-term urea fertilization alters the composition and increases the abundance of soil ureolytic bacterial communities in an upland soil', *FEMS Microbiol. Ecol.* 95(5), fiz044. https://doi.org/10.1093/femsec/fiz044.

Termorshuizen, A., et al. (2004), 'The value of compost. Re-views in Environ. *Sci. Bio-tech.* 3, 343–347. https://doi.org/10.1007/s11157-004-2333-2.

Tomasiewicz, D. and Henny, J. (1985), 'The effect of anhydrous ammonia applications on the solubility of soil organic carbon', *Can. J. Soil Sci.* 65, 737–747.

US Salinity Laboratory Staff. (1954), *Diagnosis and improvement of saline and alkali soils.* US Department of Agriculture Handbook, Washington, DC.

Vega, O. (2007), 'A review on beneficial effects of rhizosphere bacteria on soil nutrient availability and plant uptake', *Rev. Fac. Nal. Agr. Medellín.* 60(1), 3621–3643.

Wood, C., et al. (1992), 'Tillage and crop rotation effects on soil organic matter in a typic hapludult of Northern Alabama', *J. Sustain. Agric.* 2(2), 31–41. https://doi.org/10.1300/J064v02n02_04.

Yeasmin, S., et al. (2024), 'Poultry manure-urea fertilizer potentially influences nitrogen use efficiency, nutritional quality, and yield of baby corn (Zea mays L.)', *J. Agric. Food Res.* 18, 101409. https://doi.org/10.1016/j.jafr.2024.101409.

Yu, L., et al. (2024), 'An optimal combined slow-release nitrogen fertilizer and urea can enhance the decomposition rate of straw and the yield of maize by improving soil bacterial community and structure under full straw returning system', *Front Microbiol.* 15, 1358582. https://doi.org/10.3389/fmicb.2024.1358582.

Zereyesus, Y., et al. (2023), '*International food security assessment, 2023–2033* (Report No. GFA-34). U.S. Department of Agriculture, Economic Research Service. https://doi.org/10.32747/2023.8134166.ers.

Chapter 15

Understanding and optimizing the use of algae in promoting crop nutrient use efficiency and quality

Domenico Ronga, Annamaria Di Serio and Tommaso Nappo, University of Salerno, Italy

1 Introduction

Agriculture is currently facing a range of climatic and economic challenges. This includes the rising cost of fertilizers which does not translate into a proportional increase in nutrient uptake and utilization by crops. A fraction of the applied nitrogen and phosphorus may be lost through processes such as leaching, volatilization, and immobilization (Massantini et al., 1983), resulting in economic and environmental consequences that can no longer be overlooked as they were in the past. Nitrogen use efficiency (NUE) has become a key indicator for assessing the sustainability and competitiveness of agricultural production systems, with the aim of achieving a balance between productivity and applied inputs.

http://dx.doi.org/10.19103/AS.2025.0166.18

The use of biostimulants has recently become established as a complementary approach to fertilization. Rather than 'adding' nutrients, these products aim to enhance the efficiency of plant nutritional processes by improving nutrient uptake, translocation, and assimilation, reducing the impact of abiotic stresses, and improving crop quality. At the European level, the regulatory definition of biostimulants explicitly reflects these characteristics. In the new European Fertilising Products Regulation (EU) 2019/1009, approved by the European Parliament on 27 March 2019, 'plant biostimulants' are defined as any product that stimulates plant nutritional processes independently of its nutrient content, with the sole aim of improving one or more of the following characteristics of the plant or the plant rhizosphere:

- 1 Nutrient use efficiency;
- 2 Tolerance to abiotic stress;
- 3 Qualitative traits; and
- 4 Availability of nutrients confined in the soil or rhizosphere.

Biostimulants are therefore defined on the basis of their agronomic effects rather than their nature or mode of action. Consequently, a biostimulant is a product that stimulates nutritional processes regardless of its nutrient content, with the objective of improving one or more of NUE, tolerance to abiotic stress, crop quality, and the availability of nutrients confined in the soil or rhizosphere (du Jardin, 2015; Chabili et al., 2024).

Among the different categories, algal biostimulants (macroalgal extracts and, increasingly, microalgal derivatives) are of particular interest because they combine chemical complexity, wide industrial availability, and a growing body of research describing their effects on crop nutrition and performance (El Boukhari et al., 2020).

The fundamental principle underlying the action of algal biostimulants is the stimulation of plant nutritional processes. This concept implies an improvement in the efficiency with which available nutrients are absorbed, transported, and converted into biomass and yield, rather than a direct increase in nutrient availability. Nutrient use efficiency in plants depends on the integration of root uptake, biochemical assimilation, and nitrogen allocation to productive tissues, as extensively described in the physiological literature on nitrogen nutrition (Xu et al., 2012).

The aim of this chapter is to provide an integrated framework for assessing algal biostimulants, starting from the characterization of algal biostimulants, their mechanisms of action and composition. Using a series of case studies on the use of both microalgal and macroalgal biostimulants in various crops, the chapter then explores ways of optimizing the use of these products to optimize nutrient use efficiency, product quality, and yields.

2 Types of algal biostimulant

Algae used as sources of biostimulants can be broadly divided into two main groups:

- Macroalgae and
- Microalgae.

Macroalgae are classified into three major groups:

- Rhodophyta (red algae);
- Phaeophyceae (brown algae); and
- Chlorophyta (green algae).

Most scientific research and industrial applications focus on brown algae within the biostimulant sector. These biomasses are characterized by a complex structural organization and a high concentration of minerals, vitamins, phenolic compounds, structural and soluble polysaccharides, and sterols, which represent key components underlying their biological activity and effects on plants (El-Beltagi et al., 2022).

Microalgae, consisting of unicellular or colonial organisms, represent an emerging source of biostimulants. Compared with macroalgae, they exhibit a distinct biochemical composition, often richer in proteins, amino acids, photosynthetic pigments, and secondary metabolites. Moreover, the possibility of cultivating microalgae in controlled systems allows for greater standardization of the starting biomass (Chabili et al., 2024).

The application of algal biostimulants has been associated with significant improvements in major agronomic and physiological parameters, including germination rate, vegetative growth, biomass accumulation, productivity, and tolerance to abiotic stresses across a wide range of cultivated species (Sabatino et al., 2023). These effects enable algal biostimulants to be used as functional tools for optimizing production processes and for integrating nutritional and physiological management strategies within complex agricultural systems, while promoting low environmental impact. Furthermore, numerous studies have shown that the use of algal-derived biostimulants can modulate key stages of the crop cycle, contributing to the reduction of seed dormancy, enhancement of root system architecture, and improvement of reproductive processes, including flowering (Ali et al., 2019). Additional evidence also indicates positive effects on qualitative parameters of agricultural production, such as the nutritional and market traits of harvested products (Scarabattoli et al., 2023). The integration of these physiological responses ultimately results in an overall increase in crop production efficiency.

NUE represents a key indicator of the sustainability and effectiveness of modern agricultural systems. It can be defined as the ability of a crop to absorb nutrients available in the soil and convert them into biomass and, consequently, into economic yield. From an operational perspective, NUE results from the interaction between two main components: uptake efficiency and utilization efficiency (Salim & Raza, 2020; Xu et al., 2012). Improving one or both of these components allows for a reduction in fertilizer inputs while maintaining or increasing productivity. Algal biostimulants contribute to the improvement of NUE not through the direct supply of nutrients, but rather through the regulation of physiological and morpho-functional processes governing plant nutrition. Their action therefore operates at the level of nutritional processes, in accordance with the definition of biostimulants, rather than at the level of the quantitative availability of mineral elements. Indeed, metagenomic analyses (Yang et al., 2025) suggest that certain microalgae may stimulate nitrification while suppressing denitrification, thereby potentially reducing nitrogen losses and stabilizing nitrogen stocks in the soil.

3 Mechanisms of action: direct and indirect effects

For conceptual clarity, the mechanisms of action of algal biostimulants can be broadly divided into two categories (du Jardin, 2015):

- Direct plant-mediated effects and
- Indirect soil–microbiome-mediated effects.

This distinction is essential to clarify whether the observed agronomic responses arise primarily from modulation of plant physiological processes or from alterations in rhizosphere dynamics that indirectly influence nutrient availability and acquisition.

3.1 Direct plant-mediated effects

Direct plant-mediated effects involve the interaction of algal-derived bioactive compounds with plant tissues, leading to the modulation of hormonal balance, gene expression, root system architecture, and nutrient assimilation pathways (du Jardin, 2015; Shukla et al., 2019). Phytohormones, amino acids, peptides, and polysaccharides contained in algal extracts can directly influence nitrogen transporters, key enzymes of nitrogen metabolism (such as nitrate reductase and glutamine synthetase), and photosynthetic performance (Benítez García et al., 2020; Xu et al., 2012). These mechanisms ultimately enhance both nitrogen uptake efficiency (NUpE) and nitrogen utilization efficiency (NUtE) by improving the plant's capacity to absorb available nutrients and convert them into biomass and yield (Salim & Raza, 2020).

One of the main mechanisms through which algal biostimulants influence NUE is the enhancement of the root system (Khan et al., 2012, 2013). Increases in root length, lateral branching, and fine root density expand the root–soil contact surface, allowing a larger volume of soil to be explored and improving nutrient interception. This effect is particularly relevant for nutrients with low mobility in the soil, such as phosphorus and potassium, but it also contributes to more efficient uptake of nitrogen and micronutrients.

Moreover, the activation of physiological responses at the root level can enhance the plant's ability to exploit nutrient fractions that are normally poorly available, thereby reducing losses and increasing the overall efficiency of the soil–plant system.

The stimulation of root system development by biostimulants is mainly attributable to the presence of phytohormones in algal-based products. Phytohormones, commonly referred to as plant hormones, are organic molecules synthesized endogenously by plants that play a key role in regulating physiological processes, controlling growth, development, and plant adaptation to different environmental conditions (Fahad et al., 2014).

Several scientific studies have demonstrated the presence in algae of all major classes of phytohormones, including auxins (such as indole-3-acetic acid, IAA), cytokinins (CK), gibberellins (GA), abscisic acid (ABA), salicylic acid (SA), jasmonates (JA), ethylene, and brassinosteroids (BR) (Benítez García et al., 2020; Rathod et al., 2024).

Beyond hormonal regulation, algal-derived biostimulants also exert significant effects on nutrient uptake and metabolism, directly contributing to improvements in NUE. In particular, polysaccharides present in algal dry matter, such as alginates with chelating properties, increase the availability of cations in the soil and promote a more efficient rhizospheric environment for nitrogen uptake. At the same time, the high content of mineral elements (K, Ca, Mg, and micronutrients) supports the physiological processes involved in nitrogen assimilation and translocation within the plant. These effects result in improved root system architecture, characterized by increased branching and enhanced development of root hairs, thereby increasing soil exploration capacity and the efficiency of water and nitrogen acquisition, ultimately leading to higher NUE (Shukla et al., 2019).

3.2 Indirect soil microbiome-mediated effects

In addition to direct physiological effects, algal biostimulants may exert indirect effects mediated by the soil microbiome. Changes in rhizosphere microbial communities can influence nitrogen cycling processes, including nitrification and denitrification, thereby affecting nitrogen availability and losses (Yang et al., 2025). The stimulation of beneficial plant growth–promoting rhizobacteria (PGPR) and improved soil structural properties may enhance nutrient mobilization and stabilize soil nitrogen pools (Khan et al., 2012, 2013; Shukla et al., 2019). In this framework, improvements in NUE result not only from intrinsic plant responses but also from enhanced soil–plant–microbiome interactions that optimize nutrient dynamics within the agroecosystem (El Boukhari et al., 2020).

4 Composition of algal extracts

4.1 Polysaccharides

Polysaccharides represent one of the most abundant and functionally relevant fractions of algal extracts. These compounds, which include both structural and soluble polysaccharides, can account for up to 76% of the dry weight depending on the species, environmental conditions, and time of harvest (Usman et al., 2017). Polysaccharides are present in all major algal groups, within which they differ in specific structural and chemical properties. These molecules are often associated with effects on root development, modulation of stress responses, and interactions within the rhizosphere, thereby contributing to improved nutrient uptake efficiency.

4.2 Proteins and amino acids

Another important component of algae is represented by proteins. Protein content in algae shows high variability among different taxonomic groups. In particular, some red algae, such as *Palmaria palmata*, can reach values close to 47% of dry weight, whereas in green algae protein content generally ranges between 9.6% and 23.3%. In brown algae, by contrast, protein levels are lower, typically ranging from approximately 3% to 16% (Echave et al., 2022; O'Brien et al., 2022). Algal proteins occur in various forms, including structural proteins, glycoproteins, phycobiliproteins, and lectins, as well as peptides generated through hydrolytic processes and a pool of free amino acids such as proline, glycine, alanine, and betaine (Thiviya et al., 2022).

Several algal species, including *Palmaria palmata*, *Ulva lactuca*, *Gracilaria verrucosa*, *Ascophyllum nodosum*, and species belonging to the genus *Sargassum*, are characterized by a particularly rich amino acid profile, which contributes substantially to their biostimulant activity. The most abundant amino acids generally include glutamic acid, aspartic acid, alanine, glycine, proline, and arginine; although concentrations can vary widely depending on species, environmental conditions, and extraction techniques, values reported in the literature often range between approximately 2 mg g^{-1} and 8 mg g^{-1} of dry weight (Thiviya et al., 2022). Arginine is a particularly important molecule for nitrogen storage, thereby enhancing nitrogen use efficiency (Jiménez-Arias et al., 2021).

Beyond providing structural support, these molecules act as metabolic precursors, contribute to osmoregulation, and participate in signaling networks that regulate plant physiological responses. From a mechanistic perspective, algal-derived amino acids and peptides have been shown to enhance nutrient uptake through the induction of gene expression encoding nitrogen and sulfur transporters. In addition, these compounds are capable of stimulating the biosynthetic pathways of plant hormones, including auxins, cytokinins, and gibberellins, as well as activating gene regulatory networks associated with stress responses, including gene families such as DREB and LEA (Wi et al., 2018).

4.3 Polyphenols

From a functional perspective, polyphenols extracted from marine macroalgae play a significant role as biostimulants by regulating key biochemical processes in plants (Craigie, 2011; Ertani et al., 2018). In particular, compounds such as phlorotannins, which are typical of brown algae, exhibit strong antioxidant activity, contributing to increased plant tolerance to abiotic stresses (Aina et al., 2022). Similarly, bromophenols, which are more abundant in red algae, as well as various phenolic acids and flavonoids widely detected in species such as *Gracilariopsis longissima*, have been associated with the activation of defense mechanisms and the stimulation of plant growth processes.

In addition to their direct effects on plant physiology, algal polyphenols exert a positive influence on soil functionality by promoting the development of beneficial microbial communities and contributing to improvements in soil structure. Their chelating properties further enhance the availability of essential micronutrients, thereby indirectly supporting nitrogen uptake and assimilation processes. Overall, these mechanisms contribute to more efficient nutrient use, with potential benefits for NUE and the sustainability of cropping systems.

4.4 Vitamins and minerals

Marine algae also represent a relevant source of micronutrients, including fat-soluble vitamins (such as vitamins A and E) and water-soluble vitamins (including the B-complex and vitamin C), as well as essential mineral elements such as iodine, potassium, magnesium, zinc, iron, and calcium. These components contribute to the proper functioning of plant metabolic processes and to the strengthening of antioxidant defense systems (Hagan & Anyangwe, 2023). In synergy with other bioactive compounds present in algal extracts, vitamins and minerals enhance the overall biostimulant activity of marine algae, supporting plant growth, mineral nutrition, and the capacity to respond to environmental stresses.

5 Case study 1: microalgal extracts as biostimulants to increase lettuce yields

5.1 Context and objectives

An experimental study was conducted to evaluate the effectiveness of aqueous extracts of marine microalgae as biostimulants for improving the growth of horticultural crops and nitrogen use efficiency (NUE) (Di Serio et al., 2025). The investigation focused on two species of marine microalgae, *Nannochloropsis gaditana* and *Porphyridium* sp., selected for their high content of bioactive compounds potentially involved in biostimulatory processes affecting plant nutrition and physiological regulation.

The main objective of the study was to assess whether foliar application of microalgal extracts could optimize nitrogen uptake and assimilation in a short-cycle crop such as lettuce (*Lactuca sativa* L.), while maintaining unchanged nitrogen fertilizer inputs.

5.2 Experimental methods

Microalgal extracts were obtained from freeze-dried biomass subjected to aqueous extraction by sonication, followed by centrifugation and dilution. This procedure allowed the preservation of a wide range of bioactive metabolites, including polysaccharides, soluble proteins, free amino acids, photosynthetic pigments, and phenolic compounds.

Prior to greenhouse experimentation, the extracts were evaluated through germination and phytotoxicity tests on garden cress (*Lepidium sativum* L.), an annual Brassicaceae species commonly used for such assays due to its rapid germination within 24 h at 27°C and its sensitivity to substances present in the growth substrate. These tests were conducted to identify the most suitable extract concentrations in terms of biostimulant efficacy and the absence of phytotoxic effects. The selected dilutions were subsequently applied as weekly foliar treatments to the greenhouse-grown crop.

The experimental design included a comparison between plants treated with microalgal extracts, an untreated control, and a treatment with a commercially available microalgal biostimulant. Nitrogen fertilization was applied to meet the crop requirements and was identical across all treatments, allowing the observed differences to be attributed to variations in nitrogen use efficiency rather than to differences in nitrogen supply.

5.3 Agronomic and physiological results

Treatments with microalgal extracts resulted in significant improvements in several agronomic parameters, including fresh and dry biomass, leaf area, and overall plant architecture, compared with the untreated control. A more balanced distribution of assimilates was also observed, as indicated by the harvest index (HI).

From a physiological perspective, treated plants showed an increase in chlorophyll index, improved stomatal conductance, and a reduction in leaf temperature, indicating a more efficient nutritional status and enhanced photosynthetic capacity. These physiological responses are consistent with a more efficient utilization of available nitrogen.

In particular, nitrogen agronomic efficiency (NAE) was higher in microalgal treatments than in the controls, suggesting increased biomass production per unit of applied nitrogen and, consequently, an improvement in overall NUE.

5.4 Molecular investigations

Gene expression analysis revealed the modulation of genes involved in nutrient uptake processes, the regulation of primary metabolism, and stress responses. These molecular-level changes support the hypothesis that microalgal biostimulants act by improving the efficiency of physiological processes related to nitrogen use, rather than by directly increasing nutrient availability.

5.5 Practical implications and transferability

Overall, the case study demonstrates that the use of microalgal extracts as biostimulants can represent an effective strategy to increase nitrogen use efficiency in horticultural systems, while maintaining high productivity levels with unchanged fertilizer inputs. The increase in NAE observed in treated plants can be interpreted as the combined outcome of enhanced nitrogen uptake efficiency and improved nitrogen utilization efficiency (Salim & Raza, 2020). Enhanced stomatal conductance and chlorophyll content suggest improved nitrogen assimilation and photosynthetic integration, while the modulation of root activity and physiological status likely contributed to greater nitrogen acquisition from the substrate (Xu et al., 2012). The adopted approach, based on foliar applications and conventional agronomic practices, makes the results readily transferable to real production contexts and compatible with sustainable and climate-smart agriculture strategies.

6 Case study 2: integration of microalgal biostimulants with nitrogen fertilization in durum wheat cultivation

6.1 Context and objectives

In cereal agroecosystems, optimizing nitrogen fertilization is essential to increase productivity while simultaneously reducing the environmental impact associated with nitrogen losses. To this end, an experimental field study was conducted on durum wheat (*Triticum durum* cv. 'Iride') under rainfed conditions, evaluating the effects of integrating different biostimulants, including marine algal extracts, with differentiated levels of mineral nitrogen (Rossini et al., 2025).

The experimental design included three nitrogen fertilization rates (50, 100, and 150 kg N ha^{-1}) combined with the application of biostimulants based on algal extracts, betaine, and plant growth–promoting products.

6.2 Agronomic results: yield, quality and nitrogen use

The application of biostimulants resulted in significant improvements compared with the untreated control. Experimental evidence showed statistically significant effects on root development: treated plants exhibited greater root length and root biomass, with increases exceeding twofold compared with the untreated control, indicating an enhanced capacity for soil exploration and nutrient acquisition. Leaf chlorophyll content was significantly higher in biostimulant-treated plants, reflecting improved photosynthetic physiological efficiency associated with enhanced nitrogen nutrition.

A particularly relevant result concerns grain yield. When nitrogen was applied at an intermediate rate (100 kg N ha^{-1}) in combination with biostimulant treatments containing algal extracts, durum wheat grain yield was comparable to that obtained with the higher standard nitrogen rate (150 kg N ha^{-1}). This outcome indicates a fertilizer saving of approximately 33% while maintaining high yield levels.

In addition to yield quantity, product quality was also improved or maintained. Yield components and quality indices, such as grain weight and protein deposition parameters, showed positive responses to biostimulant application, suggesting a more balanced nutritional status and a potential positive impact on the marketable quality of durum wheat.

6.3 Mechanisms involved

The results of this study are consistent with the physiological mechanisms attributed to the use of algal biostimulants. Specifically:

- increased root biomass promotes the uptake of available soil nitrogen;

- higher chlorophyll content reflects improved plant health status and greater efficiency in light capture and nitrogen utilization within photosynthetic processes; and
- under intermediate fertilizer application rates, the integration of algal biostimulants enables yield and quality outcomes comparable to those achieved with higher fertilization levels, indicating an overall increase in NUE.

The yield equivalence observed at intermediate nitrogen rates (100 kg N ha^{-1}) compared with the higher fertilization level (150 kg N ha^{-1}) suggests that algal biostimulants enhanced both nitrogen uptake efficiency (through improved root system development) and nitrogen utilization efficiency (through optimized nitrogen assimilation and allocation to grain protein synthesis) (Salim & Raza, 2020; Xu et al., 2012). By acting simultaneously on acquisition and metabolic conversion processes, algal extracts allowed plants to maintain productivity with reduced external nitrogen inputs.

6.4 Practical implications and transferability

This second case study (Rossini et al., 2025), conducted under real field conditions, demonstrates how the integration of algal biostimulants into nitrogen fertilization programs can contribute to:

- reducing nitrogen fertilization costs without compromising crop yield;
- improving or maintaining grain quality, a key factor for commercial value; and
- increasing the sustainability of cropping systems through more efficient use of nutritional resources.

These findings are particularly relevant in sustainable agriculture scenarios, where reducing chemical inputs is a priority in order to mitigate nutrient losses to the environment and decrease greenhouse gas emissions associated with nitrogen fertilization, thereby contributing to a more environmentally sustainable agricultural framework.

7 Case study 3: integration of macroalgal biostimulants with nitrogen fertilization in malting barley cultivation

7.1 Context and objectives

In malting barley production chains, nitrogen represents a decisive factor for both yield and technological quality, particularly grain size and protein content. Under Mediterranean conditions, climatic variability increases the risk of producing thin kernels and protein levels outside the optimal range, making the adoption of practices that enhance NUE and yield stability especially strategic.

A field study evaluated the effectiveness of integrating a brown macroalgal extract (*Ascophyllum nodosum* extract, ANE) with different nitrogen fertilization strategies, with the aim of simultaneously improving yield, quality, and sustainability in malting barley under Mediterranean conditions (Loukakis et al., 2025).

7.2 Experimental design and treatments

The experiment was conducted across four environments in the Thessaly region (Greece) over two growing seasons, including two genotypes and two phenological application windows. The treatments compared included urea alone, urea plus algal biostimulant, urea plus a urease inhibitor, urea plus urease inhibitor combined with algal biostimulant, and an unfertilized control.

7.3 Agronomic results, quality and implications

The combined urease inhibitor plus algal biostimulant treatment (UI + B) proved to be the most consistent in improving both productive and qualitative performance. Statistical analyses showed yield increases of up to +71% compared with the unfertilized control, along with increases in thousand grain weight (TGW) of up to +27%, indicating fuller, better-filled kernels.

Statistically significant improvements were also observed in malting quality parameters. In particular, the proportion of grains larger than 2.8 mm increased by up to +22%, while protein concentration remained within the optimal range for malting (9.5–11.5%).

The outcomes of this research highlight how the use of algal biostimulants represents a valuable opportunity for agricultural enterprises, enhancing resilience and efficiency by reducing fertilizer inputs while achieving higher yields and improved quality, especially when these products are integrated with innovative agronomic techniques.

These improvements in yield and grain filling may reflect enhanced nitrogen uptake efficiency through improved root activity and better nitrogen utilization efficiency associated with optimized assimilation and allocation processes.

8 Case study 4: using macroalgal biostimulants in olive cultivation to improve nutrient use efficiency and quality

8.1 Context and objectives

In perennial tree crops such as olive (*Olea europaea*), the management of water and nutritional resources is crucial to ensure both yield and fruit quality, including oil quality, particularly under water stress conditions typical of Mediterranean systems. This case study explored the potential use of algal biostimulants to improve plant performance under different irrigation regimes, with indirect implications for nutrient use efficiency and the sustainability of olive cultivation (Dias et al., 2026).

The study evaluated the effects of a commercial *Ascophyllum nodosum* extract applied to olive trees subjected to two contrasting irrigation regimes: full irrigation (100% of field capacity) and moderate water deficit (50% of field capacity), with and without biostimulant application.

8.2 Experimental design

The algal biostimulant was applied as a foliar treatment to the treated plants, while control plants received no application. The experiment lasted 118 days, during which leaf water status, fruit growth, chemical quality, and the metabolic composition of olives were monitored.

This approach made it possible to assess whether algal extract application could mitigate the negative effects of water deficit and indirectly optimize plant physiological and nutritional responses.

8.3 Main results

Application of *Ascophyllum nodosum* extract showed positive effects on olive quality, even under water stress conditions. In particular, leaf water status improved in biostimulant-treated plants, indicating enhanced water availability within tissues, especially under moderate water deficit (50% of field capacity). Statistically significant effects were also observed on fruit size, and in both cultivars tested (*Arbequina* and *Galega*), the extract promoted increases in fruit length and width compared with the control.

Moreover, plants subjected to full irrigation combined with biostimulant application showed increased levels of nutritionally relevant compounds, including fatty acids (oleic, palmitic, linoleic, and stearic acids), squalene, sitosterol, and myo-inositol, indicating improved quality attributes. In some cultivars, the treatment also positively influenced glucose accumulation.

8.4 Practical implications

This case study demonstrates that the use of *Ascophyllum nodosum*-based biostimulants can be effectively integrated into the management plans of tree crops in Mediterranean environments, which are among the regions most exposed to abiotic stress, particularly water deficit. Treated plants maintained improved quality parameters, suggesting a potential reduction in input requirements (soil fertility and water) while preserving quality and productivity.

Overall, the application of biostimulants can contribute to sustainable and climate-smart agriculture strategies, in which the enhancement of physiological processes reduces pressure on nitrogen fertilizers and water resources.

8.5 Contribution of research to more sustainable agricultural production

Research on algal biostimulants highlights their potential to support a transition toward more efficient, resilient, and sustainable agricultural systems. The high diversity of bioactive compounds present in marine algae – including phytohormones, polysaccharides, amino acids, peptides, polyphenols, vitamins, and minerals – enables integrated modulation of plant physiological and metabolic processes, improving nutrient acquisition, assimilation, and utilization.

In particular, the optimization of NUE has emerged as one of the most relevant aspects. The use of algal biostimulants may increase crop productivity per unit of applied fertilizer by promoting better assimilate allocation, higher photosynthetic activity, and more efficient integration of nitrogen into plant metabolism. This results in the maintenance or improvement of yields with unchanged or potentially reduced fertilizer rates.

From an environmental perspective, increased NUE helps limit nitrogen losses to the environment, reducing leaching, volatilization, and greenhouse gas emissions.

In this sense, algal biostimulants fully align with climate-smart agriculture approaches, supporting production strategies with lower environmental impact and more efficient resource use.

Furthermore, the positive effects of algal compounds on root physiology, the rhizosphere, and soil microbial activity suggest a relevant role in regenerative agriculture systems. Improvements in soil structure, microbial biodiversity, and nutrient availability contribute to long-term soil health and enhance crop system resilience to abiotic stresses.

Overall, scientific evidence demonstrates that integrating algal biostimulants into crop management and fertilization programs represents a concrete strategy to reconcile agricultural productivity with economic and environmental sustainability.

An approach focused on optimizing physiological processes, rather than increasing inputs, is particularly relevant in the context of current challenges related to climate change, food security, and the reduction of agriculture's environmental footprint. Nevertheless, further studies are needed to accurately describe the physiological and genetic processes triggered by algal biostimulant use and to more precisely identify the molecules responsible for the functional effects observed to date.

9 Conclusion and future trends

Research on algal biostimulants is rapidly evolving and is increasingly moving toward a more integrated, multidisciplinary approach focused on optimizing nutrient use efficiency. One of the main future directions concerns the identification and characterization of the bioactive compounds responsible for the observed biostimulant effects, with particular attention to their synergistic interactions rather than the action of individual molecules. Advanced studies in metabolomics, proteomics, and functional genomics will be essential to elucidate the molecular mechanisms underlying improvements in NUE. Moreover, improving nitrogen use efficiency represents a global priority for the sustainability of agricultural systems, as highlighted by international guidelines on sustainable nitrogen management, which emphasize the need for strategies capable of maintaining productivity while reducing environmental losses and the excessive use of mineral fertilizers (FAO, 2019).

Another key area of development concerns the standardization of algal raw materials and extraction processes. The biochemical composition of algae is influenced by species, geographic origin, harvesting season, and environmental conditions such as temperature, salinity, and light exposure (El-Beltagi et al., 2022; El Boukhari et al., 2020). Furthermore, different extraction methods, including alkaline hydrolysis, enzymatic extraction, and cold mechanical processing, produce extracts with distinct biochemical fingerprints and bioactive profiles (El Boukhari et al., 2020; Chabili et al., 2024).

Such variability complicates cross-study comparisons and may affect consistency in agronomic performance, thereby posing challenges for large-scale adoption and regulatory acceptance (du Jardin, 2015). Future research should therefore prioritize chemical fingerprinting approaches (e.g. metabolomic profiling), batch-to-batch standardization, harmonized agronomic testing protocols, and clearer guidelines regarding application rates and timing across different crops (Chabili et al., 2024).

The formulation of increasingly targeted products represents another emerging trend. In this context, attention is shifting toward 'tailor-made' algal biostimulants designed for specific crops, phenological stages, or stress conditions, with the aim of maximizing nitrogen uptake and assimilation efficiency while simultaneously reducing fertilizer inputs and increasing biostimulant efficiency. The integration of biostimulants with precision agriculture systems and decision-support tools could further amplify these benefits.

Algal biostimulants represent a promising tool for improving nutrient use efficiency and supporting more sustainable agricultural production systems. Their complex biochemical composition enables the modulation of plant physiological processes involved in nutrient uptake, assimilation, and allocation, with positive effects on crop productivity and quality. Evidence from recent experimental studies highlights how these products can enhance both nitrogen uptake efficiency and nitrogen utilization efficiency, contributing to improved crop performance even under reduced fertilizer inputs. At the same time, their interaction with soil microbial communities may further optimize nutrient dynamics within the soil–plant system. The integration of algal biostimulants into crop management strategies therefore offers a valuable opportunity to increase resource-use efficiency while reducing environmental impacts. Continued research aimed at clarifying mechanisms of action, improving standardization, and optimizing application strategies will be essential to fully exploit biostimulant potential in sustainable and climate-smart agriculture.

10 Where to look for further information

Researchers and practitioners interested in deepening their understanding of the use of algae and algal-derived products in agriculture may consult the following sources.

10.1 Professional and scientific organizations

Relevant information is available through organizations focused on algal science, plant nutrition, and agronomy, such as International Society for Applied Phycology, International Fertilizer Association, and European Biostimulants Industry Council.

10.2 International research programs and databases

Databases such as Food and Agriculture Organization, European Commission repositories, and major agricultural research networks provide reports, project outcomes, and policy documents related to algae-based agricultural innovations.

10.3 Industry technical reports

Technical bulletins and field trial reports from companies developing algal biostimulants can offer practical insights into product formulation, application methods, and crop-specific responses. Such information should be interpreted alongside peer-reviewed scientific evidence. Given the rapidly evolving nature of algal biotechnology and biostimulant research, readers are encouraged to prioritize recent publications and interdisciplinary studies that integrate plantphysiology, soil science, microbiology, and agronomic management.

11 References

Aina, O., et al. (2022) Seaweed-derived phenolic compounds in growth promotion and stress alleviation in plants. *Life*, 12(10), 1548. https://doi.org/10.3390/life12101548.

Ali, O., et al. (2019) Biostimulatory activities of *Ascophyllum nodosum* extract in tomato and sweet pepper crops in a tropical environment. *PLoS One*, 14(5), e0216710. https://doi.org/10.1371/journal.pone.0216710.

Benítez García, I., et al. (2020) Identification and quantification of plant growth regulators and antioxidant compounds in aqueous extracts of *Padina durvillaei* and *Ulva lactuca*. *Agronomy*, 10(6), 866. https://doi.org/10.3390/agronomy10060866.

Chabili, A., et al. (2024) A comprehensive review of microalgae- and cyanobacteria-based biostimulants for agricultural uses. *Plants*, 13(2), 159. https://doi.org/10.3390/plants13020159.

Craigie, J.S. (2011) Seaweed extract stimuli in plant science and agriculture. *Journal of Applied Phycology*, 23, 371–393. https://doi.org/10.1007/s10811-010-9560-4.

Di Serio, A., et al. (2025) Marine microalgae extracts as plant biostimulant to boost baby leaf lettuce production. *Scientific Reports*, 15, 32825. https://doi.org/10.1038/s41598-025-18104-9.

du Jardin, P. (2015) Plant biostimulants: definition, concept, main categories and regulation. *Scientia Horticulturae*, 196, 3–14. https://doi.org/10.1016/j.scienta.2015.09.021.

Dias, M. C., et al. (2026). Ascophyllum nodosum extract enhances olive fruit quality under contrasting irrigation regimes. *Journal of Applied Phycology*, 1–13.

Echave, J., et al. (2022) Seaweed-derived proteins and peptides: promising marine bioactives. *Antioxidants*, 11(1), 176. https://doi.org/10.3390/antiox11010176.

El-Beltagi, H.S., et al. (2022) Phytochemical and potential properties of seaweeds and their recent applications: a review. *Marine Drugs*, 20(6), 342. https://doi.org/10.3390/md20060342.

El Boukhari, M.E.M., et al. (2020) Trends in seaweed extract-based biostimulants: manufacturing process and beneficial effect on soil–plant systems. *Plants*, 9(3), 359. https://doi.org/10.3390/plants9030359-.

Ertani, A., et al. (2018) Evaluation of seaweed extracts from *Laminaria* and *Ascophyllum nodosum* spp. as biostimulants in *Zea mays* L. using a combination of chemical, biochemical and morphological approaches. *Frontiers in Plant Science*, 9, 428. https://doi.org/10.3389/fpls.2018.00428.

Fahad, S., et al. (2014) Crop plant hormones and environmental stress. *Sustainable Agriculture Reviews*, 15, 371–400. https://doi.org/10.1007/978-3-319-09132-7_10.

FAO (2019) *The International Code of Conduct for the Sustainable Use and Management of Fertilizers*. Rome: Food and Agriculture Organization of the United Nations. Disponibile su: https://openknowledge.fao.org/server/api/core/bitstreams/a8edc4fb-1614-426b-b21c-440fff19c46b/content (Accessed:14 January 2026).

Hagan, M. & Anyangwe, N. (2023) Vitamin content in seaweeds: a systematic review on water-soluble and fat-soluble vitamins for adult daily intake. *Functional Food Science*, 3(12), 305. https://doi.org/10.31989/ffs.v3i12.1273.

Jiménez-Arias, D., et al. (2021) A beginner's guide to osmoprotection by biostimulants. *Plants*, 10(2), 363. https://doi.org/10.3390/plants10020363.

Khan, W., et al. (2012) Commercial extract of *Ascophyllum nodosum* improves root colonization of alfalfa by its bacterial symbiont *Sinorhizobium meliloti*. *Communications in Soil Science and Plant Analysis*, 43(18), 2425–2436. https://doi.org/10.1080/00103624.2012.708079.

Khan, W., et al. (2013) *Ascophyllum nodosum* extract and its organic fractions stimulate rhizobium root nodulation and growth of *Medicago sativa* (alfalfa). *Communications in Soil Science and Plant Analysis*, 44(5), 900–908. https://doi.org/10.1080/00103624.2012.744032.

Loukakis, L.O., et al. (2025) Integration of biostimulants alongside various advanced nitrogen fertilization practices improve the yield, quality, and sustainability of malting barley in Mediterranean conditions. *Agronomy*, 15(10), 2417. https://doi.org/10.3390/agronomy15102417.

Massantini, F., et al. (1983) Efficienza di assorbimento e perdite per lisciviazione dell'azoto da fertilizzanti liquidi e solidi: prove lisimetriche sul frumento. *Agrochimica*, 27, 385–398.

O'Brien, R., et al. (2022) Macroalgal proteins: a review. *Foods*, 11(4), 571. https://doi.org/10.3390/foods11040571.

Rathod, S.G., et al. (2024) Phytohormones and pheromones in the phycology literature: benchmarking of dataset and developing critical tools of biotechnological implications for commercial aquaculture industry. *Phycology*, 4(1), 1–36. https://doi.org/10.3390/phycology4010001.

Rossini, A., et al. (2025) Combining nitrogen fertilization and biostimulant application in durum wheat: effects on morphophysiological traits, grain production, and quality. *International Journal of Agronomy*, 100027. https://doi.org/10.1016/j.ijagro.2025.100027.

Sabatino, L., et al. (2023) *Ecklonia maxima*-derived seaweed extract supply as a mitigation strategy to alleviate drought stress in chicory plants. *Scientia Horticulturae*, 312, 111856. https://doi.org/10.1016/j.scienta.2022.111856.

Scarabattoli, L., et al. (2023) Efficacia di un idrolizzato proteico e suoi derivati su crescita e qualità della lattuga. *Intervento presentato al 14° Convegno "Giornate Scientifiche SOI"*, Torino, Italia. Disponibile su: https://air.unimi.it/handle/2434/1113235 (Accessed: 14 January 2026).

Salim, N. & Raza, A. (2020) Nutrient use efficiency (NUE) for sustainable wheat production: a review. *Journal of Plant Nutrition*, 43(2), 297–315. https://doi.org/10.1080/01904167.2019.1676907.

Shukla, P.S., et al. (2019) *Ascophyllum nodosum*-based biostimulants: sustainable applications in agriculture for the stimulation of plant growth, stress tolerance, and disease management. *Frontiers in Plant Science*, 10, 655. https://doi.org/10.3389/fpls.2019.00655.

Thiviya, P., et al. (2022) Seaweeds as a source of functional proteins. *Phycology*, 2(2), 216–243. https://doi.org/10.3390/phycology2020012.

Usman, A., et al. (2017) Algal polysaccharides: novel application and outlook. *Carbohydrate Polymers*, 166, 173–185. https://doi.org/10.1016/j.carbpol.2017.02.090.

Wi, J., et al. (2018) A nuclear fucosyltransferase-like protein, PtFUT, from marine red alga *Pyropia tenera* (Rhodophyta) confers osmotic stress tolerance. *Journal of Applied Phycology*, 30, 717–727. https://doi.org/10.1007/s10811-017-1241-0.
Xu, G., et al. (2012) Plant nitrogen assimilation and use efficiency. *Annual Review of Plant Biology*, 63, 153–182. https://doi.org/10.1146/annurev-arplant-042811-105532.
Yang, T., et al. (2025) Integrative multi-omics elucidates the impact of microalgae on growth, quality, phytohormones, and rhizosphere microbiome of *Angelica sinensis*. *Microbiological Research*, 278, 128418. https://doi.org/10.1016/j.micres.2025.128418.

Index

www.ingramcontent.com/pod-product-compliance
Ingram Content Group UK Ltd.
Pitfield, Milton Keynes, MK11 3LW, UK
UKHW021825100926
8556IPUK00003B/153

9 781835 451861